On the Nature of Continental Shelves

On the Nature of Continental Shelves

John J. Walsh
Department of Marine Science
University of South Florida
St. Petersburg, Florida

Academic Press, Inc.
Harcourt Brace Jovanovich, Publishers
San Diego   New York   Berkeley   Boston
London   Sydney   Tokyo   Toronto

ACADEMIC PRESS, INC.
1250 Sixth Avenue
San Diego, California 92101

United Kingdom Edition published by
ACADEMIC PRESS INC. (LONDON) LTD.
24-28 Oval Road, London NW1 7DX

Library of Congress Cataloging-in-Publication Data

Walsh, John Joseph, Date
    On the nature of continental shelves.

    Bibliography: p.
    Includes index.
    1. Continental shelf.   2. Marine biology.   3. Marine
pollution.   Biogeochemical cycles.   I. Title.
GC85.W35   1987   551.4′1   87-1484
ISBN 0-12-733775-X   (alk. paper)

PRINTED IN THE UNITED STATES OF AMERICA
88 89 90 91   9 8 7 6 5 4 3 2 1

Contents

Preface

During the period 1912–1932, Henry Bigelow conducted a series of cruises to study the physical oceanography (Bigelow, 1927), phytoplankton (Bigelow, 1926), zooplankton (Bigelow and Sears, 1939), and fishes (Bigelow and Welsh, 1925) of the continental shelf from Cape Hatteras to the Gulf of Maine. Over the next 40 years, few shipboard studies of continental shelves were more comprehensive than this initial effort, because time series of more frequent sampling intervals were required to resolve aliasing introduced by infrequent cruises over a period of years. Major research expeditions tended to occur within the deep sea, where horizontal heterogeneity of the physical habitat was minimal, allowing one to focus on one-dimensional problems of vertical exchange in the open ocean.

Early development of current meters first allowed field scientists to capture physical scales of variability at a few points on the continental shelf, while development of digital computers allowed theoreticians to run numerical models of sufficient complexity to estimate the flow fields between these few points. During the 1970s, construction of other moored instruments, satellite and aircraft sensors, as well as continuous, towed profiling systems from ships, allowed chemists and biologists to sample at time and space scales similar to the measurements of their physicist colleagues.

Within the last decade, a series of multidisciplinary studies on continental shelves adjacent to both the eastern and western boundary currents provided a wealth of information, which finally surpassed Bigelow's previous efforts. Based on these data sets, this monograph attempts to explore the results of man's perturbation experiments on the food webs of continental shelves, those of overfishing and eutrophication, in relation to the release of "greenhouse" gases within the context of natural variability of carbon and nitrogen cycling.

A number of descriptive texts on biological processes of the sea have been written, with little attempt to introduce the underlying concepts of the dynamics of the fluid within which marine organisms live. Even fewer texts on physical oceanography, however, are oriented toward the marine ecologist or fisheries manager. In an attempt to bridge the gap, this book is designed to explore the rich data sets on continental shelves with the techniques of systems analysis, which can now easily be implemented on any minicomputer. Formal knowledge of mathematics at the level of one year of calculus, of statistics, and of computer programming is beneficial, but not mandatory, since an intuitive rather than rigorous mathematical approach is used in this monograph.

The results of the research programs described in this monograph were funded mainly by the U.S. National Science Foundation, Department of Energy, National Aeronautics and Space Administration, and National Oceanic and Atmospheric Administration. Many colleagues of these and other international programs over the past 20 years have generously provided the data used in the text. Of particular importance, however, was my wife Orvita's patience, support, and insight, without which an analysis of continental shelves would have been an impossible task.

1 Introduction

... The high piled scow of garbage, bright-colored, white-flecked, ill smelling, now tilted on its sides, spills off its load into the blue water, turning it a pale green to a depth of four or five fathoms as the load spreads across the surface, the sinkable part going down and the flotsam of pond fronds, corks, bottles, and used electric light globes, seasoned with an occasional condom or a deep floating corset, the torn leaves of a student's exercise book, a well-inflated dog, the occasional rat, the no-longer distinguished cat; all this well shepherded by the boats of the garbage pickers who pluck their prizes with long poles, as interested, as intelligent, and as accurate as historians; they have the viewpoint; the stream with no visible flow, takes five loads of this a day when things are going well in La Habana and in ten miles along the coast it is clear and blue and unimpressed as it was ever before the tug hauled out the scow; and the palm fronds of our victories, the worn light bulbs of our discoveries and the empty condoms of our great loves float with no significance against one single, lasting thing—the stream.

(Hemingway, 1935; © Charles Scribner's Sons)

Situated between land and the open sea, the continental shelves may become the refuse pits of developed nations and the overfished graveyards of underdeveloped nations. For example, over the last 400 yr the human population of the northeast coastal zone of the United States has grown from a few Indian settlements to the present megalopolis of 45 million people, housed in an almost continuous urban development from Norfolk, Virginia, to Portland, Maine. About 15 million people live just within the coastal counties of the New York Bight from Cape May, New Jersey, to Montauk Point, New York. The annual energy consumption of the northeast United States was 13.6×10^{15} Btu in 1975, with only 2.7×10^{15} Btu produced in this region. A net energy import of 10.9×10^{15} Btu was thus required for the northeast, 60% of which was in the form of oil.

1

1.1 Anthropogenic Impacts

During 1974, 1.3×10^8 tons of crude petroleum and petroleum products were imported between Hampton Roads, Virginia, and Portland, Maine, of which 0.4×10^8 tons were off-loaded in storage tanks or refineries along the Hudson and Raritan Rivers. After processing by society and partial conversion to CO_2, a residue of 3×10^5 tons of oil and grease ($\sim$1%) is usually returned each year to waters of the New York Bight through dumping and wastewater runoff of the coastal communities in New York and New Jersey. With grounding of the tanker *Argo Merchant* on Nantucket Shoals during December 15, 1976, however, 2.7×10^4 tons of number 6 fuel oil were released to the coastal ecosystem in this single incident: that is, 30 times the usual daily discharge of oil to this continental shelf.

Similarly, a decade earlier on June 16, 1966, the tanker *Texaco Massachusetts* rammed another tanker, *Alva Cape*, in New York Harbor, a cargo of naphtha was discharged, and the ensuing explosion claimed 33 lives on the two ships and nearby tugs. Other oil spills, such as in the Santa Barbara, Ixtoc, and Ekofisk oil fields and from the tankers *Torrey Canyon*, *Amoco Cadiz*, *Metula*, *Florida*, and *Arrow*, constitute an estimated annual input of 2 million metric tons of petroleum to the continental shelves with an unresolved ecological impact. Another 2 million tons of petrochemicals are added each year to the coastal zone from river and sewer runoff (Goldberg, 1976). A combination of $\sim$200 fossil fuel power plants, 12 major oil refineries, and 20 nuclear power plants, in operation or under construction, is located within this northeast U.S. coastal region, where part of the annual energy supply is returned inadvertently or by planned dumping. During 1951–1967, for example, about 34,000 containers of radioactive waste (including the pressure vessel of the *Seawolf* reactor) were dumped off the U.S. northeast coast, the fate of which is still being studied near the deep-water dump site at 3800 m. Until recently, the Windscale nuclear reprocessing plant discharged radionuclides directly into the Irish Sea, from where this effluent could be traced into the Arctic Ocean. Power plants, as well as industrial and military complexes, are often sited in such coastal areas because of the proximity of cities, transportation, and available cooling water. We have come to realize, however, that dilution of these anthropogenic effluents by seawater can no longer be considered a simple, or permanent, removal process within either the open ocean or nearshore waters, such as off the Cuban coast, as described by Ernest Hemingway in the chapter opening.

The continental shelves adjacent to large human populations are subject, in fact, to a large array of both atmospheric and coastal input of pollutants in the form of heavy metals, synthetic chemicals, petroleum hydrocarbons, radionuclides, and nutrients from sewage, agrarian runoff, and deforestation. Physical transport of these pollutants, their modification by the coastal food web, and their transfer to humans have become problems of increasing complexity on the continental shelf. For example, as a result of burning fossil fuel, each year 1×10^4 tons of the toxic heavy metals cadmium, mercury, lead, and copper are released via river runoff to just the New York Bight; the concentration of cadmium in this estuarine discharge is close to the limits set by the Environmental Protection Agency (EPA). In the process of consuming the estuarine input of nutrients, the near-shore phytoplankton can also assimilate these heavy metals, accumulate them as much as 40,000-fold for copper and lead, and pass this material up the local food chain.

After 30 yr of discharge of mercury into the sea, the Minimata neurological disease of Japan was finally traced to consumption of fish and shellfish containing methyl mercuric chloride. Their *itai itai* disease is now attributed to high cadmium ingestion, and an abalone kill off California was attributed to copper pollution from the condensers of a coastal power plant. Discharge of chlorinated hydrocarbons, such as DDT (dichlorodiphenyltrichloroethane) off California, PCB (polychlorobiphenyls) in the Hudson River and within Escambia Bay, Florida, mirex in the Gulf of Mexico, and vinyl chloride in the North Sea, have led to inhibition of photosynthesis, large mortality of shrimp, and reproductive failure of birds and fish.

Overfishing is an additional human-induced stress on the food web of the continental shelf. The present commercial finfish stocks on the U.S. northeast continental shelf have been reduced by 50% over the past 10 yr as a result of heavy fishing by foreign nations and the United States, with a peak yield of fish, including menhaden, between Cape Hatteras and Nova Scotia of 1.4×10^6 tons yr^{-1} during 1970 to 1973. An order of magnitude larger yield of anchovy of 11×10^6 tons yr^{-1}, about 20% of the world's fish catch, was briefly obtained off Peru, before a collapse of this fishery in 1976 (Fig. 1). Most clupeid fisheries now yield about 10% of their maximal past harvests. As the population of a commercially exploited species drops below the carrying capacity of a particular habitat, its ability to withstand other stresses declines as well. Because of natural variability in abundance of organisms in the coastal food web, overfishing, *and* the release of pollutants to coastal waters, determination of cause and effect within a perturbation of the coastal food web is thus a difficult matter.

The 1976–1977 closures of local coastal fisheries of the Mid-Atlantic Bight can be directly traced to PCB contamination of the Hudson River and Kepone contamination of the James River. However, it was not possible to detect changes in offshore primary productivity after a shipload of the pesticide Mopac was discharged within these slope waters during March 1979. Furthermore, the $60 million loss of the shellfish industry in 1976, as a result of anoxia off the New Jersey coast, has been attributed to natural interannual oscillations in seasonal abundance of phytoplankton species at the base of the food chain rather than to pollutant impact. There is also some question about whether chlorinated hydro-

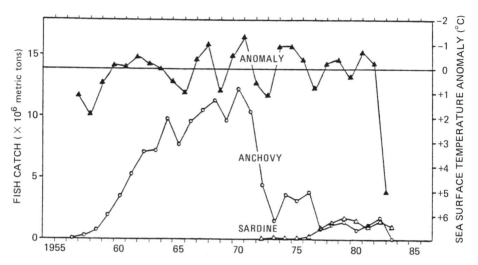

FIG. 1. Association of annual SST anomalies with annual catch of anchovies and sardines off Peru. The anomaly temperature scale is inverted: upward indicates cooler temperatures; downward indicates warmer temperatures. [After Barber and Chavez (1983); © 1983 AAAS, with permission.]

carbons are just absorbed or bioaccumulated by marine organisms and how long petrochemicals are toxic in the coastal zone, but at present, even the trajectory of these pollutants within a shelf ecosystem is poorly understood.

Knowledge of the movement and dispersive capability of the waters in a continental shelf region is thus essential to describing the response of the shelf's ecosystem and the fate of waste discharges or contaminants introduced into its waters. The mechanisms causing water movements along an entire shelf must be understood before the impact of contaminants on biota and water properties can be predicted. This depends on the ability to explain the finer-scale processes that, in sum, form the larger circulations, including the transient and nonlinear events that impair general knowledge and predictive capability. Variability of the physical habitat occurs at all time scales, from turbulent fluctuations to glaciations (Table I). Judicious filtering of the high-frequency variability, however, allows description of the time scale of interest, with appropriate parameterizations of other scales of motion—such processes are discussed in detail within Chapter 2.

With specification of the physical habitat, one approach to a quantitative assessment of the above pollutant impacts is the construction of a series of simulation models of the coastal food web in a systems analysis of the continental shelf (Walsh, 1972). Ideally, such a systems analysis of a continental shelf ecosystem would involve the sequential development of models of varying complexity, such as conceptual ecosystem models, circulation submodels, and dosage-response submodels, in a logical sequence of (1) dosage-response functions of the organisms to each class of pollutants, (2) a quantitative description of the "normal" food-web interactions of a coastal zone, (3) knowledge of where and when the introduced pollutants interact with the coastal food web, and (4) the residence time of dissolved and particulate-associated pollutants on a specific continental shelf. The generic toxicity levels in terms of median lethal concentrations (LC^{50}) of metals, pesticides, biofouling agents, industrial chemicals, and petroleum hydrocarbon fractions have, in fact, been determined for a number of marine organisms. A definition of the assimilative capacity of the continental shelves and adjacent continental slopes to detoxify or store synthetic substances without altering the food web leading to humans is hampered, however, by an

Table I

Schematic Representation of Habitat Variability on the Continental Shelf[a]

Time scale	Frequency (sec^{-1})	Spectral windows (highlighted processes)	Smaller-scale fluctuations (filtered-out processes)
1 sec	1	Microscale processes: three-dimensional "eddy" turbulence (+ surface waves)	Molecular diffusion
1 min	10^{-2}	Mesialscale processes: Internal waves Vertical microstructure Inhibited "bliny"[b] turbulence	Eddy turbulence
1 h	10^{-4}	Mesoscale processes: Inertial oscillations	"Bliny turbulence"
1 day	10^{-5}	Tides, storm surges Diurnal variations	
1 week	10^{-6}	Synoptic-scale processes: Frontal currents Meanders, "Rossby"[c] turbulence	Mesoscale variability
	10^{-7}	Seasonal-scale processes: Convective overturn Stratification	"Rossby turbulence"
1 yr	10^{-8}	Global scale processes: Climatic interactions	Seasonal variability

[a] After Walsh *et al.* (1986a); © E. J. Brill, with permission.

[b] A "bliny" (from the Russian "blini") is a pancake-shaped eddy contributing to an energy cascade to smaller scales via epidermic instabilities and internal waves.

[c] A "rossby" (from the scientist Rossby) is a pseudo-two-dimensional eddy column of scale of the order of the Rossby radius of deformation.

incomplete knowledge of the cycling rate of natural elements, such as carbon and nitrogen. The state of our understanding of the "normal" flow of carbon and nitrogen through the food webs of the continental shelves is thus the subject of this text.

During the last decade, multidisciplinary research programs have been conducted on coastal upwelling regions off Peru, Northwest Africa, and Oregon–California, on the boreal Mid-Atlantic and South Atlantic Bights, on the high-latitude North and Bering Seas, on discharge of the Yangtze, Mississippi, and Amazon Rivers to the South China, Gulf of Mexico, and Brazil shelves, and on coastal nutrient inputs to the Baltic Sea, the Sea of Japan, and the Mediterranean Sea. Using examples of these continental shelf ecosystems, structured by varying intensities of vertical exchange rates such as coastal upwelling, spin-off eddies, and tidal mixing, the natural cycles of carbon and nitrogen will be described. Humanity's present and possible future perturbations of shallow seas are then discussed in the context of both natural variability of these cycles and the likely impact of processes such as overfishing, eutrophication, and release of carbon dioxide.

1.2   Spatial Extent

The potential of the carbon cycle of the sea to either yield fish or store atmospheric CO_2 is a subject of continuing controversy as the human ability to modify the marine environment increases. Because the actual amount of CO_2 fixed annually during marine photosynthesis is unknown, the fate of phytoplankton, serving as a precursor either to fish carbon or to sediment carbon, is also unknown. Debates over the amount of potential fish harvest (Ryther, 1969; Alversen et al., 1969) and CO_2 storage capacity (Broecker et al., 1979; Walsh et al., 1981) of the ocean thus hinge on the amount and fate of marine primary production. Current estimates of annual marine primary production of the entire ocean range from 20 to 55×10^9 tons of carbon per year (Ryther, 1969; Koblentz-Mishke et al., 1970; Platt and Subba Rao, 1975; DeVooys, 1979; Walsh, 1980). This range accounts for ~25–50% of the total net global carbon fixation (Woodwell et al., 1978), and is 400–1000% of either present fish yield or fossil fuel emissions.

Over the last century, the human ability to extract nitrogen from the atmosphere has also begun to rival that of N_2 fixation by plants. Between 1950 and 1975, for example, world production of agricultural fertilizers increased 10-fold. Anthropogenic nutrient input from agrarian runoff, deforestation, and urban sewage has already impacted local streams and ponds, some large lakes, major rivers, and perhaps even the continental shelves, with possibly a 10-fold increase of nutrient loading since 1850 (Walsh, 1984). Nitrogen is now only routinely measured in 25% of the world's 215 largest rivers, however, such that few biological time series are available to document the coastal zone's past response to fluvial nutrient transients on even a decadal time scale. The annual primary production of the Dutch Waddensea, for example, has apparently increased three-fold from 80 g C m^{-2} yr^{-1} in 1950 to 240 g C m^{-2} yr^{-1} in 1970 (Postma, 1978), and presumably other shelves have responded to anthropogenic nutrient loadings as well.

Large areas of the ocean, such as the central gyres (Fig. 2), have, of course, not been impacted by eutrophication (Peng and Broecker, 1984) and still have relatively low rates of production per unit surface area ($\sim$60 g C m^{-2} yr^{-1}). They account for a major fraction of the total marine carbon fixation, however, because of their great areal extent (Table II). Nevertheless, their long food chains and the 90% recycling processes of respiration in the offshore regimes provide insignificant fish harvest (Ryther, 1969) and little net biotic storage of CO_2 (Eppley and Peterson, 1980). In contrast, the highly productive ($\sim$200 g C m^{-2} yr^{-1}) coastal and upwelling regions account for only 10% of the ocean by area, but at least 25% of the ocean's primary productivity. The coastal zooplankton populations (Fig. 3) provide the basis for more than 95% of the world's estimated fishery yield, moreover, and a large part of the proposed organic carbon sink (Walsh et al., 1981) of the atmospheric CO_2 may be located in adjacent slope sediments (Fig. 4).

The mean physical extent of today's shelf ecosystem is about 75 km in width, reaching 130 m depth at the shelfbreak (Shepard, 1963), and has a most recent origin of about 10,000 years (Milliman and Emery, 1968; Edwards and Merrill, 1977). The various shelf provinces now exhibit pronounced differences in their annual production and species assemblages as the evolutionary consequence of their different physical habitats, structured in response to global wind (Fig. 5), current, and rainfall (Fig. 6) patterns. At low latitudes over 0–30°N or 0–30°S, for example, incident radiation is sufficient to allow daily photosynthesis to proceed at a maximal rate over the whole year. Lack of much nutrient input from either the land or open sea boundaries to the oligotrophic West Florida Shelf leads to an annual primary production of only $\sim$30 g C m^{-2} yr^{-1}, however, compared to 1000–2000 g C m^{-2} yr^{-1} within the coastal upwelling region off Peru (Table III).

These major upwelling systems of $\sim$10 $\times$ 10^5 km^2 extent at low latitudes (Table III) within the Humboldt, Benguela, Canary, and California Currents of the eastern boundaries of the ocean (Fig. 6) are some of the most productive marine ecosystems, rivaling the annual primary production of $\sim$1275–1500 g C m^{-2} yr^{-1} from salt marshes (DeVooys, 1979), coral reefs (Andrews and Gentien, 1982), and seaweed beds (S. V. Smith, 1981). The seasonal, monsoon-induced upwelling regions off the African and Indian coasts and the midlatitude eastern boundary currents, such as the Canary and California Currents off Portugal and

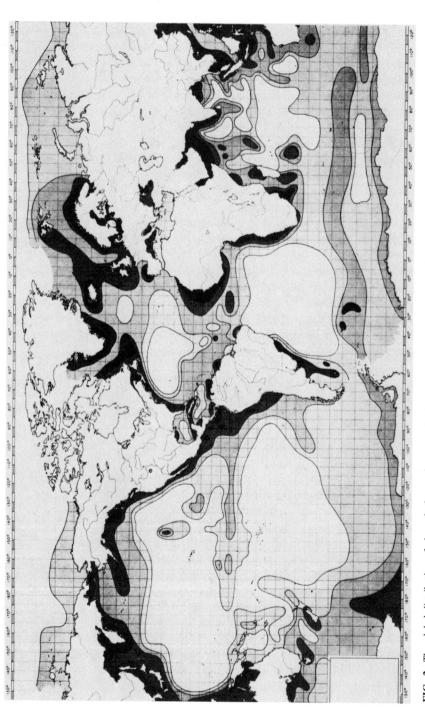

FIG. 2. The global distribution of phytoplankton primary production (mg C m⁻² day⁻¹) in five categories of >500, 250–500, 150–250, 100–150, and <100. [After Koblentz-Mishke *et al.* (1970); © Food and Agricultural Organization of the United Nations, by permission.]

Table II.

Aquatic Photosynthesis and Nitrogen Fixation in Relation to Losses of Sediment Storage of Organic Carbon, of Denitrification, and of Methane Production[a]

Region	Area (km^2)	Net primary production ($\times 10^9$ tons C yr^{-1})	Nitrogen fixation ($\times 10^7$ tons N$_2$ yr^{-1})	Sediment organic carbon sink ($\times 10^9$ tons C yr^{-1})	Denitrification loss ($\times 10^7$ tons N$_2$ yr^{-1})	Methane emission ($\times 10^7$ tons CH$_4$ yr^{-1})
Open Ocean	3.1×10^8	18.60	0.43	0.20	0	0.36
Continental shelf	2.6×10^7	5.20	0.27	0	2.97	0.04
Continental slope	3.2×10^7	2.24	0.06	0.50	5.50	0.03
Freshwater marshes	1.6×10^6	1.51	2.21	0.15	6.40	3.10
Estuaries/deltas	1.4×10^6	0.92	0.06	0.20	1.04	0.60
Salt marshes	3.5×10^5	0.49	0.48	0.05	1.40	0.80
Rivers/lakes	2.0×10^6	0.40	1.88	0.13	0.26	5.10
Coral reefs	1.1×10^5	0.30	0.28	0.01	0	0.32
Seaweed beds	2.0×10^4	0.03	0	0	0	0.08
Total aquatic area	3.8×10^8	C INPUT: 29.7	N$_2$ INPUT: 5.7	C OUTPUT: 1.2	N$_2$ OUTPUT: 17.6	CH$_4$ OUTPUT: 10.4

[a] After Walsh (1984); © *BioScience*, with permission.

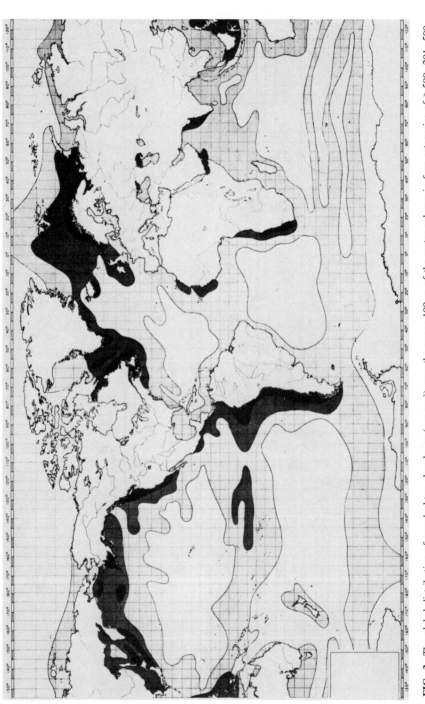

FIG. 3. The global distribution of zooplankton abundance (mg m^{-3}) over the upper 100 m of the water column in four categories of >500, 201–500, 51–200, and <50. [After Bogorov *et al.* (1968); © Food and Agricultural Organization of the United Nations, with permission.]

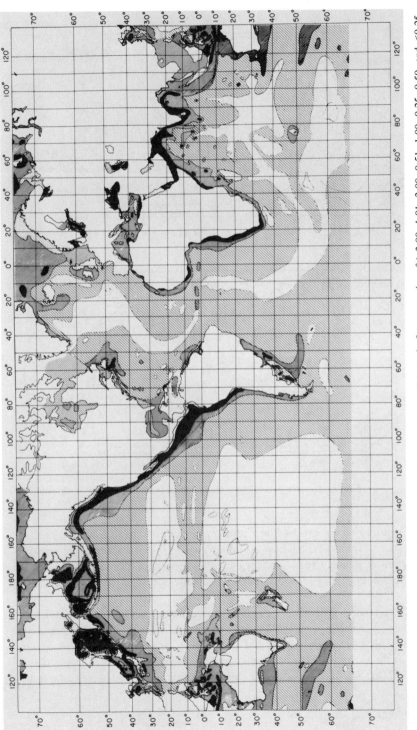

FIG. 4. The global distribution of organic carbon (% dw) within surface sediments in five categories of >2.00, 1.01–2.00, 0.51–1.00, 0.25–0.50, and <0.25. [After Premuzic *et al.* (1982); © 1982 Pergamon Journals Ltd., with permission.]

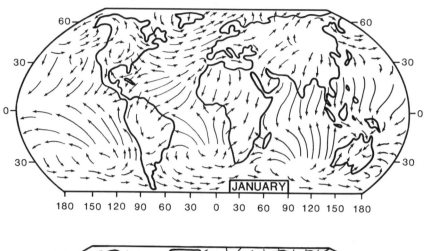

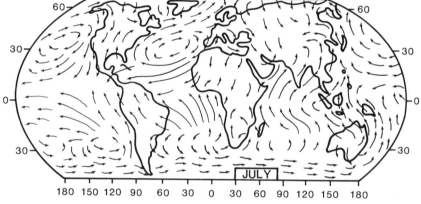

FIG. 5. The global distribution of wind fields in January and July.

Washington (Fig. 6), exhibit lower primary production than the major upwelling ecosystems (Table III), adjacent to desert coasts (Fig. 6). Because of seasonal limitation of incident radiation, for example, the annual primary production of the Oregon upwelling ecosystem, at 45°N is only ~200 g C m^{-2} yr^{-1} (Small *et al.*, 1972).

Minor coastal upwelling regions also occur nearshore at low latitudes off Vietnam, Venezuela, and the Yucatan Peninsula, in the Gulfs of Panama and Tehuantepec, within the Flores and Banda Seas, and in some offshore regions such as the Costa Rica Dome, the Angola Dome, and the Guinea Dome (Cushing, 1971a). The major portion of carbon fixation on these shelves adjacent to

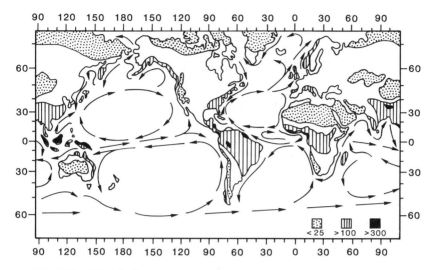

FIG. 6. The global distribution of terrestrial rainfall (cm yr^{-1}) and major ocean currents.

the western boundary currents of the sea appears, however, to be supported by eddy-induced upwelling at the shelf-break. The daily rate of nutrient input by upwelled intrusions of western boundary currents is 10-fold less than those from coastal upwelling within eastern boundary currents, with little apparent seasonal variation, such that the annual primary production of these shelf ecosystems is also 10-fold less than that off Peru or Southwest Africa.

As a consequence of the annual rainfall pattern (Fig. 6), approximately 62% of the earth's gauged freshwater runoff discharges into these western boundary shelves at 0–30° latitudes, compared to 3% (Table III) over low-latitude and midlatitude upwelling ecosystems. Five of the world's 10 largest rivers, the Amazon, Congo, Orinoco, Brahmaputra, and Mekong, for example, account for 60% of this freshwater input at low latitudes, but contain <10 μg-at. NO$_3$ l^{-1}, such that estuarine stimulation of shelf production is thought to be minimal here (Ryther *et al.*, 1967; Van Bennekom *et al.*, 1978). The marine signature of river discharge and melting ice can be seen in the global distribution of the 33 isopleth of salinity, such as on the Patagonian and Amazonian shelves, in the Mid-Atlantic Bight, the Gulfs of Guinea and Alaska, the Bay of Bengal, and the South China, East China, Labrador, Bering, Okhotsk, and Baltic Seas (Fig. 7). High surface salinities of 35‰ are mostly found in coastal regions of little freshwater discharge at lower latitudes, such as the major upwelling regions, the Arafura–Coral Seas, the Mediterranean Sea, and part of the Gulf of Mexico. Northward penetration of the Gulf Stream (Fig. 6) does introduce high-salinity water

Table III.

Geographic Location of World Shelves with Respect to River Discharge, Areal Extent, and Annual Primary Production

Latitude	Region	Major Rivers	(10^2 m^3 sec^{-1})	Area (10^5 km^2)	Unit production (g C m^{-2} yr^{-1})
		Eastern boundary currents			
0–30	Ecuador–Chile	—		2.7	1000–2000
	Southwest Africa	—		1.7	1000–2000
	Northwest Africa	—		2.6	200–500
	Baja California	—		1.1	600
	Somali Coast	Juba	5.5:5.5	0.6	175
	Arabian Sea	Indus	75.5:86.0	3.8	200
30–60	California–Washington	Columbia	79.6:97.4	1.6	150–200
	Portugal–Morocco	Tagus	3.1:8.9	0.8	60–290
		% Discharge (3)[a]		14.9	Mean: 644.2
		Western boundary currents			
0–30	Brazil	Amazon	1750.0:2005.1	6.0	90
	Gulf of Guinea	Congo	396.4:589.9	3.5	130
	Oman/Persian Gulfs	Tigris	14.5:14.5	4.3	—
	Bay of Bengal	Ganges	116.0:371.8	3.0	110
	Andaman Sea	Irrawaddy	135.6:150.6	4.0	50
	Java/Banda Seas	Brantas	4.0:8.6	6.7	—
	Timor Sea	Fitzroy	1.8:1.8	4.0	100
	Coral Sea	Fly	24.5:62.5	3.2	20–175
	Arafura Sea	Mitchell	3.6:40.3	14.2	150
	Red Sea	Awash	<0.4	1.6	34
	Mozambique Channel	Zambezi	70.8:93.3	2.6	100–150
	South China Sea	Mekong	149.0:363.3	15.7	215–317
	Caribbean Sea	Orinoco	339.3:339.3	3.7	66–139
	Central America	Magdalena	75.0:143.3	4.5	180
	West Florida Shelf	Appalachicola	6.9:6.9	2.1	30
	South Atlantic Bight	Altamaha	3.9:7.2	1.4	130–350
		% Discharge (62)[a]		80.5	Mean: 137.4
		Mesotrophic systems			
30–60	Australian Bight	Murray	7.4:7.7	8.2	50–70
	New Zealand	Waikato	4.1:25.8	2.9	—
	Argentina–Uruguay	Parana	149.0:220.1	10.4	—
	Southern Chile	Valdivia	4.5:44.3	4.0	—
	Southern Mediterranean	Nile	9.5:9.5	2.9	30–45
	Gulf of Alaska	Fraser	35.4:80.8	3.1	50
	Nova Scotia–Maine	St. Lawrence	141.6:178.7	6.7	130
	Labrador Sea	Churchill	15.8:147.5	21.2	24–100
	Okhotsk Sea	Amur	103.0:103.0	7.1	—
	Bering Sea	Kuskokwim	12.8:39.3	11.7	170
		% Discharge (12)[a]		78.2	Mean: 81.9

Table III.

(Continued)

Latitude	Region	Major Rivers (10^2 m³ sec⁻¹)		Area (10^5 km²)	Unit production (g C m⁻² yr⁻¹)
		Phototrophic systems			
60–90	Beaufort Sea	Mackenzie	97.2:97.2	2.6	10–20
	Chukchi Sea	Yukon	62.0:62.0	6.1	40–180
	East Siberian Sea	Kolyma	22.4:39.8	7.8	—
	Laptev Sea	Lena	163.09:180.2	6.9	—
	Kara Sea	Ob	122.0:300.0	10.1	—
	Barents Sea	Pechora	33.6:78.7	7.3	25–96
	Greenland–Norwegian Seas	Tjorsa	3.6:13.4	2.8	40–80
	Weddell–Ross Seas	—		3.9	12–86
		% Discharge (11)[a]		47.5	Mean: 58.9
		Eutrophic systems			
30–60	Mid-Atlantic Bight	Hudson	3.7:24.6	1.3	300–380
	Baltic Sea	Vistula	10.1:71.1	3.9	75–150
	East China Sea	Yangtze	220.0:240.3	10.7	—
	Sea of Japan	Ishikari	5.0:22.8	1.7	100–200
	North–Irish Seas	Rhine	25.4:47.8	8.7	100–250
	Northern Mediterranean	Po	14.7:49.0	2.9	68–85
	Caspian Sea	Volga	75.8:75.8	1.5	—
	Black Sea	Danube	65.3:97.9	1.6	50–150
	Bay of Biscay	Loire	8.3:16.7	2.7	—
	Texas/Louisiana	Mississippi	178.0:200.1	2.1	100
		% Discharge (12)[a]		37.1	Mean: 154.2

[a] Total percentage of the freshwater input from the earth's 215 largest rivers (>30 m³ sec⁻¹) within each of these shelf regions.

(Fig. 7) into the Greenland–Norwegian Seas, where deep-water formation subsequently occurs, that is, the sinking of surface water of both high salinity and low temperature.

High-latitude shelf ecosystems constitute 18% of the area of shallow seas, where 11% of the freshwater input is discharged (Table III), and are generally considered to be light-limited, rather than controlled by nutrient availability. Estimates of arctic annual production range, however, from 5 to 290 g C m⁻² yr⁻¹ (Subba Rao and Platt, 1984), while daily carbon fixation of 2 g C m⁻² day⁻¹ (Sambrotto *et al.*, 1984) and supersaturated oxygen conditions (Codispoti and Richards, 1971) have been measured during summer in, respectively, the Bering Strait and East Siberian Sea. Major rivers, such as the Mackenzie and Yukon, now contain nitrate contents of <10 μg-at. NO_3 l⁻¹ (Meybeck, 1982),

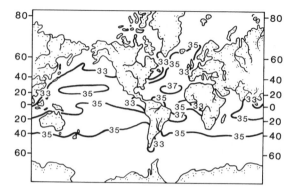

FIG. 7. The global distribution of salinity (‰) in surface waters. [After Bruns (1958); © 1963 Pergamon Books Ltd., with permission.]

with only $10-20$ g C m^{-2} yr^{-1} produced within the Beaufort Sea at 70°N (Horner, 1984). The immediate impact of human activities at high latitudes may be felt in the possible southward diversion of the outflow from the Ob and Yenisey (Kelly *et al.*, 1983), rather than from eutrophication.

In terms of a global carbon budget, however, the organic carbon content of the shelf sediments of the Chukchi Sea is $1-2\%$ dry weight (dw), an order of magnitude greater (Fig. 4) than most other shelves (Walsh *et al.*, 1985), and may constitute a major sink of organic matter. Approximately $25-50\%$ of the dissolved CO_2 in surface Atlantic water, removed each year from the Greenland–Norwegian Seas by deep water formation, may also be returned to the Arctic Ocean from North Pacific water, entrained within the 0.5- to 1.5-Sv (sverdrup; 1 Sv $= 10^6$ m^3 sec^{-1}) northward transport through Bering Strait (Walsh *et al.*, 1986a). Furthermore, a decline of marine primary production, as a result of increased light limitation at high northern latitudes on geological time scales, is suspected to be an important component of atmospheric CO_2 rise and thus global control of warming trends $\sim11,000$ yr ago (Knox and McElroy, 1984). If the West Antarctic Ice Sheet were to slide off into the Ross Sea in response to an accelerated anthropogenic temperature rise, for example, sea level would quickly rise ~6 m, drowning most coastal cities. These small polar regions, like the shelves adjacent to industrialized nations, may be of greater importance in present global cycling of elements than their spatial extent would suggest.

At midlatitudes, the remaining continental shelves extend over a 20-fold larger area than upwelling systems of the same seasonal light regime and receive 24% of the world's river discharge (Table III). Rivers such as the Hudson, Vistula, Yangtze, Rhine, Po, Loire, and Mississippi have nitrate contents of >70 μg-at. NO$_3$ l^{-1}, however, whereas the St. Lawrence, Churchill, and Kuskokwim have concentrations of <10 μg-at. NO$_3$ l^{-1}, such that half of the freshwater input to the mid-latitude shelves has been eutrophied by humans. On an average basis, the annual production of these eutrophic shelves (154.2 g C m^{-2} yr^{-1}) is evidently twice that of the unimpacted, mesotrophic shelves at the same latitudes (81.9 g C m^{-2} yr^{-1}) and perhaps larger than the western boundary shelves at low latitudes (137.4 g C m^{-2} yr^{-1}).

Before the Industrial Revolution and major changes in the terrestrial inputs of nutrients from fertilizers, sewage wastes, and landscape conversions, the input of nutrients from the open sea boundary to all the shelves was the major factor, after light, in determining daily primary production. The vertical transfer of nutrients within the open ocean at the diffusive time scale, for example, is one to two orders of magnitude less than the upward flux of nutrients on continental shelves due to coastal upwelling at the land boundary, spinoff eddies of slope currents at the shelf break, or tidal mixing (Walsh, 1976). To provide a basis for later consideration of the impacts of eutrophication, a qualitative comparison is first made here of an upwelling system, a tidally mixed system, and a seasonally stratified system, upstream of estuarine input at the same midlatitudes, that is, similar seasonal light regimes. A comparison of the seasonal cycles of nutrients, phytoplankton, and higher trophic levels on low-latitude shelves, as well as in the open sea, is then presented.

1.3   Boreal Shelf Structure

A static description of three shelf ecosystems at $40-42°$N off Oregon, on Georges Bank, and south of New York can be obtained by averaging observations over a number of years to depict seasonal cycles at the 60-m isobath. We ignore, at present, the consequences of water motion and alongshore inhomogeneity, that is, patchiness, of both nutrient inputs and marine organisms. Land-derived nutrient sources have steadily increased since the onset of the Industrial Revolution in 1850, for example, and are now partly responsible for large point sources of phytoplankton blooms, fixing $300-500$ g C m^{-2} yr^{-1} at the mouths of major U.S. rivers, such as the Hudson (Malone and Chervin, 1979), the Altamaha (Thomas, 1966), and the Mississippi (Fucik, 1974). Since the onset of the Holocene period 10,000 yr ago, however, onshore intrusions of slope water, at the wind-event (Walsh *et al.*, 1978; Hicks and Miller, 1980), eddy (Morgan and Bishop, 1977; Bishop *et al.*, 1980), and diffusive (Riley, 1967a) time scales, have also provided nutrients as a line source to the coastal zone, by subsurface shoreward movement of nutrient-rich water over long distances along horizontal isopycnal surfaces (Iselin, 1939).

In an attempt to simplify an analysis of shelf-break nutrient exchange in the absence of cultural eutrophication, seasonal cycles of temperature, nutrients, phytoplankton, zooplankton, and larval fish are provided for three shelf areas of different vertical mixing regimes, but of similar remoteness from estuarine influence, for example, Georges Bank and the New York shelf located to the east of the Hudson River and the Oregon shelf to the south of the Columbia River. As an example of the present wealth of coastal data sets, during 14 cruises in 1977–1980 on Georges Bank, as well as during 76 cruises in 1957–1961 and in 1974–1982 on the New York shelf south of Long Island, temperature, salinity, nutrients, and chlorophyll were routinely measured by personnel from the Woods Hole Oceanographic Institution, the Brookhaven National Laboratory, and the NMFS-Sandy Hook Laboratory. Additional hydrographic and chlorophyll data are available from Georges Bank during six cruises in 1939–1940 (Riley, 1941), eight cruises in 1964–1966 (Colton *et al.*, 1968), and nine cruises in 1975–1976 (Pastuszak *et al.*, 1982); nutrient data were also available from the 1939–1940 and 1975–1976 cruises. A comparable data set, at a similar latitude, was also obtained during 18 cruises by the University of Washington staff in 1961–1963 to study coastal waters off the Oregon coast (Stefansson and Richards, 1964).

Away from estuarine influences, the seasonal distribution of nutrients in the Mid-Atlantic Bight (Cape Hatteras to Georges Bank) exhibits some variability from year to year (Walsh *et al.*, 1986b), depending on the severity of the winter, as well as differences in intensity and frequency of storms during the rest of the year. Nevertheless, as in the case of a 15-yr time series of nutrients in the English Channel (Cooper, 1938), it is possible to describe the general trends of supply and utilization by combining the data sets from several years in the Mid-Atlantic Bight to estimate the yearly cycle of nutrients. Nitrogen will be used as the example nutrient, because it is both the most complex, with gaseous forms, and the most likely limiting nutrient on the continental shelf at recent (Ryther and Dunstan, 1971) or interglacial (Walsh, 1984) time scales; the other nutrients also tend to follow the same patterns. Consideration of the NO_3^{2-} and NH_4^+ species of nitrogen also allows differentiation of "new" and "recycled" production —(Dugdale and Goering, 1967) in terms of that supplied from the aphotic zone (NO_3) and that from excretory release by bacterioplankton and metazoans (NH_4), assuming low rates of local nitrification, that is, bacterial conversion of ammonium to nitrate.

1.3.1 *Temperature*

Wind events are an important source of habitat variability on the continental shelf, in contrast to the open ocean (Walsh, 1976; Beardsley *et al.*, 1976), and are responsible both for the generation of transient currents, that is, upwelling, and for vertical mixing. Because of the alignment of the North American continent with respect to the planetary wind field at 40°N (Fig. 5), a southerly wind tends to favor offshore surface flow, as a result of the apparent Coriolis force, on the east coast of the United States (i.e., to the right), compared to a northerly wind on the west coast. Nutrient-rich, cold subsurface water moves onshore and upwells at the coast to replace the warmer, nutrient-impoverished surface water, advected seaward by winds favorable to upwelling (Walsh, 1975). A downwelling circulation (onshore flow of surface water and sinking at the coast) occurs when winds are from the respective opposite directions off New York and Oregon. Unlike spin-off eddies, coastal upwelling is a terrestrial boundary process, in which most of the water is upwelled within only 10–20 km from the coast, with secondary cross-shelf flows set up as a function of the shelf width (Walsh, 1977). Under conditions of weak stratification and strong winds, vertical mixing of the water column also occurs in addition to the upwelling/downwelling cross-shelf circulation patterns (Walsh *et al.*, 1978); a quantitative description of these processes will be deferred to Chapters 2 and 3.

The average wind speed at the coast, between Cape Hatteras and Georges Bank, drops from ~9 m sec^{-1} in January to ~5 m sec^{-1} in July, similar to changes in the frequency of wind events, that is, from 5 to 6 in January to half that number by September (Walsh *et al.*, 1978). Such changes in the intensity and frequency of the wind forcing, as well as those of the vernal heating cycle and river runoff, lead to a homogeneous water column from December to March, with intense stratification of surface waters by August on the New York shelf. The annual temperature range is ~20°C (Fig. 8) at the mid-shelf region. Destratification occurs by November, with fall overturn of the water column and a return to well-mixed conditions.

Because of the persistent westerly wind fields over North America (Fig. 5), temperatures of the well-mixed water column on the mid-Atlantic shelf are lowered by polar outbreaks of wind from the land on the east coast during winter, while the shelf waters remain warm from winds off the open ocean on the west coast, for example, Oregon. Higher winter temperatures, and the summer upwelling over most of the Oregon shelf, together lead to an annual temperature range of only ~5°C (Fig. 8B) within the upper 30 m at the midshelf. Consequently, the annual temperature cycle at the 60-m isobath, 60 km off New York (Fig. 8A) is that of a typical boreal shelf, while 10 km off Oregon, at the 60 m isobath, it is that of a summer upwelling regime (Fig. 8).

On some shelves, tidal motion is an important factor of vertical mixing in addition to the wind forcing. Large tidal velocities of 55 cm sec^{-1} maximum amplitude, for example, can lead to roughly the same vertical mixing as a wind at 13 m sec^{-1} (Pingree *et al.*, 1978); tidal mixing energy is applied from below the water column, as opposed to the wind energy from above. Maximal tidal velocities at <60 m depths are ~15–25 cm sec^{-1} in most of the New York Bight, compared to ~55–110 cm sec^{-1} on Georges Bank. These latter velocities are similar to those of 1 m sec^{-1} around the British Isles (Simpson and Pingree, 1978), where an index of the tidal mixing (Simpson and Hunter, 1974) has been formulated as the ratio of depth (h) to the cube of the amplitude (u) of the tidal

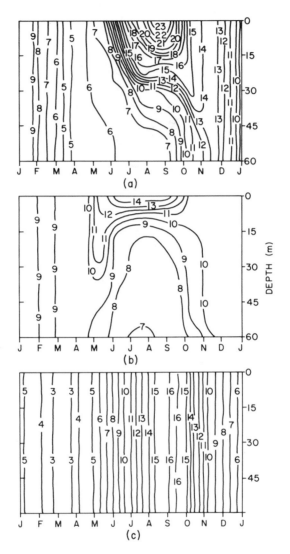

FIG. 8. A composite of the annual temperature (°C) cycle at the 60-m isobath off (a) New York, (b) Oregon, and (c) on Georges Bank. [After Walsh (1981a); © Academic Press.]

stream, that is, hu^{-3}. The reciprocal of this ratio and a constant drag coefficient, $Ch^{-1}u^3$, is the mean tidal energy dissipation rate per unit mass (Pingree *et al.*, 1978). Because of the large range of values in both parameters, a log scale is used to estimate areas of stratification (log $hu^{-3} \geq 2$ or log $h^{-1}u^3 \leq -2$), transitional areas or fronts (1.5 or -1.5), and tidally well mixed areas (≤ 1 or

$\geqslant -1$). Larger tidal velocities and shallower depths, of course, favor increased tidal mixing. At depths of 40–60 m in the New York Bight, a tidal velocity of 25 cm sec^{-1} suggests values of 3.41–3.59 for log hu^{-3}, that is, stratification of the water column. A tidal velocity of 1.10 m sec^{-1} at similar depths on Georges Bank, however, suggests values of 1.48–1.65, that is, an area of possible tidal mixing. The seasonal temperature cycle at 60 m on Georges Bank does, in fact, reflect tidal mixing with an isothermal vertical structure at all times of the year (Fig. 8). Because of both tidal and wind mixing on Georges Bank, the winter temperature minimum is about the same as that of the New York shelf, but the summer temperature maximum is similar to that off Oregon.

1.3.2 Nutrients

The effective rate of a continuous vertical flux by tidal mixing on Georges Bank has been estimated (Falkowski, 1983) to be $\sim 10^{-1}$ cm sec^{-1}, in contrast to a seasonal upwelling velocity of $\sim 10^{-2}$ cm sec^{-1} off Oregon, and $\sim 10^{-3}$ cm sec^{-1} on the New York shelf after the onset of stratification. Despite these differences in vertical exchange, as indicated by the thermal structure, between Oregon, Georges Bank, and the New York shelf (Fig. 8), the seasonal nitrate pattern in the upper 30 m on Georges Bank (Fig. 9) is remarkably similar to those of shelf waters farther south and west. Until the onset of the spring bloom on Georges Bank (Fig. 10), for example, the nitrate content of isothermal winter water in January 1940 (Riley, 1941) and February 1976 (Pastuszak *et al.,* 1982) was 6.5–7.5 μg-at. NO$_3$ l^{-1} (Fig. 9), the same as off both Oregon (Fig. 9) and New York (Fig. 9).

In January and February, the winter isothermal conditions of the New York and Oregon shelves are typified by both the relatively high concentrations of nitrate and the low amounts of ammonium (Fig. 11) at the 60-m isobath. These dissolved-nitrogen patterns are the result of reduced uptake rates by phytoplankton, nutrient renewal by onshore entrainment of slope water, and vigorous vertical mixing processes on the shelf. Concentrations of nitrate at the 60-m isobath (Fig. 9) are as high as 7 μg-at. NO$_3$ l^{-1} at all depths, reflecting the homogeneous conditions of the well-mixed water column and the low phytoplankton biomass of each ecosystem (Fig. 10).

1.3.3 Light

Because light penetration decays exponentially with depth in the ocean, phytoplankton that are mixed over the whole water column on Georges Bank experience less light than those in surface layers of stratified deeper waters (Riley, 1942). The relationship between vertical mixing, light intensity, and phytoplankton growth was first quantified as a critical depth concept (Sverdrup, 1953),

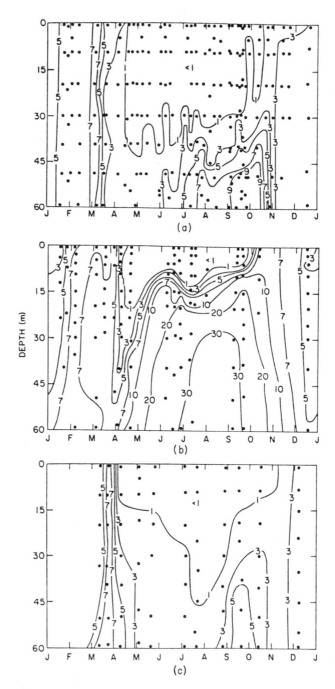

FIG. 9. A composite of the annual nitrate (μg-at. l^{-1}) cycle at the 60-m isobath off (a) New York, (b) Oregon, and (c) on Georges Bank. [After Walsh *et al.* (1986b); © MIT Press, with permission.]

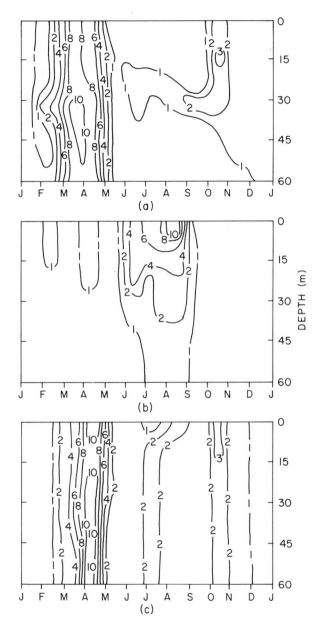

FIG. 10. A composite of the annual chlorophyll ($\mu g\ l^{-1}$) cycle at the 60-m isobath off (a) New York, (b) Oregon, and (c) on Georges Bank. [After Walsh (1981a); © Academic Press.]

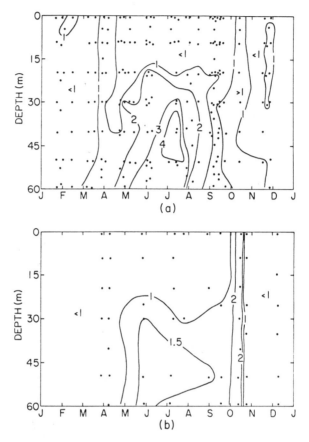

FIG. 11. A composite of the annual ammonium (μg-at. 1^{-1}) cycle at the 60-m isobath (a) off New York and (b) on Georges Bank. [After Walsh *et al.* (1986b); © MIT Press, with permission.]

below which the 24-hr respiration of the water column exceeds the integrated daily photosynthesis. The critical depth is $h_c \approx 0.2\, I_0\, (kI_c)^{-1}$, where I_0 is the incident radiation, k is the extinction coefficient, and I_c is the compensation light intensity at which algal photosynthesis equals respiration ($\sim$0.3 ly hr^{-1}). Sverdrup's concept was that if $h_c < h_m$, the depth to which the phytoplankton were mixed as a result of wind and/or tidal stirring, then no bloom would occur, even in the presence of high nutrient content; additional variants of this theme will be discussed in Section 3.1.1.

A rough estimate of the extinction coefficient can be obtained by lowering a flat white plate of 30 cm diameter, that is, a Secchi disk, into the sea; the depth at which it just disappears is about 10% of surface radiation and is termed the Secchi depth h_S. In turbid, seasonally well-mixed shelf waters with chlorophyll (chl) concentrations of ~ 2.0 μg chl l^{-1} (Fig. 10), the Secchi disk depth is ~ 11 m at the 60 m isobath in the New York Bight. Empirically, k is related to either the Secchi depth, h_S, by $k = 1.44$ h_S^{-1} (Holmes, 1970), that is, $k = 1.44(11)^{-1} = 0.131$ m^{-1}, or alternatively to chlorophyll content by $k = 0.04 + 0.0088$ [chl] $+ 0.054$ [chl]$^{2/3}$ (Riley, 1956), that is, $k = 0.143$ m^{-1}. If the average illumination of a mixed water column, $\bar{I}_h = I_0 (kh_m)^{-1}$, increases to an intensity at which the gross photoplankton production is greater than the loss rates from respiration, grazing, and sinking, then a bloom will ensue. Riley (1967b) and Hitchcock and Smayda (1977) found that above an illumination threshold of $\bar{I}_h = 40$ g-cal cm^{-2} day^{-1} (ly day^{-1}) (g-cal = gram calorie) phytoplankton blooms occurred in Long Island Sound and Narragansett Bay.

At the beginning of March off New York, Georges Bank, and Oregon, the incident radiation is <250 g-cal cm^{-2} day^{-1} and $\bar{I}_h$ is thus 32 g-cal cm^{-2} day^{-1} within the well-mixed 60-m water column; the critical depth is also too shallow at 53 m. However, incident illumination increases to more than 300 g-cal cm^{-2} day^{-1} by the end of March off New York (Walsh *et al.*, 1978) and Oregon (Small *et al.*, 1972), with associated increases in critical depth, average *in situ* light intensity, and phytoplankton biomass (Fig. 10). Similar changes in spring light intensity and critical depth have been calculated for the Irish Sea, with mean I_0 and h_c increasing from 250 g-cal cm^{-2} day^{-1} and 60 m in mid-March to 430 g-cal cm^{-2} day^{-1} and 100 m in mid-April (Pingree *et al.*, 1976).

The major phytoplankton bloom occurs in March and April on Georges Bank and off New York, compared to August and September off Oregon. With onset of water-column stratification by May and a decline in the number of wind-induced, slope intrusions of nutrient-rich water off New York, the surface waters are stripped of nitrate. Post-bloom spring surface concentrations are, at first, about 0.5 μg-at. NO$_3$ l^{-1} over the New York shelf. By late summer, however, undetectable nitrate concentrations are often observed as deep as 30 m, that is, the bottom of the euphotic zone. In contrast, concentrations of nitrate within the summer, subsurface cold pool (Fig. 8) at 50- to 60-m depths off New York can be as high as 10 μg-at. NO$_3$ l^{-1} (Walsh *et al.*, 1978), compared to the same concentration at the bottom of the euphotic zone off Oregon, and half this amount on Georges Bank (Fig. 9).

As a result of the surface nutrient depletion, high chlorophyll concentrations of summer are restricted to subsurface maxima, located just above the nutricline and at the bottom of the euphotic zone, on the New York shelf (Fig. 10). The summer phytoplankton off New York are thus mainly dependent on recycled nitrogen, for example, amino acids and ammonium, derived from zooplankton, bacterial, or benthic remineralization of particulate nitrogen from the previous spring bloom. Phytoplankton uptake of nitrogen, determined by ^{15}N incubation techniques, consists of as much as 80% ammonium during the summer off New

York (Harrison *et al.*, 1983), compared to ~50% at the end of the spring bloom (Conway and Whitledge, 1979). Ammonium concentrations are highest (~4–5 μg-at. NH_4 l^{-1}) in summer near the bottom (Fig. 11), amounting to more than 15 μg-at. NH_4 l^{-1} during August 1976, and precede the buildup of nitrate in the late summer aphotic zone (Fig. 9). Uptake studies of ^{15}N by phytoplankton on Georges Bank indicate that the persistence of homogeneous, moderate levels of chlorophyll (1–2 μg chl l^{-1}) throughout the summer water column (Fig. 10) results from tidal mixing of ammonium (Fig. 11), with less buildup of nitrate in subsurface waters (Fig. 9). Ammonium is the dominant form of nitrogen on Georges Bank between July and September, constituting more than 90% of the inorganic nitrogen pool in both the surface and bottom layers. Higher phytoplankton uptake rates of ammonium on Georges Bank in summer, rather than additional nitrifying bacterial use of this substrate in the New York aphotic zone, contribute to the greater annual primary production of 380 g C m^{-2} yr^{-1} on Georges Bank, compared to 300 g C m^{-2} yr^{-1} on the New York shelf, and 190 g C m^{-2} yr^{-1} on the Oregon shelf.

During the fall, a phytoplankton bloom (Fig. 10) responds to destratification of the water column in October and November, with a resultant nitrate concentration of ~2 μg-at. NO_3 l^{-1} throughout the New York water column after the nutrient-impoverished surface water and rich bottom water become mixed (Fig. 9). The 5 μg-at. NO_3 l^{-1} minimal increment of nitrate from fall to winter on the New York shelf is presumably supplied from either longshore transport of shelf water from Georges Bank, or onshore advection of slope water (Gordon *et al.*, 1976) during fall and winter upwelling events. With cessation of seasonal upwelling off Oregon, the greater input of nitrate to this ecosystem is curtailed in the fall, until similar amounts of nitrate are found by winter off both New York and Oregon. At the end of October, biological influence on the nitrogen cycle is declining, since incident radiation is less than 300 g-cal cm^{-2} day^{-1} in all three shelf systems and the critical depth becomes <50 m. The seasonal declines in primary production, nutrient uptake, and light then continue toward their winter minima. With appropriate time delays, this seasonal signal of carbon fixation is passed up the food web on these shelves.

1.3.4 Herbivores

At a level of ecological complexity corresponding to trophic levels (Lindeman, 1942), these shelves appear to have the same biological structure, despite their apparent differences in primary productivity. The diversity of nearshore zooplankton species is the same within 18–25 km of the coast; using a similar size of net mesh, 29 and 26 species of copepods, respectively, were found to be abundant in the New York Bight (Judkins *et al.*, 1980) and off Oregon (Petersen and Miller, 1975). The abundance of estuarine-dependent fish is also similar

(McHugh, 1976), constituting 44.1% by weight of the U.S. commercial catch in the Mid-Atlantic Bight, in contrast to 45.2% off Oregon–Washington, both in 1970. Finally, the number of common fish species on the slope and shelf is the same ($\sim$200) off Oregon (Pearcy, 1972; Alton, 1972) and New York (Hennemeuth, 1976), while the annual fish yield is similar, $\sim$10 tons km^{-2} yr^{-1}, as well. With the same output of fish, these east coast shelf systems may produce 100–200 g C m^{-2} yr^{-1} of more uneaten phytodetritus, available for export, than the west coast ecosystem. The seasonal pulses of copepod biomass off New York, Georges Bank, and Oregon (Fig. 12) reflect the regional differences in both temperature cycles and the timing of phytoplankton blooms. For example, *Pseudocalanus minutus* and *Centropages typicus* are the dominant calanoid copepods off New York and Georges Bank (Bigelow and Sears, 1939; Clarke, 1940; Sears and Clarke, 1940; Grice and Hart, 1962; Sherman, 1978; Judkins *et al.*, 1980), with the former found across the shelf in the winter–spring and the latter inshore during summer–fall; *Calanus finmarchicus* is also abundant offshore after the spring bloom. The New York offshore peak of zooplankton ($>$50 m) occurs during May–June, as offshore waters warm about one month after the major diatom bloom in March–April (Fig. 10). Inshore zooplankton ($<$50 m) become abundant during July–August, following the seasonal succession of phytoplankton from netplankton to nanoplankton after stratification of the water column.

The zooplankton biomass on Georges Bank and in the Gulf of Maine (Fig. 12) exhibits spatial and seasonal patterns (Bigelow, 1926; Redfield, 1941; Riley, 1947) similar to the respective inshore and offshore copepod communities within the New York Bight (Fig. 12). A similar zonal distribution of the same genera of copepods also occurs off Oregon, with *Pseudocalanus* sp. found across the shelf, *Calanus marshallae* offshore, and *Centropages abdominalis* inshore (Fig. 12) during the summer upwelling period (Petersen *et al.*, 1979). However, *Pseudocalanus* sp. dominates both the offshore ($>$50 m) summer peak of zooplankton in July–August, which coincides with the midshelf phytoplankton bloom, and the smaller cross-shelf peaks of copepods that occur in the absence of high chlorophyll during October (Petersen and Miller, 1975, 1977).

Pseudocalanus is a cold-water form (Corkett and McLaren, 1978), that grows well at low temperatures (Vidal, 1978) and intermittent food supply (Dagg, 1977), in contrast to *Centropages*. The summer difference in temperature between Oregon and New York waters is $\sim$10°C, a full Q_{10} range, such that *Pseudocalanus* would be metabolically favored both during summer upwelling off Oregon and in the colder spring off New York. *Pseudocalanus* is the dominant copepod on the middle Bering Sea shelf as well, and plays an important role in the interannual fluctuations of energy within this ecosystem, as discussed in Section 4.5.2. Off New York, *Pseudocalanus* females lay fewer eggs than *Centropages*, yet adult abundance is similar (Dagg, 1978), suggesting that predation

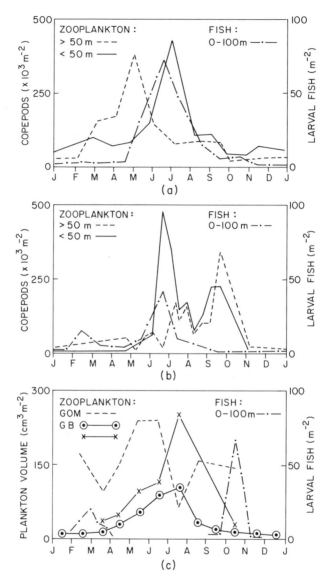

FIG. 12. Annual cycle of copepods on the inner (<50 m) and outer (>50 m) continental shelf and of larval fish over the upper 100 m off (a) New York, (b) Oregon, and (c) Georges Bank–Gulf of Maine. [After Walsh (1981a); © Academic Press.]

may be less during the spring growth phase of this cold-water form, compared to predator–prey interactions of the summer food web.

1.3.5 Invertebrate Predators

The usual increase of inshore (<50 m) chaetognath biomass after the spring peak of *Pseudocalanus minutus* may, in fact, allow the slower-growing *Centropages* to succeed *Pseudocalanus* off New York during summer. Moreover, *Centropages* is also an omnivore, in contrast to *Pseudocalanus*, and may thus partially subsist on its own or other invertebrate nauplii within the summer food web off New York. Clarke *et al.* (1943) have suggested that two to three generations of *Sagitta elegans* are produced each year on Georges Bank. There are, in fact, indications of maturation in May–June, September–October, and November–December of three broods, that is, cohorts, of these predators in all three of the shelf systems.

The main chaetognath peak is smaller and later in the year off Oregon, however, in contrast to those off New York and Georges Bank (Redfield and Beale, 1940). Similarly, major pulses of ctenophore predators occur in the fall off both California (Hirota, 1974) and New York (Malone, 1977a), after the peak of *Centropages typicus*, compared to the summer bloom of ctenophores off Oregon (Petersen and Miller, 1976). The low temperature, lack of fall bloom, and different timing of predator stress may thus all act to prevent *Centropages* from succeeding *Pseudocalanus* as the dominant copepod off Oregon. Conversely, the continued existence of a herbivore off Oregon during summer and fall may not allow a fall bloom to develop in October, just as the lack of a spring bloom in the North Pacific was thought to be due to constant grazing pressure of *Neocalanus cristatus* (McAllister *et al.*, 1960).

1.3.6 Vertebrate Predators

Seasonal cycles of larval fish (Fig. 12) tend to follow those of the zooplankton herbivores. The May–June peak of larval fish in the New York Bight comprises mainly yellowtail flounder and mackerel, while the smaller winter peak on Georges Bank consists of cod, haddock, and sand lance, following that of herring in the fall. Data were not available for the summer period on Georges Bank, but yellowtail and mackerel spawn in this area as well (Colton *et al.*, 1979). Similarly, off Oregon, the winter peak of larval fish represents sand lance, butter sole, and English sole, while the summer maximum comprises capelin (Richardson and Pearcy, 1977). Although the larval survival of fish may be a critical factor in determining fluctuations of adult abundance (Lasker, 1975; Walsh *et al.*, 1980), ichthyoplankton predators probably have little direct impact on energy flow through the lower trophic levels that support them.

For example, daily respiration of larval fish populations within the summer water column of the New York Bight represents a maximum oxygen demand of only 1.6 ml O_2 m^{-2} day^{-1}, in contrast to 48 ml O_2 m^{-2} day^{-1} for the chaetognaths, 360 ml O_2 m^{-2} day^{-1} for the benthos, and 500 ml O_2 m^{-2} day^{-1} for the copepods (Falkowski et al., 1980); see Section 6.4.4. Furthermore, a nitrogen budget for the New York shelf (Walsh et al., 1978) suggests that 53% of the nearshore phytoplankton biomass in August is consumed daily by the copepods, when predators are abundant, whereas only 6% of the inshore bloom is grazed in May, and 7% of the midshelf bloom (Fig. 10) in March, when few chaetognaths and larval fish are present. A quantitative discussion of carbon and nitrogen fluxes through these coastal food webs is deferred, however, to Chapter 4.

1.4 Tropical Shelf Rectification

At the seaward edge of the shelves of tropical seas, the daily rate of nutrient input at the shelfbreak by western boundary currents is an order of magnitude less than those introduced by coastal upwelling of the eastern boundary currents. Eddy-induced upwelling by the Guiana and Loop Currents at the edge of the Amazon and West Florida shelves, for example, may lead to annual production rates of perhaps only 30–90 g C m^{-2} yr^{-1}. Such estimates are based on extrapolation of daily measurements from just one or two cruises, however, without monthly time series to resolve possible seasonal fluctuations of primary production and energy transfer to the rest of the food web. One frequently cited time series of phytoplankton biomass during 1928–1929 off the Great Barrier Reef (Marshall, 1933) was unable to detect much seasonal variation. Theories were then advanced 10–30 yr later (Bogorov, 1941; Cushing, 1959), to be perpetuated 50 yr later (Valiela, 1984), on the time-invariant nature of primary production in the tropics. Subsequent data sets (Sournia, 1977) suggest, in fact, that 10-fold seasonal signals of production on tropical shelves do follow time-dependent inputs of nutrients, differing only in phase from boreal shelves, and in contrast to intrinsic differences between these and the food webs of the open ocean (Heinrich, 1962a; Walsh, 1976) at tropical or boreal latitudes.

1.4.1   Terrestrial Signal

With a minimal incident radiation of 250–300 g-cal cm^{-2} day^{-1} in July–August at 10°N off the Malabar coast of India, 10-fold seasonal increases of phosphate and chlorophyll at the 20-m isobath of the Laccadive shelf instead follow salinity minima (Fig. 13) during the southwest monsoon months of April to September (Shah, 1973). The low salinity of this region (Fig. 7) is derived from land drainage, which peaks in the rainy season of the southwest monsoon, providing even

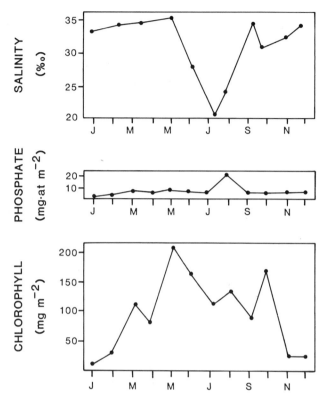

FIG. 13. The annual cycle of salinity (‰), phosphate (mg-at. m⁻²), and chlorophyll (mg m⁻²) on the Laccadive shelf. [After Shah (1973); © Springer-Verlag, with permission.]

stronger signals of salinity, nitrate, and chlorophyll (Fig. 14) within the adjacent Cochin backwater (Qasim, 1973). As in boreal coastal waters, such as 42°N in the New York Bight (Ryther and Dunstan, 1971), nitrogen rather than phosphorus limits shelf primary production at 10°N. The seaward gradient of nitrate in June–August was from 30.0 μg-at. NO_3 l^{-1} within the estuary to 0.3 μg-at. NO_3 l^{-1} 10 km offshore, while estuarine phosphate levels were 1.5 μg-at. PO_4 l^{-1}, with still 1.0 μg-at. PO_4 l^{-1} found at the offshore stations. Similar to an estimate of 371 g C m⁻² yr⁻¹ for Apalachicola Bay at 30°N (Livingston *et al.*, 1974), the annual primary production of the Cochin backwater might be at least 195 g C m⁻² yr⁻¹, of which 30 g C m⁻² yr⁻¹ may be consumed by zooplankton herbivores (Qasim, 1970), allowing significant export of the unconsumed phytodetritus.

Stimulation of tropical shelf production from local drainage basins during the rainy season has also been observed at 13°S off Madagascar, with a 10-fold

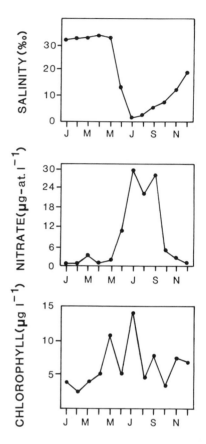

FIG. 14. The annual cycle of salinity (‰), nitrate (μg-at. l^{-1}), and chlorophyll (μg l^{-1}) in the Cochin backwater. [After Qasim (1973); © Springer-Verlag, with permission.]

increase of nitrate and chlorophyll in May–June (Sournia, 1972). Over a larger scale, before construction of the Aswan Dam in 1967, the Nile River used to discharge ~75 × 10^2 m^3 sec^{-1} in September after the summer rainy season within the 3 × 10^6 km^2 drainage basin from 10° to 20°N. This seasonal input of fresh water was about 10-fold that of the present discharge (Table III); it impacted the salinity at the 40-m isobath of the adjacent shelf 10–20 km offshore (Fig. 15), contained as much as 6.4 μg-at. PO_4 l^{-1} (Halim, 1960)(i.e., two orders of magnitude more phosphate than shelf waters), and used to support a local fall fishery for sardine, *Sardinella aurita*, of 1 × 10^4 tons yr^{-1} (Aleem and Dowidar, 1967).

In response to this fluvial nutrient input, phytoplankton biomass is still one to two orders of magnitude higher off the Nile mouth, 0.6–6.0 μg chl l^{-1} (Dowidar, 1984), than in the offshore Levantine Basin, 0.02–0.08 μg chl l^{-1} (Berman *et al.*, 1985). In the past, daily primary production rates of 7.7 g C m^{-2} day^{-1}

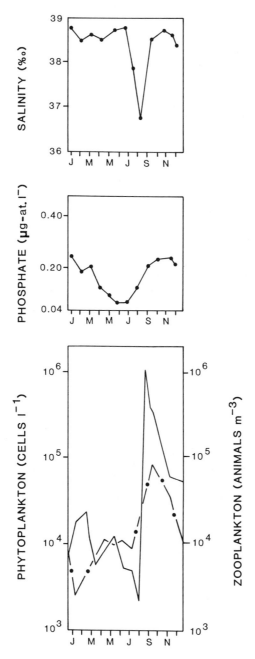

FIG. 15. The annual cycle of salinity (‰), phosphate (μg-at. l^{-1}), phytoplankton (cells l^{-1}), and zooplankton (number m^{-2}) on the Egyptian shelf. [After Aleem and Dowidar (1967); © The University of Miami, with permission.]

were observed here (Halim, 1960), as high as off Peru, and in contrast to <20 g C m^{-2} yr^{-1} off the coast of Israel (Berman *et al.*, 1984); accordingly, the zooplankton prey of the Egyptian sardine used to increase more than 10-fold (Fig. 15). Similarly, after the rainy season within the 6 × 10^6 km^2 drainage basin of the Amazon River, an estuarine nitrate content of 9.3 μg-at. NO$_3$ l^{-1} in June was 100 times that of either the same estuary in November, or of the adjacent inner shelf waters throughout the year (Ryther *et al.*, 1967). Like the Nile-stimulated pulse of marine production, phytoplankton populations within the plume of the Amazon River at 10°N are also two orders of magnitude larger than outer shelf waters (Hulburt and Corwin, 1969).

1.4.2 Oceanic Signal

On some tropical shelves, the fluvial signal of the rainy season is either masked by other inputs of nutrients or shifted in time by light limitation. Although major rivers such as the Niger and the Congo, with a combined drainage basin of 5 × 10^6 km^2, discharge into the Gulf of Guinea, the seasonal plankton dynamics of the shelves at 5°N to 5°S off the Ivory Coast (Reyssac, 1966), Ghana (Houghton and Mensah, 1978), and the Congo (Berrit, 1964) are all apparently dominated by summer upwelling. A 10-fold higher primary production is evidently found in the Gulf of Guinea during October than during May (Bessenov and Fedosov, 1965). With onset of upwelling demarcated by cold temperatures in June, for example, the now-familiar 100-fold increase of tropical phytoplankton abundance and 10-fold increase of zooplankton numbers occurs off the coast of Ghana by September (Fig. 16). An attenuated form of this seasonal signal of tropical phytoplankton production was also found within slope waters, 40 km off the Ivory Coast (Reyssac, 1966). This time series suggests that, in addition to utilization by herbivores, an export of uneaten phytoplankton may occur here, as well as within the major upwelling ecosystems, e.g., Peru (Walsh, 1981b).

Another seasonal fishery for *Sardinella aurita*, of similar yield as the previous Nile-induced harvest, is located within the coastal waters of the Gulf of Guinea (Fig. 16), with most of the harvest and spawning of the sardine taking place during July–August (Houghton and Mensah, 1978). Interannual variations in the yield of this clupeid stock off Ghana from 1964 to 1968, a period of increasing fishing pressure, were attributed to large catches during years of strong upwelling *and* low rainfall (Bakun, 1978). The impact of overfishing and the importance of estuarine sources of food for the juvenile fish must also be considered, however.

Curtailment of river discharge into the Gulf of Guinea during the 1972–1978 Sahelian droughts, for example, led to significant reductions in zooplankton biomass and *Sardinella* yield within the adjacent waters of the Ivory Coast (Longhurst, 1983), where upwelling during August is less intense than it is off Ghana

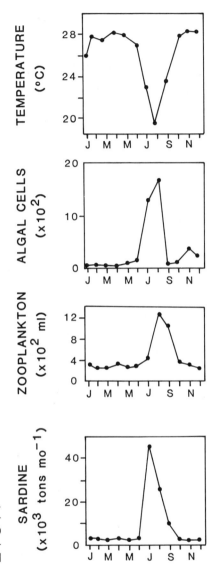

FIG. 16. The annual cycle of temperature (°C), phytoplankton (cells ml⁻¹), zooplankton (number ml⁻¹), and sardine landings (tons month⁻¹) on the Ghanian shelf. [After Houghton and Mensah (1978); © Springer-Verlag, with permission.]

(Bakun, 1978). The yields of all fish off Ghana and the Ivory Coast in 1968 were respectively 7.0×10^4 tons yr⁻¹ and 6.9×10^4 tons yr⁻¹, that is, the same, while by 1978 the yields had diverged, with 22.2×10^4 tons yr⁻¹ landed off Ghana and only 7.5×10^4 tons yr⁻¹ taken off the Ivory Coast. Farther north, another West African coastal ecosystem, which has been significantly impacted

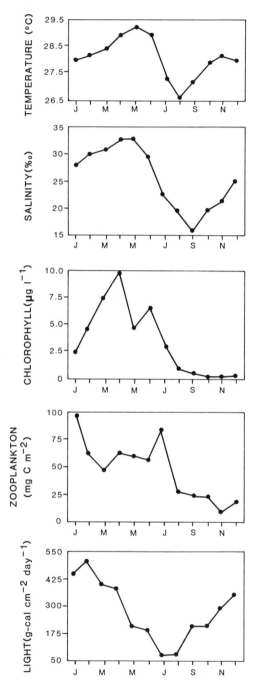

FIG. 17. The annual cycle of temperature (°C), salinity (‰), chlorophyll (μg l^{-1}), zooplankton (mg C m^{-2}), and light (g-cal cm^{-2} day^{-1}) on the Sierra Leone shelf. [After Longhurst (1983); © Academic Press, Inc. (London) Ltd.]

by human fishing, is the much wider continental shelf (~125 km) off the Sierra Leone River (Longhurst, 1983).

1.4.3   Atmospheric Signal

The annual temperature range of coastal waters at 8°N off Sierra Leone is only 2°C (Fig. 17), compared to the 10°C range off Ghana (Fig. 16), such that up-welled nutrients are not a major source of seasonal variance in primary production of this shelf. The impact of the summer rainy season is again manifested in both low salinity at the 20-m isobath (Fig. 17), and nitrate concentrations of 5–10 μg-at. NO_3 l^{-1} (Bainbridge, 1960; Watts, 1958), which should induce a July–August bloom of phytoplankton, as occurs at about the same latitude off the Malabar coast (Fig. 13). Ubiquitous cloud cover during the rainy season off Sierra Leone reduces the incident radiation, however, to 69 g-cal cm^{-2} day^{-1} in July–August, from a maximum of 480 g-cal cm^{-2} day^{-1} in February (Longhurst, 1983), which is an 86% reduction—recall that the spring bloom off New York began at ~250 g-cal cm^{-2} day^{-1}. Cloudiness off the Malabar coast also reduces the seasonal maximum of 580 g-cal cm^{-2} day^{-1} in February to 250 g-cal cm^{-2} day^{-1} during June (Qasim, 1973), which is a 57% reduction, but this decline of light energy is not sufficient to limit primary production off India.

A 10-fold seasonal increase of chlorophyll biomass and zooplankton abundance off Sierra Leone is thus instead delayed to January through June, a period of relatively high salinity (Fig. 17), that is, the dry season (Longhurst, 1983). A carbon fixation of as much as 1 g C m^{-2} day^{-1} is estimated for this period, when only 10–30% of the production is then removed by herbivore populations to support either a fishery for another species of sardine, *Sardinella eba*, or one for the menhaden, *Ethmalosa fimbriata*, feeding directly on phytoplankton (i.e., an analog of the Peruvian anchovy, *Engraulis ringens*) (Walsh *et al.,* 1980). Carbon budgets of the Sierra Leone shelf (Longhurst, 1983) and the remains of terrestrial vegetation in the deep sea off West Africa (Wolf, 1979) both suggest that export of organic matter to the continental slope may be a feature of both tropical and boreal shelves (Walsh *et al.,* 1981).

1.5   Food Web Coupling

In contrast to the apparent excess primary production of the coastal zones of the Gulf of Guinea, oceanic herbivores of the offshore waters of this gulf appear to require more phytoplankton food than is produced (Le Borgne, 1975). These offshore waters of the tropical Atlantic are oligotrophic, with an apparent seasonal range of only two- to threefold changes in nitrate content and primary production of the euphotic zone (Voituriez and Herbland, 1979). Invoking a simplistic Liebig's (1840) "law of the minimum" and the evidence for light or

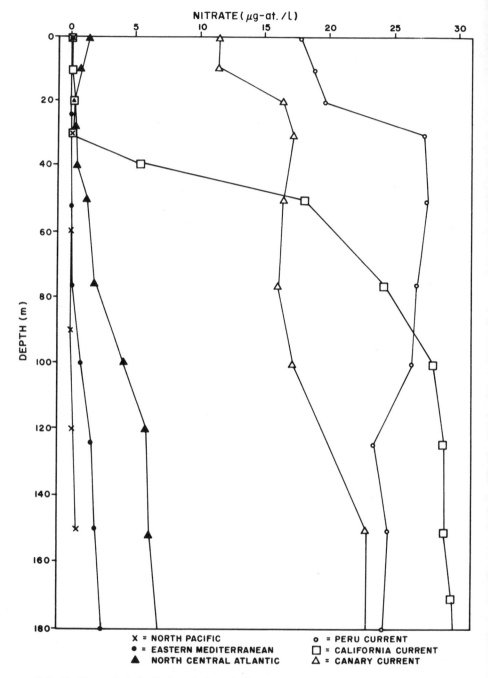

FIG. 18. The vertical distribution of nitrate in the open ocean and within the eastern boundary currents. [After Walsh, 1976; © American Society of Limnology and Oceanography, Inc., with permission.]

Table IV.

Comparison of the Consequences of the Rate of Nutrient Input to
Three Types of Subtropical/Tropical Ecosystems[a]

	Coastal upwelling	Offshore divergence	Central gyre
Vertical velocity (cm sec^{-1})	10^{-2}–10^{-3}	10^{-3}–10^{-4}	10^{-4}–10^{-5}
NO$_3$ (μg-at. l^{-1}) at 100 m	20–30	20–30	0–5
Surface horizontal gradient (km) from >10 to 0.5 μg-at. l^{-1}	30–60	300–600	None
Primary production (g C m^{-2} day^{-1})	1–10	0.5–1.0	0.1–0.5
Phytoplankton standing crop (mg chl a m^{-2})	60–180	45–90	15–30
Assimilation index (mg C mg chl a^{-1} hr^{-1})	1–10	1–10	1–5
Herbivore diversity	Low	Intermediate	High
Temporal variability of herbivore populations	High	Lower	Low
Herbivore mean weight (μg C animal^{-1})	40–400	4–40	4–40
Steps in the food chain (excluding mammals)	1–2	3–4	5–6
Terminal yield	High	Intermediate	Low

[a] After Walsh (1976); © American Society of Limnology and Oceanography, Inc., with permission.

nutrient limitation of algal production, exhibited in the above time series of these variables on shelves off the United States, Ireland, India, Madagascar, Egypt, Brazil, Sierra Leone, Ivory Coast, Ghana, and the Congo, addition of nutrients to the open ocean at low latitudes should result in increased primary production and excess food for herbivores.

A comparison of the nutrient content of the euphotic zone in some subtropical oligotrophic and eutrophic regions of the ocean (Fig. 18), for example, might suggest at first glance that nutrients such as nitrate, and their rate of supply, may be the most limiting factors (Dugdale, 1967; Eppley et al., 1973). In such coastal upwelling areas, primary production and phytoplankton standing crops are higher than in the offshore oligotrophic areas by almost an order of magnitude, as are the high nutrient input and content (Table IV). Since light is usually abundant in the subtropics, a causal relationship between nutrient abundance and productivity may be assumed.

1.5.1 Equatorial Divergences

However, such an idealized analysis of just the chemical and physical control mechanisms of primary production in the sea unfortunately breaks down when

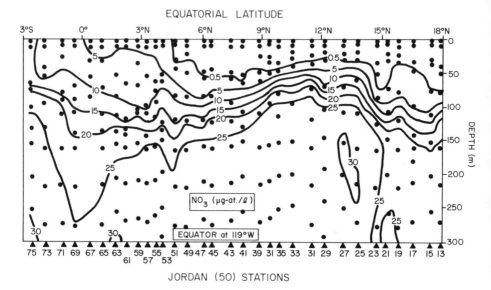

FIG. 19. The gradient of nitrate (μg-at. l^{-1}) in an offshore divergence of the Cromwell Current. [After Love (1974), Walsh (1976); © American Society of Limnology and Oceanography, Inc., with permission.]

the offshore divergences of the low latitudes are considered (Fig. 19). The horizontal nitrate gradient from 10 to 0.5 μg-at. NO$_3$ l^{-1}, north of the Pacific equatorial divergence at 119°W, is 10-fold wider than those of the coastal upwelling systems, yet the supply rate is 10-fold less. The spatial extent of this equatorial gradient of unused nutrients appears to be a persistent feature, analogous in origin to a spreading from a line source as far west as 119°W (Love, 1974), 140°W (Sverdrup *et al.*, 1942), and 160°W (Cromwell, 1953; Reid, 1965).

Light and nutrients thus appear to be sufficient for phytoplankton growth in this equatorial divergence system. The assimilation index [mg C(mg chl *a*)$^{-1}$ hr^{-1}], or estimate of potential growth, of individual phytoplankton cells in the equatorial divergence (Barber and Ryther, 1969) appears similar to those usually found for organisms in both coastal upwelling systems (Barber *et al.*, 1971; Walsh *et al.*, 1974; Estrada, 1974) and oligotrophic gyres (Thomas, 1970a; Eppley *et al.*, 1973). Furthermore, nitrogen enrichment experiments have been performed on phytoplankton from oligotrophic water in the California Current (Eppley *et al.*, 1971) and near the equatorial divergence (Thomas, 1970b), and both gave about the same maximal algal division rate of 0.7–1.5 doublings day^{-1}, that is, similar to those of the rich Peru and Baja California upwelling systems (Walsh, 1975; Walsh *et al.*, 1974).

Analysis of variance of integrated primary production with latitude in the

EASTROPAC observations (Owen and Zeitzschel, 1970) demonstrated no significant spatial differences across the equatorial divergence. Examination of the seasonal cycles, averaged over 3°S to 16°N, of chlorophyll, zooplankton, and primary carnivores (e.g., 1- to 10-cm fish and squid) (Blackburn *et al.*, 1970), indicates less than twofold variation in their standing crops (Fig. 20). Such small seasonal changes were the same as those of the pelagic ecosystem within the equatorial divergence of the Atlantic Ocean (Voituriez and Herbland, 1979). A more recent analysis (Dessier and Donguy, 1985) of chlorophyll and copepod abundance across the Pacific equatorial divergence from 1978 to 1982 also shows only a twofold seasonal variation, similar to the EASTROPAC data in 1967–1968 (Blackburn *et al.*, 1970).

One is thus faced with an anomaly of offshore ecosystems with relatively high nutrients, abundant light, perhaps no intrinsic differences in potential growth between these phytoplankton and those of coastal communities, yet evidently low autotrophic utilization of the nutrients occurs within oceanic upwelling areas. There is no evidence here of spatial or temporal blooms of microalgae on seasonal or decadal time scales as encountered within the food webs of continental shelves. Differential importance of herbivory in marine ecosystems may explain the contrast of horizontal nutrient gradients among the three types of

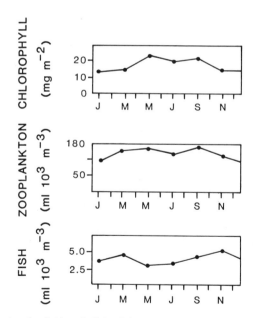

FIG. 20. The annual cycle of chlorophyll (μg l^{-1}), zooplankton (ml m^{-3}) and fish (ml m^{-3}) within the Pacific equatorial divergence. [After Blackburn *et al.* (1970); © Springer-Verlag, with permission.]

pelagic systems (Table IV), for the loss rate of a phytoplankton population, as opposed to an individual cell, may also set its growth rate (Dugdale, 1967).

The central biotic provinces of the oceans have been characterized as high-diversity systems with little temporal variability (McGowan, 1974) of their low zooplankton standing crop (Table IV). In contrast, the eastern boundary currents appear to have zooplankton populations of low diversity, high standing crop, and high variability (Longhurst, 1967a; Wickett, 1967). A similar gradient of lower diversity and higher standing crop of herbivores also appears to be a function of the increased upwelling rate between coastal and offshore divergences (Timonin, 1971). The size of individual herbivores (Taniguchi, 1973) declines in oceanic ecosystems, while the number of steps in the central gyre food chains (Ryther, 1969) and of those of the equatorial divergence becomes larger (Sette, 1955). One must ask, however, which of these smaller, more diverse herbivores of the open ocean food webs is capable of growth rates equivalent to those of the phytoplankton, to prevent an algal bloom from taking place.

Large netplankton, such as diatoms, require an energy subsidy in the form of wind or tidal mixing (Margalef, 1978) to prevent these nonmotile organisms from sinking out of the water column at velocities of as much as 10 m day^{-1} (Smayda, 1970; Eppley et al., 1978). Even at low sinking rates within experimental enclosures (Eppley et al., 1978; Oviatt, 1981), other phytoplankton, that is, flagellates with their own energy subsidy of motility, replace diatoms as the dominant phytoplankton. Larger phytoplankton cells lead to shorter food chains and are usually found in the eastern boundary currents and coastal regions of the sea rather than in the open ocean (Ryther, 1969; Malone, 1971a; Parsons and Takahashi, 1973; Margalef, 1974). With summer stratification of temperate coastal waters, a seasonal succession of the netplankton diatoms by smaller phytoplankton usually occurs as well (Yentsch and Ryther, 1959; Malone et al., 1983).

In initial studies of size fractionation within tropical oceanic phytoplankton communities, algal cells passing through a 107-μm mesh net constituted 99.9% of phytoplankton biomass within the Florida Straits (Bsharah, 1957). Subsequent studies of nanoplankton of <20–35 μm diameter indicated that these small organisms (e.g., flagellates) constituted as much as 75% of the biomass and 80% of the primary production of tropical waters (Malone, 1971b). More recent studies in the eastern tropical Pacific (Li et al., 1983) and south equatorial Atlantic (Herbland and LeBouteiller, 1981) indicate that the even smaller picoplankton of <1–3 μm size (i.e., coccoid cyanobacteria) may constitute 25–90% of the chlorophyll biomass and ~20–90% of the primary production in the open ocean. Most metazoan herbivores, such as copepods, are not capable of grazing the picoplankton, but protozoans of the microzooplankton size class are candidates for a biological control mechanism, which can prevent a bloom of picoplankton from developing in the equatorial divergences. Within the EASTRO-

PAC studies, protozoans were the second dominant group of zooplankton, after copepods, in the equatorial Pacific, forming 10% of the total abundance (Blackburn et al., 1970). Further, the microplankton consumed only 23% of the daily phytoplankton production of the California Current, but 70% of the production across the equator at 105°W (Beers and Stewart, 1971). In offshore habitats, the herbivores may have evolved evolutionary strategies, such as seasonal migration in high latitudes (McAllister et al., 1960; Voronina, 1972), and increased speciation in low latitudes, to anticipate phytoplankton blooms. An expansion of the number of niches within a relatively stable physical habitat would allow a wide diversity of oceanic herbivores to crop all size classes of phytoplankton (Sheldon et al., 1972).

In contrast to the continental shelves, grazing may be an extremely efficient control mechanism within the gyres and, perhaps, offshore divergences, such that most phytoplankton are always consumed by herbivores in both systems. Based on moored sediment traps in the equatorial Atlantic and the north Pacific gyre, for example, less than 2% of the surface primary production survives the descent to 5000 m (Honjo, 1980). During coastal upwelling within the California Current, however, as much as 53% of the primary production is caught within sediment traps on the slope at 700 m (Knauer et al., 1979).

1.5.2   Subarctic Gyres

Seasonally invariant, open-ocean food webs of the tropics are tightly coupled such that 90% of the nitrogen demand of daily primary productivity is met by recycled forms of nitrogen, e.g., metazoan or bacterial excretory products of ammonium, urea, and amino acids (McCarthy and Goldman, 1979). In such situations, weak or undetectable seasonal pulses of nutrient injection, of phytoplankton growth, and of secondary production make it very difficult to isolate the processes of autotrophic fixation of carbon from those of heterotrophic consumption. Within a boreal oceanic ecosystem, such as the subarctic North Pacific at 50°N, the interaction of phytoplankton and the herbivores should be uncoupled seasonally, since there is a light pulse in the spring. After 30 yr of observations near Station P at 50°N, 145°W, however, little seasonality of chlorophyll biomass is evident (Fig. 21).

The cyanobacteria of the picoplankton are ubiquitous (Johnson and Sieburth, 1979; Waterbury et al., 1979) and, together with cryptomonads, small diatoms, coccolithophores, and other eukaryotes, dominate the flora of the North Atlantic (Murphy and Haugen, 1985) and North Pacific (Booth et al., 1982) as well as in the tropical open ocean. The <5 μm size fraction contributes as much as 98% of the phytoplankton abundance and 74% of their biomass near Station P. A time-invariant chlorophyll structure of 0.3–0.5 μg chl l^{-1} at Station P thus implies herbivores that are capable of continuously grazing the picoplankton, either mi-

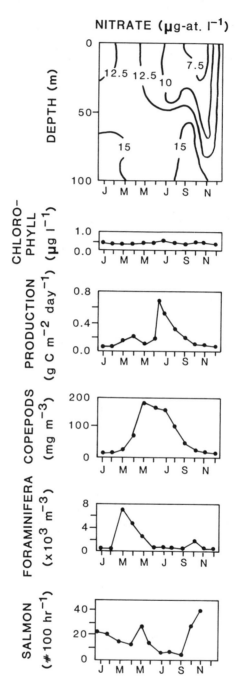

FIG. 21. The annual cycle of nitrate (μg-at. l^{-1}), chlorophyll (μg l^{-1}), primary production (g C m^{-2} day^{-1}), copepods (mg m^{-3}), foraminifera (number m^{-3}), and salmon *landings* (hr^{-1}) within the subarctic Pacific at Station P. [After LeBrasseur (1965), Stephens (1968), Anderson *et al.* (1969), LeBrasseur and Kennedy (1972), Fulton (1978), and Valiala (1984); © 1969 Pergamon Journals, Ltd., and © 1984 Springer-Verlag, with permission.]

crozooplankton and/or ontogenetic migrators of this ecosystem, such as *Neocalanus plumchrus*. Based on pigment budgets, for example, microzooplankton may graze 95% of the primary production in the central Pacific gyres, 64% at Station P, and 33% in a coastal fjord off the Washington coast (Welschmeyer and Lorenzen, 1985; C. Lorenzen, personal communication). The large copepods, *N. plumchrus* and *N. cristatus*, are capable of grazing particles of 2 μm diameter, but feed more efficiently on particles >5 μm at Station P. Their feeding potential is three- to fourfold greater than has been actually measured in this ecosystem at ambient algal concentrations of 0.3 μg chl l^{-1} (Frost *et al.*, 1983). These same herbivores, feeding on the diatom spring bloom of 10–20 μg chl l^{-1} on the Bering shelf (Dagg *et al.*, 1982; Vidal and Smith, 1986), for example, have a threefold larger body size than their counterparts at Station P. Although there is little indication of food limitation of these grazers on the Bering shelf, they remove only 18–75% of the daily primary production here. It is possible that these copepods are instead omnivores at Station P, feeding on both the microzooplankton and the picoplankton.

After overwintering as adults in deep waters of the North Pacific (Heinrich, 1962a; C. B. Miller *et al.*, 1984), these calanoid copepods undergo a seasonal migration in spring to surface waters. They remain here until July–August, when once more they enter a diapause phase and exit the euphotic zone. Such a regular seasonal signal of copepod biomass at Station P, observed from 1956 to 1976 (LeBrasseur, 1965; Fulton, 1978), may impact the summer productivity of their plant prey, the abundance of their microzooplankton prey, and the availability of their fish predators (Fig. 21).

A seasonal increase of daily primary production at Station P from ~0.2 g C m^{-2} day^{-1} in April to ~0.6 g C m^{-2} day^{-1} in August (Fig. 21) may reflect possible successional changes in specific growth rates (per day) of different components of the picoplankton with no alteration of plant biomass (i.e., g chl m^{-2} or g C m^{-2}). Shade-adapted cyanobacteria of slower growth rates may be replaced by light-tolerant diatoms or coccolithophores, after a reduction of vertical mixing by shoaling of the thermocline, from ~50 m in the spring to 30 m by August–September (Dodimead *et al.*, 1963). In any case, nitrate stocks of the upper 30 m (Fig. 21) are actually reduced from ~5.7 g NO$_3$ m^{-2} in April to 2.7 g NO$_3$ m^{-2} by September (Anderson *et al.*, 1969). The amount of dissolved silicate in surface waters is strongly depleted by August–September as well (Honjo, 1984)—that is, it is an index of diatom production. The total algal fixation of dissolved C, N, and Si as particulate matter thus increases, with relaxation of grazing pressure by August. Since the calanoid grazers of the macrozooplankton leave the upper 30 m of the water column by August and the microzooplankton abundance may also be less then, compared to the spring (Fig. 21), the sinking flux of the picoplankton must increase if their chlorophyll biomass remains constant. An estimate of the seasonal increase of these inferred phytoplankton sinking losses can be obtained from computation of "new" production, derived both from nitrate depletion within the upper 30 m and from its resupply by vertical diffusion of subsurface stocks; the implied partial differential equations of such a mass balance are described in Section 3.1.4. The observed depletion rate over time, from 13.6 μg-at. NO_3 l^{-1} in April to 6.4 μg-at. NO_3 l^{-1} in September within 0–30 m of surface water (Fig. 21), is 3 g NO_3 m^{-2} per 180 days. This suggests a mean nitrate uptake by phytoplankton of 16.7 mg NO_3 m^{-2} day^{-1}, or a production of 100.2 mg C m^{-2} day^{-1} using a C/N ratio of 6 : 1, for example, with no renewal of surface nitrate from the aphotic zone each day.

The vertical exchange rate for the nitrate flux across the thermocline in two regions of the Pacific Ocean has been estimated to be about 1.0 cm^2 sec^{-1}, that is, 8.6 m^2 day^{-1} (King and Devol, 1979; Eppley et al., 1979). The observed vertical nitrate gradient was ~6 μg-at. NO_3 l^{-1} over the 30 m and 50 m depths of the September water column at Station P (Fig. 21), that is, 4.2 mg NO_3 m^{-4}. Multiplying the estimate of vertical transfer and of the nitrate gradient yields a diffusive resupply for this time of year of an additional 36.1 mg NO_3 m^{-2} day^{-1}, or a production of 216.6 mg C m^{-2} day^{-1} with a C/N ratio of 6 : 1. A computed total carbon fixation of 316.8 mg C m^{-2} day^{-1} from the temporal and spatial changes of nitrate, compared to the measured average production of 600 mg C m^{-2} day^{-1} (Fig. 21), suggests that "new" production is ~50% of the total production at Station P. This result is similar to other measurements on the continental shelves (Walsh, 1983) and even within the Sargasso Sea when nitrate is abundant (Platt and Harrison, 1985). This amount of ~300 mg C m^{-2} day^{-1} of new production would be available for summer export within the water column at Station P, either as sinking particles or as migratory fish.

Based on one 10-day-long mooring of seven sediment traps over the 4200-m water column near Station P in June 1978, perhaps 7% of the daily production might arrive at 250 m after leaving the euphotic zone, with 4% surviving the rest of the descent to the sea floor (Lorenzen et al., 1983a). Deployment of two other time-series sediment traps at 1000-m and 3800-m depths of the Station P water column from March to October 1983 (Honjo, 1984) captured the seasonal signal of primary production (Fig. 22), similar to the results of long-term trap studies in the Sargasso Sea (Deuser and Ross, 1980). An order of magnitude more organic carbon was caught at Station P with the July–September trap samples than with the March–April samples.

A summer input of 30–40 mg C m^{-2} day^{-1} to the traps at 1000–3800 m (Fig. 22) would be 9–13% of the algal export, estimated above from the nitrate budget to be perhaps sinking out of this North Pacific euphotic zone. The May peak of biogenic matter in the sediment traps was also associated with both zooplankton remains, such as fecal pellets and radiolarian shells, and a large flux of carbonate from the coccolithophorids (Honjo, 1984). The August peak of organic matter, opal, and carbonate consisted instead of mainly phytoplankton

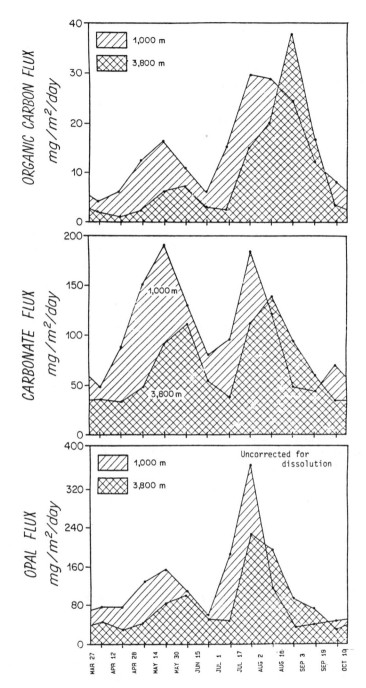

FIG. 22. The annual cycle of organic carbon, carbonate, and opal fluxes (mg m⁻² day⁻¹) caught with sediment traps moored at 1000 m and 3800 m near Station P. [After Honjo (1984).]

detritus, corroborating a seasonal cycle of algal export from the euphotic zone, induced here by biotic factors rather than light and nutrients. A bulk settling velocity of 150 m day^{-1}, derived from time lags of particle arrival at these 1000-m and 3800-m traps (Honjo, 1984), raises intriguing questions, however, about the mode of delivery of the phytodetritus when grazing is presumed minimal; we shall return to this subject in Section 5.1.

Preservation of a seasonal signal of primary production also occurs within sediment traps at 3200 m near Bermuda at 32°N, 65°W in the Sargasso Sea (Deuser, 1986). The daily detrital flux is instead only 1% of the surface carbon production and requires a similar rapid-descent mechanism (Deuser and Ross, 1980). The seasonal range of primary production at Station S in the Sargasso Sea (Fig. 23) is 0.2–0.8 g C m^{-2} day^{-1} (Menzel and Ryther, 1960), about the same as the Station P variability in the North Pacific (Fig. 21), but the maximum occurs during the winter–spring rather than summer–fall. Furthermore, the seasonal maximum of organic carbon caught within sediment traps at 3200 m near Station S in the Sargasso Sea is only 3–4 mg C m^{-2} day^{-1} (Fig. 23), compared to 30–40 mg C m^{-2} day^{-1} at 3800 m near Station P (Fig. 22). Seasonality of primary production at Station S is attributed to nutrient injection by convective overturn of the water column (Fig. 23), with direct transfer of the resultant carbon fixation to zooplankton biomass (Menzel and Ryther, 1961a) and production of fecal pellets. Transfer of phytodetritus through herbivores at Station S, with the usual ecological efficiency of 10% (Slobodkin, 1961), would provide 10-fold less input to sediment traps here than at Station P in the absence of summer grazing stress. As one moves toward the oligotrophic waters of the tropics, nutrient reserves of the oceanic gyres become less and grazing pressure by microzooplankton increases. The winter maximum of nitrate at Station S in the Sargasso Sea (Fig. 23) is only 1–2 μg-at. NO_3 l^{-1} (Menzel and Ryther, 1961b), for example, compared to 10–15 μg-at. NO_3 l^{-1} at Station P (Fig. 21). Just a 10% reduction in grazing stress for a few days at Station S might thus allow phytoplankton growth equal to the usual spring blooms (Sheldon et al., 1973), if nutrients were available on an intermittent basis (Platt and Harrison, 1985).

1.5.3   Subtropical Shelves

Although the food webs at Stations P and S are based on picoplankton, removal of grazing stress may lead to 10-fold greater seasonal export of algal carbon in the North Pacific, similar to boreal, tropical, and, as we shall see, subtropical shelves. Within the oligotrophic Gulf Stream in the 800-m depth of the Florida Straits at 25°N, ~10 km off the Bahama Banks, for example, the annual range of surface temperature is only ~3°C, and the phosphate content is less than 0.1 μg-at. PO_4 l^{-1} within the upper 100 m throughout the year (Fig. 24). Nitrate

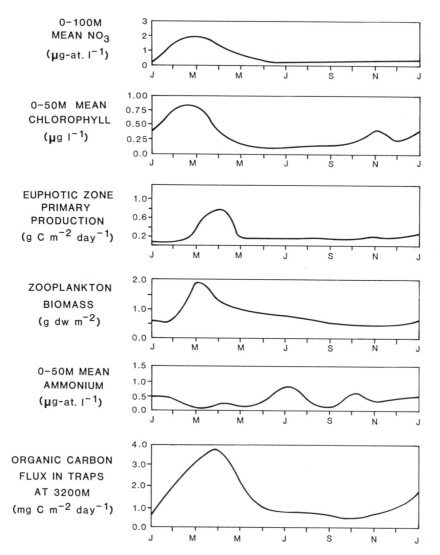

FIG. 23. The annual cycle of nitrate (μg-at. l^{-1}), chlorophyll (μg l^{-1}), primary production (g C m^{-2} day^{-1}), zooplankton (g dw m^{-2}), ammonium (μg-at. l^{-1}) in the euphotic zone, and of organic carbon flux (mg m^{-2} day^{-1}) at 3200 m within the north Atlantic at Station S near Bermuda. [After Menzel and Ryther (1960), Menzel and Ryther (1961a), Menzel and Spaeth (1962), and Deuser and Ross (1980); © 1960 Pergamon Journals, Ltd., © 1961 International Council for the Exploration of the Sea, © 1962 American Society of Limnology and Oceanography, Inc., © 1980 Macmillan Journals, Ltd., with permission.]

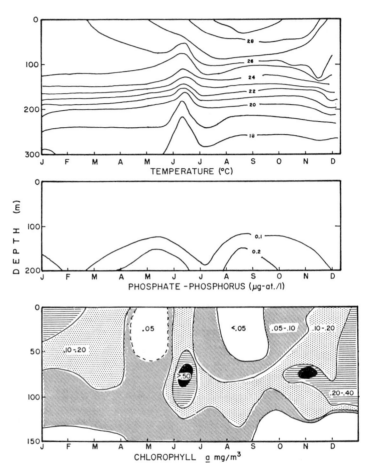

FIG. 24. Seasonal profiles during 1960 of temperature, phosphate, and chlorophyll *a* at Cat Cay, Bahamas. [After Alexander *et al.* (1962).]

concentrations of >10 μg-at. NO_3 l^{-1} are usually found deeper than 300 m (Bsharah, 1957), similar to other oceanic regimes (Fig. 18). A small seasonal maximum of chlorophyll (Alexander *et al.,* 1962) might occur in winter (Fig. 24), similar to observations in the Sargasso Sea (Menzel and Ryther, 1960). However, a seasonal maximum of dry weight of nanoplankton, the dominant organisms (Miller and Moore, 1953), was instead observed in summer during a second study (Bsharah, 1957), while seasonal maxima of algal cell numbers were found during spring in a third study (Vargo, 1968). Such results either suggest no persistent seasonal cycles, or imply that fluctuations of algal population dynamics in this subtropical region are induced by phenomena at shorter time scales, that is, unresolved, or aliased, by the above three sampling plans of monthly intervals.

Farther to the north, at the edge of the Florida–Georgia shelves between 29 and 32°N, spin-off eddies of the same Gulf Stream, in fact, upwell more than 10 μg-at. NO_3 l^{-1} of the deep subtropical water to the bottom of the shelf euphotic zone. These nutrient intrusions induce diatom blooms, similar to those of the seasonal cycles discussed above for boreal and tropical shelves, but over time intervals of 1–3 weeks rather than 1–3 months. Near-bottom intrusions of cold water on the Florida shelf had been detected over 40 yr ago (Green, 1944), but an explanation of their occurrence had to await development of appropriate theories. Classical explanations of the origin of western boundary currents such as the Gulf Stream (Stommel, 1948; Munk, 1950) had suggested that positive (counterclockwise) vorticity was developed at the shelf-edge sides of these fast-moving currents ($\sim$200 cm sec^{-1}), to provide a frictional balance to the negative vorticity of the basin circulation, induced by the curl of the wind stress.

Spin-off cyclonic eddies in the Florida Current (Lee, 1975) do, in fact, rotate counterclockwise, and their occurrence is positively correlated with the curl of the wind stress (Duing et al., 1977)—that is, the greater the curl, the more intense is the positive vorticity generated at the edge of the Florida–Georgia shelf; precise definitions of vorticity and wind curl will be provided in Section 2.6. Ekman suction is created in the center of these eddies (Lee and Mayer, 1977), with an upwelling velocity of $\sim 10^{-3}$ cm sec^{-1} at the edge of the shelf (Hsueh and O'Brien, 1972), in contrast to $\sim 10^{-2}$ cm sec^{-1} found nearshore during coastal upwelling (Walsh, 1975).

As much as 10 μg-at. NO_3 l^{-1} can be brought into the oligotrophic outer shelf euphotic zone within the center of these eddies (Lee et al., 1981), with less than 1 μg-at. NO_3 l^{-1} found at their boundaries (Fig. 25). Diatom blooms of 10 μg chl l^{-1} and daily production rates of 2.0 g C m^{-2} day^{-1} have been associated with these eddies, in contrast to 0.1 g C m^{-2} day^{-1} when they are absent from the outer shelf (Yoder et al., 1985). The spin-off eddies are present about 50% of the time on the outer Georgia shelf, with little seasonal variation, such that present estimates of annual primary production are 360 g C m^{-2} yr^{-1} (Yoder et al., 1985), compared to previous estimates of 130 g C m^{-2} yr^{-1} (Haines and Dunstan, 1975).

The poleward-flowing East Australian Current is another major western boundary current, which induces upwelled intrusions of nutrient-rich water off both the New South Wales coast at 27–32°S (Rochford, 1975) and off the Great Barrier Reef at 18°S (Andrews and Gentien, 1982). Unlike the seasonal cycle of coastal upwelling induced by the spring Southeast Trade Winds off the northwest coast of Australia in the Arafura Sea (Rochford, 1962), however, upwelled intrusions of the East Australian Current near the Great Barrier Reef are not correlated with alteration of the wind regime from 4 months of northeast monsoon to

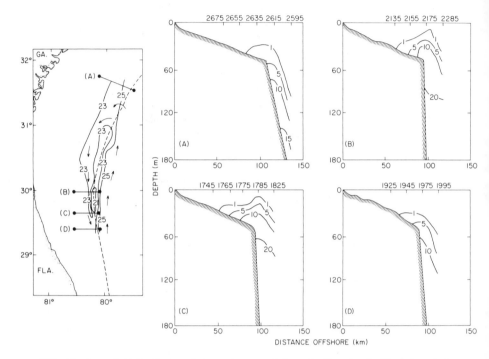

FIG. 25. Cross-shelf distribution of nitrate (in μg-at. l^{-1}) on the Georgia–Florida shelves during April 1977 in relation to surface temperatures of a spin-off eddy. [After Lee *et al.* (1981); © 1981 Pergamon Journals, Ltd., with permission.]

8 months of the Trade Winds (Andrews and Gentien, 1982). Fluctuations of this boundary current occur instead at about 1- to 3-month intervals as topographic Rossby waves (Hamon and Kerr, 1968; Garrett, 1979).

With an intensification of the East Australian Current, an upwelled intrusion leads to a biological response at the shelf break of the Coral Sea (Fig. 26), similar to that within the South Atlantic Bight (Fig. 25). With introduction of up to 10 μg-at. NO_3 l^{-1} at the shelf break, 10- to 100-fold increases of chlorophyll occur on the outer shelf of the Coral Sea after each upwelled intrusion. As much as 175 g C m^{-2} yr^{-1} of primary production is thought to be derived from these nutrient injections (Andrews and Gentien, 1982), which, in fact, were first detected during the original Great Barrier Reef Expedition of 1927–1928 (Orr, 1933). Similarly, increases of chlorophyll at the shelf break of the Tasman Sea near Sydney were attributed in earlier studies to nutrient-rich intrusions of slope water (Humphrey, 1963); subsequent studies found that they were associated with pulses of diatom production here during intensification events of the East Australian Current (Hallagraeff, 1981).

Outwelling of nutrients from estuaries and marshes of the South Atlantic

Bight (Turner *et al.*, 1979) provides an additional source of nutrients, with little apparent seasonal variation (Bishop *et al.*, 1980), except at the mouths of major rivers, such as the Altamaha (Thomas, 1966). Annual primary production within the plume of the Altamaha River is ~600 g C m^{-2} yr^{-1} (Thomas, 1966), while carbon fixation within other regions of the inner shelf (0–20 m isobaths) is ~285 g C m^{-2} yr^{-1} (Haines and Dunstan, 1975). In contrast, land runoff is negligible near the Great Barrier Reef at 18°S (Andrews and Gentien, 1982), such that the phytoplankton production within the open waters of the Reef is only 10–20 g C m^{-2} yr^{-1} (Sorokin, 1973), an order of magnitude less than that of the inner shelf of the South Atlantic Bight. The fate of the primary production on the outer shelves of the Coral and Tasman Seas is unknown, although part may be consumed off Queensland by the Reef communities (Glynn, 1973), while the fate of the outer shelf production of the South Atlantic Bight is probably export.

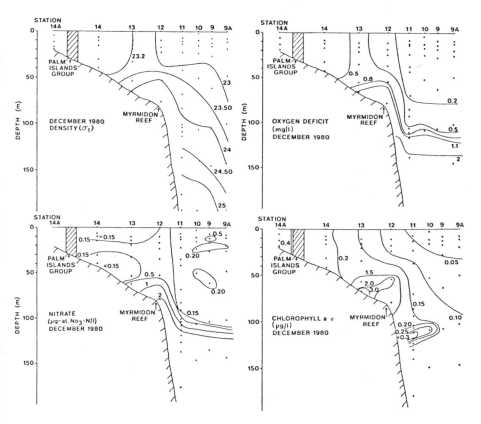

FIG. 26. Vertical sections of density (σ_t), oxygen deficit, nitrate, and chlorophyll *a* during an East Australian Current intensification. [After Andrews and Gentien (1982); © Inter-Research, with permission.]

Nutrients and presumably chlorophyll do not have a long residence time on the outer shelf of the South Atlantic Bight, since this area is flushed after every upwelling event of 1–3 week duration (Lee and Mayer, 1977). Although fast-growing gelatinous zooplankton are found at the shelf break, their abundance is not correlated with the spin-off eddies (Atkinson et al., 1978). Most of the eddy-induced primary production may sink to the bottom and/or be flushed offshore, rather than be incorporated within higher trophic levels. The seasonal oscillations of copepod and larval fish biomass of the Georgia–Florida shelf are instead correlated with nearshore increases both of estuarine-derived nutrients and of landward penetration of the upwelled intrusions during increased stratification of the water column in late summer–early fall (Turner et al., 1979; Yoder et al., 1985). Although the annual primary production of the Mid-Atlantic and South Atlantic Bights may be the same, ~300 g C m^{-2} yr^{-1} (Table III), the fish biomass and yield of the southern ecosystem is ~10% of the northern one, implying most of the primary production from the spin-off eddies is not passed up this shelf food web.

Interaction of a western boundary current with a continental shelf leads to a transition from the nanoplankton community of the open ocean (e.g., that of the Florida Straits) to a netplankton community of the shallow seas (e.g., the South Atlantic Bight) on a time scale of 1–3 weeks with subsequent change in the fate of fixed carbon, from consumption by herbivores to export. Similar changes in species composition of phytoplankton and their associated food webs occur seasonally within an eastern boundary current off Baja California (Walsh et al., 1974, 1977), within the Gulf of Guinea (Reyssac, 1966), on boreal shelves subject to convective overturn (Malone et al., 1983), and on tropical shelves subject to land drainage (Sournia, 1972; Shah, 1973). Increased upwelling within another eastern boundary current off Peru also apparently leads to a switch in the dominant composition of the shelf primary producers, from diatoms to flagellates. The herbivores then crop only 25% of the productivity during strong upwelling off Peru instead of 80% during weak upwelling (Beers et al., 1971; Walsh, 1975), with a terminal fish yield of perhaps only 1% of the primary production in the former case, instead of 10% in the latter (Parsons, 1976). A comparison of a total of 15 continental shelf ecosystems with those of six open ocean regimes suggests, in fact, that on an annual basis only 16% of the primary production of these shelves was utilized by herbivores, in contrast to 80% of that above the deep sea (Joiris et al., 1982). Understanding the coastal ecosystem processes has far greater significance than their overall contribution to total marine carbon fixation (Table II) would suggest because (1) The fate of carbon and nitrogen fixed in these highly productive shelf regions is quite different from the oceanic areas of the sea, with part of the shelf export sinking to slope depocenters instead of most of the primary production being grazed within the water column in the open sea (Eppley and Peterson, 1980; Walsh *et al.*, 1981), and (2) The impacts of human activity are greater in the coastal region.

Thus, there is a strong motivation to provide a quantitative basis for an analysis of synoptic biomass information, coupled with rate process data, required to study these highly dynamic oceanographic regions over longer periods at annual and decadal time scales, as opposed to the above seasonal scale, and in addition to a much higher sampling frequency of hours for resolution of their basic biological processes. Sampling constraints and the statistical requirements of such time series analyses will be discussed in Section 3.2, after the following description of water motion on continental shelves.

2 Circulation

And the professor, who was a very remarkable man, didn't tell them not to be silly, . . . but believed the whole story. "No," he said, "I don't think it will be any good trying to go back through the wardrobe door . . . you won't get into Narnia again by *that* route . . . Indeed, don't *try* to get there at all. It'll happen when you're not looking for it." . . . The Editor has asked me to tell you how I came to write The Lion, The Witch and the Wardrobe, I will try, but you must not believe all that authors tell you about how they wrote their books. . . . At first I had very little idea how the story would go. But then suddenly Aslan came bounding into it. I think I had been having a good many dreams of lions about that time. Apart from that, I don't know where the Lion came from or why He came. But once He was there He pulled the whole story together, and soon He pulled the other six Narnian stories in after him. . . . That is, I don't know where the pictures came from. And I don't believe anyone knows exactly how he "makes things up." Making up is a very mysterious thing. When you "have an idea" could you tell anyone exactly how you thought of it?

(Lewis, 1950, 1960)

Progress in oceanography, like other sciences, may indeed proceed at a seren-dipitous pace, as originally suggested by Horace Walpole in 1754, but invocation of the scientific method of investigation implies a rigorous testing of one's ideas, that is, hypotheses, at some point in time. Without a mathematical definition of Aslan's motion, a quantitative knowledge of his trajectory is impossible, and the case is similar for those particles graphically described by Ernest Hemingway in 1935 off the Cuban coast. Instead, the description of a continental shelf ecosystem is reduced to vague mumbling about an ill-defined black box (i.e., a poorly lit wardrobe) of coastal water. Variability of the physical environment of the continental shelves occurs, furthermore, at all time scales (Fig. 27), from turbulent motion to glaciations, over which marine organisms have adapted various life history strategies. During the Cretaceous, for example, when shelf areal extent may have been three- to fourfold that of today, the present shelf food chain of diatoms, copepods, and teleosts evidently evolved. Succinct records of their population fluctuations since the last Wisconsin glaciation can be seen in phytoplankton varves and fish scales preserved within shelf and slope sediments.

At the time scales of living plankton (weeks to months), drifting with the mean flow on continental shelves, seasonal changes in light, temperature, and stability of the water column determine their species succession. Other motions of water, in response to wind forcing at storm frequencies of <1 week, provide an additional source of energy for injection of nutrients, resuspension of diatoms, and dispersal of fish larvae on continental shelves. Finally, although tidal mixing is important in local shelf areas such as Georges Bank, motions of these and higher frequencies are usually parameterized by mixing coefficients in most ecological studies. To provide a quantitative basis for simulation analysis of carbon and nitrogen fluxes through the food webs of continental shelves, the important scales of motion and their forcing functions will first be described.

Dominant motions on the continental shelves are certainly the rotary currents, which are induced by the tides and amount to more than 50% of the kinetic energy. Theoretical consideration (Csanady, 1982) of the movement of water parcels during a tidal cycle involves rotation of their current vector over 360° in a period close to the Earth's rotation rate, $\Omega = 0.73 \times 10^{-4}$ radians sec^{-1}, with no net displacement. Over a number of these cycles, however, residual motion does occur, with water parcel displacements greater than the diameter of the tidal ellipse. Except for local enhancement of vertical exchange processes from tidal mixing these longer-term residual flows, at lower frequency than the tides, constitute the main transport pattern on shelves, effecting the distribution of physical, chemical, and biological variables, such as temperature, nutrients, and plankton. Transient wind forcing at time and space scales of 2–10 days and 100–1000 km extent is the major stimulus of time-dependent flow, or acceleration of water in shallow seas. We will thus not explicitly discuss tides and waves, the subject of numerous other texts, such as Lamb (1879), Neumann and Pierson (1966), LeBlond and Mysak (1978), and Nihoul (1982).

2.1 Equations of Motion

A quantitative description of water motion began in 1686 with Isaac Newton's second law of motion,

$$A = F/M \tag{1}$$

where A is the acceleration of the fluid, that is, the time derivative, dV^*/dt of the total velocity field (V^*); M is its mass; and F is the applied force. Knowledge of all of the forces (F_i) acting per unit mass on the waters of the continental

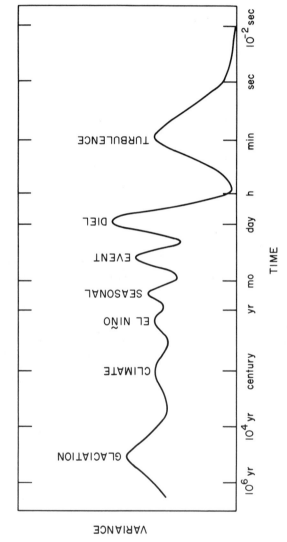

FIG. 27. A hypothetical spectrum of habitat variability from scales of 10^6 yr to 10^{-2} sec. [After Monin (1970).]

shelf, F_i/M, would then allow exact specification of the time rate of change of the flow field, $dV*/dt$, or A. Analytical or numerical solution of $dV*/dt$ would then provide values of $V*$, a description of the fluid motion. The validity of such a Galilean–Newtonian hypothesis, model, or law, however, eventually depends on verification data, which in turn imply measurements and a frame of reference for such observations.

Equation (1) does not hold for an observer who is also accelerated, i.e., it assumes a nonaccelerated, or inertial, frame of reference. A reference frame of observations fixed to the Earth, for example, is not a Newtonian one, due to the rotation of the Earth. A somewhat existentialistic adjustment to Eq. (1) was added 150 yr later by Gaspard Coriolis, allowing transformation of the "idea" from an inertial coordinate system to one that rotates with the Earth (Coriolis, 1835) by

$$A = F_i/M + F_{i+1}/M \tag{2}$$

where the apparent deflecting force of the Earth's rotation, or Coriolis force, is the second term on the right. It is approximated by $F_{i+1}/M = fV*$, in which $f = 2\Omega \sin \theta$, the Coriolis parameter at latitude θ, in the horizontal dimension; $f = 2\Omega \cos \theta$ in the vertical; and Ω was already defined as the Earth's rotation rate.

The Coriolis force is *not* a physical force, but enters the equations of motion as a transformation parameter from a "fixed" to a rotating coordinate system, allowing description of the relative motion of the fluid. The Rossby number, $\text{Ro} = V*L^{-1}\Omega^{-1}$, in which L is the horizontal length scale, provides an index of when this transformation is required, that is, $\text{Ro} \ll 1$. At an along-shore flow of 10 cm sec^{-1} over a 1000-km-long shelf, or at a cross-shelf flow of 1 cm sec^{-1} over a 100-km-wide shelf, for example, $\text{Ro} \approx 1.4 \times 10^{-3}$. Such motions are sufficiently slow that the effects of the Earth's rotation beneath the water column must be considered in further applications of Eq. (1).

2.1.1 Navier–Stokes Equations

By 1850, a more detailed description of the F_i/M terms of Eq. (2) had been formulated as the Navier–Stokes equations of motion. They are a series of non-linear expressions, requiring judicious deletion, or scaling of the forces, to allow solution. Ignoring both tidal motion and the curvature of the Earth over scales of motion less than 1000 km, these equations can be written in a local Cartesian coordinate system of x, y, z axes with the respective u, v, w components of the velocity field ($V*$) taken as positive eastward, northward, and upward from the sea surface. In the Lagrangian framework of an observer following the motion, they are

$$A = \frac{dV*}{dt} = \frac{du}{dt} + \frac{dv}{dt} + \frac{dw}{dt} = \frac{F_i}{M} + \frac{F_{i+1}}{M} \tag{3}$$

The total time derivatives of each component of motion can also be expanded in their Eulerian description past a stationary observer as

$$\frac{du}{dt} = \frac{\partial u}{\partial t} + u\frac{\partial u}{\partial x} + v\frac{\partial u}{\partial y} + w\frac{\partial u}{\partial z} = -\frac{1}{\rho}\frac{\partial P}{\partial x} + F_x + fv \tag{4}$$

$$\frac{dv}{dt} = \frac{\partial v}{\partial t} + u\frac{\partial v}{\partial x} + v\frac{\partial v}{\partial y} + w\frac{\partial v}{\partial z} = -\frac{1}{\rho}\frac{\partial P}{\partial y} + F_y - fu \tag{5}$$

$$\frac{dw}{dt} = \frac{\partial w}{\partial t} + u\frac{\partial w}{\partial x} + v\frac{\partial w}{\partial y} + w\frac{\partial w}{\partial z} = -\frac{1}{\rho}\frac{\partial P}{\partial z} + F_z - g \tag{6}$$

where the partial derivatives of u, v, and w with time are the local change of the flow field; their partial derivatives with space involve the nonlinear advection of momentum; the spatial derivatives of pressure P are the gradient forces arising from differences in density ρ and sea-surface slope; F_x, F_y, and F_z are the frictional forces arising from external stresses of wind forcing and bottom friction, as well as internal turbulent stresses; fv and fu are the horizontal components of the Coriolis pseudoforce, performing no work at right angles to the velocity field; and the gravity term g both is much larger than the vertical component of the Coriolis transformation, which is thus not included in Eq. (6), and incorporates the centrifugal effects of the Earth's rotation in determining the vertical position of the free surface e.

2.1.2 Hydrostatic Approximation

Effective use of this form of the Navier–Stokes equations requires additional simplifying assumptions, begun with the approximate vertical momentum balance of Eq. (6). Vertical velocities in the sea, w, are small, $\sim10^{-1}-10^{-3}$ cm sec^{-1}, compared to horizontal motions, u and v, on shelves, $\sim10^0-10^1$ cm sec^{-1}, and within the western boundary currents, $\sim10^2$ cm sec^{-1}, such as the Gulf Stream or Kuroshio. As a result, the vertical acceleration, dw/dt, is usually small, except for surface wind waves with periods of seconds. A change in w of $10^{-1}-10^{-3}$ cm sec^{-1} over 1 day ($\sim10^5$ sec) after the onset of upwelling or downwelling would lead to an acceleration of $10^{-6}-10^{-8}$ cm sec^{-2}. Similarly, the largest vertical stress gradients, F_z, are usually $\sim10^{-7}$ cm sec^{-2} (Csanady, 1982). In contrast, the gravitational acceleration g is 980 cm sec^{-2}, such that Eq. (6) essentially reduces to

$$\frac{dw}{dt} \approx 0 \approx -\frac{1}{\rho}\frac{\partial P}{\partial z} - g \qquad \text{or} \qquad \frac{\partial P}{\partial z} = -\rho g \tag{7}$$

for residual motions, where Eq. (7) is the hydrostatic approximation of a fluid at rest applied to vertical motion in the sea.

With no net force in the vertical dimension, the pressure P at any depth z can then be found by taking the depth integral of Eq. (7) from the height of the free surface elevation e to depth z, where the hydrostatic equilibrium position of the free surface is assumed to be $z = 0$. We then obtain the vertical pressure

$$P(z) = P(z=0) + \int_{-z}^{e} \rho g \, dz \tag{8}$$

as a function of the density field, where $P(z = e = 0)$ is the atmospheric pressure and the second term is the integrated pressure of the water column, which basically increases linearly with depth, since ρ only varies by a few percent in the sea. Using this simple dynamic height calculation (Sverdrup *et al.*, 1942), those parts of the horizontal pressure gradients, $\partial P/\partial x$ and $\partial P/\partial y$ of Eqs. (4) and (5), attributable to changes in the vertical density field can then be obtained from Eq. (8).

2.1.3   Pressure Fields

The horizontal pressure forces can be separated into those due to spatial changes in the vertical mass field, that is, the relative pressure or buoyancy field, and those due to piled-up mass, that is, the slope of the sea surface, by

$$\frac{\partial P}{\partial x} = \int_{-z}^{e} \frac{\partial \rho}{\partial x} g \, dz + \rho g \frac{\partial e}{\partial x} \tag{9}$$

and

$$\frac{\partial P}{\partial y} = \int_{-z}^{e} \frac{\partial \rho}{\partial y} g \, dz + \rho g \frac{\partial e}{\partial y} \tag{10}$$

where the first terms of Eqs. (9) and (10) are the products of the constant acceleration of gravity and spatial gradients in buoyancy forces, arising from differences in the depth integral of the fluid's density along each axis. The second terms contain the product of the acceleration of gravity and the sea-surface slope, as if the water parcels were rolling downhill in both situations. As a result of the land boundary at the coast, currents due to the slope of the sea surface are particularly important on continental shelves. Their associated acceleration is independent of depth, constituting a barotropic mode of circulation, in which the currents are the same at all depths: that is, $\partial u/\partial z$, $\partial v/\partial z = 0$. The vertical density fields can only be used to estimate the buoyancy field and its associated horizontal acceleration. These latter currents constitute the baroclinic circulation mode, in which the isobaric (equal pressure) and isosteric (equal density) surfaces intersect—that is, the vertical shear has a depth dependence such that $\partial u/\partial z$, $\partial v/\partial z \neq 0$.

To place these two pressure forces in perspective, a wind event lasting one day, $\sim 10^5$ sec, might generate a horizontal velocity of ~ 10 cm sec^{-1} throughout the whole water column, that is, a transient acceleration of 10^{-4} cm sec^{-2} on the continental shelf. Tidal forces, in general, are of the same order of 10^{-4} cm sec^{-2} (Smith, 1976). A sea-surface slope of 1 cm elevation between the coastline and the shelf break 100 km offshore contributes the same acceleration to the horizontal momentum balance as well (Csanady, 1982). To achieve the same acceleration from horizontal changes in the buoyancy field, a lighter surface layer, of $1.0\sigma_t$ difference in density from bottom water, need be only 30 m thick near shore and 20 m deep at the shelf break, that is, a density gradient found during summer on the continental shelf (Csanady, 1982). Thus, at times, both of the pressure forces are equivalent to an acceleration induced by storm events or tides.

2.1.4   Turbulence

The remaining terms of Eqs. (4) and (5) are the friction forces derived from turbulence, wind forcing, and bottom drag. In addition to high frequency accelerations as a result of tidal and inertial forces, apparently random accelerations, at time scales <1 sec, of the water occur. They are complex flows, or eddies, superimposed and interacting with the more repeatable, mean flow patterns at lower frequency. Turbulence provides mixing and exchange of momentum, dissolved substances, and particles, deriving its kinetic energy either from the potential energy of the fluid as a convective instability, or from the kinetic energy of the mean motion as a shear flow instability. Rigorous treatment of turbulent acceleration is beyond the scope of this monograph, with more detailed information found in other texts (Phillips, 1966; Nihoul, 1980). The discussion is limited to a brief description of its parameterization by eddy coefficients, used in the subsequent simulation models.

The acceleration terms of Eq. (4), such as $u \, \partial u/\partial x$, can be written as the sum of mean and turbulent fluctuations of each component of motion, $(\bar{u} + u') \, \partial(\bar{u} + u')/\partial x$, where $\bar{u}$ is a time average of minutes to hours, sufficient to remove fluctuations of the random turbulent flow u' measured at shorter time scales than the averaging period for $\bar{u}$. By definition, the turbulent flow has zero mean, that is, $\bar{u}' = 0$, similar to residuals in a statistical treatment of variance, while the instantaneous flow is $\bar{u} + u'$; time series of similar terms for currents and fluorescence are presented in Section 3.4. Expansion of the above expression leads to

$$u \frac{\partial u}{\partial x} = (\bar{u} + u') \frac{\partial(\bar{u} + u')}{\partial x} = \bar{u} \frac{\partial \bar{u}}{\partial x} + \bar{u} \frac{\partial u'}{\partial x} + u' \frac{\partial \bar{u}}{\partial x} + u' \frac{\partial u'}{\partial x} \quad (11)$$

Taking the time average of Eq. (11), we then obtain

$$\overline{\left(\frac{u\,\partial u}{\partial x}\right)} = \overline{(\bar{u} + u')\frac{\partial(\bar{u} + u')}{\partial x}} = \frac{\bar{u}\,\partial\bar{u}}{\partial x} + \overline{\left(\frac{u'\,\partial u'}{\partial x}\right)} \tag{12}$$

since $\bar{u}' = 0$, but the average of the second term on the right, containing products of the fluctuating quantities, is not necessarily zero. Taking a similar mean of the left side of Eq. (4), we obtain

$$\frac{\overline{du}}{dt} = \frac{\partial\bar{u}}{\partial t} + \bar{u}\frac{\partial\bar{u}}{\partial x} + \bar{v}\frac{\partial\bar{u}}{\partial y} + \bar{w}\frac{\partial\bar{u}}{\partial y} = -\frac{\overline{u'\,\partial u'}}{\partial x} - \frac{\overline{v'\,\partial u'}}{\partial y} - \frac{\overline{w'\,\partial u'}}{\partial z} \tag{13}$$

In collecting the turbulent terms on the right side of Eq. (13), they appear as a force acting on the mean acceleration and are termed the Reynolds stress.

It is difficult to measure Reynolds stress on a small enough time scale of milliseconds to resolve turbulent accelerations; the required sampling protocol is discussed in Section 3.2. Our applications of the Navier–Stokes equations, in any case, concern longer time scales, such that it is convenient to describe the Reynolds stresses in terms of the averaged variables, such as $\bar{u}$, which are measured routinely by current meters every 10–20 min. The turbulent stresses are thus parameterized by gradient transport expressions involving turbulent viscosity, or eddy, coefficients and the mean flow, that is,

$$\frac{\overline{u'\,\partial u'}}{\partial x} = \frac{\partial}{\partial x}\left(K_x\frac{\partial\bar{u}}{\partial x}\right), \quad \frac{\overline{v'\,\partial u'}}{\partial y} = \frac{\partial}{\partial y}\left(K_y\frac{\partial\bar{u}}{\partial y}\right),$$

$$\text{and} \quad \frac{\overline{w'\,\partial u'}}{\partial z} = \frac{\partial}{\partial z}\left(K_z\frac{\partial\bar{u}}{\partial z}\right) \tag{14}$$

We will not continue to use the overbar notation of $\bar{u}$, but it is assumed that all further discussion of motion now refers to suitably averaged velocity components. These eddy coefficients will be applied later to the exchange of nutrients, abiotic particles, and plankton in examples within Chapters 3–6.

If motion in the sea were not turbulent, but laminar, where individual layers of different velocity slide over each other in a vertical shear of $\partial u/\partial z$, for example, the vertical exchange of momentum would only result from random molecular motion. This force on the fluid could then be analogously described by $(\partial/\partial z)(\eta\,\partial u/\partial z)$, where the coefficient η is the molecular dynamic viscosity. It is considered to be a characteristic property of the fluid, resisting angular deformation and independent of the motion, such that the term can be rewritten as $\eta\,\partial^2 u/\partial z^2$. Similar stresses can be described along the other spatial axes, $\eta\,\partial^2 u/\partial y^2$ and $\eta\,\partial^2 u/\partial x^2$. Assuming that seawater is an incompressible fluid, that is, that the change in density of a water parcel is small compared with its density, then $\eta\,\partial^2 u/\partial z^2$ can be converted to the viscous shear stress, $(\eta/\rho)\,\partial^2 u/\partial z^2$, with division by ρ. The coefficient η/ρ is assumed to be a constant, called the kinematic viscosity, and has the same units, cm² sec⁻¹, as the eddy coefficients, K_x, K_y, and K_z. Typical values of $\eta/\rho : K_z : K_x, K_y$ are $10^{-2} : 10 : 10^6$ cm² sec⁻¹ such that the Reynolds stresses are much more important than the viscous shear stresses in the sea.

Of the three Reynolds stresses in Eqs. (13) and (14), the horizontal terms are ignored in some physical problems. The accelerations derived from $\partial/\partial x$ ($K_x \, \partial u/\partial x$) and $\partial/\partial y$ ($K_y \, \partial u/\partial y$) are of the order 10^{-5} cm sec⁻², 10-fold less than the impact of tides, storms, and the horizontal pressure gradients. Since the eddy coefficients are *not* a characteristic property of the fluid, they vary as a function of the averaging period, the state of the motion, and the stratification of the water column; in particular, K_z changes seasonally. For example, K_z is estimated to range from ~30.0 cm² sec⁻¹ during winter in the New York Bight (Stommel and Leetma, 1972) to ~0.3 cm² sec⁻¹ during the summer (Harrison *et al.*, 1983).

During the periods of strong vertical stratification of the water column, the force of the third Reynolds stress, $\partial/\partial z$ ($K_z \, \partial u/\partial z$), across the pycnocline is typically two orders of magnitude less than either the surface wind stress, τ_s, or the bottom drag stress, τ_b, within the two mixed layers at these boundaries. When the vertical density gradient is small, however, the vertical flux of horizontal momentum, $\partial/\partial z$ ($K_z \, \partial u/\partial z$), becomes very important in shallow seas, communicating the surface and bottom stresses of these two mixed layers to the interior of the water column. This process is described by the Richardson number Ri as

$$\text{Ri} = N^2 \left(\frac{\partial u}{\partial z} \right)^{-2} + N^2 \left(\frac{\partial v}{\partial z} \right)^{-2} \tag{15}$$

where N^2 is the Brunt–Väisälä frequency, that is, an index of the "stability" of the fluid due to density stratification, and $\partial u/\partial z$ and $\partial v/\partial z$ are the components of the vertical shear.

On continental shelves, the Brunt–Väisälä frequency is approximated by

$$N^2 \approx -g \frac{\partial \sigma_t}{\partial z} \times 10^{-3} \tag{16}$$

where σ_t is the shorthand notation of density at atmospheric pressure and a specific temperature and salinity, that is, $\sigma = (\rho - 1) \times 10^3$. A sufficient condition for instability of the water column, or enhanced vertical turbulent transfer of momentum, occurs when Ri < $\frac{1}{4}$ (Phillips, 1966). During May 1976 in the Mid-Atlantic Bight after a 2-dyn cm⁻² wind event, for example, Ri was as small as 0.15, and parts of the shelf water column were almost isothermal, indicating well-mixed conditions (Walsh *et al.*, 1978). During August on this shelf, Ri >

2.00 under weak wind conditions, the water column is strongly stratified (see Fig. 42), and little vertical transfer of either momentum or dissolved substances, such as nitrate, takes place (Walsh *et al.*, 1978).

2.1.5 External forcing

At the surface ($z = 0$) and bottom ($z = -H$) boundaries, $K_z \, \partial u/\partial z$ represents the frictional acceleration imparted to the water column by wind forcing at the top (τ_s) and deceleration by drag forces at the bottom (τ_b). These are external Reynolds stresses, involving turbulent vertical transfer of horizontal momentum between air, water, and sediment, but, like the interior Reynolds stresses of Eq. (13), are difficult to measure. As in the case of the eddy coefficients, these two stresses are parameterized as well, in particular by quadratic drag laws

$$\tau_{s,x} = \rho_a C_{10} W W_x, \quad \tau_{s,y} = \rho_a C_{10} W W_y \tag{17}$$

and

$$\tau_{b,x} = \rho C_d u (u^2 + v^2)^{1/2}, \quad \tau_{b,y} = \rho C_d v (u^2 + v^2)^{1/2} \tag{18}$$

where, for example, $\tau_{s,x}$ is the x component of the wind stress in cm^2 sec^{-2} (dyn cm^{-2}), i.e., the force per unit area; ρ_a is the density of air, 10^{-3}; W is the wind speed; W_x is the x component of the wind velocity; and C_{10} is a dimensionless drag coefficient, nominally estimated at a height 10 m above the free surface (it is taken to be 1.6×10^{-3} when $W < 7$ m sec^{-1} and 2.5×10^{-3} when $W \leqslant 10$ m sec^{-1}); while $\tau_{b,x}$ is the x component of the bottom stress and C_d is the bottom drag coefficient, at usually 1 m above the bottom, of $\sim 2 \times 10^{-3}$.

At a "typical" wind speed of 7 m sec^{-1}, Eq. (17) yields a τ_s of ~ 1 cm^2 sec^{-2}, or a force of ~ 1 dyn cm^{-2}, as the surface stress. Conversely, with a "typical" bottom current velocity of 20 cm sec^{-1}, Eq. (18) suggests a similar bottom stress of ~ 1 cm^2 sec^{-2}. On certain shelves, such as the North Sea, Georges Bank, and the Bering Sea, tidal streams of 100 cm sec^{-1} are not uncommon, such that a bottom stress of ~ 30 cm^2 sec^{-2} might occur. The magnitude of this force would only be matched by a similar wind stress during an occasional hurricane or typhoon.

The depth of penetration of these external frictional perturbations, from their respective boundaries toward the interior of the water column, determines their contribution to the acceleration and deceleration of the whole fluid. For example, with a 1 cm^2 sec^{-2} surface or bottom stress, the changes of velocity with time might be an acceleration of 10^{-3} cm sec^{-2} over a 10 m surface mixed layer and a similar deceleration within a bottom one of the same thickness, compared to perhaps 10^{-4} cm sec^{-2} over a 100-m water column. The acceleration of water, in fact, is not restricted to a surface mixed layer in response to a wind stress, but the kinetic energy is transferred to the interior of the shelf water column instead by readjustment of the sea-surface slope and associated pressure gradients of Eqs. (9) and (10). Further details of this process are discussed in Sections 2.2 and 2.3.

In summary, Eqs. (19) and (20) describe the assumptions, parameterizations, and omitted terms of the original Navier–Stokes equations, Eqs. (4)–(6):

$$\frac{\partial u}{\partial t} + u\frac{\partial u}{\partial x} + v\frac{\partial u}{\partial y} + w\frac{\partial u}{\partial z} = -\frac{1}{\rho}\int_{-z}^{e}\frac{\partial \rho}{\partial x}g\,dz - g\frac{\partial e}{\partial x}$$

$$+ \frac{\partial}{\partial z}\left(K_z\frac{\partial u}{\partial z}\right) + fv \tag{19}$$

$$\frac{\partial v}{\partial t} + u\frac{\partial v}{\partial x} + v\frac{\partial v}{\partial y} + w\frac{\partial v}{\partial z} = -\frac{1}{\rho}\int_{-z}^{e}\frac{\partial \rho}{\partial y}g\,dz - g\frac{\partial e}{\partial y}$$

$$+ \frac{\partial}{\partial z}\left(K_z\frac{\partial v}{\partial z}\right) - fu \tag{20}$$

We have deleted Eq. (6) for the vertical acceleration, using the hydrostatic approximation and the dynamic height calculation to estimate the buoyancy part of the horizontal pressure gradient. The horizontal turbulent stress terms, like the tidal forces, have been ignored, with parameterization of the vertical transfer of the shear stress by squared drag coefficients at the surface and bottom boundaries. To describe the acceleration and deceleration of seawater in a coordinate system attached to a rotating Earth, the vector product of the Earth's rotation velocity and that of the fluid have been added as a Coriolis pseudoforce.

2.1.6 Continuity

Upon integration of Eqs. (19) and (20) to obtain u and v, we can recover w as well from the law of conservation of mass, expressed by

$$\frac{\partial \rho}{\partial t} + \frac{\partial(\rho u)}{\partial x} + \frac{\partial(\rho v)}{\partial y} + \frac{\partial(\rho w)}{\partial z} = 0 \tag{21}$$

or by rewriting Eq. (21)

$$\frac{1}{\rho}\frac{d\rho}{dt} + \left(\frac{\partial u}{\partial x} + \frac{\partial v}{\partial y} + \frac{\partial w}{\partial z}\right) = 0 \tag{22}$$

Again assuming that seawater is an incompressible fluid, Eq. (22) becomes the continuity equation of momentum conservation, that is

$$\frac{\partial u}{\partial x} + \frac{\partial v}{\partial y} + \frac{\partial w}{\partial z} = 0 \quad\text{or}\quad -\frac{\partial w}{\partial z} = \frac{\partial u}{\partial x} + \frac{\partial v}{\partial y} \tag{23}$$

The kinematic boundary condition at the sea surface, $z = 0$, is $w = de/dt$, which is the vertical displacement of the sea surface with respect to its equilibrium position; this is analogous to $u = dx/dt$ and $v = dy/dt$.

At the bottom, $z = -H$, the other boundary condition is

$$-w = u \frac{\partial H}{\partial x} + v \frac{\partial H}{\partial y} \tag{24}$$

which is of great importance in the flux of energy through different food webs on continental shelves of varying width. In this Cartesian coordinate system, with the x axis directed offshore and the y axis along shore, a negative u means onshore flow and a positive w denotes upwelling. In a situation of only shoreward, cross-isobath flow, that is, $\partial H/\partial y = 0$, the same u leads to a greater w if the gradient $\partial H/\partial x$ increases, that is, greater upwelling occurs on a narrow shelf compared to a broad one. We shall see later in Sections 4.1.2 and 6.2.1, for example, that the structure of individual shelf food webs is very different as a consequence of this phenomenon—for example, the 500-km-wide Bering Sea shelf compared to the 10-km-wide shelf off Peru.

2.1.7 Inertial Oscillations

Although Eqs. (19) and (20) contain fewer terms than Eqs. (4)–(6), they are usually not solved in this form as well, with other terms dropped in formulation of different concepts such as inertial oscillations or geostrophic balance. Various numbers, or ratios of these terms, have been developed to characterize flow conditions in which the different forces predominate; recall that the Rossby number was a ratio of the inertial and Coriolis terms, indicating when the Coriolis transformation cannot be ignored. Similarly, the ratio of the inertial and pressure terms is the Euler number; of inertial and gravitational terms, the Froude number; of inertial and frictional terms, the Reynolds number; and of Coriolis and frictional terms, the Ekman number. A model, or analysis, based on one of these dimensionless numbers is usually referred to as a Froude model, Reynolds model, etc., and they are described in detail by Von Arx (1962).

Away from the boundaries of the ocean, for example, the constraints of a coastline no longer effect changes in sea level, that is, $\partial e/\partial x = \partial e/\partial y = 0$. If we also assume that the ocean is homogeneous or well mixed, with no changes in density, $\int_{-z}^{e} \partial \rho/\partial x = \int_{-z}^{e} \partial \rho/\partial y = 0$, then this other horizontal pressure term also disappears. An assumption that the advection of momentum is small, a step called "linearization" of the Navier–Stokes equation, deletes those terms, $u\, \partial u/\partial x$, $u\, \partial v/\partial x$, etc., as well. To complete this escape from reality, the wind and bottom stresses of the deep sea are also assumed to be negligible, that is, $\partial/\partial z\, (K_z\, \partial v/\partial z)$ and $\partial/\partial z\, (K_z\, \partial u/\partial z) = 0$, and we are left with the definition of an inertial oscillation as

$$\frac{\partial u}{\partial t} - fv = 0 \tag{25}$$

and

$$\frac{\partial v}{\partial t} + fu = 0 \tag{26}$$

where the inertial terms of $\partial u/\partial t$ and $\partial v/\partial t$ are balanced by the Coriolis terms. These Eqs. (25) and (26) have a solution of the form $u, v \approx e^{ift}$ and lead to oscillating motion of water parcels, with period $T = 2\pi/f$, that is, about 17 hr at midlatitudes. The description of this motion is not consistent with the existence of a coast, however, and is thus quite unrealistic as a shelf circulation model.

2.1.8   Geostrophy

Replacement of only the pressure gradient due to the slope of the sea surface in Eqs. (25) and (26), but assuming a steady state, $\partial u/\partial t = \partial v/\partial t = 0$, introduces a more useful shelf circulation model, that of geostrophic balance between the Coriolis and pressure forces:

$$fv = g \frac{\partial e}{\partial x} \qquad \text{or} \qquad v = \frac{g}{f} \frac{\partial e}{\partial x} \tag{27}$$

and

$$-fu = g \frac{\partial e}{\partial y} \qquad \text{or} \qquad u = -\frac{g}{f} \frac{\partial e}{\partial y} \tag{28}$$

Alternatively, away from the coast within an inhomogeneous ocean, we can add instead the buoyancy term and delete the pressure force of the sea-surface slope from Eqs. (27) and (28). These steady-state geostrophic balances then become

$$v = \frac{1}{\rho f} \int_{-z}^{e} \frac{\partial \rho}{\partial x} g \, dz \tag{29}$$

and

$$u = -\frac{1}{\rho f} \int_{-z}^{e} \frac{\partial \rho}{\partial y} g \, dz \tag{30}$$

which are known as the thermal wind equations of meteorology, and form the basis for dynamic height calculations (Sverdrup et al., 1942) within a frictionless sea of the geostrophic currents u, v.

2.1.9 Shallow-Water Equations

Retention of the time-dependent inertial terms of u and v in Eqs. (27) and (28) and inclusion of the Reynolds stress term constitute, together with the continuity equation [Eq. (23)], the linearized set of "shallow-water equations,"

$$\frac{\partial u}{\partial t} = -g \frac{\partial e}{\partial x} + \frac{\partial}{\partial z}\left(K_z \frac{\partial u}{\partial z}\right) + fv \tag{31}$$

$$\frac{\partial v}{\partial t} = -g \frac{\partial e}{\partial y} + \frac{\partial}{\partial z}\left(K_z \frac{\partial v}{\partial z}\right) - fu \tag{32}$$

In principle, the response of a stratified shelf sea to external forcing can be viewed as consisting of two additive components: a homogeneous, barotropic circulation mode and a density-related, baroclinic mode. The shallow water equations lack the buoyancy term and may thus be a reasonable representation of only the well-mixed, winter flow conditions on the continental shelf. Addition of the baroclinic mode to Eqs. (31) and (32) is a significant complication, however, requiring knowledge of the density field at appropriate time intervals, which is deferred to Section 2.7. We will first consider three examples of the use of these simple shallow-water equations in case studies of the Ekman dynamics, spatial resolution, and time dependence of shelf currents.

The changing slope of the sea level on continental shelves is the major mechanism for transmitting the external wind forcing to the interior of the water column. It is thus appropriate to adopt a technique of solution (Welander, 1957) for the above shallow-water equations that treats these gradients of sea level as an external input as well. Depth integration of the velocity components, for example, yields their horizontal transports of

$$U = \int_{-H}^{e} u \, dz \quad \text{and} \quad V = \int_{-H}^{e} v \, dz \tag{33}$$

Similarly, depth integration of Eqs. (31), (32), and (23) gives the transport equations

$$\frac{\partial U}{\partial t} = -gH \frac{\partial e}{\partial x} + F_x - B_x + fV \tag{34}$$

$$\frac{\partial V}{\partial t} = -gH \frac{\partial e}{\partial y} + F_y - B_y - fU \tag{35}$$

$$\frac{\partial U}{\partial x} + \frac{\partial V}{\partial y} + (w_{z=e} - w_{z=-H}) = 0; \tag{36}$$

where F_x and F_y are now just the components of the external wind forcing, rather than the total frictional forces of Eqs. (4) and (5), while B_x and B_y are the components of the bottom stress. In coastal regions,

$$\frac{\partial U}{\partial x} + \frac{\partial V}{\partial y} = -\frac{\partial e}{\partial t} \tag{36a}$$

while offshore $\partial U/\partial x + \partial V/\partial y$ is zero. With appropriate boundary conditions, the x, y dependence of sea level e can be obtained from Eqs. (34)–(36a). Using an approximation of $\partial K_z/\partial z$, and values of $e(x, y)$ as external input, then Eqs. (31) and (32) can be solved for $u(z)$ and $v(z)$.

2.2 Ekman Dynamics

Intrigued by Fridtjof Nansen's observations, in 1893–1896 during the polar cruise of the *Fram*, that ice drift was apparently 20–40° to the right of the wind forcing, Vagn Walfrid Ekman developed one of the first theories (Ekman, 1905) of a wind-driven ocean over 80 yr ago. Away from the continental shelves, Ekman assumed that the steady-state solution, $\partial u/\partial t = \partial v/\partial t = 0$, of the flow was a balance of the Coriolis acceleration, fv, fu, and that of the vertical shear of momentum, $\partial/\partial z(K_z \, \partial u/\partial z)$, $\partial/\partial z(K_z \, \partial v/\partial z)$, that is, the derivation of the Ekman number or model. Any movement in the coastal ocean, however, must eventually adjust to geostrophic equilibrium, involving the pressure gradient of Eqs. (27) and (28). Rossby (1938) and Charney (1955) provided the additional theoretical basis of this momentum transfer from the surface mixed layer to the interior of the fluid. Many analytical models now exist of the various aspects of this phenomenon, such as an impulsive wind, rectangular basins, circular basins, longshore and cross-shelf winds, or variable depth. A simple one, of a wind setup in a closed basin of constant depth, is presented in Section 2.3 as an example of this approach; see Csanady (1982) for more detail.

We start in the deep sea, by assuming that a wind stress, $\tau_y = \rho F_y$, is applied only along the y axis in an open ocean, that is, $\partial e/\partial x = \partial e/\partial y = \partial e/\partial t = 0$, of steady flow, $\partial U/\partial t = \partial V/\partial t = 0$, where the bottom stress is also negligible. Equations (34)–(36) of the transport then become

$$-fV = 0 \tag{37}$$

$$fU = F_y \tag{38}$$

$$\frac{\partial U}{\partial x} + \frac{\partial V}{\partial y} = 0 \tag{39}$$

The analytical solutions for Eqs. (37)–(39) are $V = 0$ and $U = F_y/f$; that is, the Ekman transport is 90° to the right of the wind in the northern hemisphere, where f is positive. For typical values of $F_y = 1$ dyn cm^{-2}, and $f = 10^{-4}$ sec^{-1}, the transport is $U = 1 \times 10^4$ cm^2 sec^{-1}.

A steady-state, depth-averaged current u of 10 cm sec^{-1} might occur over the upper 10 m of the water column. Recall, however, that in Section 2.1.5 the same impulsive wind forcing of 1 cm^2 sec^{-2} (1 dyn cm^{-2}) might lead to an acceleration of 10^{-3} cm sec^{-2} within the upper 10 m of the water column, or a transient current of 10^2 cm sec^{-1} over 1 day (10^5 sec). Either $\sim 90\%$ of the wind input of energy is dissipated over time by generation of bottom and interfacial frictional stresses, or the wind-driven flow, even the time-averaged mean, extends deeper than 10 m. To decipher the possible depth dependence of the currents, summing to the transport, we adopt the assumptions of Ekman's model.

Solutions for $u(z)$ and $v(z)$ in the interior of this idealized open ocean are thus obtained by balancing the steady-state vertical transfer of momentum with the Coriolis acceleration and assuming that K_z is a constant, that is,

$$-fv = K_z \frac{\partial^2 u}{\partial z^2} \tag{40}$$

$$fu = K_z \frac{\partial^2 v}{\partial z^2} \tag{41}$$

with the boundary conditions $\partial u/\partial z = \partial v/\partial z = 0$ as $z \rightarrow -\infty$, while $K_z \, \partial u/\partial z = 0$ and $K_z \, \partial v/\partial z = u_*^2$ at $z = 0$, where $u_* = \sqrt{\tau_y/\rho}$, a friction velocity.

The z coordinate can be scaled by the Ekman depth of frictional resistance, $D = \pi\sqrt{2K_z f^{-1}}$, at which the current vector is opposite the surface direction. At depths of $-z \gg D$, both the velocity components and the frictional stress vanish, with, of course, no mechanism to either further dissipate or transfer momentum to depths much below the Ekman layer. Under these constraints, the analytical solutions (Csanady, 1982) are

$$u(z) = \left(\frac{u_*^2}{fD}\right) e^{z/D} \left(\cos \frac{z}{D} - \sin \frac{z}{D}\right) \tag{42}$$

$$v(z) = \left(\frac{u_*^2}{fD}\right) e^{z/D} \left(\cos \frac{z}{D} + \sin \frac{z}{D}\right) \tag{43}$$

Empirically, the depth of a turbulent Ekman layer is usually $D \approx 0.1 \, u_*/f$ and K_z is estimated by $u_* D/20$. With $\tau_y = 1$ cm^2 sec^{-2} and $f = 10^{-4}$ sec^{-1}, u_* is 1 cm sec^{-1}, D is 10 m, and K_z is 50 cm^2 sec^{-1}, similar to the wintertime estimate for the Mid-Atlantic Bight (Stommel and Leetma, 1972). The velocity components at the surface, $z = 0$, are $u = v = 10u_*$, or ~ 10 cm sec^{-1} for this case; they are equal to $e^{-\pi}u$, $e^{-\pi}v$ in the Ekman spiral at $z = -D$, that is, $(1/23)u, v$ or ~ 0.5 cm sec^{-1}; and they are negligible at

$z = -3D$, or 30 m. Addition of a term for the slope pressure gradient, however, in Eqs. (37)–(41) would allow deeper penetration of the kinetic energy of the wind throughout the water column on the continental shelf.

2.3   Vertical Coupling

We now consider a closed basin of a deep constant depth H where the spatial and temporal derivatives of the slope pressure field are not zero, but the wind stress, $F_y = u_*^2$, again acts only in the y direction and the bottom stresses are negligible. The steady-state transports U, V, are also both zero, since it is a closed basin with coasts everywhere: that is, the flow perpendicular to every coast is zero by definition of a coastal constraint on all sides of the basin. This situation is described by the transport equations

$$0 = -gH \frac{\partial e}{\partial x} \tag{44}$$

$$0 = -gH \frac{\partial e}{\partial y} + F_y \tag{45}$$

$$0 = -\frac{\partial e}{\partial t} \tag{46}$$

There are no Coriolis terms in these transport Eqs. (44)–(46), and the simple solution

$$e = \frac{F_y y}{gH} = \frac{u_*^2 y}{gH} \tag{47}$$

reflects just a balance of transport between the y components of wind stress and of the pressure gradient of water, piled up on one coast and drained from the other.

When we examine the nature of the interior solution to this problem (Csanady, 1982), however, the flow at depth is in geostrophic balance with this pressure gradient, because of the Coriolis force inherent in the development of a surface Ekman layer, which also forms in response to the wind forcing. Using the steady-state version of Eqs. (31) and (32),

$$\frac{\partial u}{\partial t} = 0 = -g \frac{\partial e}{\partial x} + \frac{\partial}{\partial z}\left(K_z \frac{\partial u}{\partial z}\right) + fv \tag{48}$$

$$\frac{\partial v}{\partial t} = 0 = -g \frac{\partial e}{\partial y} + \frac{\partial}{\partial z}\left(K_z \frac{\partial v}{\partial z}\right) - fu \tag{49}$$

and substituting Eq. (47) into Eqs. (48) and (49) we obtain

$$-fv = \frac{\partial}{\partial z}\left(K_z \frac{\partial u}{\partial z}\right) \tag{50}$$

$$f\left(u + \frac{u_*^2}{fH}\right) = \frac{\partial}{\partial z}\left(K_z \frac{\partial v}{\partial z}\right) \tag{51}$$

which have the same solution as Eqs. (40) and (41), if K_z is constant and $u = u + u_*^2/fH$.

However, we no longer assume that $\partial u/\partial z = \partial v/\partial z = 0$ as $z \rightarrow -\infty$. The surface boundary conditions of wind stress for Eqs. (50) and (51) are the same, but a small bottom stress is now assumed. The solutions to Eqs. (50) and (51) are

$$u = -\frac{u_*^2}{fH} + u_E \tag{52}$$

$$v = v_E \tag{53}$$

where u_E, and v_E are the Ekman solutions of Eqs. (42) and (43). At the surface, the flow is the sum of both the geostrophic velocity and that of the Ekman layer (Fig. 28). At $-z \gg D$, the solutions below the Ekman layer are, of course, just

$$u = -\frac{u_*^2}{fH} \tag{54}$$

$$v = 0 \tag{55}$$

where the flow at depth along the x dimension is said to be in geostrophic balance with the pressure gradient $\partial e/\partial y$. A similar approach is used in Section 2.6.4 for numerical obtention of the geostrophic solution and those of two Ekman layers, one at the surface and the other at the bottom.

As the wind stress, F_y, piles up the water on one of the coasts in the y dimension of this simple model, the fluid would like to move back down the inclined surface of the sea slope, $\partial e/\partial y$, due to the force of gravity. In the frictionless interior of the water column below the Ekman depth D the fluid would experience, at the same time, a Coriolis adjustment, perpendicular to this flow in the y dimension, that is, $u = -u_*^2/fH$ along the x dimension for $-z \gg D$ as expressed in Eq. (54). Since there was assumed to be no equivalent wind stress in the x direction, there is no adjustment of the flow to geostrophic equilibrium at depth in the y direction, that is, $v = 0$ for $-z \gg D$ in Eq. (55). The effect of the wind stress in the y direction is thus only found in the surface layer of the water column above D. Away from enclosed basins such as the Great Lakes, Caspian Sea, or Black Sea, a more realistic representation of the continental

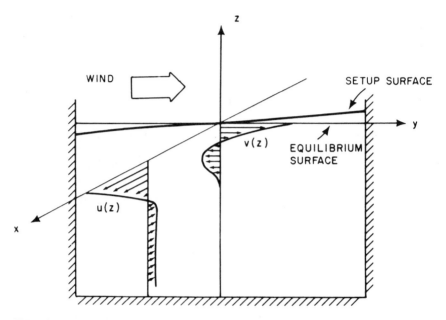

FIG. 28. Wind setup in constant-depth basin, and associated interior velocities, an Ekman layer at the surface and geostrophic flow below. [After Csanady (1982); © D. Reidel Publishing Co., with permission.]

shelf is one of open boundaries everywhere except at the coast, where the same general principles apply for transfer of momentum from the surface to the interior of the water column.

We thus now consider a straight coastline, that is, $\partial H/\partial y = 0$, where the wind stress, F_y, is still aligned in just the positive y direction (Fig. 29). Such an assumption means that $\partial e/\partial y = 0$ since there is no physical boundary to allow setup of the slope pressure field downstream in the y direction. At steady state, without a slope of the sea surface in the y dimension and ignoring bottom friction, there is also no "downhill," or subsurface return flow, in y, such that there appears to be no way for the momentum to be transferred deeper within the water column in this dimension.

An offshore, surface Ekman drift ($+u$) occurs, however, along the x dimension, in response to apparent Coriolis deflection of the longshore current (Fig. 29). In this situation, water is drained from the coast, generating a sea-surface slope, such that, $\partial e/\partial x \neq 0$, because of the land boundary at $x = 0$. An "adjustment drift" or return, cross-isobath flow occurs below the Ekman layer in the x dimension as a result of this cross-shelf pressure field of the sea-surface slope, where sea level is higher offshore than at the coast. Coriolis also acts on this adjustment drift of water moving onshore, such that motion also occurs in the y dimension at depth, below the Ekman layer.

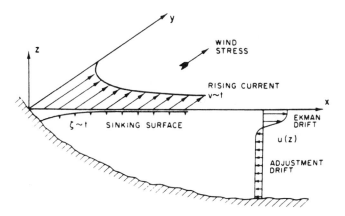

FIG. 29. Generation of coastal currents by longshore wind. Ekman drift in the surface layer onshore or offshore is compensated for by motions accompanying adjustment to geostrophic equilibrium. The fluid below the Ekman layer is accelerated in the longshore direction by the Coriolis force of (cross-shore) adjustment drift. [After Csanady (1981a); © American Geophysical Union.]

Applying Eq. (24), to this simple coastal idealization of the balance of friction and pressure forces leads to

$$-w = u \frac{\partial H}{\partial x} \tag{56}$$

at the bottom of the water column, since $\partial H/\partial y = 0$; recall the discussion of Eq. (24) where a positive w would result in this Cartesian coordinate system (Fig. 29). The consequence of such a wind stress applied to the coastal ocean is thus both transfer of momentum to the interior of the water column and generation of a positive vertical velocity, or upwelling. A wind stress, F_y, from the negative y dimension would, of course, lead to the opposite circulation pattern of onshore flow at the surface, reversal of the sea slope gradient with offshore flow at depth, and a $-w$ at the bottom, that is, downwelling of water near the coast. Various observed sequences of wind forcing and the resultant circulation pattern on the continental shelf lead us to the subject of spatial and temporal resolution of the physical habitat.

2.4 Spatial Resolution

With the advent of satellite sensors in the last decade to complement previous *in situ* instrumentation, the motivation to describe the spatial heterogeneity of the ocean in which these measurements are embedded has increased significantly. High-speed digital computers also now allow us to describe spatial properties of shelf flow fields, which is usually not possible from the above simple analytical models. By solving nonlinear partial differential equations with numerical techniques, which requires less computer time each year, both the spatial and temporal consequences of the Navier–Stokes equations can be explored. The steady-state, or time-invariant, solution of these equations over space is a special case of their time-dependent solutions.

The basis for numerical solution of these and the subsequent biochemical state equations will thus be first illustrated with the ordinary differential form of one of the Navier–Stokes equations,

$$\frac{du}{dt} = F_x + fv \tag{57}$$

where the pressure terms of Eq. (4) have been deleted. The left-hand side of this equation, du/dt, implies differentiation of a continuous function with time and, if the right-hand side of the equation were sufficiently simple, it could be solved to give an analytical solution, as we have done in the above examples. In general, however, analytical solutions are not possible for the full Navier–Stokes equations or for the nonlinear biochemical state equations. Simulation analysis becomes instead concerned with numerical solution of discrete, finite-difference equations, for which there are always solutions: spurious ones and those less so!

2.4.1 Temporal Approximation

If the changes in u with time are relatively small, the assumption, or approximation, is made that

$$\frac{du}{dt} \approx \frac{\Delta u}{\Delta t} \tag{58}$$

where Δ indicates a finite increment of u and t. Now $\Delta u/\Delta t$ can also be written as

$$\frac{\Delta u}{\Delta t} = \frac{u^{n+1} - u^n}{t^{n+1} - t^n} \tag{59}$$

where u^{n+1} is the value of u after the time interval t^{n+1}, while u^n is the value of u at a previous time, t^n.

Equation (58) can then be written as

$$\frac{du}{dt} \approx \frac{u^{n+1} - u^n}{t^{n+1} - t^n} \tag{60}$$

If $(t^{n+1} - t^n)$ is transferred to the left-hand side, one obtains

$$(t^{n+1} - t^n)\left(\frac{du}{dt}\right) \approx u^{n+1} - u^n \tag{61}$$

or, since $\Delta t = (t^{n+1} - t^n)$,

$$\Delta t \frac{du}{dt} \approx u^{n+1} - u^n \tag{62}$$

and finally,

$$u^{n+1} \approx u^n + \Delta t \frac{du}{dt} \tag{63}$$

which gives the approximate future value of u as a function of the present value plus the product of the time interval and the continuous derivative of u with respect to time.

If the finite difference approximation with time, $\Delta u/\Delta t$, were taken to be so small that $\Delta u/\Delta t = du/dt$, then Eq. (64) would follow from substitution of $\Delta u/\Delta t$ in equation (63), that is, the Δt and dt would cancel, leaving

$$u^{n+1} = u^n + \Delta u \tag{64}$$

If Δt were identical to dt, there would then be no error involved in calculation of future values of the state variables of an ordinary differential equation, one with no spatial derivatives. Thus, it is in the selection of the time step Δt, that computational errors, or instability, become first introduced in the numerical solution. The smaller the time step, or Δt, the smaller will be the error, but there is a trade off in terms of computer time and perhaps costs. With a shorter time step, there are more calculations; it takes longer for a computer to make these calculations over a fixed period of simulated time. Consequently, more central processor time of the computer is used and a larger bill arrives from your central computer center; of course, if you own your computer, then one need only pay the bill from the local electrical utility.

At each iteration, or time step, of this procedure, Eq. (63) is used to calculate the next time value of u from the last one. The time step Δt is selected as a compromise between accuracy, speed of the computer, and available funds, while du/dt is calculated by summing the terms on the right-hand side of Eq. (57),

$$u^{n+1} = u^n + (\Delta t)(F_x + fV) \tag{65}$$

The time-dependent and steady-state $(\partial u/\partial t = 0)$ solutions of this and other differential equations can be calculated in such a manner. The value of u at an asymptotic steady state is reached in the procedure, for example, when $du/dt \approx 0$, that is,

$$u^{n+1} = u^n + (\Delta t)\,(0) = u^n + 0 = u^n \tag{66}$$

The above simple method of numerical solution, or integration, is called the Euler, or forward time step, method. Other techniques, involving time-centered and implicit time steps, such as the Nystrom, Runge–Kutta, or Adams–Bashford approaches, can be found in standard texts on numerical methods (Ralston, 1965; Hamming, 1973; Hildebrand, 1974). Only elementary calculus is required for perusal of Cheney and Kincaid's (1980) text, while particular emphasis on solution of the partial differential equations of fluid mechanics is given by Roache (1976). These other methods use more than one iteration per time step, requiring additional previous derivatives of the state variables beyond the last time step, to predict future values with greater accuracy than the Euler method. The equation

$$u^{n+1} = u^n + \left(\frac{\Delta t}{24}\right) \left(55 \frac{du}{dt} - 59 \frac{du^{n-1}}{dt} + 37 \frac{du^{n-2}}{dt} - 9 \frac{du^{n-3}}{dt}\right)$$

(67)

is an example of the calculation involved in the first iteration of the Adams–Bashford method, which weights a series of previous time derivatives. These more accurate methods require greater computer storage and time, however, with again a tradeoff between computer time, or costs, and computational accuracy; furthermore, we have only considered the time derivative thus far.

2.4.2   Spatial Approximation

Similarly, the continuous spatial derivatives of a partial differential equation are approximated by a spatial finite difference rather than by a temporal finite difference, that is,

$$\frac{\partial e}{\partial x} \approx \frac{e_{i+1} - e_{i-1}}{2 \, \Delta x}$$

(68)

Equation (68) is a centered spatial difference of the gradient of the values of e over three grid points, $i + 1, i, i - 1$ of a spatial mesh in the $x(i)$ dimension. In this context, we are using the FORTRAN convention of i columns and j rows; matrix algebra uses the inverse convention of i rows and j columns. Other finite differences are used for the $y(j)$ and $z(k)$ dimensions. Using only two grid points, an alternative finite difference expression is

$$\frac{\partial e}{\partial x} \approx \frac{e_{i+1} - e_i}{\Delta x}$$

(69)

termed an upstream, or forward, finite difference, where Δx is the distance between the two adjacent grid points.

Using these finite-difference expressions of the temporal and spatial derivatives, together with appropriate initial and boundary conditions, any partial differential equation can be solved with numerous computer algorithms, or proce-

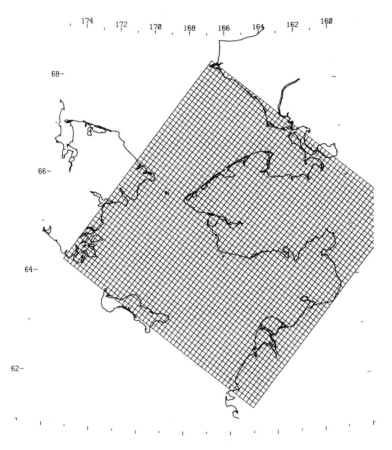

174   172   170   168   166   164   162   160

68-

66-

64-

62-

FIG. 30. Eulerian grid, with 10-km spacing, of a numerical model of the Bering Strait. [After Walsh and Dieterle (1986); © Elsevier Science Publishers B.V., with permission.]

dures, of varying accuracy and time (Roache, 1976). For example, in a model of the northern Bering Sea circulation (Walsh and Dieterle, 1986) over a staggered Eulerian grid between St. Lawrence Island and Bering Strait (Fig. 30), the numerical integration of the depth-integrated shallow water equations [Eqs. (34)–(35) and (36a)] was carried out by the following forward difference scheme in time and midpoint difference in space:

$$U_{i,j}^{n+1} = U_{i,j}^n + 0.25(\Delta t)(V_{i,j}^n + V_{i,j+1}^n + V_{i-1,j}^n + V_{i-1,j+1}^n)(f)$$

$$- \frac{g(\Delta t)}{2\,\Delta x}(H_{i,j} + H_{i-1,j})(e_{i,j}^n - e_{i-1,j}^n) - (\Delta t)(B_{i,j}^x) \tag{70}$$

$$V_{i,j}^{n+1} = V_{i,j}^n - 0.25(\Delta t)(U_{i,j}^{n+1} + U_{i+1,j}^{n+1} + U_{i,j-1}^{n+1} + U_{i+1,j-1}^{n+1})(f)$$

$$- \frac{g(\Delta t)}{2\,\Delta y}(H_{i,j} + H_{i,j-1})(e_{i,j}^n - e_{i,j-1}^n) - (\Delta t)(B_{i,j}^y) \tag{71}$$

$$e_{i,j}^{n+1} = e_{i,j}^n - \frac{\Delta t}{\Delta x}(U_{i+1,j}^{n+1} - U_{i,j}^{n+1}) - \frac{\Delta t}{\Delta y}(V_{i,j+1}^{n+1} - V_{i,j}^{n+1}) \tag{72}$$

At time $n + 1$ in their numerical solution, Eq. (70) was computed for the entire spatial grid (Fig. 30), then Eq. (71), and then Eq. (72); that is, only one time level was stored for any variable at each of 1825 grid points. Errors are also introduced by the approximations of the spatial derivatives; thus the time step Δt, in this finite difference scheme of a two-dimensional x–y spatial grid must also satisfy the requirement that $\Delta t < \Delta x/\sqrt{2gH_{max}}$ where H_{max} was set to 50 m, south of St. Lawrence Island. With the fast speed of a gravity wave,

$$\sqrt{gH} = c = \frac{\Delta x}{\Delta t} = u \tag{73}$$

the above numerical stability criterion becomes $u\,\Delta t/\Delta x < 0.7$, a more conservative constraint than the usual Courant–Friedricks–Lewy condition of $u\,\Delta t/\Delta x < 1$ for explicit time-difference algorithms. Evaluation of $\Delta x/\sqrt{2gH_{max}}$, with $\Delta x = 10$ km, $g = 10$ m sec^{-2}, and $H_{max} = 50$ m, yields Δt must be less than 5 min; a 3-min time step was used in the Bering Strait simulations. Each circulation model run, solving Eqs. (70)–(72), simulated 4800 time steps, or 10 days, in about 45 min on a super-minicomputer system to obtain steady-state solutions. We shall see in Section 2.6.5 that an alternative formulation of this problem as a vorticity balance leads to faster steady-state solutions.

2.4.3   Bering Strait

The interior solutions of u and v of the Bering Strait model were obtained by assuming $\partial u/\partial z = \partial v/\partial z = 0$, that is, dividing the transports U and V, by the local depth, $-H$, which was entered in the model as a digitized form of the bottom topography (Fig. 31). At the upstream open boundaries of the model (Fig. 30) across the Anadyr and Shpanberg Straits at the western and eastern sides of St. Lawrence Island, U was prescribed by assuming that of the total northward transport through Bering Strait, 60% passed through Anadyr Strait and 40% through Shpanberg Strait; e was then determined here at these upstream boundaries from Eq. (72). At the downstream open boundary, V and e were functions of both adjacent, interior solutions and the phase velocity $\sqrt{gH}$, that is, a radiative boundary condition (Orlanski, 1976) in which only outward energy flux occurs, without significant distortion of the solutions of Eqs.

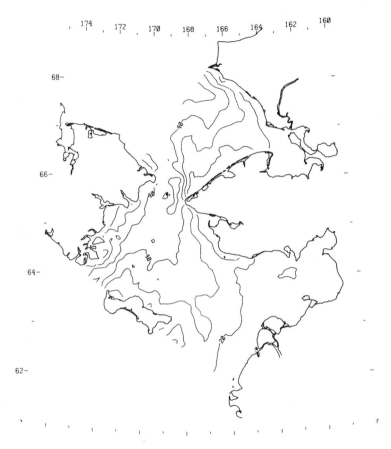

FIG. 31. Bottom topography (m) of a numerical model of the Bering/Chukchi Seas. [After Walsh and Dieterle (1986); © Elsevier Science Publishers B.V., with permission.]

(70)–(72). At the land boundaries, $U = V = 0$ such that there were 1825 active grid points (i,j) of 10 km spacing in these simulations (Fig. 30).

At the bottom boundary (Fig. 31), a mean, linearized bottom stress was used over two adjacent grid points; that is, instead of the usual $B_x = C\rho u^2$ and $B_y = C\rho v^2$, the formulation was

$$B_{i,j}^x = (2\alpha C\rho)(H_{i,j} + H_{i-1,j})^{-1}(U_{i,j}) \qquad (74)$$

$$B_{i,j}^y = (2\alpha C\rho)(H_{i,j} + H_{i,j-1})^{-1}(V_{i,j}) \qquad (75)$$

where ρ is the density of sea water, C is the drag coefficient ($\sim 2 \times 10^{-3}$), and α is assumed to be a constant of 10 cm sec^{-1}. At the surface boundary no wind stress was applied, such that $F_x = F_y = 0$, in three cases of varying transport

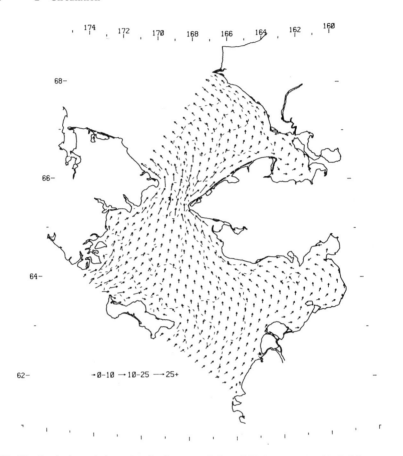

FIG. 32. Continuity solution of a depth-averaged flow field (cm sec^{-1}) with 0.6-Sv transport through the Bering Strait. [After Walsh and Dieterle (1986); © Elsevier Science Publishers B.V., with permission.]

through the Bering Strait. One could think of the three circulation patterns with 0.3-, 0.6-, and 1.2-Sv transport, however, as the flow response under varying conditions of northerly or southerly wind forcing.

To demonstrate the relative importance of the different forces in Eqs. (70)–(72), a series of cases was run with different terms of these equations set to zero. With no pressure or friction forces acting on the interior flow field of the 0.6-Sv transport case, for example, there was no setup of the surface elevation, such that only the continuity Eq. (72) applied, with Eqs. (70) and (71) not relevant. As a result, the computed currents neither followed the bottom topography, nor accelerated downstream, except within the constricted area of Bering Strait, with a maximum flow of 45.6 cm sec^{-1} (Fig. 32).

Inclusion of the Coriolis, pressure, and frictional forces of Eqs. (70) and (71)

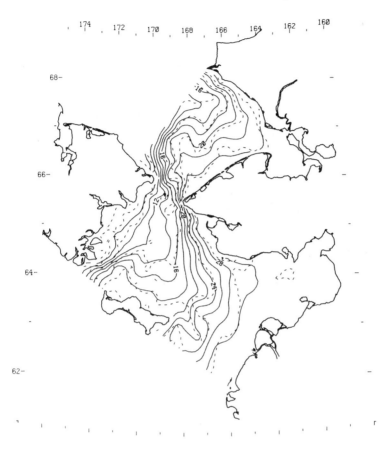

FIG. 33. The spatial distribution of sea-level elevation (cm) in a flow field with 0.6-Sv transport through the Bering Strait. [After Walsh and Dieterle, 1986); © Elsevier Science Publishers B.V., with permission.]

acting on the boundary currents that enter the model's domain in the 0.6-Sv case, generated horizontal sea-level gradients in both the x and y directions (Fig. 33). At steady state, such that $\partial e/\partial t = \partial U/\partial t = \partial V/\partial t = 0$, the largest gradient, ~ 20 cm/300 km, was between Anadyr Stream Water, off the Siberian peninsula, and Alaska Stream Water, off the mouth of the Yukon River (Fig. 33). Such a sea-surface slope of 0.07 cm cm^{-5}, and $g = 10^3$ cm sec^{-2}, would lead to a fluid acceleration of $\sim 7 \times 10^{-4}$ cm sec^{-2}; recall that accelerations of the flow of $\sim 1 \times 10^{-4}$ cm sec^{-2} might result from either a 7-m sec^{-1} wind, exerting stress at the sea surface for 1 day (10^5 sec), or from the usual tidal forces. Over 1 day, with no frictional losses at the bottom, a current of 70 cm sec^{-1} would develop from such an acceleration of the fluid in this 0.6-Sv case of the Bering Sea model.

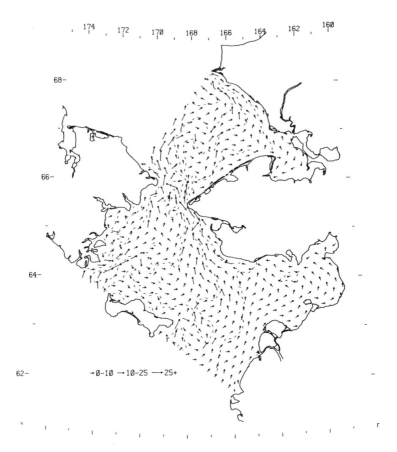

FIG. 34. The depth-averaged flow field (cm sec^{-1}) of a circulation submodel with horizontal pressure, bottom friction, and Coriolis forces after 240 hr in the 0.6-Sv case. [After Walsh and Dieterle (1986); © Elsevier Science Publishers B.V., with permission.]

With the linear bottom stress of Eqs. (74) and (75), the largest computed currents within Bering Strait were actually 35.6 cm sec^{-1} for the 0.3-Sv case and 72.9 cm sec^{-1} for the 0.6-Sv case (Fig. 34), compared to the continuity solution of 45.6 cm sec^{-1} for the same case (Fig. 32) and 142.7 cm sec^{-1} for the 1.2-Sv case. A combination of the Coriolis correction for a rotating Earth coordinate system and the acceleration of the boundary currents induced by the horizontal pressure gradient led to flow along the isobaths (Fig. 34), in sharp contrast to the simple continuity case (Fig. 32). In this model, the currents can be simplistically described as entering Anadyr and Shpanberg Straits from the south, "deflecting" to the right, or toward the east, by Coriolis, piling up against the Alaskan coast to create the sea-surface slope, running back down the slope to the west under the influence of gravity, and being "deflected" once more to the north by Coriolis, with additional acceleration of the flow at Bering Strait by the continuity constraint. The flow pattern north of Bering Strait exhibited the same characteristics as that north of Anadyr and Shpanberg Straits (Fig. 34), subject to the same simplistic combination of physical forces described above. The coastal ocean is rarely in steady state, however, and we thus proceed to an example of the time-dependent solution to the same finite-difference equations, Eqs. (70)–(72).

2.5   Time Dependence

With specification of the initial and boundary conditions for the finite-difference equations (70)–(72) of the Bering Sea circulation model, the numerical procedure constitutes an *initial-value problem,* or *marching problem,* in mathematical jargon. Starting with the initial values of the variables at each grid point, one marches forward in time with some iterative computer algorithm to integrate the equations. If one obtains the steady solution, $\partial U/\partial t = 0$, to a time-dependent problem, at an asymptotic steady state, as we have done in the above example, the transient solutions of the marching process have no physical significance. If a time-dependent forcing function were added to the initial-value problem, however, the transient solutions would then approximate the nonsteady solutions, that is, Eqs. (31) and (32) rather than Eqs. (48) and (49).

Another class of mathematical models constitutes *boundary-value problems,* or *jury problems,* in which only the boundary conditions need be specified for solution, that is, there is no time dependence in the problem. There are numerous iterative techniques for the solution of these steady partial differential equations (Roache, 1976) as well, all of which are analogous to obtaining the steady-state solutions of a time-dependent problem. Among these techniques, some direct methods have now become practical for the solution of boundary-value problems with the advent of matrix algebra and high-speed digital computers; we will discuss such a technique for solution of the vorticity equation in Section 2.6.5 on Gaussian elimination. Addition of the wind stress term to the Bering Sea model, for one instant in time, would lead to a different steady-state solution at each grid point of this marching problem than those of the above cases, without wind forcing. Changing the direction and intensity of the wind forcing (such as Fig. 43) in a number of such simulations would further lead each time to a different, transient solution; the set of transient solutions would then approximate the time-dependent response of the coastal ocean to wind forcing over some specified time scale. Using the same equations [Eq. (34), (35), and (36a)], the same finite-difference scheme, and similar boundary conditions of the Bering Sea model, a time-dependent flow field (Fig. 35) was calculated over a 5.7-km grid of the New York Bight in response to hourly observations of changes in the wind forcing (Tingle *et al.,* 1979).

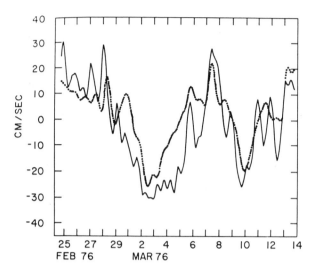

FIG. 35. The computed currents at one grid point (dotted line) versus the observed current at the BNL spar buoy south of Long Island. [After Tingle *et al.* (1979); © 1979 Plenum Press.]

The complexity of the physical processes that govern the circulation in the New York Bight is discussed by Beardsley *et al.* (1976). They include tides, estuarine exchanges, local wind forcing, topographic steering, input of low-salinity water from the upstream Gulf of Maine, an along-shore pressure gradient, and shelf-slope water exchange. It is doubtful that a single physical model can be credibly used to describe all of these circulation details. However, the present simple model does include the effects of local wind, bottom topography, and pressure gradients of the sea-surface slope. Such a circulation model is most applicable during winter in the coastal region, when the currents tend to be barotropic, that is, to have the same speed and direction at depth. The performance of this time-dependent model in the nearshore zone is shown in Fig. 35. The computed along-shore current at one of the grid points is compared against observations at a spar buoy south of Long Island in water <30 m deep during February and March 1976 [see Scott and Csanady (1976) for details of the current measurements]. Data from only one wind station on the south shore of Long Island were used to drive the model; this input was not valid during fast-moving storms, as shown, for example, during March 6 (Fig. 35)—that is, there was spatial variation in the wind forcing not considered in the model. At this time, the circulation model was computing 5 cm sec^{-1} of eastward flow, rather than the observed 10 cm sec^{-1} of westward flow. This disagreement of model and field data is probably caused by the complex geostrophic forces set up by a cold front. Note that the storm on March 10 was modeled quite well. Furthermore, the small high-frequency peaks of flow in Fig. 35 are the diurnal tides, which were not completely filtered out of the observations—see Section 3.2.3 for more detail.

Such deficiencies mean that, on one hand, a larger-scale wind field must be used in computing all details of the general shelf circulation but that, of course, on the other hand, a more detailed description of bottom friction must be considered close to the coast. The comparison of computed and observed currents (Fig. 35) does indicate, however, that there is some fidelity in the time-dependent barotropic currents, usually within 10 cm sec^{-1}, except in certain complex meteorological situations. Addition of a bottom Ekman layer and the baroclinic pressure field will increase the fidelity of the model to within 1 cm sec^{-1} for winter conditions of steady-state flow in Section 2.7: that is, some of the variance is removed by averaging the wind forcing, density, and observed currents.

Numerical solution of the linearized momentum and continuity equations [i.e., Eqs. (34), (35), and (36a)] of the Bering Sea and the New York Bight models, however, is time-consuming and expensive on digital computers. Inclusion of equations for biological and chemical state variables in Section 3.1.4 will accordingly increase the computation time for their solution, in coupled models of biochemical interaction within a steady-state flow field of the continental shelf. One could also compile time series of observed winds at a number of coastal and offshore locations to obtain solution of these coupled models for each interval of interest, obviously increasing again the computer time for each simulation run in a time-dependent mode. If the number of equations describing the flow field could be reduced, the amount of computer time and costs could also be reduced, with perhaps acquisition of additional physical insight as well, that is, thinking about fewer variables and their interactions at any one time.

2.6   Vorticity

Thus far, we have mainly described the motion of a water parcel within a translational framework in a vector combination of acceleration along each of the dimensions of the Cartesian coordinates, for example, $\partial u/\partial x$, $\partial v/\partial y$, or $\partial w/\partial z$. We introduced Ω the angular rotation of the Earth, however, as part of Eqs. (4) and (5) in the Coriolis transformations fv and fu, for description of acceleration on a rotating Earth; recall that $f = 2\Omega \sin \theta$ in the horizontal plane and $f = 2\Omega \cos \theta$ in the vertical, where θ is the latitude. The tendency of a water parcel to rotate with the Earth is defined as its planetary vorticity f, that is, twice the angular velocity of the Earth at a particular latitude. Because of the strong vertical gradients of density in the sea, compared to those in the horizontal plane, the vorticity of the ocean is suppressed around the horizontal axis, but resting water parcels have significant planetary vorticity around the vertical axis, relative, of course, to the inertial framework of the fixed stars.

If a water parcel in the sea has an additional tendency to spin around the vertical axis, beyond that of the planetary vorticity, it is termed relative vorticity ω. The vertical component of relative vorticity is defined as

$$\omega = \frac{\partial v}{\partial x} - \frac{\partial u}{\partial y} \tag{76}$$

with the convention that a positive value of ω indicates cyclonic or counterclockwise motion, that is, the sense of the rotation of the Earth in the northern hemisphere. The algebraic sum of the planetary and relative vorticities ($f + \omega$) is termed the absolute vorticity of the fluid, which is a constant in the absence of friction over a water column of unchanging depth: that is, the potential vorticity ($f + \omega)H^{-1}$ is constant. In this idealized situation over an unchanging depth of the water column, if a fluid in the northern hemisphere were to move towards the equator, f would decrease, with an associated increase in ω, such that the fluid would appear to find the Earth spinning more slowly beneath it (Longuet-Higgins, 1965), with the development of cyclonic (counterclockwise) rotation or more positive relative vorticity.

2.6.1   Vorticity Equation

Cross-differentiation of the Navier–Stokes equations [Eqs. (4) and (5)] and subtraction of Eq. (4) from Eq. (5) (i.e., taking the curl, in calculus jargon) allows derivation of the vorticity equation,

$$\frac{d(f+\omega)}{dt} = -(f+\omega)\left(\frac{\partial u}{\partial x} + \frac{\partial v}{\partial y}\right) + \left(\frac{\partial w}{\partial y}\frac{\partial u}{\partial z} - \frac{\partial w}{\partial x}\frac{\partial v}{\partial z}\right)$$

$$+ \left(\frac{\partial P}{\partial x}\frac{\partial \rho^{-1}}{\partial y} - \frac{\partial P}{\partial y}\frac{\partial \rho^{-1}}{\partial x}\right) + \left(\frac{\partial F_y}{\partial x} - \frac{\partial F_x}{\partial y}\right) \tag{77}$$

which together with the continuity equation [Eq. (23)], has been used to address various oceanographic problems, after judicious deletion of some terms. Development of early theories on the intensification of western boundary currents (Stommel, 1948), wind-driven ocean circulation (Munk, 1950), and topographic effects on coastal upwelling in eastern boundary currents (Arthur, 1965), for example, was all based on various considerations of the vorticity equation. Similar manipulation of the depth-integrated, barotropic equations of motion, described by Eqs. (34) and (35), for example, leads to a depth-integrated approximation of the vorticity equation of

$$\frac{d(f+\omega)}{dt} = -(f+\omega)\left(\frac{\partial U}{\partial x} + \frac{\partial V}{\partial y}\right) + \left(\frac{\partial H}{\partial y}g\frac{\partial e}{\partial x} - \frac{\partial H}{\partial x}g\frac{\partial e}{\partial y}\right)$$

$$+ \left[\frac{\partial(F_y - B_y)}{\partial x} - \frac{\partial(F_x - B_x)}{\partial y}\right] \tag{78}$$

where the horizontal advection of momentum and changes in density are ignored. As in the case of Eqs. (34) and (35), F_y and F_x are now components of the wind stress in Eq. (78) rather than the whole frictional terms of Eq. (77). We will consider application of Eq. (78) within large-scale ocean basins ($\sim 1-2 \times 10^7$ km^2) and local continental shelves ($1-2 \times 10^5$ km^2).

2.6.2 Basin Scale

Over the large horizontal scale of the ocean basins, ω is negligible, compared to f, such that an expansion of the total derivative of absolute vorticity leads to

$$\frac{d(f+\omega)}{dt} \approx \frac{\partial f}{\partial t} + u \frac{\partial f}{\partial x} + v \frac{\partial f}{\partial y} \tag{79}$$

similar to the Lagrangian description of Eqs. (4)–(6). There is little change of the planetary vorticity with time, $\partial f/\partial t \approx 0$, and it varies with latitude, not longitude, such that $\partial f/\partial x = 0$. The change of f with latitude can be assumed to be linear, that is, $\partial f/\partial y = \beta \approx 2 \times 10^{-13}$ cm^{-1} sec^{-1}, and is termed the β-plane approximation. With these simplifications, Eq. (79) reduces to

$$\frac{d(f+\omega)}{dt} \approx \beta v \tag{80}$$

for large-scale basins.

From Eq. (33), the depth-integrated form of Eq. (80) is

$$\frac{d(f+\omega)}{dt} \approx \beta \int_{-H}^{0} v \, dz = \beta V \tag{81}$$

Furthermore, in a deep sea with a negligible bottom stress ($B_y, B_x \approx 0$), a constant depth of integration ($\partial H/\partial y, \partial H/\partial x = 0$), and of nondivergent flow ($\partial U/\partial x + \partial V/\partial y = 0$), Eq. (78) then becomes

$$\beta V = \left(\frac{\partial F_y}{\partial x} - \frac{\partial F_x}{\partial y} \right) \tag{82}$$

where the meridional (north–south) transport of water is balanced by the wind curl, as first noted by Sverdrup (1947). The marked asymmetry of flow between western (~ 200 cm sec^{-1}) and eastern (~ 20 cm sec^{-1}) boundary currents of the major ocean basins (Fig. 6) can then be explained (Stommel, 1948; Munk, 1950) by the implied vorticity balance of Eq. (82). The planetary wind field is not asymmetric as well (Fig. 5). The predominant westerly wind fields at $\sim 40°$N/S latitudes and easterly forcing along the equator lead instead to a clockwise, or negative, wind curl for an annual mean between ~ 20 and $50°$N in the Atlantic and Pacific basins, between 50 and $70°$S in the west-wind drift, and along the equator (Fig. 36). Based on a seasonal analysis of wind speeds, averaged over

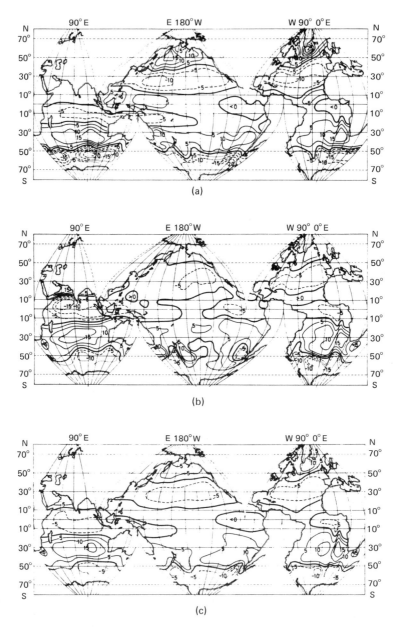

FIG. 36. The global distribution of surface wind stress curl (10^{-9} dyn cm^{-3}) for (a) northern winter (December–February), (b) northern summer (June–August), and (c) the annual mean. [After Hantel (1972); Goudon and Breach Science Publishers Inc., with permission.]

~2.5 × 10⁵ km² blocks of the ocean and converted by an algorithm similar to Eq. (17) to wind stress (Hellerman, 1967), the maximum range of the curl of the wind stress is from + 10 × 10⁻⁹ dyn cm⁻³ at 55°N and 35°S to − 10 × 10⁻⁹ dyn cm⁻³ at 30°N (Hantel, 1972).

Within the northern hemisphere, where f is positive (i.e., β is also positive), the negative wind curl over the subarctic North Atlantic (Fig. 36), for example, and Eq. (82) indicate that the water transport must be negative, or toward the south, over almost all of this basin. We note that conservation of mass and vorticity occur, such that the water is not drained and the winds do not spin up this ocean basin into a faster and faster clockwise rotating, liquid top. The intensified northward flow of a western boundary current, such as the Gulf Stream, instead must both return water to the North Atlantic basin and generate counterclockwise (positive) vorticity, by frictional drag against the continental margins, that is, the spin-off eddies discussed in Section 1.5.3.

Within a deep sea of divergent flow, where the water transport is not integrated to the bottom, or to a depth of no motion, a spatial variation in the wind stress (i.e., a non-vanishing curl) could cause upwelling (positive curl) or downwelling (negative curl) in the northern hemisphere. Below a surface mixed layer of negligible meridional transport, the vertical velocity at the base of the surface layer, w_b, can be estimated (Yoshida and Mao, 1957) by

$$w_b \approx \frac{1}{\rho f} \left(\frac{\partial F_y}{\partial x} - \frac{\partial F_x}{\partial y} \right) \tag{83}$$

where the wind curl is now balanced by vertical motion. For example, away from the equator, where f is zero, Eq. (83) and the negative curl at ~4°S (Fig. 36) (i.e., $f < 0$) lead to upwelling rates of ~0.5 × 10⁻³ cm sec⁻¹ (Hantel, 1972); recall Fig. 19, in which the upwelled input of nitrate appears to be greatest at ~3°S.

2.6.3 Shelf Scale

Near the coast, the depth-integrated form of continuity equation, Eq. (23), is that of Eq. (36a), $\partial U/\partial x + \partial V/\partial y = -\partial e/\partial t$, that is, another case of divergent flow, such that Eq. (78) becomes

$$\frac{d(f+\omega)}{dt} = (f+\omega)\left(\frac{\partial e}{\partial t} \right) + \left(\frac{\partial H}{\partial y} g \frac{\partial e}{\partial x} - \frac{\partial H}{\partial x} g \frac{\partial e}{\partial y} \right)$$
$$+ \left(\frac{\partial F_y}{\partial x} - \frac{\partial B_y}{\partial x} - \frac{\partial F_x}{\partial y} + \frac{\partial B_x}{\partial y} \right) \tag{84}$$

At local scales of ~50 km on the continental shelf, the relative vorticity can no longer be ignored, such that the Lagrangian expansion of the total derivative of absolute vorticity is

$$\frac{d(f + \omega)}{dt} = \beta v + \frac{\partial \omega}{\partial t} + u \frac{\partial \omega}{\partial x} + v \frac{\partial \omega}{\partial y} \tag{85}$$

instead of Eq. (80). The spatial derivatives of ω become important when the coastal flow has significant curvature (i.e., near points and capes), where for a radius of curvature of 50 km, the relative vorticity will be about twice that of the planetary vorticity (Arthur, 1965).

In the simple case of steady-state flow, the time derivatives of both vorticity and sea level are zero, however, such that the depth-integrated vorticity equation for a continental shelf then becomes

$$0 = \left(\frac{\partial H}{\partial x} g \frac{\partial e}{\partial y} - \frac{\partial H}{\partial y} g \frac{\partial e}{\partial x} \right) + \left(\frac{\partial B_y}{\partial x} - \frac{\partial B_x}{\partial y} \right) = \left(\frac{\partial F_y}{\partial x} - \frac{\partial F_x}{\partial y} \right) \tag{86}$$

With rearrangement of the partial derivatives of the surface and bottom friction components, the first term on the left is now the description of vortex stretching, or shrinking, of the water column as it crosses isobaths, the second term on the left is the curl of the bottom stress, and the term on the right is the curl of the wind stress.

In the absence of wind forcing (i.e., a vanishing curl), Eq. (86) becomes

$$\left(\frac{\partial H}{\partial x} g \frac{\partial e}{\partial y} - \frac{\partial H}{\partial y} g \frac{\partial e}{\partial x} \right) = \left(\frac{\partial B_x}{\partial y} - \frac{\partial B_y}{\partial x} \right) \tag{87}$$

where the generation of relative vorticity as the water column moves onshore, or offshore, is then balanced by bottom friction. In this situation of no wind forcing, conservation of vorticity implies that the steady-state circulation will tend to flow parallel to the isobaths; recall Figure 34 of the Bering Sea model.

In contrast to three equations [e.g., Eqs. (34), (35), and (36a), or Eqs. (70)–(72)] necessary for the description of barotropic, steady-state flow on the Bering Sea and New York shelves, use of the vorticity balance results in only one equation; the same flow field can be described by Eq. (87). With knowledge of $\partial e/\partial y$ and $\partial e/\partial x$, u and v can then be obtained by vertical decomposition and the geostrophic relationship [i.e., Eqs. (27) and (28)] upon solution of Eq. (87). Human nature being what it is, however, a reduction in the number of state equations, that is, momentum and continuity, leads to an increase in complexity of the vorticity equation. Analytical (Csanady, 1981b) or numerical (Hsueh and Peng, 1978; Hopkins and Dieterle, 1983) solutions of Eq. (86), for example, usually involve more complicated parameterizations of the bottom stress than the simple quadratic form of Eq. (18).

2.6.4   Bottom Ekman Layer

Analogous to the flow within a surface Ekman layer of Eqs. (42) and (43), induced by the wind stress, the currents within a bottom Ekman layer, induced by friction at the sediment interface, can be approximated by a trigonometric function of the veer angle and the height of this layer. In the previous quadratic formulation of bottom stress as Eqs. (74) and (75), such that

$$B_x = C\rho\alpha u \qquad B_y = C\rho\alpha v$$

the three parameters of this approximation are now replaced by a function termed the bottom resistance coefficient r, that is,

$$B_x = ru, \qquad B_y = rv \tag{88}$$

where $r = (f/g)h \cos \phi$, in which h is the height of the bottom Ekman layer and ϕ is the veer angle (Hsueh and Peng, 1978). Since u and v can be expressed as functions of sea-level elevation from Eqs. (27) and (28), substitution of their geostrophic relation into Eq. (88) leads to

$$B_x = -\frac{f}{g}(h \cos \phi)\left(\frac{g}{f}\frac{\partial e}{\partial y}\right) \tag{89}$$

$$B_y = \frac{f}{g}(h \cos \phi)\left(\frac{g}{f}\frac{\partial e}{\partial x}\right) \tag{90}$$

Upon cancellation of the gravity and Coriolis terms, and differentiation to form the x, y components of the bottom curl, we obtain

$$\frac{\partial B_x}{\partial y} = -(h \cos \phi)\frac{\partial^2 e}{\partial y^2} \tag{91}$$

$$\frac{\partial B_y}{\partial x} = (h \cos \phi)\frac{\partial^2 e}{\partial x^2} \tag{92}$$

Substitution of these expressions into Eq. (86) leads to a new vorticity equation of

$$\left(\frac{\partial H}{\partial x}g\frac{\partial e}{\partial y} - \frac{\partial H}{\partial y}g\frac{\partial e}{\partial x}\right) + (h \cos \phi)\left(\frac{\partial^2 e}{\partial x^2} + \frac{\partial^2 e}{\partial y^2}\right) = \left(\frac{\partial F_y}{\partial x} - \frac{\partial F_x}{\partial y}\right) \tag{93}$$

where in calculus jargon the first term on the left is the Jacobian of H, e, representing the vortex generation; part of the second term is the Laplacian of e, multiplied by the trigonometric function, with this term representing the bottom friction; and recall that the right-hand term is the curl of the wind stress. This equation can be solved with specification of the time-dependent wind forcing, of the local bottom topography, of the height and veer of the bottom Ekman layer, and of boundary conditions for the sea elevation; values of h have ranged from 5 to 15 m and ϕ from 0 to 45° in previous analytical (Csanady, 1981b) and numerical (Hsueh and Peng, 1978; Han *et al.*, 1980; Hsueh, 1980) studies. Hopkins and Dieterle (1983) extended the functional dependence of the bottom stress

to include the effects of surface wind stress in waters of $H < 30$ m; this parameterization of the friction terms is used in the circulation model of Section 2.6.6.

Few wind measurements are available from a grid of buoys moored on continental shelves. Consequently, time series of the actual values of the wind curl
are scant, but for a typical wind stress of 1 dyn cm^{-2}, wind curl estimates are
$\sim 5 \times 10^{-9}$ dyn cm^{-3} over an ocean basin (Hantel, 1972). Presumably the mean
wind curl is less over a shelf area, which is two orders of magnitude smaller than
that of an ocean basin; in any case, little information is available on the time
dependence of the wind curl on a shelf. We thus assume that Eq. (93) can be
reduced to

$$\left(\frac{\partial H}{\partial x} g \frac{\partial e}{\partial y} - \frac{\partial H}{\partial y} g \frac{\partial e}{\partial x}\right) + (h \cos \phi)\left(\frac{\partial^2 e}{\partial x^2} + \frac{\partial^2 e}{\partial y^2}\right) \approx 0 \qquad (94)$$

without introducing significant errors in the model.

A time-dependent wind forcing can still enter the solution of this equation,
however, through the usual coastal boundary condition that the depth-integrated
transport, normal to the coastline, must be zero. With $U = 0$ in Eq. (35), the
steady-state balance is

$$F_y = B_y + gH \frac{\partial e}{\partial y} \qquad (95)$$

and, substituting Eq. (90) in Eq. (95), we obtain

$$F_y = (h \cos \phi)\left(\frac{\partial e}{\partial x}\right) + gH \frac{\partial e}{\partial y} \qquad (96)$$

for the coastal boundary condition at $x = 0$. Using current meter data, or other
rationalizations (Hsueh and Peng, 1978; Csanady, 1981b; Hopkins and Dieterle,
1983), expressions are derived for the upstream, downstream, and offshore
boundaries as well, allowing solution of Eq. (94). If the value of e is specified
at the boundary, it is termed a Dirichlet condition; if the value of $\partial e/\partial x$ or $\partial e/\partial y$
normal to the boundary is specified, it is a Neumann condition; and if both e and
the spatial derivatives are specified, it is termed a Cauchy condition. Examples
of the solution and application of Eq. (94) are presented as Sections 2.6.5 and
2.6.6.

2.6.5   Gaussian Elimination

Finite-difference representation of a second-order derivative (i.e., $\partial^2 e/\partial x^2$) is
analogous to that of the first-order derivative $\partial e/\partial x$ represented by the upstream
finite difference of Eq. (69). Since $\partial^2 e/\partial x^2$ is equivalent to $(\partial/\partial x)(\partial e/\partial x)$, we can
expand Eq. (69) as

$$\frac{\partial}{\partial x}\frac{(e_{i+1} - e_i)}{\Delta x} \approx \frac{1}{\Delta x}\left[\frac{(e_{i+1} - e_i)}{\Delta x} - \frac{(e_i - e_{i-1})}{\Delta x}\right] \tag{97}$$

or

$$\frac{\partial^2 e}{\partial x^2} \approx \frac{(e_{i+1} + e_{i-1} - 2e_i)}{\Delta x^2} \tag{98}$$

Rewriting Eq. (94) as an upstream finite difference then leads to a series of algebraic equations at each grid point i, j, in the form

$$a(H_{i+1,j} - H_{i,j})(e_{i,j+1} - e_{i,j}) - a(H_{i,j+1} - H_{i,j})(e_{i+1,j} - e_{i,j})$$
$$+ b(e_{i+1,j} + e_{i-1,j} - 2e_{e,j}) + c(e_{i,j+1} + e_{i,j-1} - 2e_{i,j}) = 0 \tag{99}$$

where $a = g/(\Delta x\,\Delta y)$, $b = h\cos\phi/\Delta x^2$, and $c = h\cos\phi/\Delta y^2$. Since Eq. (94) constitutes a boundary-value problem (recall Section 2.5), its finite-difference form of Eq. (99) can be solved numerically either by iterative methods (Roache, 1976) or by a number of direct methods involving a technique known as Gaussian elimination (Varga, 1962).

Detailed discussion of the matrix algebra used on digital computers (Eisenstat et al., 1976) to speed this previously slow method (Roache, 1976), is beyond the scope of this text. A simple example (Cheney and Kincaid, 1980) will be presented, however, to illustrate the method of solution, similar to Section 2.4.1. Consider the set of three linear equations in three unknowns,

$$3e_1 + 2e_2 - e_3 = 7 \tag{100}$$

$$5e_1 + 3e_2 + 2e_3 = 4 \tag{101}$$

$$-e_1 + e_2 - 3e_3 = -1 \tag{102}$$

In the forward sweep, or elimination procedure, of Gaussian elimination, successive multiplications and subtractions are used to eliminate the variables e_1 and e_2 from Eqs. (101) and (102). For example, if we first multiply Eq. (100) by $\frac{5}{3}$ and subtract the result from Eq. (101), and then multiply Eq. (100) by $-\frac{1}{3}$ and subtract this result from Eq. (102), we obtain the equivalent set of equations of

$$3e_1 + 2e_2 - e_3 = 7 \tag{103}$$

$$-\frac{1}{3}e_2 + \frac{11}{3}e_3 = -\frac{23}{3} \tag{104}$$

$$\frac{5}{3}e_2 - \frac{10}{3}e_3 = \frac{4}{3} \tag{105}$$

If we continue, by multiplying Eq. (104) by -5 and subtracting the result from Eq. (105), we obtain the next equivalent set of

$$3e_1 + 2e_2 - e_3 = 7 \tag{106}$$

$$-\tfrac{1}{3}e_2 + \tfrac{11}{3}e_3 = -\tfrac{23}{3} \tag{107}$$

$$15e_3 = -37 \tag{108}$$

The backward sweep, or substitution part, of the Gaussian elimination procedure consists of solving Eq. (108) for e_3, substituting this value in Eq. (107) to solve for e_2, and substituting both values for e_2 and e_3 in Eq. (106), or Eq. (103) and Eq. (100), to solve for e_1. We arrive at $e_3 = -\tfrac{37}{15}$, $e_2 = -\tfrac{62}{15}$, and $e_1 = \tfrac{64}{15}$. Eqs. (100)–(102) can also be represented in matrix form as

$$\begin{bmatrix} 3 & 2 & -1 \\ 5 & 3 & 2 \\ -1 & 1 & -3 \end{bmatrix} \times \begin{bmatrix} e_1 \\ e_2 \\ e_3 \end{bmatrix} = \begin{bmatrix} 7 \\ 4 \\ -1 \end{bmatrix} \tag{109}$$

in which the above procedure could be performed with matrix algebra, implemented on a digital computer, to provide the same solutions for e_1, e_2, and e_3.

Use of just the Laplacian of e in Eq. (94) and the appropriate boundary conditions on a rectangular grid i, j, however, generates $(i-2)(j-2)$ simultaneous equations, requiring of the order $[(i-2)(j-2)]^3$ multiplications in the usual Gaussian elimination procedures (Roache, 1976). Significant roundoff error occurs if $(i-2)(j-2) > 50$ (Hamming, 1973). Application of sparse matrix algorithms obviates some of these problems; Eq. (94) was solved (Eisenstat et al., 1976) on a grid of ~6000 elements (Fig. 37) over the Mid-Atlantic Bight in ~15 minutes, compared to ~45 minutes over a mesh of ~2000 elements in the Bering Sea model (Fig. 30), using Eqs. (70)–(72). Replacing the momentum equations with the vorticity equation, and using Gaussian elimination techniques of solution rather than time-dependent techniques, thus reduced the computer time about 10-fold for the same size spatial grid.

2.6.6   Mid-Atlantic Circulation

Because of complex bottom topography and an irregular coastline, a curvilinear grid (Fig. 37) was used over the Mid-Atlantic Bight rather than the rectangular grids (Fig. 30) of the above marching problems, embodied in the previous Bering Sea and New York Bight circulation models. In their study of the New York Bight, Han et al. (1980) also departed from a rectangular grid, using a triangular one instead; their results are discussed in Section 6.4.5. Equation (99) of the boundary-value problem was solved with the Gaussian elimination technique over a spatial domain of the shelf, extending between Cape Hatteras and Martha's Vineyard, of 886 km along shore at the coast and of 676 km at the shelf break.

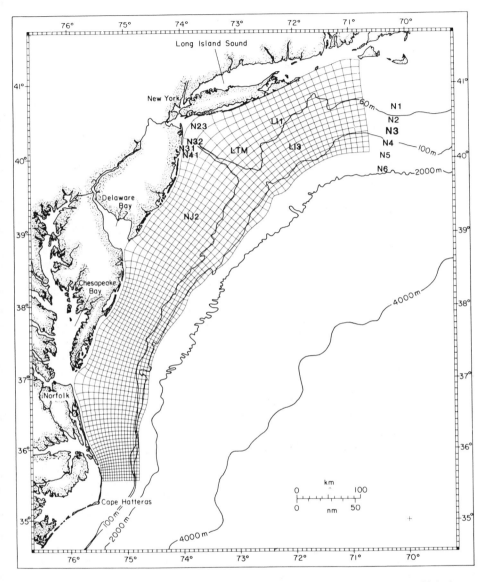

FIG. 37. The spatial grid of a numerical model for comparison of simulated currents with both WHOI/NMFS current-meter observations (N1–N6) at the upstream boundary and AOML observations (LI1, LI3, LTM, N23, N31, N32, N41, NJ2) in the New York Bight. [After Walsh *et al.* (1987a); © 1987 Pergamon Journals Ltd., with permission.]

The curvilinear grid chosen to represent this shelf domain maps the smoothed coastal and oceanic (200 m) boundaries into constant coordinate lines in a transformed (η, ξ) coordinate system. Each nearshore cell of the grid (Fig. 37) is thus orthogonal to the coastline, such that the transport boundary condition, $U = 0$, can be unambiguously specified here. The mapping procedure (Reid and Vastano, 1966; Reid et al., 1977) uses a truncated Fourier series to transform the geographic region of the x–y plane into a rectangular domain of $-\frac{1}{2}x < \eta < \frac{1}{2}x$ and $0 < \xi < y$, without introduction of new terms in Eq. (99)—see Hopkins and Dieterle (1983) or Reid et al. (1977) for more detail. Alternatively, a rectangular grid and stream functions, that is, $U = -\partial\psi/\partial y$ and $V = \partial\psi/\partial x$, could have been used in this boundary-value problem (Chapman et al., 1986), with a constant ψ, or streamline, at the coast to ensure the boundary condition of zero water transport in or out of the coast, that is, $\partial\psi/\partial y = 0$.

The sea elevation at the upstream boundary of the grid (Fig. 37) was computed from the observed currents at six moorings (N1–N6) south of Martha's Vineyard (Beardsley et al., 1985). At the offshore boundary, e was set to a constant 5 cm, such that mass exchange was confined to the surface and bottom Ekman layers at the shelf break; at the downstream boundary, the along-shore gradient of sea elevation was set to zero (Hsueh, 1980). For each steady-state solution of Eq. (94), the mean wind forcing, from observations at John F. Kennedy airport, was entered in the model as the along-shore component of the wind stress at the land boundary, where only along-shore transport of water occurred. The simulated currents at various depths were then compared with Atlantic Oceanographic Meteorological Laboratory (AOML) current meter observations (Mayer et al., 1982) near eight interior grid points of the model (Fig. 37) by analytical solution (Hopkins and Slatest, 1986) of the depth-dependent Ekman equations (e.g., Section 2.3).

After a mean northwest wind forcing (296°T) of 1.07 dyn cm^{-2}, applied as F_y at the land boundary of the model over April 5–12, 1979 (Fig. 38a), the depth-averaged currents of this circulation case were ≤5 cm sec^{-1}. Offshore flow occurred between the 20- and 40-m isobaths, with southwesterly flow between the 40- and 60-m isobaths at midshelf, south of the Hudson Canyon and north of Norfolk. In response to such northwest wind events, surface waters are pushed offshore, and the predominantly westward alongshore flow is slowed down (Beardsley and Butman, 1974). An upwelling circulation pattern of 7 m day^{-1} was created near the coast, in which shelf water of the surface Ekman layer was advected offshore at >10 cm sec^{-1} in the model. At 20 m depth, the flow was onshore at ~5 cm sec^{-1}; that is, subsurface slope water can be returned within a bottom Ekman layer.

During northeast wind events in the Mid-Atlantic Bight, the westward flow is instead intensified, however, and weak onshore flow usually occurs at the surface, with offshore flow of subsurface water (Beardsley et al., 1985). By changing the boundary conditions and obtaining another steady-state solution, we note that under a wind forcing of only 0.31 dyn cm^{-2} from the northeast (068°T), the mean flow of the model's water column during April 12–16, 1979, was >10 cm sec^{-1} to the southwest over most of the shelf, except for offshore flows south of Long Island, near the Hudson Canyon, off Delaware Bay, and south of Norfolk (Fig. 38b). Consistent with a downwelling circulation pattern, the offshore flow at 20-m depth was twice that of the surface flow south of Norfolk (Fig. 39); ubiquitous offshore flow occurred in the bottom layer at 45-m depth.

Vertical decomposition of the model's flow field during this northeast wind forcing provides insight for both the veracity (Fig. 40) and complexity of the simulated currents (Fig. 39). With steady-state solutions of $e_{i,j}$ at each grid point and a constant K_z, continuous values of $u_{i,j}$ and $v_{i,j}$ with depth can be analytically obtained from Eqs. (48) and (49). This complete depth-dependent decomposition of the transport can be viewed as the linear superposition of three terms:

$$u(z) = \frac{U}{H} + u_E + u_B \tag{110}$$

$$v(z) = \frac{V}{H} + v_E + v_B \tag{111}$$

that is, analogous to Eqs. (52) and (53) in Section 2.3, where u_E and v_E are appropriate solutions of flow within the surface Ekman layer, and u_B and v_B are the solutions within the bottom Ekman layer.

The terms U/H and V/H represent those components of the geostrophic flow field at middepth of the water column, $z = H/2$, while the relative contributions of the second and third terms are decaying functions of D and h, the respective depths of the Ekman layers, away from the boundaries at the surface and bottom of the water column. Simons (1980) discusses the many approaches to analytical solutions of current variation with depth. Figure 39 represents the vertical decomposition of the flow field at 0-m, 20-m, and 45-m depths of the northeast wind case, using one particular analytical procedure (Hopkins and Slatest, 1986).

A comparison of the computed and observed currents during the same April time period at 16 locations of varying depth in the New York Bight (Walsh et al., 1987a) is presented in Fig. 40. Except for the nearshore current meter moorings, N31 and N41, where a hydrographic survey (Hazelworth and Berberian, 1979) indicated a freshwater plume downstream of the Hudson/Raritan estuaries, the direction and speed of the model's flow field match fairly well the observed currents. Moreover, addition of the term for horizontal changes in the buoyancy force to Eq. (94) (i.e., the baroclinic component of flow) alters the estimate of April 1979 shelf transport by only ∼10% (Hopkins and Dieterle, 1987).

As we shall see in Section 5.1.3, temporal changes of chlorophyll biomass during the spring bloom appear to be aptly described by time-dependent alter-

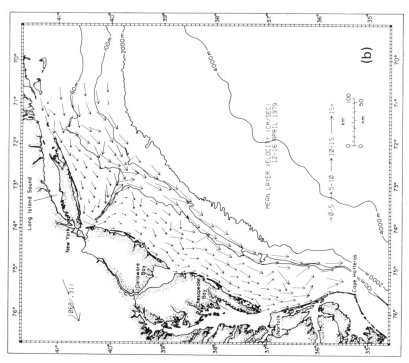

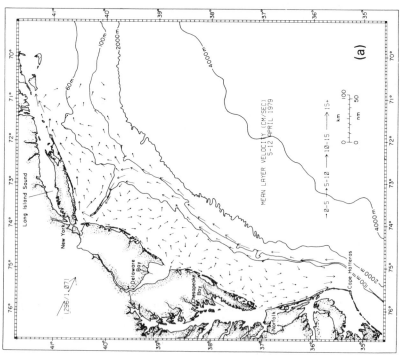

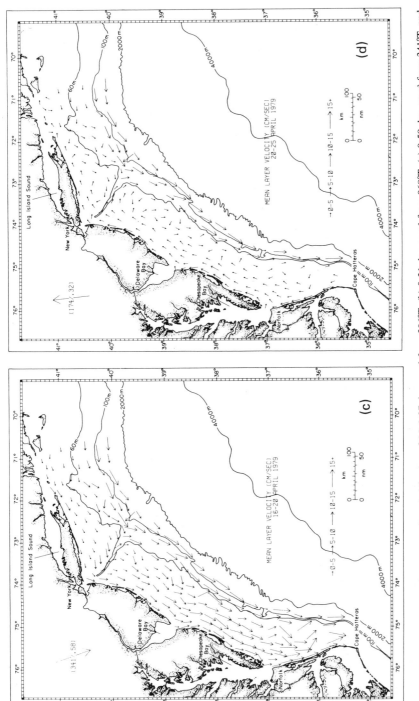

FIG. 38. The depth-averaged currents under wind forcing of (a) 1.07 dyn cm⁻² from 296°T, (b) 0.31 dyn cm⁻² from 068°T, (c) 0.58 dyn cm⁻² from 341°T, and (d) 0.32 dyn cm⁻² from 174°T. [After Walsh *et al.* (1987a); © 1987 Pergamon Journals Ltd., with permission.]

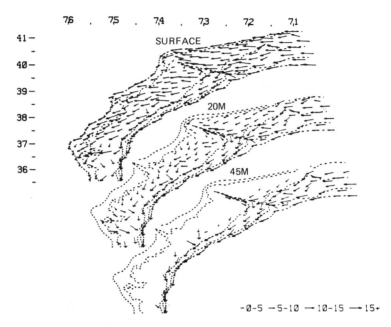

FIG. 39. The simulated currents (in centimeters per second) at three depth levels (0 m, 20 m, 45 m) over the Mid-Atlantic Bight during April 12–16, 1979. [After Walsh *et al.* (1987a); © 1987 Pergamon Journals Ltd., with permission.]

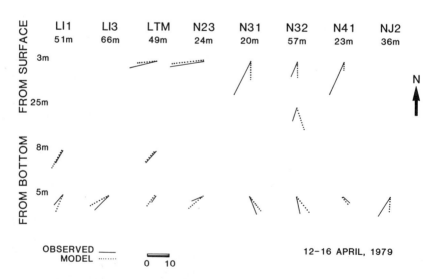

FIG. 40. A comparison of observed and computed currents at 16 locations within the Mid-Atlantic Bight during April 12–16, 1979. [After Walsh *et al.* (1987a); © 1987 Pergamon Journals Ltd., with permission.]

ation of barotropic circulation patterns and the associated changes of plankton dynamics in response to these habitat fluctuations. Following this northeast wind event, another mean wind forcing from the northwest (341°T) occurred, for example, during April 16–20, 1979 (Fig. 38c), but with half the intensity (0.58 dyn cm^{-2}) of the first period (Fig. 38a). In response to this shift in wind forcing at the boundary, the third steady-state solution led to a mean depth-averaged along-shore flow, which was now weaker north of the Hudson Canyon and stronger south of Delaware Bay during April 16–20, compared to the April 12–16 time period (Fig. 38b). In the model, there was now little or no offshore flow south of Long Island and New Jersey, but continued offshore movement of water occurred off Delaware Bay and Norfolk (Fig. 38c). Even the flow within the surface Ekman layer was directed along-shore rather than offshore in this third case, except for waters south of Virginia.

Finally, with southerly wind forcing in the real world, surface flow is at times offshore and to the east, reversing the predominantly westward currents, if a storm is of sufficient intensity within the Mid-Atlantic Bight. During the fourth case for April 20–25, 1979, a mean wind forcing of 0.32 dyn cm^{-2} from the south (174°T) was sufficient to drive weak, depth-averaged currents (<5 cm sec^{-1}) to the northeast within the 10- to 40-m isobaths (Fig. 38d), in contrast to the three previous flow fields of the model (Figs. 38a–c). This simulated circulation pattern led to an upwelling of ~5 m day^{-1} at 5-m depth on the inner shelf south of Delaware, similar to that observed north of New Jersey during the case of April 5–12, 1979 (Fig. 38a). Within the surface Ekman layer, offshore flow of 5–10 cm sec^{-1} occurred only on the outer shelf, however, south of Delaware Bay during April 20–25, 1979. The last wind forcing is typical of summer conditions (Fig. 5), with a seasonal switch in wind origin from the northwest to the southwest, but the stratified water-column conditions of this period invalidate our assumption of a barotropic sea.

2.7   Baroclinicity

Variation of a current field in the vertical dimension can be caused either by frictional effects, which we have parameterized as vertical changes in eddy viscosity, or by baroclinic effects, that is, vertical changes in density. Of perhaps equal importance are horizontal changes in density, induced by river runoff on the continental shelf, which we have thus far ignored in the equations of both motion and vorticity. We briefly contrast the results of barotropic and baroclinic calculations for the simulated currents of the New York Bight (Fig. 41) during April 1979 and August 1978, to explore the importance of the density field in both diagnostic circulation models (Sarkisyan, 1977) and shelf food-web models—see Section 6.4.2.

The baroclinic analog of the depth-integrated, barotropic vorticity equation

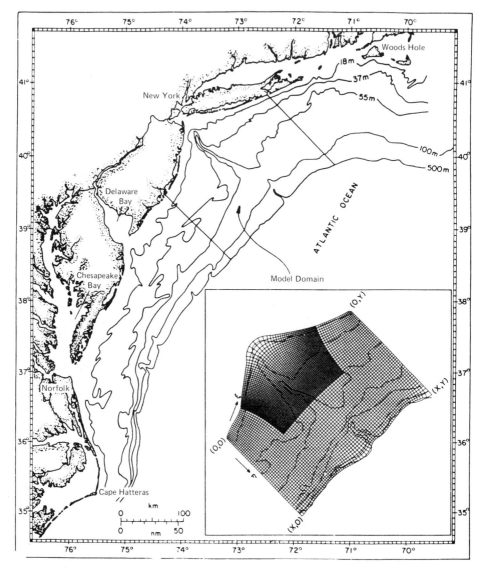

FIG. 41. The New York Bight and the location of the model domain. The insert shows the quasi-bathymetric shelf grid and the smaller nested, apex grid. [After Hopkins and Dieterle (1986); © International Council for the Exploration of the Sea, with permission.]

[Eq. (93)] involves the addition of another term, the Jacobian of H, R, to represent the horizontal variation of the buoyancy field, where $R = \int^{e}_{-H} \rho g \, dz$. Equation (93) then becomes

$$\left(\frac{\partial H}{\partial x} g \frac{\partial e}{\partial y} - \frac{\partial H}{\partial y} g \frac{\partial e}{\partial x}\right) + \left(\frac{\partial H}{\partial x} \frac{\partial R}{\partial y} - \frac{\partial H}{\partial y} \frac{\partial R}{\partial x}\right)$$
$$+ (h \cos \phi)\left(\frac{\partial^2 e}{\partial x^2} + \frac{\partial^2 e}{\partial y^2}\right) = \left(\frac{\partial F_y}{\partial x} - \frac{\partial F_x}{\partial y}\right) \quad (112)$$

which may be solved numerically (Hopkins and Dieterle, 1987), or analytically after appropriate simplification (Csanady, 1984). Assuming a negligible wind curl, a known bottom topography ($\partial H/\partial x$, $\partial H/\partial y$), and specification of the veer angle (ϕ), height (h) of the bottom Ekman layer, and boundary conditions for e, the steady-state solution of Eq. (112) still requires definition of the depth-integrated density field in $\partial R/\partial y$, $\partial R/\partial x$. One can either compute the density field, from models of air–sea interactions of buoyancy exchange (Nihoul, 1984), or prescribe the density field from data (i.e., a diagnostic circulation model).

The planetary wind field (Fig. 5) and the global distribution of salinity (Fig. 7) and temperature (i.e., density) all arise from differential heating of the Earth, with latitude, by the sun. The resultant motions induced by these friction and pressure forces are coupled on at least seasonal time scales, but it is convenient to distinguish between wind-driven and thermohaline circulation, for example, that within the upper kilometer of the ocean in contrast to the abyssal circulation. The local thermohaline circulation of a shelf is assumed to be the result of independent, time-invariant buoyancy forces, where the horizontal density gradients are determined by global processes of precipitation, evaporation, and insolation. Similar to momentum [Eqs. (4)–(6)] and relative vorticity [Eq. (85)], however, density can also be advected, such that cross-shelf transport of salt by shear diffusion in the Mid-Atlantic Bight (Fischer, 1980) is of the same order as the along-shore transport of salt by self-advection (Csanady, 1982). Diagnostic circulation models thus assume not only that an observed density field in $\partial R/\partial y$, $\partial R/\partial x$ is synoptic, but that these input data can be independently prescribed for at most 1–2 weeks (Han et al., 1980; Lee et al., 1982).

Observed density fields at ~75 stations during two cruises on August 2–8, 1978, and April 9–16, 1979, were interpolated (Hopkins and Dieterle, 1987), for example, to a model grid (Fig. 41) of ~4000 elements of 3 km spacing, extending ~140 km offshore and ~250 km alongshore between Cape May, New Jersey, and Montauk Point, Long Island, that is, over the New York Bight. Curvilinear coordinates were again employed, as in the above case of the Mid-Atlantic Bight circulation (Fig. 37), with another nested grid of approximately fourfold greater spatial resolution used to approximate the apex of the New York Bight (Fig. 41)—see Section 5.1.1. The seasonal discharge of fresh water was

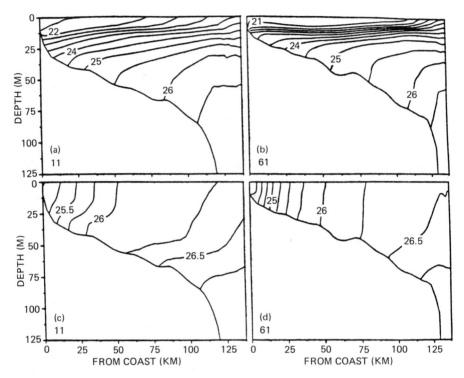

FIG. 42. The smoothed density fields for input, at the upstream and downstream boundaries of a baroclinic model, during August 1978 off (a) Long Island (b) and New Jersey and (c, d) over the same transects during April 1979. [After Hopkins and Dieterle (1987); © 1987 Pergamon Journals Ltd., with permission.]

simulated by a mass flux of 2300 m^3 sec^{-1} in April and 300 m^3 sec^{-1} in August across the estuarine boundaries of the apex grid. The inflow of fresh water from both Long Island Sound and the Hudson River can be seen in the smoothed August σ_t fields off the Long Island (Fig. 42a) and New Jersey (Fig. 42b) coasts, and to a lesser extent in the April σ_t fields (Figs. 42c, d). The cross-shelf gradients of density near the north and south boundaries of the grid were only 1–2 σ_t between lighter coastal and heavier slope waters in April 1979, compared to 5–6 σ_t in August 1978 (Fig. 42), such that the buoyancy force of the summer water column was greater than that of the spring.

The bathymetry was entered in the model at 5-km spacing, and the wind forcing for April and August (Fig. 43) was obtained from a tower at Tiana Beach, Long Island, for input of the wind stress at the coastal boundary [i.e., Eq. (96)]. The mean wind stress was 0.31 dyn cm^{-2} from the south during August 1–9, 1978, and 0.50 dyn cm^{-2} from the northeast during April 8–16, 1979, which is over twice the averaging period of the previous model (Figs. 38a–d). With

AOML current meters at 30, 80, and 150 km off Long Island along the northern boundary of the grid (Mayer *et al.*, 1982), the sea elevation *e* was again estimated at the upstream boundary by subtracting the computed baroclinic component of the observed flow [i.e., from the thermal wind equations, Eqs. (29) and (30)] to obtain the barotropic component of Eq. (36a). The offshore and downstream boundary conditions for *e* were the same as in the above Mid-Atlantic circulation model, and Gaussian elimination (Section 2.6.5) was used to solve for seasonal steady states of the New York Bight currents in April and August.

Under northeasterly wind forcing and southwest flow (Fig. 39), water piles up against the coast of the New York Bight. The gradient of sea level during the April case, for example, increased shoreward from 5 cm at the offshore boundary condition to 24–25 cm at the coast (Hopkins and Dieterle, 1987), which is ~20 cm/100 km or threefold the sea-level gradient of the Bering Sea model (Fig. 33). In a purely barotropic model, the water would then run "downhill," or offshore along the gradient of sea level, and turn to the right, or toward the southwest, under the influence of the Coriolis term. The density gradient of the buoyancy term increased seaward (Fig. 42), however, in the opposite direction of the sea-surface slope under northeast wind forcing, such that the portion of geostrophic flow induced by the buoyancy force would tend to oppose the other component of geostrophic flow, induced by the pressure force of the sea-surface slope. In

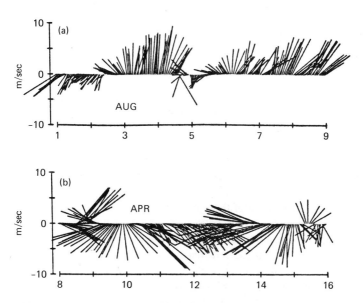

FIG. 43. Wind stick diagrams, where the length of the vector is the speed (m sec^{-1}) and the north direction is toward the top of the page, during (a) August 1978 and (b) April 1979. [After Hopkins and Dieterle (1987); © 1987 Pergamon Journals Ltd., with permission.]

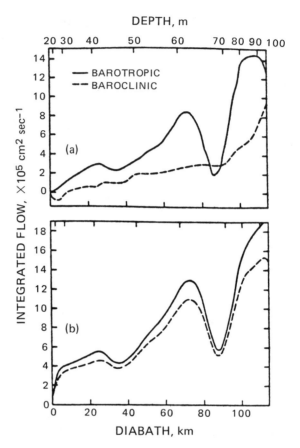

FIG. 44. The depth-integrated transport (cm² sec⁻¹) for the barotropic solution (solid) and full, baroclinic solution (dashed) during (a) August 1978 and (b) April 1979. [After Hopkins and Dieterle (1987); © 1987 Pergamon Journals Ltd., with permission.]

terms of the vorticity balance of Eq. (112), the negative relative vorticity in-duced by the baroclinic term is partially offset by the positive vorticity of the barotropic term.

The seasonal importance of these two pressure forces is examined in Fig. 44 of the depth-integrated transports across the northern boundary of the model in April and August. The along-shore transport increased seaward during each sea-son, of course, because the depth increased, except for the local minimum at ~65 m where the Jacobian of H, e changed sign, such that the bathymetry flat-tened offshore and shoaled to the west (Hopkins and Dieterle, 1987). At each isobath, the transport V from just the barotropic term was consistently larger than the same parabathic transport computed from the sum of the barotropic and baroclinic pressure terms. During April, the cross-shelf density gradient was larger near the coast than at depths >50 m (Fig. 42), but the depth integral also increased seaward, such that the reduction of transport, due to the baroclinic shear, remained a uniform 10% out to the 80-m isobath. The April currents (Fig. 45), computed from both the barotropic and baroclinic terms, compared well with 18 AOML current meters (Hopkins and Dieterle, 1987), yielding an average vector error of only ~1 cm sec^{-1} and 10° from the circulation model.

Under the southerly wind forcing of the August case (Fig. 43), the cross-shelf gradient of sea elevation still increased toward the coast, but was reduced to ~12 cm/100 km, or half that of the April case. The resultant barotropic transport across the northern boundary of the model was similarly reduced by ~40% (Fig. 44). The August buoyancy force was also greater than that of April, such that the relative reduction of the transport, due to the baroclinic shear, was then ~60% in August out to the 80-m isobath. The computed along-shore transport from the combined barotropic and baroclinic terms, in conjunction with a southerly wind forcing in August, was ~25% of the previous V under the April northeaster. Vertical resolution of this flow field (Fig. 45) allows further assessment

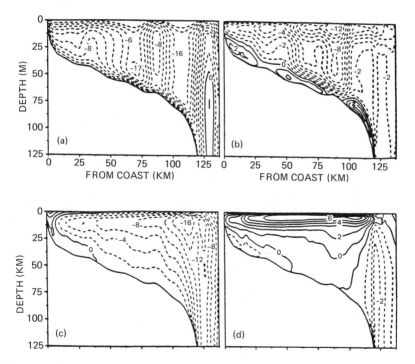

FIG. 45. The v and u flow components during (a, b) April 1979 and (c, d) August 1978. [After Hopkins and Dieterle (1987); © 1987 Pergamon Journals Ltd., with permission.]

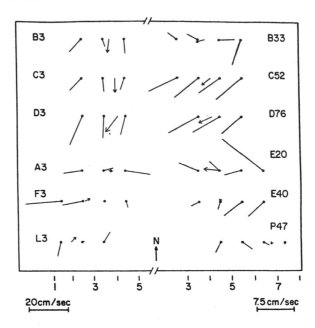

FIG. 46. Stick diagrams of the observed flow during August 1978. The vectors represent daily averages. The modeled flow is inserted one inertial period after the local density field was sampled, where the letter code corresponds to current-meter moorings and the number following the letter indicates the depth in meters. The scale of the velocity in cm sec^{-1} is given at the bottom of each panel. [After Hopkins and Dieterle (1987); © Pergamon Journals Ltd., with permission.]

of the impact of both the ageostrophic forcing and seasonal changes in the pressure field. During the August case of southerly winds, offshore flow (positive u) occurred within the upper 25 m of the surface Ekman layer, in contrast to the onshore flow (negative u) over most of the water column in the April case of northeasterly winds (Fig. 45). Appropriate reversal of cross-shelf flow was computed for the bottom Ekman layer under the two wind forcings, with a downwelling velocity w of ~2.5 m day^{-1} computed in the bottom layer of the April case, for example. In the August upwelling situation, the combined wind stress and buoyancy force were sufficient to actually reverse the along-shore flow field (positive v) within 40–50 km of the coast. At midshelf, the southward (negative v) geostrophic velocities in the middle of the column were reduced by ~12 cm sec^{-1} during the seasonal transition from a well-mixed to a stratified water column; the smallest seasonal change of the parabathic flow field occurred at the shelf break.

A comparison of the more complex flow field of the August case with 12 AOML current meters (Fig. 46) leads to less fidelity of the simulated circulation to the "real" world. Data from moorings B, C, and D over the upstream boundary agree fairly well with the computed currents, while F and L in the apex and the rest in the interior of the grid exhibit greater divergence from the model output. Inaccurate boundary conditions, ignored terms of the Navier–Stokes equations, and inadequate time averages of wind and density forcing are all candidates for refinement of the above diagnostic model (Hopkins and Dieterle, 1987). Realistic sampling constraints suggest, however, that there are limits to our ability to continuously resolve the physical variance of the shelf habitat. An alternative approach is to next consider chemical and biological tracers, which constitute information at varying scales of integration of shelf dynamics.

3 Production

It would almost seem as if water, especially sea water, had some slumbering force within itself, a dormant sympathy for organic life which needed merely the slightest stimulus to awaken and take its share in dynamic animation . . . only at the surface can vegetable life exist and develop, changing sunlight into edible materials . . . so let us consider sunlight and darkness, or even light and shadow. . . . .

The boundary of air and water above me appeared perfectly solid. . . . The sunlight shifted down in long, oblique rays as if through some unearthly beautiful cathedral window. . . . When less than 50 feet beneath the surface I happened to glance at a large deep-sea prawn . . . to my astonishment it was no longer scarlet, but a deep velvety black. . . . On this and other dives I carefully studied the changing colors . . . at 20 feet there was only a thread of red and at 50 the orange was dominant. This in turn vanished at 150 feet. Three hundred feet found . . . the yellow almost gone . . . at 450 feet, no blue remained, only violet, and green too faint for naming. . . . At 800 feet . . . when I looked outside I saw only the deepest, blackest-blue imaginable. . . . One had to sit quietly and absorb these beauties before one could remember to be an ichthyologist.

(Beebe, 1926, 1934)

Biological organisms are improbable, unusual structures that thwart the Second Law of Thermodynamics by using solar energy to enzymatically speed up chemical reaction rates, maintaining a nonequilibrium distribution of the elements over the Earth. The ratios of nutrient concentrations of the deep sea are thus similar to those in living plankton (Redfield *et al.*, 1963), reflecting a biological imprint on chemical reactions in the sea. Unlike CO_2, which is abundant within seawater in the form of HCO_3^{2-}, major nutrients such as nitrate, phosphate, and silicate limit the population growth of phytoplankton and must be returned to the sunlit regions of the sea. The availability of trace nutrients, such as molybdenum or iron, may further determine the rate of biological extraction of nitrogen from the atmosphere, such as nitrogen fixation by microalgae, such that phosphorus is the limiting major nutrient in fresh water, as is nitrogen on the continental shelves (Howarth and Cole, 1985). Consequently, the regional differences in shelf primary production (Table III) partially reflect the varying rates of nitrogen input to the euphotic zone from subsurface slope waters.

Averaging over the 40 shelf regions where estimates of primary production were known to me, an annual fixation of 215 g C m^{-2} yr^{-1} is obtained from >100 field studies listed in the references. This is somewhat larger than the earlier shelf means of 175 g C m^{-2} yr^{-1} (Cushing, 1971b) and 183 g C m^{-2} yr^{-1} (Platt and Subba Rao, 1975), based on observations taken prior to 1972. Within 23 of these shelf regions (Table III), successive observations suggest a range of 168–351 g C m^{-2} yr^{-1} for the annual shelf primary production. Such variation reflects different methodologies, spatial heterogeneity, and eutrophication over time. A mathematical description of marine photosynthesis allows us to assess discrepancies in field estimates of the specific rate of autotrophic production (day^{-1}). Recent coastal zone color scanner (CZCS) and moored fluorometer time series also provide sufficient spatial and temporal resolution of the local phytoplankton biomass fields (g C m^{-2}) for analysis of the second component of daily primary production; these are the subjects of this chapter. Consideration of the interannual changes of primary production as a consequence of eutrophication will be deferred to Chapter 6.

3.1 Photosynthesis

Within the various versions of the Second Law of Motion [e.g., Eqs. (1), (4)–(6), and (77)], the physical forces causing positive accelerations, such as wind, sea-surface slope, and buoyancy, all depend eventually on the input of electromagnetic radiation to the planet in the form of heat. Differential latitudinal heating of the Earth between the poles and the equator leads to the planetary wind fields, which in turn pile water up against the coast. Recall that the buoyancy force is a function of spatial changes in the vertical distribution of light, warm water above cold, heavy water. Similarly, circumvention of the Second Law of Thermodynamics by organisms, that is, avoidance of an increase in entropy, or molecular disorder, requires input of electromagnetic radiation to Earth in the form of light. In energetic terms, the average energy input to the sea from solar radiation is ~10^4 more than that derived from the wind or tides.

Biological utilization of solar energy as autotrophic growth of phytoplankton can be described by a simple stoichiometric expression of concurrent carbon fixation and protein synthesis in

$$4CO_2 + 2H_2O + 2NO_3 \xrightarrow{\text{light}} 2CHO + 2CHN + 7O_2, \tag{113}$$

where carbon dioxide, water, and nitrate are converted by sunlit plants to precursors of carbohydrates, proteins, and free oxygen. Additional major nutrients, such as phosphorus, silicon, and sulfur, as well as trace elements, are required, of course, for growth of marine microalgae and represent a set of interacting limiting factors. The photosynthesis part of Eq. (113) results in the production of carbohydrates and oxygen and consists basically of two processes: a photochemical, or light, reaction, in which photons of light are trapped by the plant's chlorophyll molecules to be converted into biochemical potential energy [i.e., adenosine triphosphate (ATP) and nicotinamide adenine dinucleotide phosphate (NADP)], and an enzymatic, or dark, reaction, in which the potential energy stored within these compounds of the algal cell is then used to synthesize organic material. The light-limited reaction is independent of ambient temperatures, whereas enzymatic transfers of energy are temperature-dependent, such that the regulation of primary production in the sea by light, nutrients, and temperature must be considered.

3.1.1   Light Regulation

The portion of the spectrum of total incident solar energy, in the form of light utilized by marine phytoplankton, is restricted to the visible wavelengths from violet (350 nm) to red (750 nm). During transit of the atmosphere, most of the ultraviolet radiation (<350 nm) is absorbed by ozone, while water vapor and CO_2 absorb some of the infrared (>750 nm) radiation; the ability of atmospheric CO_2 to absorb radiation at the thermal end of the spectrum will be discussed in more detail in Section 6.1.1. Almost half of the solar energy at the surface of the sea is in the form of infrared radiation, which is absorbed within the first meter of the water column. The photosynthetically active radiation (PAR) is considered to range from 400 to 720 nm and to amount to 50% of the solar energy at the air–sea interface (Strickland, 1958), with the light of >600 nm wavelength mainly absorbed by chlorophyll a and <600 nm by the other accessory algal pigments.

   The depth penetration of light in the sea ranges from a few hundred meters in coastal waters to about 1000 m in the open ocean (Clarke and Denton, 1962) and can be described with Beer's law as

$$I(z) = \frac{\delta}{I_0 e^{-kz}} \tag{114}$$

where $I(z)$ is the PAR at depth z; I_0 is the total incident radiation; δ is the ratio of $I(0)/I_0$, usually assumed to be 0.5; and k is the diffuse attenuation, or extinction, coefficient. This coefficient can be partitioned into those components of the downwelling irradiance absorbed by water, phytoplankton, and other substances as

$$k = k_w + k_p + k_s \tag{114a}$$

where $k_w = 0.015$ m^{-1} for pure water, $k_s = 0.065$ m^{-1} for coastal water (K. Carder, personal communication), and $k_p = 0.03$ chl a (Smith and Baker, 1982). The attenuation length, or depth, is the reciprocal of this coefficient, k^{-1}, and represents the depth of penetration of 37% of surface irradiance, while the bottom of the euphotic zone (1% light level) is 4.61 attenuation lengths. Using the above estimates of k_w and k_s, a range in mean chlorophyll concentrations of 14.0–0.6 μg l^{-1} would lead to attenuation lengths of 2–10 m and euphotic zones of 9.2–46.1 m depth on the continental shelf.

As Beebe observed in his bathysphere off Bermuda 50 yr ago, the spectral quality of submarine light changes with depth; in pure water, the attenuation depth of red light (600 nm) is 5 m, compared to 25 m for yellow (520 nm) and 28 m for blue–green (480 nm) light (Duntley, 1963). The incident radiation has a spectral peak near the blue–green wavelength, which penetrates the deepest in the sea (Jerlov, 1976), and all taxonomic groups of marine phytoplankton have carotenoid pigments to absorb radiation at this wavelength (Parsons et al., 1977). Recent modification of Eq. (114) involves spectral decomposition of $I(z)$ into blue–green light and other wavelengths (Paulson and Simpson, 1977). Blue light is selectively scattered by particulate matter in turbid coastal waters, however, shifting the spectral peak towards the red (Jerlov, 1976), such that the average extinction coefficient over the PAR spectrum is sufficient for most shelf applications.

Evaluation of Eq. (114) requires specification of I_0, which varies as a function of albedo, cloudiness, day length, time of day, and season of the year. The amount of light reflected at the sea surface is usually only a few percent for angles of incidence less than 45° from the normal, for example, ~7% in April at 51°N (Fasham et al., 1983). As discussed in Section 1.4.3, seasonal changes in cloud cover can have a major impact on reduction of I_0 within some coastal areas, and local adjustments must be made to global estimates of clear sky radiation (Kimball, 1928). Approximations of the seasonal changes in day length and altitude of the sun, of daily changes in cloudiness, and of hourly changes in solar radiation can all be obtained with appropriate cosine and sine functions.

Over a period of 4 yr (1974–1977) on Long Island, for example, the mean incident radiation measured at 41°N (Fig. 47) during the winter solstice (12/22), I_0^w, was 120 g-cal cm^{-2} day^{-1}, compared to a mean of 500 g-cal cm^{-2} day^{-1} during the summer solstice (6/22), I_0^s (i.e., when the sun is farthest north of the

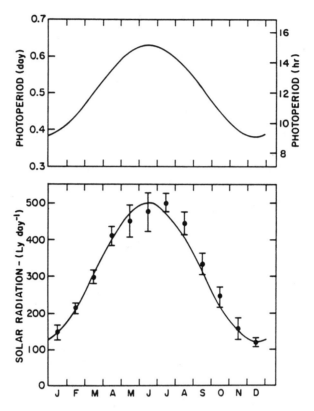

FIG. 47. Seasonal variation of photoperiod and incident solar radiation at 41°N, 73°W. [After Nagle (1978); with permission under DOE Contract No. DE-AC02-76CH00016.]

equator in this hemisphere). These seasonal light data can be described (Fig. 47) by a cosine function of I_0^w, I_0^s, and d, the Julian date (Nagle, 1978), as

$$I_0(d) = I_0^w + 0.5(I_0^s - I_0^w)\left[1 - \frac{\cos 2\pi(t - 356)}{365}\right] \tag{115}$$

where 356 is the Julian date of the winter solstice. Similarly, knowledge of the day length, or photoperiod, i, at the winter (i^w) and summer (i^s) solstices allows calculation of the seasonal change by

$$i(d) = i^w + 0.5(i^s - i^w)\left[1 - \frac{\cos 2\pi(t - 356)}{365}\right] \tag{116}$$

At 41°N, the mean photoperiod of the winter solstice is 9.1 hr (Fig. 47), compared to 15.6 hr at the summer solstice (Nagle, 1978). At 70°N near the Bering

Strait, the photoperiod is 24 hr during the summer solstice, with an incident radiation of 200 g-cal cm^{-2} day^{-1} (Sambrotto et al., 1984).

With specification of the day length and input of total radiation to the sea surface, the hourly change of solar irradiance with time, t, can then be estimated (Ikushima, 1967) during a cloudless day as

$$I_0(t) = I_m \sin^3\left[\frac{\pi(t - a)}{i(d)}\right] \tag{117}$$

where I_m is the maximum light intensity (g-cal cm^{-2} hr^{-1}) at local noon of the sine function, that is, $I_0(d) = \int_{j=1}^{i(d)} I_j\, dt$ for $j = 1, \ldots m, \ldots i(d)$, while $a = 12 - 0.5i(d)$ sets the time of sunrise, a. With $i(d) = 8$ hr, for example, Eq. (117) becomes at noon ($t = 12$)

$$I_0(12) = I_m \sin^3\left(\frac{\pi}{2}\right) \tag{118}$$

for a day length of 8 a.m. to 4 p.m. The maximum light intensity is at noon, since sin $\pi/2$ is 1.0, as, of course, is the cube of this sine function. From 1600 hr to 0800 hr, the cube expression becomes negative, reflecting no light. It must be set to zero in numerical models, however, to avoid spurious errors of subtraction at night (Walsh, 1975). For cloudy days or significant albedo, other refinements can be added to Eq. (117) to reduce the value of $I_0(t)$.

Substitution of Eq. (117) into Eq. (114) results in

$$I(z, d, t) = \delta I_m \sin^3\left[\frac{\pi(t - a)}{i(d)}\right] e^{-(k_w + k_p + k_s)z} \tag{119}$$

which we shall use to estimate the light regulation of marine photosynthesis. In the absence of temperature effects on the dark reaction of photosynthesis, the rate of photosynthesis in the sea should be maximal at the surface and should decline with depth as a function of the availability of light expressed by Eq. (119). Examination of the observed change of the photosynthetic rate with light intensity in the sea, however, suggests a nonlinear response of marine phytoplankton to various levels of solar radiation (Fig. 48).

As previously discussed in Chapter 1, I_c of Sverdrup's critical depth theory is the compensation light intensity, on a daily basis, at which algal photosynthesis equals the respiration λ of these plants. The positive portion, above the compensation point, of Fig. 48 is thus a plot of the rate of net photosynthesis against light intensity. At shallower depths of the water column (i.e., higher light intensities), a saturation light intensity I_s can be found at which occurs the maximum net photosynthesis per unit chlorophyll biomass, μ_m, of the plant or the assimilation index (mg C mg chl^{-1} hr^{-1}).

Conversion of the chlorophyll biomass to its carbon equivalent by a C/chl

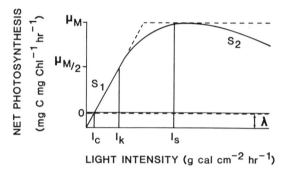

FIG. 48. Net photosynthesis as a function of light intensity.

ratio yields the light-saturated, maximum-potential, specific growth rate ε_m of the phytoplankton (hr^{-1}), at a particular temperature, and assuming no limitation by nutrients or other factors. Multiplication of the specific growth rate by the carbon biomass of the phytoplankton would yield the primary production of these algae (mg C m^{-3} hr^{-1}). At light intensities above I_s, there appears to be an inhibition of the rate of photosynthesis by natural phytoplankton populations. This may result from ultraviolet radiation at the sea surface, since laboratory cultures do not demonstrate an inhibitive response at intensities above I_s after spectral removal of short wavelengths (Harris, 1980).

The slope, s_1, of the light-limited portion of photosynthesis—that is, before the electron transfers among plastoquinone and cytochrome of photosystems I and II become saturated (Fujita, 1970) at I_s—is an index of how efficiently algae can use quanta of light (Fig. 48). Early studies (Ryther, 1956) of marine phytoplankton suggested that s_1 might be less for dinoflagellates than for green algae and diatoms. The dinoflagellate photosynthetic rate may be saturated at higher light intensities, which is consistent with the migratory pattern of dinoflagellates to surface waters of greater $I(z)$, after nutrient uptake near the nutricline (Walsh et al., 1974). Light or shade adaptation of algae by changing the size of the P_{700} light-collecting antenna of the photosynthetic unit (Falkowski, 1981; Perry et al., 1981) can take place within 12 hr, however, such that prior assumptions of a constant s_1 (Bannister, 1974a) are somewhat invalid.

Based on considerations of a maximum quantum yield (Platt and Jassby, 1976), the upper limit for s_1 is 0.86 mg C mg chl^{-1} hr^{-1} (g-cal PAR cm^{-2} $hr^{-1})^{-1}$, or 0.115 mg C mg chl^{-1} hr^{-1} (μE m^{-2} $sec^{-1})^{-1}$ on a quantum, or photon, basis. Four quanta of visible light are required to produce 1 mol of ATP and NADPH, while 6×10^{23} quanta are equivalent to either 1 einstein (E), or 5×10^4 g-cal over the visible spectrum, that is, about 1×10^5 g-cal of total incident radiation. In a seasonal study of various shelf habitats within the Mid-Atlantic Bight (Malone and Neale, 1981), s_1 ranged from 0.011 to 0.153 mg C mg chl^{-1} hr^{-1} (μE m^{-2} $sec^{-1})^{-1}$ for nanoplankton chlorophytes and from 0.002

to 0.154 mg C mg chl^{-1} hr^{-1} (μE m^{-2} sec^{-1})$^{-1}$ for netplankton diatoms, approaching at times the maximum quantum yield. The mean s_1 for these two size classes of phytoplankton was respectively 0.06 and 0.05 mg C mg chl^{-1} hr^{-1} (μE m^{-2} sec^{-1})$^{-1}$, however, similar to previous observations (Platt and Jassby, 1976; Taguchi, 1976). These field and culture data suggest that, over appropriate time scales longer than those of light-shade adaptation, the quantum yield s_1 can be considered a time-invariant parameter of models of homogeneous phytoplankton populations, that is, those with chlorophyll or carbon as their state variables rather than individual species.

Building upon Liebig's (1840) concept of a limiting nutrient, or "foodstuff," Blackman (1905) first considered multiple limiting factors of photosynthesis, in which the rate of photosynthesis increased linearly with increments of the limiting factor, until another component became operative. This concept is represented graphically in Fig. 48 by both the initial slope s_1 of the photosynthesis–irradiance ($P–I$) curve and the dotted line, described by

$$\mu = \mu_m \frac{I}{I_s} \qquad \text{for} \quad I \leq I_s$$

and   (120)

$$\mu = \mu_m \qquad \text{for} \quad I > I_s$$

where μ is the realized, or resultant, rate of photosynthesis, and μ_m is the maximal rate defined previously at a saturating light intensity, I_s.

Subsequent data on higher plants (Baly, 1935), algal cultures (Tamiya et al., 1953), and marine phytoplankton (Ryther and Yentsch, 1957) indicated, however, that μ approached μ_m asymptotically (Fig. 48). Such curvature of the $P–I$ relationship could be described as a rectangular hyperbola by

$$\mu = \frac{\mu_m(I/I_k)}{1 + (I/I_k)} \qquad \text{or} \qquad \frac{\mu_m I}{I_k + I} = \mu \qquad (121)$$

where I_k is the light intensity at which $\mu = \mu_m/2$, and the initial slope s_1 is μ_m/I_k. This same expression is used to describe the dependence of photosynthesis, nutrient uptake, and algal growth on the availability of external and internal nutrients (Fig. 49). Variants of Eq. (121) have been used (Smith, 1936) to optimize the fit of photosynthetic rate data from algal cultures (Winokur, 1948) and field populations (Talling, 1957).

Additional field data suggested that μ_m actually occurred at a depth where the light intensity was 30–50% of the surface value, that is, within the first attenuation depth. Since light inhibition is not described by either Eq. (120) or Eq. (121), Steele (1962) proposed a third expression,

$$\mu = \left(\mu_m \frac{I}{I_s}\right) e^{(1 - I/I_s)}, \qquad (122)$$

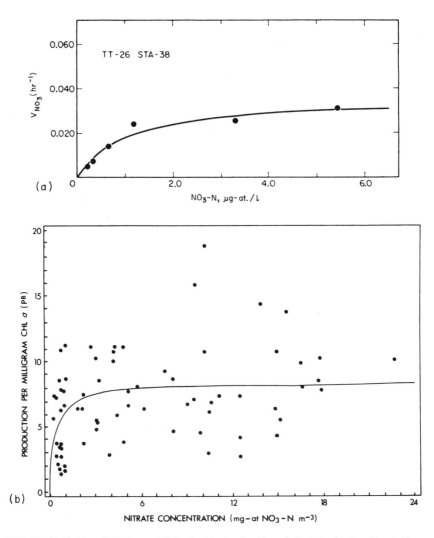

FIG. 49. (a) Uptake of nitrate as a Michaelis–Menten function of nitrate in shipboard incubations of ^{15}N-labeled substrate. [After MacIsaac and Dugdale (1969); © 1969 Pergamon Journals Ltd., with permission. (b) Primary production per unit chlorophyll as a Michaelis–Menten function of ambient sea surface nitrate off southwest Africa. [After Brown and Field (1986); © IRL Press, with permission.]

where s_1 is now $\mu_m e/I_s$, or μ_m is $s_1 I_s e^{-1}$ (Bannister, 1974a), and μ, μ_m are specific photosynthetic rates (hr^{-1}), normalized to carbon rather than to chlorophyll. Of course, further variations of Eq. (122) were subsequently introduced, involving specification of more parameters (Vollenweider, 1965; Fee, 1969; Parker, 1974; Platt et al., 1977, 1980; Fasham and Platt, 1983). For a constant s_1

and μ_m, Eq. (122) can be analytically integrated with depth (Vollenweider, 1965) to yield the rate of photosynthesis per unit surface area as

$$\int_z \mu = \mu_m[e^{(1 \; - \; I(z)/I_s)} - e^{(1 \; - \; I_0/I_s)}]k^{-1} \tag{123}$$

where $I(z)$ is obtained from Eq. (119) and k from Eq. (114a). After consideration of nutrient limitation, examples of these submodels will be presented in Sections 3.1.2, 3.2.1, 5.1.3, and 6.4.2.

3.1.2 Nutrient Regulation

Early studies of nutrient assimilation by marine phytoplankton (Ketchum, 1939) demonstrated, as in the case of light intensity, that nutrient utilization increased with experimental increments of dissolved nutrients. Initial mathematical formulation of such nutrient uptake by microalgae in the sea (Riley, 1946; Riley et al., 1949; Steele and Menzel, 1962) followed a linear Blackman relationship. One of the first descriptions of multiple interaction of limiting factors of photosynthesis in the sea (Riley, 1946) included the Blackman nutrient term as

$$\mu = p_1 I_0 (1 - e^{-kh_e})(kh_e)^{-1}\left(1 - \frac{0.55 - N}{0.55}\right)\left(\frac{h_e}{h_c}\right) \tag{124}$$

where p_1 describes the light dependence of μ, and h_e and h_c are the depths of the euphotic zone and mixed layer, such that the first three terms of Eq. (124) constitute the mean photosynthesis of the euphotic zone, analogous to Eq. (123). The last term is an index of turbulent mixing, while N of the fourth term is the nutrient concentration, phosphate in this case (Riley, 1946), with a linear limitation of photosynthesis from $N = 0.55$ μg-at. PO_4 l^{-1} to zero.

By 1913, Leonor Michaelis and Maud Menten had derived a kinetic model for the action of a single inducible enzyme on a substrate, which was first invoked to describe nutrient dynamics of continuous cultures of bacteria in the 1940s (Monod, 1949). A further extrapolation of this classical kinetic model to whole marine organisms (Caperon, 1967) led to the parameterization (Fig. 49) of nutrient uptake by phytoplankton (Dugdale, 1967) as

$$\chi = \frac{\chi_m N}{n + N} \tag{125}$$

where χ and χ_m are now the resultant and maximum rates (hr^{-1}) of nutrient assimilation, rather than photosynthesis; N is again the external concentration (μg-at. l^{-1}) of the limiting nutrient; and n is the Michaelis, or half-saturation, constant, at which $\chi = \chi_m/2$. The values of n are greater for coastal phytoplankton than for microalgae from oceanic habitats (Eppley et al., 1969; MacIsaac and Dugdale, 1969; Carpenter and Guillard, 1971), where nutrients are less abundant. They are also greater for large cells than for small cells (Eppley *et al.*, 1969; Goering *et al.*, 1973), which, because of their higher surface-to-volume ratio, have more uptake sites for the permease enzymes of the cell wall to act on the external substrates; smaller cells may thus be favored in oligotrophic habitats.

Additional studies of silicon (Paasche, 1973) and nitrogen (Caperon and Meyer, 1972) uptake indicated that there was a threshold concentration N_0 at which χ was zero, that is, the rectangular hyperbola of Fig. 49 did not have a zero intercept. A modified version of Eq. (125) was used to describe both this phenomenon and patchiness of prey for grazing organisms (Walsh, 1975) by

$$\chi = \frac{\chi_m (N - N_0)}{n + (N - N_0)} \qquad (126)$$

[See Eq. (171) as well.] The same experimental work with continuous cultures of phytoplankton (Caperon and Meyer, 1972) also indicated that cell growth depended more on the history of nutrient assimilation than on the nutrient concentration at time t. A further modification of Eq. (125) was introduced (Winter *et al.*, 1975) to simulate the time history of nutrient uptake by

$$\varepsilon(t) = \frac{\varepsilon_m(t)N(t - t_L)}{n(t) + N(t - t_L)} \qquad (127)$$

where ε_m is the maximum algal growth rate, ε is the resultant growth rate, and t_L is a time lag of 3 days.

Nutrient uptake can occur in the dark (Ketchum, 1939; Dugdale and Goering, 1967) and without cell division (Fitzgerald, 1968), such that nutrient uptake, photosynthesis, and growth can be uncoupled over smaller time scales than that of the daily population increment. The half-saturation constants of algal growth and nutrient uptake may, at times, be similar in field populations (Eppley and Thomas, 1969), however. Furthermore, at steady state, $\varepsilon = \chi$; that is; the amount of dissolved nitrogen that enters the cell wall during nutrient uptake, before cell division, exits as particulate nitrogen, after growth and division of the parent cell to form a daughter cell.

Nonsteady transients of growth can be described (Caperon, 1968; Droop, 1973) by a relationship similar to Eq. (125) with

$$\varepsilon = \frac{\varepsilon_m(Q - q)}{\theta + (Q - q)} \qquad (128)$$

where the rates (hr^{-1}) ε_m and ε are defined above; Q is the cell quota, or biomass, of a particular nutrient (pg-at. cell^{-1}); q is the minimum cell content of a nutrient before cell division can proceed (Eppley and Strickland, 1968); and θ is another half-saturation constant, but of particulate nitrogen in the algal cell, for example, rather than of dissolved nitrogen within the water column. In transient situations of laboratory cultures, θ can be much less than n (Droop, 1968; Caperon and Meyer, 1972). Since Q is a unit of biomass per cell, the nutrient flux at steady state can also be related to growth of either the algal cell or its population by

$$m = \varepsilon Q, \text{ or } m = (\varepsilon Q)(\text{cells}) \tag{129}$$

where m is the mass flux (μg-at. 1^{-1} hr^{-1}).

The fidelity of Eqs. (119), (122), (123), (125), (128), and (129) for description of population increments in terms of either chlorophyll, particulate carbon, or particulate nitrogen was examined in a simulation analysis (Howe, 1979) of an isolated population of phytoplankton. The experimental study involved the enrichment of natural microalgal populations off California with nitrate (Eppley et al., 1971) in an ocean without advection or diffusion, that is, 200-1 batch cultures on the deck of a ship. With a PAR of 361 g-cal cm^{-2} day^{-1}, or 69 E m^{-2} day^{-1}, a photoperiod of 14 hr on July 11 at 32°N, an $I(z)$ of 20% PAR, a depth integral of 1 m for the light intensity within a batch culture under a neutral density filter, an I_s of 11 g-cal cm^{-2} hr^{-1}, $k_s = 0$, and an s_1 of 60 mg-at. C m^{-3} E^{-1}, the simulated algal carbon, produced from photosynthesis in Eq. (123), matched fairly well the observed increase of particulate carbon (Fig. 50a). The sea water had only been filtered through a 183-μm mesh to remove large zooplankton (Eppley et al., 1971), however, such that grazing of microzooplankton may have resulted in lower particulate carbon in the batch culture after 100 hr, compared to the model with no grazing losses.

Since N_0 is small, ~0.06 μg-at. N 1^{-1} (Caperon and Meyer, 1972), Eq. (125) rather than Eq. (126) was used to model nitrate uptake, with a χ_m of 0.12 hr^{-1} and an n of 1.0 μg-at. NO$_3$ 1^{-1}. A cell quota Q of 9.2 pg-at. N cell^{-1}, that is, 9.2×10^{-12} mol N cell^{-1}; a minimum content q of 2.3 pg-at. N cell^{-1}; an ε_m of 0.06 hr^{-1}, that is, half of χ_m; and a half-saturation constant θ of 2.3 pg-at. N cell^{-1}, the same as q, were used to model nitrogen growth of the batch culture. At initial cell concentrations of 0.5×10^6 cells 1^{-1} in the batch culture, the model's q of 2.3 pg-at. N cell^{-1} was equivalent to 1.2 μg-at. N 1^{-1}, that is, about the same as n. The dissolved nitrate stimuli, N, consisted of addition of 5 μg-at. NO$_3$ 1^{-1} at $t = 0$, of 20 μg-at. NO$_3$ 1^{-1} at $t = 28$ hr, and of 50 μg-at. NO$_3$ 1^{-1} at 77 hr, which were depleted after nitrate uptake (Fig. 50b) and production of particulate nitrogen by the phytoplankton (Fig. 50c).

The simulated C/N ratio of the model (Howe, 1979) was ~5.2 by weight, or 6.1 by atoms, in agreement with the shipboard measurements (Eppley et al., 1971) and field estimates of phytoplankton populations with adequate nitrogen supplies (Walsh et al., 1981). The most important component of this C/N ratio is nitrogen. Although CO$_2$ may be limiting in alpine (Gale, 1972), forest (Odum, 1970), lake (Shapiro, 1973), and tide-pool (Byers, 1963) ecosystems, it is unlikely to be limiting in any well-buffered marine ecosystem, where the dissolved carbon concentration is at least two orders of magnitude greater than phytoplank-

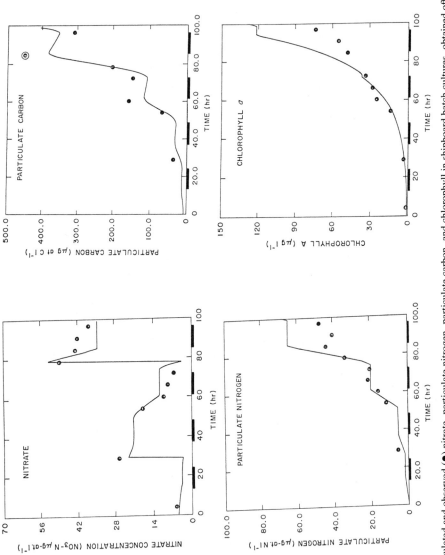

FIG. 50. Simulated and observed (●) nitrate, particulate nitrogen, particulate carbon, and chlorophyll in shipboard batch cultures, obtained off California. [After Eppley *et al.* (1971) and Howe (1979).]

ton carbon (Wangersky, 1965). In contrast, nitrogen compounds are recycled more slowly than those of phosphorus in oligotrophic (Thomas, 1970a,b) and eutrophic (Ryther and Dunstan, 1971) marine waters, and thus may determine the overall productivity of an area through regulation of protein synthesis. Silicate availability may additionally determine phyletic composition of the phytoplankton—see Section 6.2 on eutrophication. The degree to which the carbohydrate/protein food quality of any individual plant species is altered by nitrogen deficiency will affect its suitability as an appropriate diet for either terrestrial (Witkamp, 1966) or marine (Menzel, 1960) herbivores.

Laboratory-induced nitrogen deficiency has generated high phytoplankton C/N atomic ratios of 10–45 (Hobson and Pariser, 1970; Caperon and Meyer, 1972), while *in situ* phytoplankton populations have lower ratios of ~5–7 in nearshore upwelling areas off Peru (Walsh, 1975), Southwest Africa (Hobson, 1971), California (Holm-Hansen *et al.*, 1966), and Oregon (Small and Ramberg, 1971). Problems of detrital contamination of field measurements of particulate matter in the sea are presumably obviated in computing the relative particulate C/N ratio. This carbohydrate and fat-to-protein index (the usual practice is to multiply phytoplankton N by 6.25 to obtain protein content) of natural populations reflects the coupling of photosynthesis and nutrient uptake. The low particulate C/N ratios of Figs. 50a and c would suggest that nitrogen depletion is not sufficiently severe in this experiment to uncouple photosynthesis from nutrient uptake and protein synthesis, with consequent higher carbon storage in carbohydrates and fats (Fogg, 1959; Holm-Hansen *et al.*, 1959).

Based on nitrate depletion (Fig. 50b) and a ^{14}C estimate of carbon assimilation, however, the weight ratio of hourly C/N uptake of the batch culture was ~14 (Eppley *et al.*, 1971). Similarly, comparisons of independent ^{14}C and ^{15}N estimates of *net* photosynthetic and nutrient fluxes over the depth of the euphotic zone for field populations on a daily basis have not yielded a $^{14}C/^{15}N$ uptake ratio of 5–7. Instead a ratio of at least 14 to 65, or more than twice that of the particulate material, has been found (MacIsaac and Dugdale, 1972; McCarthy, 1972), with at times a ratio of as much as 105 (Eppley *et al.*, 1973). Such depth-integrated carbon or nitrogen fluxes, computed from daily estimates of algal growth, contained artifacts induced by incubation of phytoplankton in small bottles.

Carbon-14 methodology, $^{15}NO_3^-$ uptake, and nitrate depletion from the water column can give consistent estimates of daily primary production during the early spring bloom, when nitrate is the major source of nitrogen in the absence of grazing. In the Bering Sea, for example, from April 26 to May 27, 1979, April 23 to May 24, 1980, and April 13 to May 14, 1981, NO_3^- depletion within the mean euphotic zone (19 m) and a C/N ratio of 5 : 1 by weight yield a mean primary production of $1.06 \text{ g C m}^{-2} \text{ day}^{-1}$ over the outer and middle shelf (Whitledge *et al.*, 1986). Such a calculation does not consider either the diffusive resupply of NO_3 from shelf-break waters or the vertical excursion of an algal cell in the surface mixed layer.

Incubations of phytoplankton with ^{15}N-labeled nitrate in small bottles, where *no* diffusive resupply of nutrients occurs and light conditions are static (i.e., no vertical mixing), yield a similar mean estimate of 0.91 g C m^{-2} day^{-1} ($N = 35$), with a C/N ratio of 5 : 1. The average surface mixed layer in spring of 1979 to 1981 ($N = 330$), however, was 27.2 m: that is, the phytoplankton could remove nitrate from the upper 27 m, not just 19 m, while being mixed in and out of the euphotic zone. Observed nitrate depletion within this surface mixed layer yields, instead, a primary production of 1.50 g C m^{-2} day^{-1} from the local time rate of change of nitrate (Whitledge *et al.*, 1986). Finally, using a diffusive resupply rate of 0.29 μg-at. NO_3^- l^{-1} day^{-1} summed over the upper 27 m (Coachman and Walsh, 1981), and a C/N ratio of 5 : 1, the calculated average primary production would then be a total of 2.05 g C m^{-2} day^{-1} for the above 31-day periods in 1979, 1980, and 1981.

An individual phytoplankton cell is not NO_3^- limited under the observed and simulated conditions of Fig. 50b, with typical diatom half-saturation constants of 1.0–1.5 μg-at. NO_3^- l^{-1}. Photosynthesis is light-limited, however, such that carbon fixation can only take place in the euphotic zone, in contrast to nitrogen uptake over the whole mixed layer. The ^{14}C estimates of daily primary production in the Bering Sea, integrated down to the mean 1% light depth (19 m), over the same times in 1979, 1980, and 1981, for example, yield an average ($N = 49$) of 2.05 g C m^{-2} day^{-1} (R. Iverson, personal communication). This is the same production as that derived from the above calculation of nitrogen utilization over the mixed layer depth, with a C/N uptake ratio of 5 : 1. Extrapolation of a description of plankton dynamics for Eqs. (119)–(129) from a small bottle to the continental shelf thus requires explicit addition of terms for advection and mixing.

Chlorophyll is a biomass parameter that can be measured more frequently over time and space by remote and *in situ* sensors than particulate carbon or nitrogen. The chlorophyll increment with time in the batch culture was accordingly modeled by relating chlorophyll synthesis to nitrogen uptake and growth (Howe, 1979). A molecule of chlorophyll *a* is 6.27% nitrogen by weight (Shuman and Lorenzen, 1975), and the ratio of cell nitrogen to chlorophyll is 16 : 1 under nitrogen limitation (Strickland, 1965), such that 0.4% of the net nitrogen uptake of the cell may be devoted to chlorophyll synthesis. With these assumptions, the model's chlorophyll synthesis matched that of the batch culture within the first 80 hr (Fig. 50d). The computed ratios of carbon to chlorophyll ranged from 40 : 1 to 60 : 1, after nutrient addition and a reduction of the initial ratio from 120 : 1, as observed in the batch culture (Eppley *et al.*, 1971). Like the cases of the simulated particulate carbon and nitrogen, however, the final chlorophyll biomass of the model was higher than that of the batch culture, suggesting that terms for grazing, as well as a description of the physical habitat, must be added to Eqs. (119)–(129). After parameterization of temperature effects on phytoplankton metabolism, Section 3.1.4 considers such an ecological context of autotrophic growth in the sea.

3.1.3   Temperature Regulation

Experimental studies of temperature effects on the chemical reaction rates of homogeneous gases in the 1850s, which suggested that the rate about doubled for every 10°C increment, led Savante Arrhenius in 1889 to describe this process by

$$j = Je^{-S/\xi}T_k \qquad (130)$$

where j is the specific rate of the chemical reaction; J is the number of collisions of molecules in the ideal gas per unit time; S is the activation energy required to produce sufficient collisions of molecules to break stable chemical bonds of the reactants, allowing a chemical change to occur at a particular temperature T_k (in kelvin); and ξ is the ideal gas constant, ~ 2 kcal T_k^{-1} g-at.$^{-1}$, relating Boyle's Law of 1662 and Charles's Law of 1787 as a ratio of the product of the pressure and volume of a gas to its temperature. The Arrhenius temperature equation, like the Michaelis–Menten paradigm of enzyme kinetics, is most applicable to simple cases of chemical reactions in solutions or homogeneous gases, but fails for more complicated chain reactions.

By plotting the natural logarithm (ln) of j against $1/T_k$, the value of S can be obtained from the slope of the resultant straight line, since it is $-S/\xi$, or the Q_{10} temperature coefficient of chemical reactions. Using this procedure, the activation energy of the respiratory electron transport system of phytoplankton in the north Pacific Ocean, for example, is 15.8 kcal g-at.$^{-1}$ (Packard et al., 1975). The Arrhenius equation was proposed (Goldman and Carpenter, 1974) as an empirical fit to temperature and growth-rate data from continuous cultures of marine and freshwater phytoplankton (Fig. 51). They suggested that the maximum growth rate ε_m of the microalgae can be substituted for j in Eq. (130), with J as an empirical parameter, 5.4×10^9 day^{-1}, resulting in

$$\varepsilon_m = (5.4 \times 10^9)e^{-6472/T_k} \qquad (131)$$

with a Q_{10} of 2.08. Since $\varepsilon_m = \chi_m$ at steady state, another multiplicative model of temperature and nutrient interaction, under light-saturated conditions, can then be written as

$$\chi = (5.43 \times 10^9)\, e^{-6472/T_k}\left(\frac{N}{n + N}\right) \qquad (132)$$

by substituting Eq. (131) into Eq. (125).

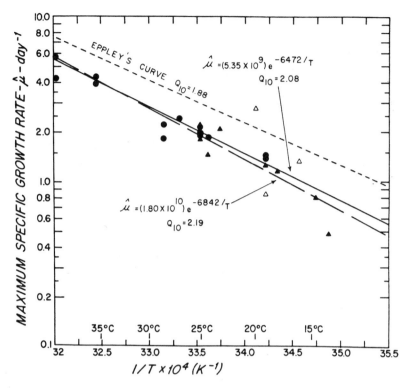

FIG. 51. Effects of temperature on the maximum growth rate (day⁻¹) of marine and freshwater algae grown in continuous cultures. ●, Freshwater species; ▲ marine species; △ *Thalassiosira pseudonana* (3H) and *Monochrysis lutheri*. [After Goldman and Carpenter (1974); © 1974 American Society of Limnology and Oceanography, Inc., with permission.]

An earlier analysis (Eppley, 1972) of temperature and growth rates from batch cultures of microalgae (Fig. 52) yielded a similar Q_{10} of 1.88, but used an empirical formulation of an exponential function to the base 10, rather than e, as

$$\varepsilon'_m = (0.85)10^{0.0275T} \tag{133}$$

where ε'_m is the maximum division rate (doublings day⁻¹) under continuous light and T is instead Celsius temperature. The growth and division rates of phytoplankton populations are both derived from consideration of the net increase of the standing stocks, in the absence of grazing or sinking.

By assuming that an algal population is in an exponential growth phase, the size of a phytoplankton population $M(t)$ at time t can be related to the previous size $M(0)$ at time 0 by

$$M(t) = M(0)e^{\varepsilon_m t} \tag{134}$$

Dividing this equation by $M(0)$ and taking the natural logarithm results in

$$\ln\left[\frac{M(t)}{M(0)}\right] = \varepsilon_m t \quad \text{or} \quad \varepsilon_m = \left(\frac{1}{t}\right)\ln\left[\frac{M(t)}{M(0)}\right] \tag{135}$$

Replacing the base e of the natural logarithm of Eq. (135) by 2 then leads to

$$\varepsilon_m' = \left(\frac{1}{t}\right)\log_2\left[\frac{M(t)}{M(0)}\right] \tag{136}$$

as the definition of the division rate in doublings per day. Using the natural

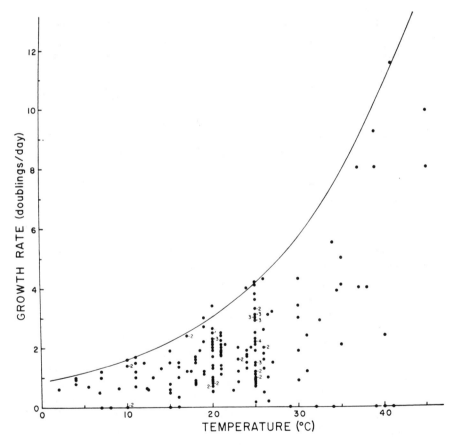

FIG. 52. Effects of temperature on the maximum division rate (doublings day^{-1}) of marine and freshwater algae grown in batch cultures. [After Eppley (1972).]

logarithm for both growth and temperature dependence, Eq. (133) becomes instead

$$\varepsilon_m = (0.59)e^{0.0633T} \tag{137}$$

yielding a maximum growth rate of 2.1 day^{-1} at 20°C, for example, assuming a 24-hr photoperiod.

Over the seasonal temperature cycle of ~20°C (Fig. 8) on the New York and Nova Scotia shelves, the assimilation index of light-saturated photosynthesis, μ_m, varied from 0.3 to 30.8 mg C mg chl^{-1} hr^{-1} (Platt and Jassby, 1976; Malone and Neale, 1981), compared to a theoretical maximum of 24.0 mg C mg chl^{-1} hr^{-1} for a light-adapted cell in surface waters (Falkowski, 1981). However, although the seasonal change of the nanoplankton μ_m could be related to changes in temperature by Eq. (131), the netplankton μ_m exhibited no significant correlation with temperature on the east coast of the United States (Malone and Neale, 1981). An analysis of the change of phytoplankton respiration with temperature off the west coast of North America similarly detected no significant relationship (Packard *et al.*, 1975).

Reduction of the Mid-Atlantic sample size by almost half and division by the C/chl ratio to further reduce the variance with a carbon-specific division rate ε_m finally led to a significant correlation with seasonal temperature changes (Malone, 1982). The maximum laboratory division rate of 4 doublings day^{-1} at 25°C (Fig. 52) was not observed at sea (1.5 for netplankton and 2.0 for nanoplankton), however, such that an adjustment of Eq. (133) for a 50% photoperiod might be appropriate (Malone, 1982). We then obtain another simple multiplicative model, but of light and temperature interaction, expressed by

$$\varepsilon'_m = (0.5)(0.85)10^{0.0275T} \tag{138}$$

A better empirical fit of data to any model can, of course, be achieved by increasing the number of "tunable" parameters as well as by reducing the sample size and level of significance. Similar to optimal light intensities, there are optimal temperatures at which microalgae grow in the laboratory (Li, 1980) and field (Li, 1985), which can be described (Ratowsky *et al.*, 1983) by

$$\mu_m = [p_2(T - T_1)]^2 [1 - e^{p_3(T - T_u)}]^2 \tag{139}$$

where T_1 and T_u are the minimum and maximum temperatures at which μ_m is zero. The two parameters p_2 and p_3, are obtained by a nonlinear regression of the data, and as such have little biological meaning. As one deviates from a mechanistic model with causal implications by adding parameters of no insight, one eventually arrives at regression analysis, involving statistical "fishing trips" for understanding in a noncausal manner.

Gordon Riley (1939) was one of the first to suggest multiple regression techniques for analysis of phytoplankton relationships, where the dependent variable Y of the regression equation

$$Y = b_0 + b_1X_1 + b_2X_2 + \cdots + b_nX_n \tag{140}$$

is related to the independent variables $X_1, \ldots, X_n$ in a linear manner. A regression equation for the dependent variable is written in a series of steps, where the independent variable that has the highest correlation with the dependent variable is entered first. Then subsequent independent variables with the sequential highest partial correlation with the dependent variable are added until the regression model exceeds the value of the F-test selected for the degree of confidence. Not all of the independent variables may be entered, and some may be removed from the multiple regression equation as indicated by a partial F-test at each step; t-tests are made to see if the regression coefficients $b_1, b_2, \ldots, b_n$ are significant.

Individual environmental factors interact synergistically to determine the species composition and fluxes of elements within phytoplankton communities with some factors of more direct, first-order importance than others in any one ecosystem. The coefficient of determination of a regression equation, r^2, is the total amount of variation of the dependent variable that is paralleled by changes in the independent variables. With sequential addition of environmental variables in such stepwise multiple regressions, r^2 becomes an index of the relative importance of each habitat variable. First- and second-order parameters are distinguished in such a study by whether or not r^2 of the equation is increased by addition of an environmental factor.

Although somewhat limited by their underlying assumptions, statistical analyses such as stepwise multiple regression can at least be used to infer which variables might be important (Walsh, 1971) and to assign priorities to those fluxes and habitat variables that could be measured for inclusion in a simulation model. In sharp contrast, however, polynomial regressions of the nonlinear dependence of Y on a single variable, such as X^2, like the equation

$$N = b_0 - b_1T - b_2T^2 + b_3T^3 \tag{141}$$

for nitrate on temperature (Kamykowski and Zentara, 1986), become exercises in curve fitting. Like the square of the temperature differences in Eq. (139), the nitrate distribution in relation to cube of the temperature in Eq. (141) has little basis in principles of fluid dynamics, physical chemistry, or enzyme kinetics. Without a causal relationship, it is not surprising that Eq. (141) yields (Kamykowski and Zentara, 1986) an r^2 of 0.99 for spatial variations of temperature and nitrate over an area of the ocean of $\sim 10^6$ km^2 centered at 15°S, 123°W, that is, near the meridional nitrate gradient of the South Pacific depicted in Fig. 19, but only an r^2 of 0.53 for a block of the ocean centered at 45°S, 115°E, south of Australia in the Indian Ocean (Fig. 53).

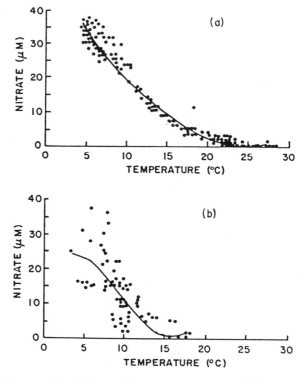

FIG. 53. A cubic regression of nitrate against temperature over an $\sim 10^6$ km² area centered at (a) 15°S, 123°W and (b) 45°S, 115°E. [After Kamykowski and Zentara (1986); © 1986 Pergamon Journals Ltd., with permission.]

Linear regression analysis may be appropriate for preliminary insight into a complex system, but it is an inadequate description of biological phenomena. Linear Blackman relations such as, for example, the regression equation

$$\text{Nutrient uptake} = b_0 + b_1(T) + b_2(N) + b_3(I) \tag{142}$$

cannot be expected to fully describe or predict biological relationships, which are basically nonlinear and consist of thresholds, time lags, and saturation and inhibition effects. Polynomial fits such as Eq. (143) may lead to greater coefficients of determination, such as

$$\text{Nutrient uptake} = b_0 + b_1(T) + b_2(T)^2 + b_3(T)^3 + b_4(N) \tag{143}$$

in regression analysis, but the biological meaning of the higher-order terms raises a serious problem in interpretation of the results.

An alternative approach is to incorporate functional nonlinear expressions of relations between state and habitat variables in simulation models, which are based on causal experimental analyses. A more appropriate formulation of Eq. (142) might be the multiplicative interaction of light and nutrients in Eq. (124), of temperature and nutrients in Eq. (132), of light and temperature in Eq. (138), or of some combination of all three variables. Blackman's (1905) concept of multiple limiting factors assumed, however, that only one habitat variable was limiting the rate of photosynthesis at any instant of time. Since multiplication of several suboptimal rates of growth, limited simultaneously by light, nutrients, and temperature, may underestimate the actual increment of the phytoplankton population, one could instead compute ε_m separately from Eqs. (127), (125), and (137) at each time step of a model (Walsh, 1975), applying the lowest ε_m over that time interval. An example of this approach is presented in Section 3.2.1.

3.1.4 Ecological Context

At the end of the nineteenth century, equations had been developed for water motion in the euphotic zone by Ekman, for photosynthesis by Blackman, and for nutrient and temperature regulation by Michaelis, Menten, and Arrhenius, while the importance of grazing and sinking had been stressed respectively by Lohmann (1908) and Ostwald (1903). Although predator–prey models were introduced by Lotka (1925) and Volterra (1926) in the mid 1920s, formulation of the marine production cycle as a predator–prey interaction was not advanced until the 1940s by Fleming (1939) and Riley (1946) as

$$\frac{dM}{dt} = M(\varepsilon - \gamma) \tag{144}$$

where M is the microalgae, or phytoplankton; ε is the growth rate; and γ is the grazing loss to herbivores. Basically, Eq. (144) was advanced by Thomas Malthus in 1798 to describe the birth and death processes, ε and γ, of a human population, which might double every 25 yr in an exponential fashion (Malthus, 1798).

Equation (134) is the analytical solution of Eq. (144) in the special case where $\gamma = 0$. In this situation, we have

$$\frac{dM}{dt} = \varepsilon M \tag{145}$$

which is a first-order, constant coefficient, ordinary differential equation and can be solved by the technique of separation of variables, that is,

$$\frac{dM}{M} = \varepsilon \, dt \tag{146}$$

Upon integration of each side of Eq. (146), we obtain

$$\int \frac{1}{M} dM = \ln M + c_1 = \varepsilon t + c_2 = \varepsilon \int dt = \int \varepsilon \, dt \qquad (147)$$

or

$$\ln M = \varepsilon t + c_3 \quad \text{where} \quad c_2 - c_1 = c_3 \qquad (148)$$

Exponentiation of Eq. (148) leads to

$$e^{\ln M} = M = c_4 e^{\varepsilon t} = (e^{\varepsilon t})(e^{c_3}) = e^{\varepsilon t + c_3} \qquad (149)$$

since e^{c_3} is an arbitrary constant c_4. Now from Eq. (149), $M(0)$ at time 0 is

$$M(0) = c_4 e^{\varepsilon \cdot 0} = (c_4)(1) = c_4 \qquad (150)$$

or the value of the arbitrary constant c_4 is that of the original population size at $t = 0$. We then obtain Eq. (134) from substitution of $M(0)$ for c_4 in Eq. (149). Recall that the finite-difference form of Eq. (134) would be

$$M_{t+1} = M_t + (\Delta t)(\varepsilon M) \qquad (151)$$

providing a numerical solution to Eq. (134) as well.

Riley *et al.* (1949) added sinking and vertical eddy diffusivity terms to Eq. (144), which already contained a description of photosynthesis and nutrient limitation from Eq. (124), as

$$\frac{dM}{dt} = M(\varepsilon - \gamma) + K_z \frac{\partial^2 M}{\partial z^2} - w_s \frac{\partial M}{\partial z} \qquad (152)$$

Upon Lagrangian expansion of the Eulerian derivative, dM/dt, as in the previous cases of water motion and vorticity, Eq. (152) becomes

$$\frac{\partial M}{\partial t} + u \frac{\partial M}{\partial x} + v \frac{\partial M}{\partial y} + w \frac{\partial M}{\partial z} = M(\varepsilon - \gamma) + K_z \frac{\partial^2 M}{\partial z^2} - w_s \frac{\partial M}{\partial z} \qquad (153)$$

By the 1950s, these predator–prey equations for changes of a phytoplankton population in the sea had become sufficiently complicated that analytical solutions like Eq. (134) were no longer possible, resulting in laborious numerical solutions on desk calculators (Riley, 1951).

Explicit consideration of nutrients and herbivores leads, furthermore, to a set of coupled partial differential equations,

$$\frac{\partial N}{\partial t} = K_x \frac{\partial^2 N}{\partial x^2} + K_y \frac{\partial^2 N}{\partial y^2} + K_z \frac{\partial^2 N}{\partial z^2}$$

$$- \frac{\partial(uN)}{\partial x} - \frac{\partial(vN)}{\partial y} - \frac{\partial(wN)}{\partial z} - \chi M + lG \qquad (154)$$

$$\frac{\partial M}{\partial t} = K_x \frac{\partial^2 M}{\partial x^2} + K_y \frac{\partial^2 M}{\partial y^2} + K_z \frac{\partial^2 M}{\partial z^2} - \frac{\partial(uM)}{\partial x}$$

$$- \frac{\partial(vM)}{\partial y} - \frac{\partial(wM)}{\partial z} + \varepsilon M - \gamma G - w_s \frac{\partial M}{\partial z} \tag{155}$$

$$\frac{\partial G}{\partial t} = - \frac{\partial(uG)}{\partial x} - \frac{\partial(vG)}{\partial y} - \frac{\partial(wG)}{\partial z} - lG - oO \tag{156}$$

where K_x, K_y, and K_z are now all of the horizontal and vertical eddy coefficients; u, v, and w are the cross-shelf, along-shore, and vertical velocities; χ is the specific uptake rate (t^{-1}) of nutrients, during photosynthesis, μ (as a function of a number of habitat variables, such as light, temperature, and nutrients) by the phytoplankton population, with ε as their growth rate (t^{-1}); γ is the specific grazing rate (t^{-1}) on phytoplankton by the herbivores (G); w_s is the sinking velocity of phytoplankton; γ' is the growth rate (t^{-1}) of the herbivores (as a function of assimilation and reproductive losses); l is the nutrient excretion rate (t^{-1}) of the herbivores; and o is the predation rate (t^{-1}) of the carnivores (O).

The major advance in the 1970s over previous plankton models, described 30 yr earlier by the same general equations [Eqs. (154)–(156)], was the introduction of spatial resolution of the velocity fields, generated from numerous circulation submodels of continental shelf ecosystems (see, e.g., Thompson, 1974; Hamilton and Rattray, 1978; Preller and O'Brien, 1980). With such flow fields, K_x and K_y of $0.5–1.2 \times 10^6$ cm^2 sec^{-1}, K_z of 1 cm^2 sec^{-1}, and χ, ε of ~ 0.1 hr^{-1}, the simulated spatial structure of nitrate (Walsh, 1975; Wroblewski, 1977; Howe, 1979) matched reasonably well the nutrient observations off Peru, Oregon, and Northwest Africa: that is, the numerous plankton growth parameters embodied in ε of Eq. (155) appear to have been correctly approximated. An example is presented in Section 3.2.1. The predicted phytoplankton fields of Eq. (155) did not match the limited set of shipboard chlorophyll observations, however, without imposition of unrealistically high grazing rates (Walsh, 1983), discussed in Chapter 4. Sinking and/or lateral export of phytoplankton, such as $\partial(uM)/\partial x$ and $\partial(vM)/\partial y$ of Eq. (155), are additional loss terms (Walsh et al., 1981), which will be explored in Section 5.1.

3.1.5 Numerical Dispersion

The advection of momentum, vorticity, and density was ignored in the various numerical studies of Chapter 2. These nonlinear terms of Eqs. (154)–(156) can no longer be ignored, however, and constitute error sources in the methods of numerical solution employed thus far. In particular, attempts to express the product of a continuous space derivative of a scalar quantity and the advection term in a partial differential equation with the previous finite-difference estimates lead to numerical errors. These approximations essentially introduce another term in the original equation, that of implicit numerical diffusion (Hirt, 1968).

Using a forward difference in time and an upstream difference in space for expression of the local change and advection of an algal population M, for example, described continuously by

$$\frac{\partial M}{\partial t} = -\frac{\partial(uM)}{\partial x}$$
(157)

results at time $n+1$ (Molenkamp, 1968) in

$$\frac{M_{i,j}^{n+1} - M_{i,j}^n}{\Delta t} = -\frac{(u_{i+1,j}^n - u_{i,j}^n)(M_{i+1,j}^n - M_{i,j}^n)}{\Delta x}$$
$$+ \frac{K_f(M_{i+1,j}^n + M_{i-1,j}^n - 2M_{i,j}^n)}{\Delta x^2}$$
(158)

where K_f is an apparent diffusion coefficient derived from the finite-difference approximation in space and time. It is equivalent to

$$K_f \approx \frac{u_{i,j}(\Delta x - u_{i,j}\,\Delta t)}{2}$$
(159)

The spatial dependence of this K_f can be removed with a centered finite difference scheme in space (Hirt, 1968) by

$$\frac{M_{i,j}^{n+1} - M_{i,j}^n}{\Delta t} = -\frac{(u_{i+1,j}^n - u_{i-1,j}^n)(M_{i+1,j}^n - M_{i-1,j}^n)}{2\,\Delta x}$$
$$+ K_f\frac{(M_{i+1,j}^n + M_{i-1,j}^n - 2M_{i,j}^n)}{\Delta x^2}$$
(160)

where K_f reduces to

$$K_f \approx \frac{u_{i,j}^2\,\Delta t}{2}$$
(161)

It is important to note, however, that the following two finite-difference expressions are both unconditionally unstable: that is, their numerical solution would not converge to an analytical one—if it could be computed!

$$\frac{M_{i,j}^{n+1} - M_{i,j}^n}{\Delta t} = -\frac{(u_{i+1,j}^n - u_{i-1,j}^n)(M_{i+1,j}^n - M_{i-1,j}^n)}{2\,\Delta x}$$
(162)

or

$$\frac{M_{i,j}^{n+1} - M_{i,j}^{n-1}}{2\Delta t} = -\frac{(u_{i+1,j}^n - u_{i-1,j}^n)(M_{i+1,j}^n - M_{i-1,j}^n)}{2\,\Delta x}$$
$$+ \frac{(K_x)(M_{i+1,j}^n + M_{i-1,j}^n + 2M_{i,j}^n)}{\Delta x^2}$$
(163)

Equation (162) can be made conditionally stable by adding another term for explicit diffusion, for example, similar to the implicit one of Eq. (160), to this forward time difference. Similarly, Eq. (163) can be adjusted by lagging the diffusive term by a time step for this difference expression centered in time, for example,

$$\frac{M_{i,j}^{n+1} - M_{i,j}^{n-1}}{2\Delta t} = -\frac{(u_{i+1,j}^n - u_{i-1,j}^n)(M_{i+1,j}^n - M_{i-1,j}^n)}{2 \Delta x}$$
$$+ \frac{(K_x)(M_{i+1,j}^{n-1} + M_{i-1,j}^{n-1} - 2M_{i,j}^{n-1})}{\Delta x^2} \tag{164}$$

These changes in Eqs. (163) and (164) only lead to conditionally stable solutions because they must still meet the Von Neumann criterion of

$$\frac{K_x \Delta t}{\Delta x^2} < \frac{1}{2} \tag{165}$$

and the Courant–Friedricks–Lewy (C–F–L) condition of

$$\frac{u\Delta t}{\Delta x} < 1 \tag{166}$$

With $u = 10$ cm sec^{-1}, $\Delta x = 10$ km, and $\Delta t = 1$ hr, Eq. (159) yields a K_f of 4.82×10^6 cm^2 sec^{-1}, compared to an order of magnitude less numerical diffusion of 0.18×10^6 cm^2 sec^{-1} from Eq. (161). Use of the finite-difference expression, centered in space, of Eq. (160) does require knowledge of $M_{i-1,j}$ and $u_{i-1,j}$ at the boundary, $i - 1$, in the x direction, however, thereby influencing the interior solution. Use of Eq. (160), or (162), also requires that an explicit diffusion term must be added to this equation to allow stable numerical solutions of the finite difference expression. At length scales of 10 km, such an explicit eddy coefficient would be of the order 10^6–10^7 cm^2 sec^{-1}, similar to the implicit numerical diffusion of Eq. (159).

At a tidal velocity of 100 cm sec^{-1}, $\Delta x = 10$ km, and Δt of 1 hr, the C–F–L conditions of Eq. (166) would still be satisfied, but Eq. (161) then would yield a K_f of 18.4×10^6 cm^2 sec^{-1}, compared to 32.2×10^6 cm^2 sec^{-1} from Eq. (159). Excessive damping or smoothing (Long and Pepper, 1981) of the numerical solution of Eq. (158) or (160) would occur by numerical diffusion, for example, at the computed current speeds of ~ 140 cm sec^{-1} in the Bering Strait model, between the Diomede Islands and Cape Wales under the 1.6-Sv flow case—recall Section 2.4.3. Adoption of a Lagrangian approach in time, with a cubic spline interpolation of the spatial gradient (Purnell, 1976), can instead reduce K_f of Eq. (161) by two more orders of magnitude (Long and Pepper, 1981).

The numerical diffusion introduced into simulation models by an Eulerian description of plankton dynamics can be avoided with use of a Lagrangian reference frame. Using a particle-in-volume Lagrangian technique (Harlow, 1963), the trajectory of individual particles can instead be simulated to overcome the problem of phase and amplitude errors within standard finite difference techniques for numerical approximation of the advective process. Recall that in a Lagrangian formulation the concentration of material refers to a coordinate system that is fixed in relation to the mean flow of the fluid, whereas in an Eulerian formulation the concentration of material refers to a coordinate system that is fixed in space through which the fluid is moving.

In this alternate approach, the physical space of the continental shelf is first divided into a number of volumes to make the fixed Eulerian grid on which the velocity fields and external sources of particles are represented, and on which the turbulent fluxes can be evaluated. The Eulerian grid for this purpose is the same grid as that of the previous circulation models described in Chapter 2. The spatial distribution of the particles is then represented as a discrete number of Lagrangian particles that undergo transport from volume to volume as they are moved by the velocity field.

The numerical integration of Eq. (157) is instead performed in two steps. In the Lagrangian step, the new M particle positions (x, y) at time $t + \Delta t$ are calculated using explicit forward differencing of

$$x(t + \Delta t) = x(t) + u_p \, \Delta t, \tag{167}$$

$$y(t + \Delta t) = y(t) + v_p \, \Delta t, \tag{168}$$

where u_p and v_p are the velocities of one phytoplankton particle as determined from the Eulerian grid calculation of a separate circulation model and interpolated to the particle position. Although there is no computational stability requirement for the time step Δt in Eqs. (167) and (168), it would be undesirable for a particle to be moved too far from the grid location of the velocities responsible for the advection. Thus a time increment that restricts particle movement to no more than one-half of a grid volume per time step is usually imposed.

The interpolation procedure consists of a linear area (A_i) weighting scheme in which each particle is arbitrarily assigned the same area as the Eulerian grid volumes. The velocity at the position of the particle is then obtained as the weighted sum of the velocities in the neighboring Eulerian cells, where the weights correspond to the amount of overlap of the particle within the respective cells, that is, in the two dimensions of a depth-integrated model,

$$u_p = \frac{u_1 A_1 + u_2 A_2 + u_3 A_3 + u_4 A_4}{\Delta x \, \Delta y} \tag{169}$$

In Eq. (169), $u_1 - u_4$ represent the u velocity components that have been averaged to the grid of the particular circulation model. Within the Eulerian part of this computation, a concentration field of M on the Eulerian grid can be determined each time step from the particle positions, using the interpolation procedure outlined above. The turbulent fluxes are still evaluated with an eddy coefficient and the local gradient of particles; however, there is no numerical diffusion in this approach, and some of the problems raised by using downstream boundary conditions in these calculations are obviated. Examples of this approach are presented in Sections 4.5.1 and 5.1.1.

3.2 Sampling Considerations

Most predators in the sea are larger than their prey, spending a large part of each day searching a dilute broth of faster-growing, smaller organisms (Fig. 54). Oceanographers spend a great deal of time and money in the same activity, but usually to sample biological populations in the ocean, with perhaps the ultimate goal of also eating them. The advective terms of Eqs. (155) and (156) and the life history of each size class of organisms lead to distinct spatial domains of each population (Fig. 55a), the resolution of which requires various sampling platforms of different endurance and range (Fig. 55b).

One must be able both to sample frequently enough to resolve the phenomenon of interest and to average the data over appropriate time and space scales to allow their interpretation, for testing of the original hypothesis that led to the measurements (Walsh, 1972). Aliasing and filtering of data describing natural processes are inherent problems for both model and "real-world" studies. Inadequate testing of a simulation analysis with shipboard data is thus presented first in this section; the synoptic, but infrequent temporal, coverage of remote sensors is discussed next; finally, a description of the temporally rich, but spatially poor, data sets derived from moored instruments completes Chapter 3.

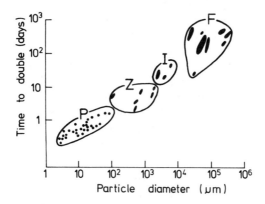

FIG. 54. The doubling times of phytoplankton (P), herbivores (Z), invertebrate carnivores (I), and fish (F) as a function of their size. [After Sheldon *et al.* (1972) and Steele (1978); © 1972 American Society of Limnology and Oceanography Inc.; © 1978 Plenum Press, with permission.]

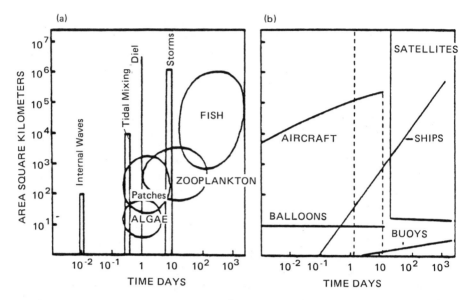

FIG. 55. Time and space domains of (a) habitat variance and biological abundance in relation to (b) duration and scale of sampling platforms. [After Steele (1978) and Esaias (1980); © 1978, 1980 Plenum Press.]

3.2.1   Testing

A two-dimensional model of the upwelling ecosystem off Punta San Juan, Peru, was used (Walsh, 1975; Walsh and Howe, 1976) to relate the input of light, ammonium, nitrate, silicate, and phosphate to the steady-state carbon and nitrogen biomass of phytoplankton, zooplankton, and fish over depths of 0–50 m from 0 to 50 km off the coast. Estimates of water circulation, nutrient uptake, photosynthesis, diel migration, grazing stress, and herbivore excretion from a March–April 1969 cruise were used to formulate a simulation model of this austral autumn ecosystem. A second independent set of data, taken in the same place in March–April 3 yr earlier, was used to test the model. Data from another June 1969 cruise to the same area were used for a second test of the response of the autumn model to winter upwelling. Consideration of the goodness of fit of such a model in the time domain, that is, nonsteady transients, then leads to an alternative representation of time series in the frequency domain.

Two independent data sets taken at the same time of year at the same place off Punta San Juan, Peru, but 3 yr apart, were used to build and test the model. The 1966 cruise (Ryther *et al.*, 1971) was a drogue study aboard the R/V *Anton Bruun*, while a grid of stations was occupied repeatedly on the second cruise in 1969 aboard the R/V *Thomas G. Thompson* (Walsh *et al.*, 1971). The drogue cruise track of *Bruun* in March–April 1966 (validation data) and of the grid and one of the underway temperature maps of the *Thompson* in March–April 1969

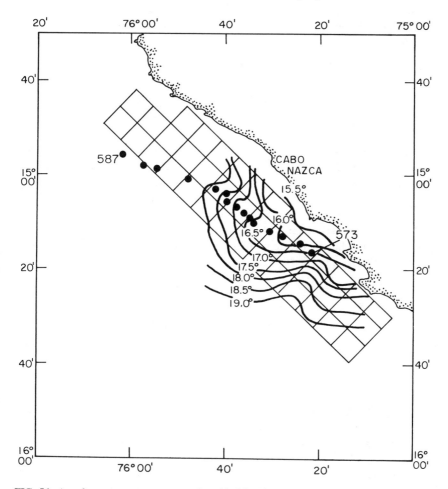

FIG. 56. A surface-temperature map over the grid of the *Thompson* input data in relation to drogue stations (●) of the *Anton Bruun* validation data off Cabo Nazca, Peru. [After Walsh (1975); © 1975 Pergamon Journals Ltd., with permission.]

(input data) are shown in Fig. 56. The spatial domain of the model, with a grid spacing of 10 km in the horizontal and 10 m in the vertical, was directed 100 km downstream along the trajectory of the drogue, with an axis at a 30° angle to the coast (45° to the left of the wind). The temporal domain involved a spin-up time scale for the model (i.e., to reach steady state) of 10 days, with diel periodicity of nutrient uptake, of photosynthesis, of grazing, and of vertical herbivore migration simulated by periodic functions.

The following set of nonlinear, coupled partial differential equations was written for the state variables light, nitrate, recycled nitrogen, phosphate, and silicate, as well as phytoplankton, zooplankton, detritus, and fish in terms of both particulate nitrogen and carbon. The spatial distribution of nitrate (NO_3) was described by

$$\frac{\partial(NO_3)}{\partial t} = -\frac{\partial[u(NO_3)]}{\partial x} - \frac{\partial[w(NO_3)]}{\partial z}$$

$$+ K_y \frac{\partial^2(NO_3)}{\partial y^2} - \frac{\chi'_m(NO_3)(PN)}{(n + NO_3)}, \tag{170}$$

where u and w were the downstream and vertical velocities specified at each grid point from a simple circulation model (Walsh, 1975); K_y was a constant lateral eddy coefficient of $\sim 10^6$ cm^2 sec^{-1}; n was the half saturation constant of 1.5 μg-at. NO_3 l^{-1} for nitrate uptake; χ'_m was the maximum uptake rate of NO_3 as a function of both ammonium and daylight, expressed as χ'_m = $[0.11 - 0.02(NH_4)](\sin 0.2618t)$; and PN was the particulate nitrogen of phytoplankton. The other form of dissolved nitrogen was recycled nitrogen, represented by ammonium (NH_4), in

$$\frac{\partial(NH_4)}{\partial t} = -\frac{\partial[u(NH_4)]}{\partial x} - \frac{\partial[w(NH_4)]}{\partial z} + K_y \frac{\partial^2(NH_4)}{\partial y^2}$$

$$- \frac{\chi_m(NH_4)(PN)}{(n + NH_4)} + 0.67 \frac{\gamma_1(PN - P_z)(ZPN)}{\theta_p + (PN - P_z)}$$

$$+ 0.67 \frac{\gamma_2(PN - P_0)(FPN)}{\theta_p + (PN - P_0)} \tag{171}$$

where χ_m was the mean uptake rate of 0.08 hr^{-1} [0.11 hr^{-1} adjusted the maximum of a sinusoid to be χ_m = 0.11($\sin 0.2618t$)] for all other nutrients; n was the 1.5 μg-at. NH_4 l^{-1} half-saturation constant for ammonium uptake; the excretion coefficient of 0.67 was essentially l of Eqs. (154) and (156); γ_1 and γ_2 were the maximum specific grazing rates of the zooplankton (ZPN) and fish (FPN) herbivores expressed as a cosine function for nocturnal grazing within the upper 30 m of the water column, that is, γ_2 = 0.008 cos(0.2618t + 1.571); to simulate diurnal migration of the herbivores, the functional forms of the grazing rates were changed with depth, that is, for zooplankton, γ_1 was

$$(0.03) \cos(0.2618t + 1.571) \quad \text{if} \quad z < 30 \text{ m}$$

$$(0.03) \sin(0.2618t) \quad \text{if} \quad z > 30 \text{ m}$$

while that of the anchovy, γ_2 , was

$$(0.008) \cos(0.2618t + 1.571) \quad \text{if} \quad z < 30 \text{ m}$$

$$(0.008) \sin(0.2618t) \quad \text{if} \quad z > 30 \text{ m}$$

P_z was the zooplankton grazing threshold of 0.5 μg-at. PN l^{-1} and 2.5 μg-at. PC l^{-1}; P_0 was the anchoveta grazing threshold of 3.0 μg-at. PN l^{-1} and 15.0

μg-at. PC 1^{-1} necessary to parameterize phytoplankton patchiness; and θ_p was the half-saturation constant for grazing of 1.5 μg-at. PN 1^{-1} and 7.5 μg-at. PC 1^{-1}.

Similarly, the nutrient flux of phosphate (PO_4) was described by

$$\frac{\partial(PO_4)}{\partial t} = -\frac{\partial[u(PO_4)]}{\partial x} - \frac{\partial[w(PO_4)]}{\partial z} + K_y \frac{\partial^2(PO_4)}{\partial y^2} - \frac{\chi_m(PO_4)(PP)}{(n + PO_4)}$$
$$+ 0.13 \frac{\gamma_1(PN - P_z)(ZPN)}{\theta_p + (PN - P_z)} + 0.13 \frac{\gamma_2(PN - P_0)(FPN)}{\theta_p + (PN - P_0)} \qquad (172)$$

where PP was particulate phosphorus of phytoplankton and n was the 1.5 μg-at. PO_4 1^{-1} half-saturation constant of phosphate uptake. The relative excretion input of phosphorus was 0.13N in relation to nitrogen recycling, based on an analysis of the possible anchoveta impact on nutrient regeneration (Walsh, 1975).

The distribution of silicate (SiO_4) was computed by

$$\frac{\partial(SiO_4)}{\partial t} = -\frac{\partial[u(SiO_4)]}{\partial x} - \frac{\partial[w(SiO_4)]}{\partial z}$$
$$+ K_y \frac{\partial^2(SiO_4)}{\partial y^2} - \frac{\chi_m(SiO_4)(PSi)}{(n + SiO_4)} \qquad (173)$$

where PSi was particulate silicon of phytoplankton and n was the 1.5 μg-at. SiO_4 1^{-1} half-saturation constant for silicate uptake. Silicate was not assumed to be regenerated through herbivore excretion.

The particulate nitrogen (PN) of phytoplankton was computed from

$$\frac{\partial(PN)}{\partial t} = -\frac{\partial[u(PN)]}{\partial x} - \frac{\partial[w(PN)]}{\partial z} + K_y \frac{\partial^2(PN)}{\partial y^2}$$
$$- \frac{\gamma_1(PN - P_z)(ZPN)}{\theta_p + (PN - P_z)} - \frac{\gamma_2(PN - P_0)(FPN)}{\theta_p + (PN - P_0)}$$
$$+ \min \left\{ \left[\frac{\chi'_m(NO_3)(PN)}{(n + NO_3)} + \frac{\chi_m(NH_4)(PN)}{(n + NH_4)} \right], \right.$$
$$\left. \frac{[\chi_m(PO_4)(PP)]}{(n + PO_4)}, \frac{[\chi_m(SiO_4)(PSi)]}{(n + SiO_4)}, \frac{\chi_m(I_z)(PN)}{(I_k + I_z)} \right\} \qquad (174)$$

where the lowest of the four uptake expressions was selected as the growth rate, ε, of the phytoplankton at each time step of the model. The term I_z was equivalent to $I(z)$ computed from a version of Eqs. (114) and (114a) and I_k was defined in Eq. (121). Note that there were no sinking losses of uneaten algae in Eq. (174), in contrast to the sinking fecal pellets of zooplankton and anchovy in Eq. (178). We shall return to the consequences of this omission in Section 6.3.2.

Table V.

Values of Parameters in a Series of Simulation Models of Marine Ecosystems[a]

Parameter	System					
	Peru	NW Africa	Baja California	Alta California	Southern Ocean	Sargasso Sea
Vertical velocity (w), cm sec^{-1}	10^{-2}	10^{-2}	10^{-2}	—	10^{-4}–10^{-5}	10^{-3}–10^{-4}
Downstream velocity (u), cm sec^{-1}	25	25	25	—	4	1
Eddy coefficient (K_y), cm^2 sec^{-1}	10^6	0	1–5×10^6	—	10^7	10^6
Maximum nutrient uptake						
(χ_m), day^{-1}	1.0	1.0	1.0	2.0	1.0	0.5
N half saturation, μg-at./l	1.5	1.5	1.5	1.0	1.0	0.5
Si half saturation, μg-at./l	0.75	0.75	0	1.0	—	—
PO$_4$ half saturation, μg-at./l	0.25	0.25	0.25	—	—	—
Zooplankton grazing rate (γ_1), day^{-1}	0.24	0.24	0.24	—	0.10	0.43
Nekton grazing rate (γ_2), day^{-1}	0.06	0.06	0.06	—	0.01	0.06
Zooplankton grazing threshold (P_z), μg-at./l	0.5	0.5	0.5	—	0.1	0.16
Nekton grazing threshold (P_0), μg-at./l	3.0	3.0	3.0	—	0	0.02

[a] From Walsh (1977); © John Wiley and Sons, Inc., with permission.

The state equation for particulate carbon (PC) of the phytoplankton was

$$\frac{\partial(PC)}{\partial t} = -\frac{\partial[u(PC)]}{\partial x} - \frac{\partial[w(PC)]}{\partial z} + K_y\frac{\partial^2(PC)}{\partial y^2}$$

$$-\frac{\gamma_1(PC - P_z)(ZPC)}{\theta_p + (PC - P_z)} - \frac{\gamma_2(PC - P_0)(FPC)}{\theta_p + (PC - P_0)}$$

$$+\frac{(PC)(\mu_m)[e^{(1 - I_z/I_s)} - e^{(1 - I_0/I_s)}]}{kz}\left[\frac{\chi_m(SiO_4)(PSi)}{n' + SiO_4}\right] \quad (175)$$

where all terms in this equation, except for the last one, were similar to those in the above phytoplankton PN equation. The last term of Eq. (175) was the product of Steele's (1962) formulation of light regulation of photosynthesis, in-depth integral form of Eq. (123), and silicon limitation with $n' = 0.5$ μg-at. SiO_4 1^{-1}. Like all the rate parameters of these equations, the μ_m of 0.10 hr^{-1}, at the peak of the daily sinusoid, was based on experiments (Barber *et al.*, 1971) made during the *Thompson* cruise.

The transfer of phytoplankton carbon and nitrogen to the herbivores was described by a steady-state budget for zooplankton particulate nitrogen (ZPN) of

$$\frac{\partial(ZPN)}{\partial t} = (1 - 0.67 - 0.13 - 0.20)\frac{\gamma_1(PN - P_z)(ZPN)}{\theta_p + (PN - P_z)} \quad (176)$$

and for anchoveta particulate nitrogen (FPN) of

$$\frac{\partial(FPN)}{\partial t} = (1 - 0.67 - 0.13 - 0.20)\frac{\gamma_2(PN - P_0)(FPN)}{\theta_p + (PN - P_0)} \quad (177)$$

where 20% of the food ingested by these herbivores was lost as fecal pellets and 80% as excretory products: that is, there was no net growth of the zooplankton or anchovy populations. The anchovy were assumed to be herbivores in this model, not zoophagous, such that l and o of Eq. (156) were respectively 0.8 and 0.0. The equations for the particulate carbon of zooplankton and anchovy were analogous to Eqs. (176) and (177).

The last state variable was detrital particulate nitrogen (DN), described by

$$\frac{\partial(DN)}{\partial t} = -\frac{\partial[u(DN)]}{\partial x} - \frac{\partial[w(DN)]}{\partial z} - w_s\frac{\partial(DN)}{\partial z}$$

$$+\frac{(0.2)(\gamma_1)(P - P_z)(ZPN)}{\theta_p + (P - P_z)} + \frac{(0.2)(\gamma_2)(P - P_0)(FPN)}{\theta_p + (P - P_0)} \quad (178)$$

where the assumed sinking velocity w_s of 150 m day^{-1} was sufficiently fast to make horizontal eddy diffusion of little consequence for these particles. Decomposition of the fecal pellets was ignored in this model, and the other terms have already been described. Table V summarizes the values of each parameter used

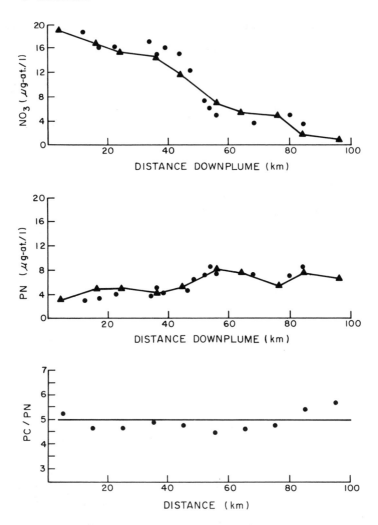

FIG. 57. Simulated and observed (●) nitrate, particulate nitrogen (PN), and particulate C/N ratio in a two-dimensional model of the upwelling ecosystem at 15°S off Peru. [After Walsh (1975) and Walsh and Howe (1976); © 1975 Pergamon Journals Ltd., © 1976 Academic Press, with permission.]

in this model and other numerical studies off Baja California, Northwest Africa, Antarctica, and Bermuda. Equations (170)–(178) were converted to upstream finite difference equations, i.e., there was numerical dispersion in the x and z dimensions) and solved with an Adams–Bashford predictor technique [recall Eq. (67)], using a time step of 1 hr to satisfy Eq. (166).

The importance of nonlinear interaction of grazing thresholds, multiple nutrient limitation, and diel periodicity in determining the distributions of nutrients

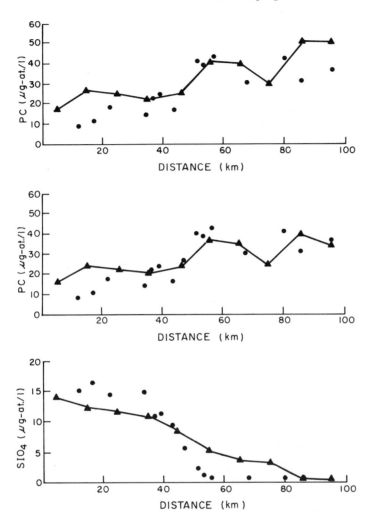

FIG. 58. Simulated and observed (●) particulate carbon and silicate, where photosynthesis is decoupled from silicon regulation in the first panel and coupled in the second, in a two-dimensional model of the upwelling ecosystem at 15°S off Peru. [After Walsh and Howe (1976); © Academic Press.]

and phytoplankton on a mesoscale of $1-10$ days and 10^1-10^2 km² (Fig. 55a) is displayed in Figs. 57 and 58. The circles in these two figures are, respectively, the nitrate, silicate, particulate nitrogen computed from particulate phosphorus (PN = 16 × PP), and particulate carbon from the chlorophyll (with a C/chl ratio of 50 : 1), measured over $0-10$ m depth at each of the *Anton Bruun* drift stations. The triangles connected by line segments of Figs. 57a,b are the predicted nitrate and particulate nitrogen for each grid point at local time of the

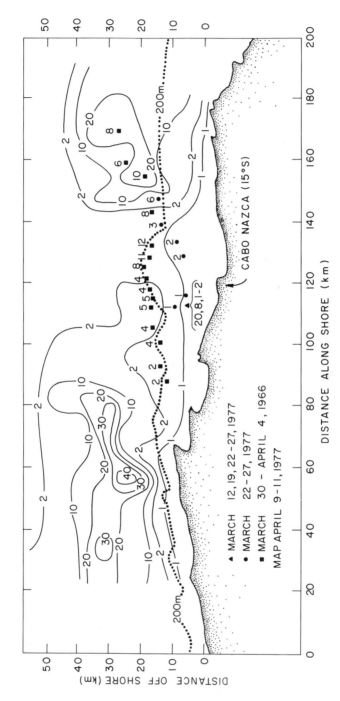

FIG. 59. The surface distribution of chlorophyll during an underway map on April 9–11, 1977, off 15°S in relation to drogue results both in March–April 1966 (□) and March 1977 (●), as well as from a time series (△) at one 1977 station. [After Ryther *et al.* (1971) and Walsh *et al.* (1980); © 1980 Pergamon Journals Ltd., with permission.]

closest *Bruun* station. The close fit of these state variables of the model with the independent set of observations suggests that the flow field and biological flux assumptions of the model may be reasonable approximations of the Peru upwelling ecosystem.

Output of some of the other variables of the model is shown in Figs. 58a–c. The first result of the particulate carbon simulation (Fig. 58a) is a case in which photosynthesis was a function of light only and was not regulated by silicate availability. The predicted carbon agreed fairly well with the measured chlorophyll and a C/chl ratio of 50 $\times$ 1 at most of the inshore stations, but it diverged from the offshore observations. The second carbon simulation (Fig. 58b) included the multiplicative coupling of silicate depression of photosynthesis, using a value for n' of 0.5 μg-at. SiO_4 l^{-1}, which was one-third the value used in the Michaelis–Menten alternative choice expression, [Eq. (174)]. The closer fit of predicted and estimated particulate carbon in Fig. 58b suggests that both nitrogen and silicon may be regulating carbon productivity off Punta San Juan, Peru. The predicted downplume silicate distribution of this second carbon simulation is shown in Fig. 58c, for comparison with the 0–10 m mean silicate concentrations observed at each of the *Bruun* stations.

The combined results of the multiplicative and alternative formulations of nutrient and light regulation are shown in Fig. 57c as the predicted downplume C/N ratios at grid points closest to each of the *Bruun* stations. The mean predicted atomic C/N ratio from the model was 4.8, compared to an observed mean of 5.6 from 85 samples collected on the 1969 *Thompson* cruise, and a mean of 4.0 from the five *Bruun* samples (Dugdale and Goering, 1970). The reasonable match of predicted and observed C/N ratios suggests that the Peru phytoplankton may usually be up on the plateau of their respective hyperbolae of growth versus ambient nitrogen, phosphorus, silicon, and perhaps light, in terms of both maximal carbon and nitrogen productivity in the upper 10 m of the inshore regions.

Such a hypothesis suggests that a C/N ratio of ~5 : 1 may be about as low as can be expected for *in situ* diatoms (the C/N ratio of pure protein is ~3.6), reflecting the maximum productivity of individual phytoplankton cells in upwelling areas. As a phytoplankton population is advected offshore, away from the source of upwelled nutrients, it crosses the ecotone to an oligotrophic system, where the rate of nitrogen input may limit the total yield of algal biomass. The algae may then begin to operate on the nonlinear part of the hyperbolic functions, with C/N ratios increasing and production declining. The amount of grazing stress imposed by the nearshore herbivores, rather than the rate of nutrient supply, may instead set the population limits of phytoplankton in the Peru upwelling ecosystem.

During another cruise of the R/V *Melville* to the Punta San Juan study area in April 1977, for example, the chlorophyll biomass was two to four times greater than that of the *Anton Bruun* study (Fig. 59), perhaps reflecting overfishing of the anchovy and a decline in their grazing pressure on the microalgae (Walsh, 1981b). The above model was not designed, however, to resolve either spatial variability at the synoptic scale ($>10^4$ km^2), or temporal fluctuations at the event (<10 days) and interannual (>1 yr) time scales. Some of the weaknesses of the above model were the parameterization of lateral diffusion in one along-shore plume of temperature (Fig. 56) and chlorophyll (Fig. 59), and the assumption of homogeneously distributed herbivores. Without a detailed physical model, no mechanism could account for 10-fold changes of chlorophyll over 5 days at the same station within the nearshore 1977 upwelling region as well (Walsh *et al.*, 1980). These problems are inherent in the representation of three-dimensional, time-dependent, "real-world" structure as a steady-state, two-dimensional $x-z$ model. The heuristic value of such an early model, however, was the realization that additional insight into coastal food webs would have to await development of new sampling techniques to compile biological time series of greater spatial and temporal resolution.

3.2.2 Aliasing

By 1822, Baron Jean Fourier had introduced an alternative description of a time-dependent function as the series expansion of a sum of periodic functions, now bearing his name (Fourier, 1822). Specifically, each term of a finite sequence of equally spaced observations, $X_1, X_2, \ldots X_n$, in the time domain can be approximated by

$$X_n = a_0 + \sum_{i=1}^{m} \left(a_i \cos \frac{2\pi ni}{T_0} \right) + \left(b_i \sin \frac{2\pi ni}{T_0} \right) \tag{179}$$

in the frequency domain, where a_0 is X, the mean of the observations; a_i and b_i are coefficients of the Fourier series, representing the amplitude of the periodic functions and thus the importance of each term at a particular frequency i/T_0 (cycles hr^{-1}), in which the last term, $m = n/2$, of this sum represents the m harmonic of a fundamental frequency $1/T_0$, with period T_0. If the values of the coefficients a_i, b_i at each discrete harmonic frequency $1/T_0, 2/T_0, \ldots m/T_0$, are plotted against frequency, a spectrum of their intensity, $\frac{1}{2}(a_i^2 + b_i^2)$, or spectral density is derived. Since a_0 is $\overline{X}$, the finite sum of the discrete Fourier transform in Eq. (179) represents an estimate of the residual, $X_n - \overline{X}$, that is, how much each X_n departs from $\overline{X}$, at different frequencies, which might provide insight into the sources of variance of X_n over the sequence of observations.

For example, a hypothetical curve for the wind velocity in a coastal region, with zero mean, could be constructed as the sum of three sinusoids, with a period of 24 hr, of perhaps 2 weeks, and the fundamental harmonic term of an annual period (Fig. 60). The variance spectrum, or Fourier transform, of this wind velocity time series would appear as three discrete lines, partitioning the total

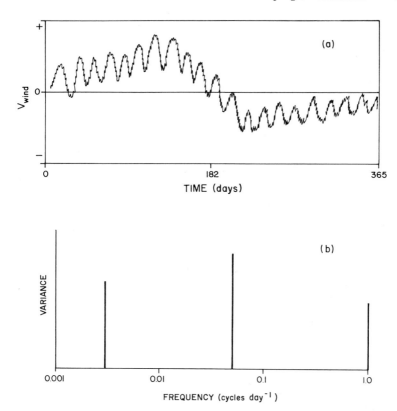

FIG. 60. (a) Schematic time series and (b) variance spectrum of wind forcing with only three sources of variability: seasonal, event, and diel scales. [After Walsh *et al.* (1977); © 1977 American Society of Limnology and Oceanography Inc., with permission.]

variance among only three frequencies (Fig. 60), since $m = 3$ in Eq. (179). The height of these line spectra (Jenkins and Watts, 1968) is proportional to the variance contribution of $\frac{1}{2}(a_i^2 + b_i^2)$ at that frequency; therefore the relative heights indicate the relative importance of the sources of variance at the three time scales.

This simple time series is band-limited in the signal-analysis jargon of electrical engineering (Cooper and McGillem, 1967; Haykin, 1983), with the lowest fundamental frequency of 0.0027 day^{-1}, the highest of 1.0 day^{-1}, and none outside these bands. The low-frequency band is usually determined by the length of the time series, while the high-frequency band is determined by constraints of the sampling device. Measuring the temperature (Fig. 61) at 17 m, 5 km off the Baja California coast, every 5 min for only 7.6 days between March 29 and April 6, 1973 (Walsh *et al.*, 1977), for example, precludes analysis of turbulent or seasonal sources of variance. Spectra determined from actual time series are

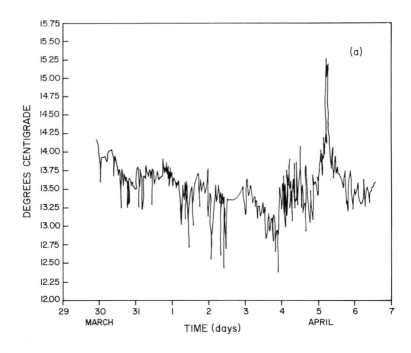

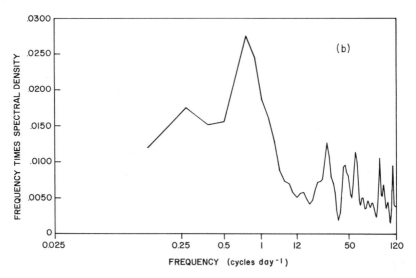

FIG. 61. (a) Time series and (b) variance spectrum of temperature data from a thermistor at 17 m depth on a mooring 5 km off the Baja California coast from March 29 to April 6, 1973. [After Walsh *et al.* (1977); © 1977 American Society of Limnology and Oceanography Inc., with permission.]

continuous curves rather than discrete lines, of course, since the variance is partitioned over all frequencies in the measured bands. Such curves have peaks at those frequencies that contribute most to the total variance. The integral of the area under the curve represents the total variance of the original time series, with the diel variability (~ 1 cycle day^{-1}) accounting for most of the variance in this particular time series of temperature (Fig. 61).

Computation of the variance spectrum of a time series requires evaluation of the Fourier coefficients a_i, b_i defined by

$$a_i = \frac{2}{n} \sum_{i=1}^{n} X_n \cos \frac{2\pi ni}{T_0} \tag{180}$$

$$b_i = \frac{2}{n} \sum_{i=1}^{n} X_n \sin \frac{2\pi ni}{T_0} \tag{181}$$

after appropriate manipulation (Panofsky and Grier, 1965) of Eq. (179). With the growth of digital computers, a fast Fourier transform (FFT) algorithm (Cooley and Tukey, 1965) was developed to supplant previous methods of computing the Fourier coefficients from the autocovariance (Blackman and Tukey, 1958). An extensive literature on FFT algorithms now exists (Haykin, 1983) and is beyond the scope of this monograph; for a time series of $n > 1024$, FFTs take about 20% as much time as the autocovariance method and preserve both the amplitude and the phase information. Thus far, we have only discussed the Fourier series description of a sequence of observations in time, but the variance of a spatial sequence Y_1, Y_2, . . . Y_n could be as easily described by

$$Y_n = a_0 + \sum_{i=1}^{m} \left(a_i \cos \frac{2\pi ni}{L_0} \right) + \left(b_i \sin \frac{2\pi ni}{L_0} \right) \tag{182}$$

where a_0 is now the spatial mean, $\overline{Y}$, of the observations; a_i and b_i are the Fourier coefficients at a particular wavelength, i/L_0; and L_0 is the wave number. Before the development of moored fluorometers (Whitledge and Wirick, 1983), most spectral analyses (Platt and Denman, 1975) of biological variables (Platt, 1972; Denman and Platt, 1976) were restricted, in fact, to data sets collected from short transects of a ship steaming over a small area. Figure 62, of temperature and chlorophyll fluorescence along a 21-km transect on the outer shelf of the northern North Sea in May (Horwood, 1978), is an example of such underway data that became available in the late 1960s (Lorenzen, 1971; Walsh et al., 1971); the map of surface chlorophyll distribution off Peru (Fig. 59) was constructed from a series of ship transects.

The variance spectra of temperature and chlorophyll on this transect (Fig. 63c) over wavelengths of 160 m to 16 km and of shorter transects in other parts of the North Sea (Horwood, 1978) over wavelengths of 28 m to 2.8 km are presented in Fig. 63. If temperature is considered a passive tracer, or scalar, embedded in a turbulent ocean, the slope of its variance spectrum (Kolmogorov,

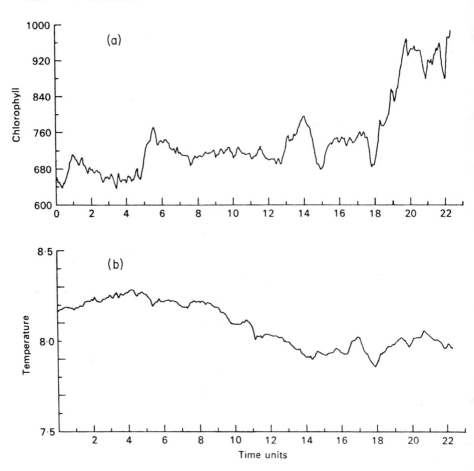

FIG. 62. A plot of (a) surface chlorophyll and (b) temperature against time during a 21-km line map in the northern North Sea. [After Horwood (1978); © 1978 British Crown, Marine Biological Association of the United Kingdom, with permission of Cambridge University Press.]

1941) should have a value of $-\frac{5}{3}$ over length scales of 100 m to 0.1 cm, in a subinertial regime of energy cascade as three-dimensional isotropic turbulence, below which viscous dissipation occurs. Over larger length scales of 100 m to 100 km, where a large portion of the ocean's kinetic energy is contained within synoptic eddies (Table I), variance spectra of temperature should have slopes of -2 to -3 within two-dimensional turbulence regimes, similar to the variance spectra of water motion (Ozmidov, 1965; Webster, 1969). Phytoplankton presumably grow at much faster time scales than those of the physical processes that alter local temperature, such that above a critical length scale the variance spectrum of chlorophyll might have a biological source.

In an early model of phytoplankton patchiness, which considered a balance of just algal growth and an eddy diffusive loss of biomass, Kierstead and Slobodkin (1953) suggested that such a critical length scale L_c might be

$$L_c \approx \sqrt{\frac{K_y}{\varepsilon}} \tag{183}$$

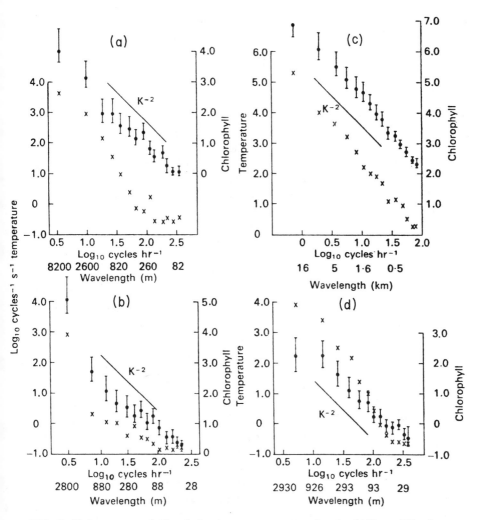

FIG. 63. Variance spectra of chlorophyll and temperature over wavelengths of 160 m to 16 km from the line map in the northern North Sea (c), and over wavelengths of 28 m to 2.8 km from shorter transects in other areas (a, b, d) of the North Sea. [After Horwood (1978); © 1978 British Crown, Marine Biological Association of the United Kingdom, with permission of Cambridge University Press.]

or about 1 km for a K_y of 10^5 cm^2 sec^{-1} and a growth rate of 1 day^{-1}. A subsequent dimensional analysis of spatial spectra of chlorophyll fluorescence within Canadian waters (Denman and Platt, 1976) also suggested that, at wavelengths less than 1 km, the slope of chlorophyll variance might be -2 to $-\frac{5}{3}$. They theorized that, at longer wavelengths, the slope might reduce to -1, reflecting a net increase, or spatial inhomogeneity, of algal biomass. Internal wave fluctuations were thought to induce the chlorophyll variance at wavelengths of 1 km to 100 m, and turbulent diffusion at scales <100 m (Platt and Denman, 1975), since a similar change in the slope of chlorophyll variance appeared to occur at ~100 m wavelength in lakes (Powell et al., 1975).

Additional field observations (Fig. 63) did not support a change in slope of the chlorophyll variance spectra at the 1-km wavelength in most of the North Sea (Horwood, 1978), however. Reformulation of the spectral model (Denman et al., 1977) suggested that the net growth rate of the algal population $(\varepsilon - \gamma - \omega_s)$, rather than ε, might lead to a larger critical length scale of ~10 km. Most of the variance and coherence of temperature and chlorophyll were, indeed, at the fundamental wavelength of 21 km in the northern North Sea experiment (Fig. 62), reflecting a possible change in water types (Richerson et al., 1978). Chlorophyll patches of 20–40 km extent were also observed in the underway map off Peru (Fig. 59).

As one samples over larger length scales with a slow research vessel at 8–12 kt, however, the frozen-turbulence hypothesis (Hinze, 1959) of these spatial variance spectra is no longer valid. It was assumed that the temperature or chlorophyll sensor was towed much faster through the ocean than possible accelerations of the velocity field, such that variations with time of the measured variable (Fig. 62) were synonymous with its spatial changes. The underway transect in the North Sea was obtained over 2–3 hr, while the underway map off Peru occupied 2–3 days of steaming time. To avoid confounding time and space in such sampling programs above length scales of 10 km, over which biological processes are now thought to dominate those of physical forcing, different sampling strategies and platforms must be used.

Representation of a time, or equivalent space, series of length n by the sum of $m = n/2$ sine and cosine terms [e.g., Eqs. (179) and (182)] involves equally spaced frequencies in this sum, ranging from the fundamental harmonic of $1/n\Delta$, where the fundamental period of the time series $T_0 = n\Delta$ and Δ is the sampling interval, to the m highest harmonic of $1/2\Delta$, that is, $(1/n\Delta)(n/2)$, termed the Nyquist frequency. The value of $\frac{1}{2}(a_i^2 + b_i^2)$ at the fundamental harmonic frequency has the greatest contribution to the variance of the time series, with the least at the Nyquist frequency, such that the variance spectrum is always red, that is, the variance decreases with increasing frequency (Panofsky and Grier, 1965). A convenient mathematical fiction of band-limited white noise is sometimes employed (Cooper and McGillem, 1967), where the variance is assumed constant over a finite band width and zero outside this frequency range.

The above definition of the Nyquist frequency as $\frac{1}{2}\Delta$ allows description of the Nyquist sampling theorem (Nyquist, 1928) in terms of the frequency at which one must sample, that is, at least twice the maximum frequency of phenomena one wishes to resolve. If a larger sampling interval of $>\Delta$ is employed, aliasing, or improper resolution of the temporal description of the processes, occurs with spurious high-frequency variance associated with spectral estimates at lower frequencies than the Nyquist frequency. Assume, for example, that the process of interest has a characteristic frequency of ϕ_0, or period λ_0, that we wish to study (Fig. 64a). Yet our sampling plan involves collection of data at a frequency ϕ_a

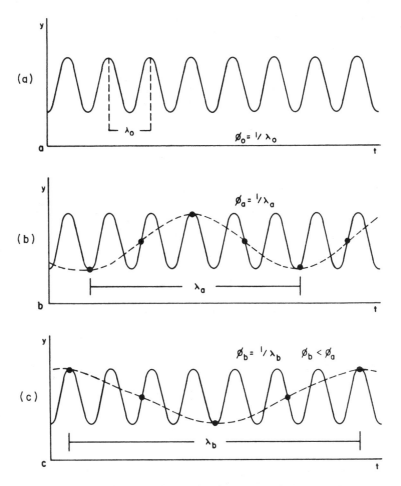

FIG. 64. A sinusoidal function y with characteristic frequency (a) ϕ_0, in which aliasing is created by sampling either at frequency (b) ϕ_a or (c) ϕ_b, to generate the dashed lines, inadequately representing the original function y. [After Kelley (1976); © 1976 Blackwell Scientific Publications Ltd., with permission.]

$> \phi_0$ as exemplified by the large dots in Fig. 64b; our interpretation of the data would be the dashed line in this panel, not the true process described by the solid line. Similarly, if the sampling plan was at even lower frequency, $\phi_b > \phi_a$, an even more distorted view of the process is presented in Fig. 64c. An example of the consequences of such aliasing is described in Section 3.3.2.

Phytoplankton species on the continental shelves can divide every 0.5–2.0 days; without significant losses, an algal population during the spring bloom could increase at the same rate. To resolve the temporal and spatial consequences of this resultant growth process, a Nyquist sampling interval of at least 0.25 day is required by sampling theory (Blackman and Tukey, 1958). If one sampled every 4 hr on the continental shelf within an along-shore flow regime of 30 cm sec^{-1} to resolve this process, at least five ships would be required to steam randomly over a grid of 20 km^2, occupying a station every hour to obtain the necessary biomass measurements (Kelley, 1976)—hopefully, there would be no fog during this experiment! We thus arrive at a major reason for moored fluorometer and satellite color data—the need to analyze distributions of biological properties at frequencies that can resolve causally the sources of their variance.

3.2.3 Filtering

Upon acquisition from the sea of a data set, which required a great deal of effort and planning to obtain, we immediately begin to discard information. Consider the signals received by your brain as you navigate a car down a busy highway—most of the data is filtered in an unknown process that allows you to consider a very small subset of the bits of information impinging on your sensory receptors. Similarly, we routinely employ simple digital filters to data sets, in order to remove complexities at time and space scales that are presumably irrelevant to the particular process under study. If one wished to ignore diurnal changes in a temperature cycle with a period of 1 day, for example, one could average hourly observations to yield the mean daily temperature.

The simplest digital filter of a time series $X_1, \ldots, X_n$ is the running mean of equal weight, w_i, where

$$Y_n = w_i X_{n-1} + w_i X_n \tag{184}$$

yields the term of the filtered time series, Y_n. For $w_i = 0.5$ and an unfiltered sequence of 8, 2, 5, 3, 10, 4, 8, for example, Eq. (184) leads to the filtered sequence of 5, 3.5, 4, 6.5, 7, 6, with one less term. In the autocovariance method of computing variance spectra (Blackman and Tukey, 1958), a smoothing scheme of unequal weights was used,

$$Y_n = 0.25X_{n-1} + 0.50X_n + 0.25X_{n+1} \tag{185}$$

yielding a filtered sequence of 4.3, 3.8, 5.3, 6.8, 6.5 from the original sequence, but with two less terms. Smoothing functions can also be designed to approximate the shape of the normal curve by making the weights proportional to binomial coefficients, for example,

$$Y_n = 0.06X_{n-2} + 0.25X_{n-1} + 0.38X_n \quad (186)$$
$$+ 0.25X_{n+1} + 0.06X_{n+2}$$

which yields a filtered sequence of 4.3, 5.3, 2.9 from the same original sequence, but with four less terms.

We would like to know both how the variance spectrum of the original time series $X_1, \ldots, X_n$ is changed by the smoothing, or filtering, procedure and how the weights w_i might be selected to delete data at particular frequencies. A description of the behavior of the smoothing function in the frequency domain is given by the ratio at each frequency of the Fourier transform of the filtered time series $Y_1, \ldots, Y_n$, or output signal in the jargon of electrical engineers (Cooper and McGillem, 1967), to the Fourier transform of the input signal, $X_1, \ldots, X_n$. This ratio of the two transforms is termed the frequency response or transfer function, $H(f)$, and can be described as a Fourier expansion of the symmetric, or even, filtering function, $w_{-i} = w_i$.

An even function can be represented as the sum of a mean and a Fourier series containing only cosines, such that the Fourier transform of the frequency response is

$$H(f) = w_0 + 2 \sum_{i=1}^{m} w_i \cos 2\pi i f \Delta \quad (187)$$

at frequency f, while w_i is the ith weight numbered outward from the central weight w_0, and Δ is the time interval between successive observations of the time series. Equation (187) is then solved (Hamming, 1977) for values of w_i that generate the desired transfer function, or filter, $H(f)$; most digital filters either attenuate data at high frequencies (low-pass), at low frequencies (high-pass), or allow passage of data only at a selected band of frequencies (bandpass). For example, assume that there was a requirement for a 4-hr low-pass filter that had the ideal transfer function

$$H(f) = \begin{cases} 1 & \text{over} \quad 0 \le f \le 0.25 \quad \text{cycles hr}^{-1} \\ 0 & \text{over } 0.25 \quad \text{cycles hr}^{-1} < f \le (2\Delta)^{-1} \quad \text{cycles hr}^{-1} \end{cases} \quad (188)$$

that is, up to the Nyquist frequency, where Δ is 1 hour.

Such a digital filter should pass all variations ≤ 0.25 cycles hr^{-1} or, equivalently, stop variations with periods <4 hr. In general, the filter with the most coefficients w_i will yield the best transfer function. As an illustration, two 4-hr low-pass filters with 10 and 100 coefficients (w_i) were calculated, with their resulting transfer functions shown in Fig. 65. The first digital filter of 10 coefficients exhibits poor behavior, with undesirable ripples in the transfer function and a relatively wide transition zone between the pass and stop frequencies. The

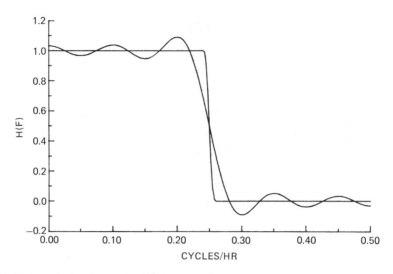

FIG. 65. Transfer functions $H(f)$ of a 4-hr low-pass digital filter, with 10 coefficients (rippled line) and 100 coefficients (C. D. Wirick, personal communication).

ripple can always be reduced by smoothing the coefficients, but this widens the transition zone. Clearly, the second digital filter of $100w_i$ has the better transfer function, but when a time series is filtered the first and last i terms are lost [recall Eqs. (185) and (186)], such that the loss of data must be weighed against the gain in performance of the filter. Conventional shipboard surveys of 1- to 3-week duration to study dominant shelf phytoplankton processes, such as the spring bloom, cannot resolve either algal population growth (0.5 day^{-1}) or wind resuspension events (0.2 day^{-1}), thereby severely aliasing estimates of primary production and its consumption (Walsh *et al.,* 1987a). As part of a multidisciplinary effort to determine the fate and transport of biogenic particles from the coastal zone to slope depocenters, a series of moored fluorometers, transmissometers, thermistors, and current meters were deployed on the continental shelf during February–April 1984 within the Mid-Atlantic Bight. The sampling rates of the four moored instruments were sufficient (0.007–0.012 day^{-1}) to resolve major scales of physical and biological variability. Fig. 66b is an example of a 2-hr, low-pass digital filter applied to a 53-day data record from a moored fluorometer at 13-m depth on the 80-m isobath, south of Martha's Vineyard. In contrast, Fig. 66a is the result of the same time series filtered by a 40-hr, low-pass filter; note that much of the high-frequency variance has been removed.

A high-passed time series, containing only information at high frequencies, could be constructed by subtracting the data record of Fig. 66a from that of Fig. 66b. Alternatively, one could remove long-term trends of a data set by smoothing a time series with a high-pass filter. A previous experiment at a sampling rate of

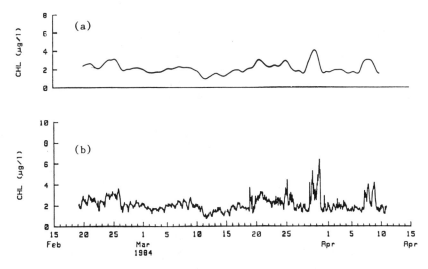

FIG. 66. (a) A 40-hr and (b) a 2-hr low-passed data record of chlorophyll fluctuations at 13 m on the 80-m isobath, south of Martha's Vineyard, from February 17 to April 12, 1984. [After Walsh *et al.* (1987b); © 1987 Pergamon Journals Ltd., with permission.]

0.012 day^{-1} during September–October 1977 had yielded an 11-day time series of chlorophyll fluctuations at 10.5-m depth on the 28-m isobath, south of Long Island (Fig. 67a). This data record was instead filtered (Wirick, 1981) with a high-pass frequency of 0.4 cycles day^{-1}, yielding a time series in which the low-frequency variance had been removed (Fig. 67b); note that the filtered data record is 1–2 days shorter than the original time series. Not surprisingly, most of the variance of the filtered chlorophyll fluorescence record was then contained within the diurnal frequency band (Fig. 68). Interpretation of these *in situ* time series from moored fluorometers is deferred to Section 3.4.1, after discussion of measurements of phytoplankton biomass from another perspective, that of remote sensing from aircraft and spacecraft.

3.3   Spatially Synoptic Chlorophyll Fields

Restricted by cost and flight time to local mesoscale areas, remote sensing by aircraft nevertheless fills a critical gap in the space–time sampling domain (Fig. 55), between conventional measurements made from ships and those currently within the capability of satellite sensors. Parameters such as salinity and chlorophyll fluorescence (Table VI) can be measured from aircraft platforms but are at least a decade away from being sensed by spacecraft. The instrument of particular uniqueness to aircraft platforms is the laser (Lidar) fluorosensor. It may eventually provide a measure of the relative abundance of different pigment

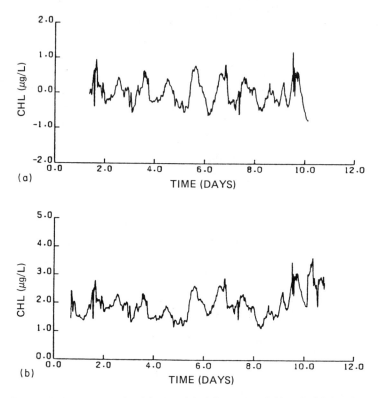

FIG. 67. (a) A high-passed record and (b) the original time series of chlorophyll fluctuations at 10.5 m on the 28-m isobath, south of Long Island, from September 23 to October 3, 1977. [After Wirick (1981).]

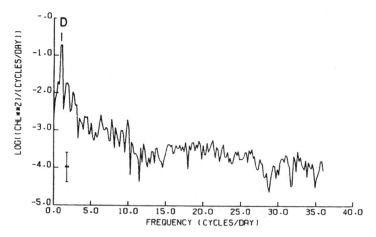

FIG. 68. The variance spectrum of chlorophyll fluctuations within the high-passed record at 10.5 m on the 28-m isobath, south of Long Island, from September 23 to October 3, 1977. [After Wirick (1981).]

Table VI.

Aircraft-Borne Remote Sensing Techniques

Name	Type of Sensor	Characteristics	Measurements
AOL	Laser (Lidar) fluorosensor	Uses single-wavelength laser to induce fluorescence; measures emission in 40 channels; has vertical profiling capability	Fluorescence of chlorophyll *a* and other pigments; light attenuation: phytoplankton color group diversity
L-Band	Microwave radiometer	Measures passive microwave radiation from water surface in single channel	Salinity (requires independent measurement of surface temperature)
PRT-5	Infrared radiometer	Measures passive thermal radiation from water surface in single channel; commercially available	Surface temperature
MOCS	Multispectral scanner	Measures backscattered sunlight in visible and near infrared spectral range; has 20 bands, 15 nm wide	Chlorophyll *a* in real time: suspended and dissolved matter that affects color
OCS	Multispectral scanner	Has 10 bands in visible and near-infrared spectral range; forerunner of CZCS instrument on *Nimbus*-7 satellite; flown on NASA Lear jet	Two-dimensional high-altitude imagery; maps of chlorophyll *a* and suspended sediments
IOS	Multichannel spectrometer	Measures upwelling radiance spectra in 256 channels in the visible and near-infrared range	Chlorophyll *a* by natural (sun-stimulated) fluorescence and broadband color effects; suspended sediments

classes or color groups of phytoplankton, such as the golden-brown species (diatoms and dinoflagellates), by using laser light of different frequencies to excite the photopigments. Moreover, the resolution and spatial scales of these different airborne sensors (Table VI) can provide ground truth and calibration data for the satellite sensors over wide areas, thus potentially extending the utility of the present generation of satellite instrumentation.

Laser fluorosensors are in a class known as "active" remote sensors because they provide their own source of energy. Laser pulses are fired into the water column from low-flying aircraft, and the induced emission spectrum is sensed in narrow spectral bands. The returning laser light varies as a function of backscatter from particulate matter within the ocean and as a function of absorption. Red-

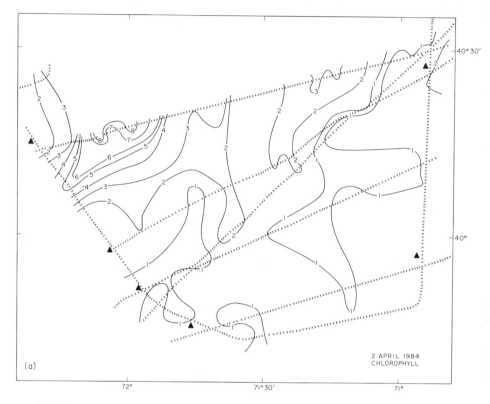

FIG. 69. The surface distribution of chlorophyll, measured with Lidar, during overpasses of a P-3A aircraft on the New England shelf during (a) April 2, 1984 and (b) April 8, 1984. [After Walsh *et al.* (1987b); © 1987 Pergamon Journals Ltd., with permission.]

shifted Raman backscatter from the water molecule itself is proportional to the number of water molecules accessed, or, equivalently, to the penetration depth of the laser beam into the water. The Raman backscatter provides a direct measure of water clarity or turbidity. Further, the strength of the Raman signal can be used to correct fluorescent signals received from photopigments for spatial variations in optical transmission properties of water (Bristow *et al.*, 1979; Hoge and Swift, 1981), thus eliminating the need for extensive surface truthing of water transmissivity. As with shipboard or moored fluorometers, the corrected chlorophyll *a* fluorescence signal centered near 685 nm is used to gauge chlorophyll *a* concentration.

Extracted chlorophyll data from shipboard observations on the New England shelf during April 1984 were used to calibrate the laser-induced estimates of chlorophyll from this airborne Lidar system (Fig. 69). An r^2 of 0.92 was obtained for the shipboard and aircraft measurements during 1984, somewhat better

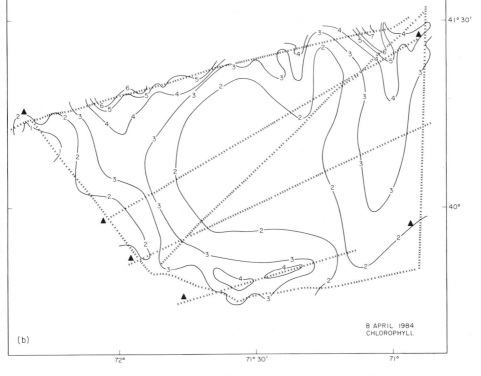

(b)

8 APRIL 1984
CHLOROPHYLL

FIG. 69. (continued).

than an r^2 of 0.77 within turbid Mid-Atlantic Bight waters off Virginia in 1980 (Hoge and Swift, 1981). Such aircraft overflights provide 20-fold more surface temperature or chlorophyll measurements within 2 hr than can be provided by a ship with over ~200 hr of routine sampling at hydrographic stations, to be later contoured as aliased surface maps of the thermal and phytoplankton fields.

The April 1984 aircraft overflights were both taken between the 60-m and 400-m isobaths, south of Long Island, after northwest wind events of 1.0–1.5 dynes cm^{-2} intensity on April 1 and 8. The aircraft chlorophyll sampling on April 8 (Fig. 69b) occurred during a northwest wind event, however, while the overflight of April 2 (Fig. 69a) took place 24 hr after a previous northwest event. The higher chlorophyll values encountered over the slope during the second overflight may thus represent phytoplankton cells, which had yet to sink out of the water column. The aircraft phytoplankton fields on April 2 (Fig. 69a) and April 8 (Fig. 69b) are apparently different, with shelf plumes of 3–4 μg chl l^{-1} extending out over the slope during the latter overflight, compared to seaward tongues of only 1–2 μg chl l^{-1} encountered during the former. We shall see similar spatial patterns from satellite overflights over much larger areas of this shelf (Figs. 75 and 78), which suggest that the aircraft data are correct in their depiction of both shelf export and rapid removal of algal biomass from surface waters.

Preliminary investigations undertaken in the 1960s by George Clarke, Gif Ewing, Carl Lorenzen, Charlie Yentsch, and others provided evidence that the quality of light or color reflected from the sea surface and remotely sensed in a "passive" mode by aircraft instrumentation might also be interpreted as phytoplankton biomass (i.e., chlorophyll) in the upper portion of the water column. These workers (see, e.g., Clarke et al., 1970) were limited by their equipment to an altitude of 3 km. However, even at that altitude, the influence of the atmospheric backscatter was quite obvious as it began to dominate the color signal reflected from the ocean surface. This raised the question of whether the rather poorly reflected ocean could be sensed through the entire atmosphere from a spacecraft, and if the contributions of the Rayleigh backscatter and aerosol backscatter could be effectively removed from the signal seen by a spacecraft. Additional studies in 1971 and 1972 with Lear jet and U-2 aircraft and a rapid scan spectrometer at altitudes of 14.9 and 19.8 km demonstrated that this concept could be used to develop spacecraft equipment for the purpose of estimating chlorophyll from Earth-orbiting satellites. This became possible through the realization that problems associated with the scattering properties of the atmosphere, as well as direct reflectance of the sun from the sea surface (glint), could be either avoided or corrected (Hovis and Leung, 1977).

3.3.1 Coastal Zone Color Scanner

The first satellite-borne ocean color sensor, the coastal zone color scanner (CZCS), was launched aboard Nimbus-7 in October 1978 with four visible and two infrared (one of which is thermal) bands, allowing a sensitivity about 60 times that of the Landsat-1 multispectral scanner. Unlike many satellite sensors of ocean properties (Table VII), the CZCS responds to more than the features of the mere surface of the sea and is sensitive to algal pigment concentrations in the upper 20–30% of the euphotic zone (Gordon et al., 1980; Hovis et al., 1980). The CZCS was specifically designed to detect upwelling radiance in blue, green, yellow, and red spectral bands (443, 520, 550, and 670 nm) selected for the purpose of describing spatial variations in the concentrations of phytoplankton pigments from an altitude of 955 km, with a resolution of 825 m at nadir. The theoretical and experimental techniques for describing the biooptical state of ocean waters and its relationship to optical parameters that can be remotely sensed have been discussed by a number of workers (Morel and Prieur, 1977; Smith and Baker, 1978a,b; Gordon and Morel, 1983).

Table VII.

Satellite-Borne Remote-Sensing Techniques

Altimeter. A pencil-beam microwave radar that measures the distance between the spacecraft and the earth. Measurements yield the topography and roughness of the sea surface from which the surface current and average wave height can be estimated.

Color scanner. A radiometer that measures the intensity of radiation emitted from the sea in visible and near-infrared bands in a broad swath beneath the spacecraft. Measurements yield ocean color, from which chlorophyll concentration and the location of sediment-laden waters can be estimated.

Infrared radiometer. A radiometer that measures the intensity of radiation emitted from the sea in the infrared band in a broad swath beneath the spacecraft. Measurements yield estimates of sea-surface temperature.

Microwave radiometer. A radiometer that measures the intensity of radiation emitted from the sea in the microwave band in a broad swath beneath the spacecraft. Measurements yield microwave brightness temperatures, from which wind speed, water vapor, rain rate, sea-surface temperature, and ice cover can be estimated.

Scatterometer. A microwave radar that measures the roughness of the sea surface in a broad swath on either side of the spacecraft with a spatial resolution of 50 km. Measurements yield the amplitude of short surface waves that are approximately in equilibrium with the local wind and from which the surface wind velocity can be estimated.

Synthetic Aperture Radar. A microwave radar similar to the scatterometer except that it electronically synthesizes the equivalent of an antenna large enough to achieve a spatial resolution of 25m. Measurements yield information on features (swell, internal waves, rain, current boundaries, and so on) that modulate the amplitude of the short surface waves; they also yield information on the position and character of sea ice from which, with successive views, the velocity of ice flows can be estimated.

Simply stated, the CZCS radiance data can be utilized to estimate chlorophyll concentrations by detecting shifts in ocean color, particularly in offshore waters. Clear open ocean waters have low chlorophyll concentrations ($0.01-1.0$ μg chl l^{-1}) and the solar radiation reflected from the upper layers of these waters is blue; conversely, waters with high concentrations of chlorophyll (>1.0 μg chl l^{-1}) are green (Morel and Smith, 1974). It has been demonstrated that this change in ocean color can now provide a quantitative estimate of chlorophyll concentration (Gordon and Clark, 1980; Smith and Baker, 1982) for oceanic regions with an accuracy of $0.3-0.5$ log chl (where chl is the chlorophyll concentration). The initial comparisons between CZCS imagery and surface pigments measured continuously along ship tracks carried out by Gordon *et al.* (1980) and Smith and Wilson (1981) had suggested that chl could be retrieved from the imagery to within about a factor of two. Subsequently, Smith and Baker (1982) and Gordon *et al.* (1982) have shown that accuracies on the order of $\pm 30\%$ in chl are possible for Morel's Case-I waters (Morel and Prieur, 1977), that is, areas of little sediment or humic matter within the water column.

Within Case-II coastal waters, such as the Mid-Atlantic Bight, *in situ* calibration data (Gordon *et al.*, 1982), lead to pigment estimates from the CZCS-derived values of water-leaving radiances that are within 15% of the chlorophyll

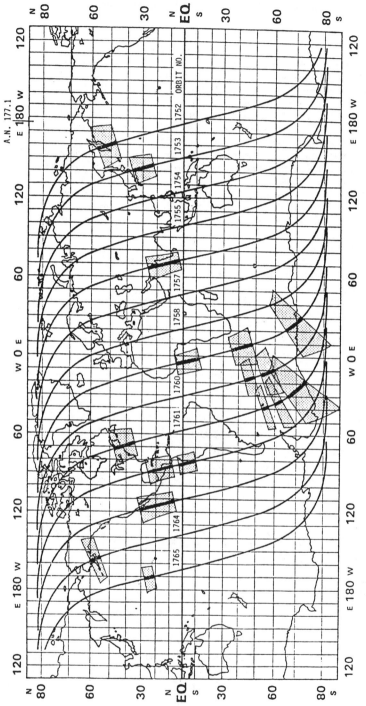

FIG. 70. The availability of CZCS data from relatively cloud-free regions (shaded areas) during the 14 orbits of Earth by the *Nimbus-7* satellite on February 28, 1979.

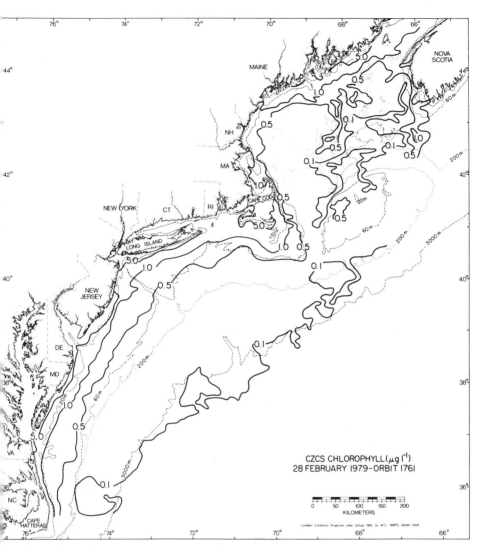

FIG. 71. The distribution of satellite-estimated chlorophyll over a grid of 4500 pixels, 5 nautical miles apart, within the Mid-Atlantic Bight on February 28, 1979.

measured aboard ship at the time of the satellite overpass. We are thus in a position to systematically exploit the rich CZCS data base, obtained somewhere on the world shelves (Fig. 70) each day over an 8-yr interval from 1978 to 1986. Fig. 71, showing the CZCS-derived (Systems and Applied Sciences Corp. [SASC], 1984) chlorophyll distribution during February 28, 1979, within the Mid-Atlantic Bight, is an example of the information available from one of the

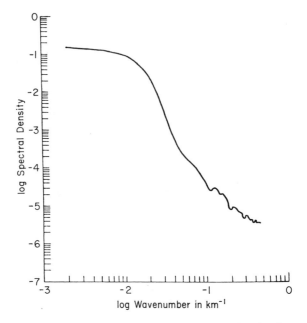

y-axis: log Spectral Density

x-axis: log Wavenumber in km⁻¹

FIG. 72. The isotropic variance spectrum of temperature over wavelengths of 1–1000 km from a June 14, 1979, overpass of the TIROS-N AVHRR sensor above slope waters of the Mid-Atlantic Bight (O. B. Brown, personal communication).

many satellite images taken daily during the *Nimbus*-7 orbits around the planet (Fig. 70). The chlorophyll isopleths of Fig. 71 were contoured from a grid of ~4500 data points between the 10- and 3000-m isobaths. Each grid point was 5 nautical miles (~9 km), apart and each chlorophyll value was thus the mean of ~100 observations since the pixel resolution of the CZCS is ~0.8 km, that is, an order of magnitude less than the possible critical length scale, L_c.

The cross-isobath and parabathic gradients of the synoptic CZCS-derived chlorophyll field (Fig. 71) are "typical" of shipboard composites of chlorophyll data taken at sea during the same season from 1975 to 1984 on this continental margin (Walsh *et al.,* 1978; O'Reilly and Busch, 1984; Walsh *et al.,* 1987a). On a cloudless planet, the relevance of the shipboard observations could simply be addressed by first assembling a time series of daily CZCS data. One could then compute the local rate of change of phytoplankton at 4500–450,000 grid points within successive 2-min images, and unravel the rate processes responsible for the amount of algal biomass left behind in the upper water column.

Remote sensing imagery from both the TIROS-N radiometer (AVHRR) and the *Nimbus*-7 coastal zone color scanner (CZCS) on June 14, 1979, in the vicinity of a warm core ring off the Mid-Atlantic Bight, for example, leads to two-dimensional variance spectra of temperature (Fig. 72) and pigment (Fig. 73).

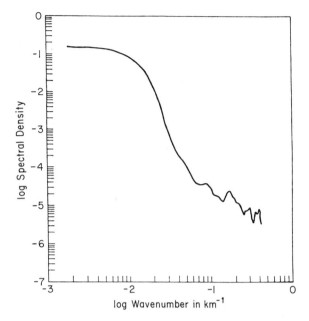

FIG. 73. The isotropic variance spectrum of chlorophyll over wavelengths of 1–1000 km from a June 14, 1979, overpass of the *Nimbus*-7 CZCS sensor above slope waters of the Mid-Atlantic Bight (O. B. Brown, personal communication).

These data sets, at the full 1-km resolution, could permit identification of time and space loci, where biological variability is dominated by physical processes and contrasting situations where biological interactions are dominant. The spectra both have a slope of approximately -2 up to about 9-km wavelength, at which the chlorophyll variance diverges towards a slope of -1, supporting a critical length scale L_c of ~10 km.

Phytoplankton standing stocks as indicated by CZCS estimates of pigment concentrations at a fixed time, within a particular spatial pattern, are the result of a complicated set of biological, chemical, and physical processes, with time scales ranging from seconds to seasonal, and space scales ranging from global to microscopic. The shapes and locations of chlorophyll patterns delineating the synoptic-scale shelf features are dominated by physical–dynamical processes, transporting those plankton populations, left in the water column as the net result of birth and death processes. Therefore, the synoptic-scale patterns ($100 < x < 1000$ km) tend to evolve over time scales, of order a few weeks to a month. The mesoscale spatial patterns ($10 < x < 100$ km), involving L_c, tend to evolve over time scales ranging from several hours of a phytoplankton cell division to a few days, however—that is, at the time scale of a wind event (Walsh, 1976)—and require resolution at a CZCS sampling protocol of at least a daily overflight.

3.3.2 Implications

Unfortunately, it has not been possible to obtain CZCS measurements of the global oceans on anything close to a daily basis. On any given day, a major fraction of our watery planet is obscured by clouds. Our experience to date suggests that present CZCS coverage would yield, on average, between 10 (at the equator) and 20 (at 40°N) usable images per month. The upper estimate represents a mean sampling interval of about every 1.5 days, for a given 1000 km × 1000 km boreal ocean domain, with the majority of usable data in patchy subscenes of typically a few hundred kilometers in extent, excepting an occasional clear view of most of the domain in one image. Coverage frequencies, however, fluctuate seasonally (and regionally) around these nominal estimates; coverage gaps of 2–3 weeks are likely to occur several times per year, with less frequent gaps of longer duration. In winter, low sun elevations will cause sampling voids of several weeks to a few months (increasing with latitude) at latitudes above 40°.

Of the nominal 2 hr of *Nimbus*-7 CZCS coverage taken and recorded each day, an average of approximately 30–40% is rejected and not processed due to total cloud cover (no significant open water areas). Furthermore, other data gaps for a particular site are derived from the inability of one satellite to provide sufficient daily overlap in swath width during the 14 orbits (Fig. 70). As a result, only 45 useful CZCS images were available for the Mid-Atlantic Bight between January 1 and June 30, 1979, for example; the 25% data recovery was also not grouped in equal time increments.

The results of eight CZCS images taken on March 18, 20, 21, and 23 (Fig. 75) and on April 10, 17, 19, and 21, 1979 (Fig. 78), are shown as representative time series of the types of data sets that are likely to be available from some continental shelves. These sampling constraints mean that the CZCS data sets violate stationary assumptions—that is, time-invariant probability density functions over the time domain of interest—inherent in most statistical approaches to time-series analyses of phytoplankton processes, such that subsets of the data do not have the same mean and variance. A simulation analysis of this data set in Section 5.1.3 does allow daily interpolation, however, among the available CZCS images.

3.3.3 Mid-Atlantic Time Series

As we begin to address questions about the amount, transport, and fate of the residues of coastal primary production, detected by CZCS time series within shelf (Smith *et al.*, 1982; Shannon *et al.*, 1984; Abbott and Zion, 1985) and adjacent slope waters (Holligan *et al.*, 1983; Brown *et al.*, 1985), we must specify where and what the CZCS radiometer senses. In terms of the relative amount of penetration of surface solar irradiance, the first attenuation depth, sensed by the CZCS, varies from ~1 to 10 m as a function of algal biomass and other suspended matter—recall Eqs. (114) and (114a). This depth only represents the 37% light level of the water column, while the bottom of the euphotic zone is 4.61 attenuation depths. At 50 stations during unstratified water column conditions of the mid-Atlantic spring bloom during April 1982, however, the near-surface chlorophyll concentrations, integrated to the first attenuation depth, had an r^2 of 0.86 with the chlorophyll integral of the whole euphotic zone (Brown et al., 1985). A set of 1979 time series of algal pigments, compiled with CZCS imagery in the Mid-Atlantic Bight, may thus reflect temporal and areal changes of phytoplankton biomass over most of the water column, rather than just vertical microstructure within the first attenuation depth.

A previous study of other regions in the central north Pacific, Sargasso Sea, Gulf of Mexico, Gulf of California, and Californian coastal waters yielded a similar r^2 of 0.91 for variation of chlorophyll biomass between the 100% and 37% light levels and the entire euphotic zone (Smith and Baker, 1978a). Within the Benguela Current, an r^2 of 0.98 was obtained for surface chlorophyll and the mean value of the euphotic zone (Shannon et al., 1984). Additional analysis of 850 vertical chlorophyll profiles from diverse Peruvian and Canadian waters (Platt and Herman, 1983) suggests that both the chlorophyll and associated primary production, potentially sensed by the CZCS, are "surprisingly constant proportions of the total chlorophyll and production integrals," respectively 2.9–5.2% and 6.1–10.7%.

Present CZCS algorithms do not distinguish between chlorophyll a and its degradation product, phaeophytin a, however, in derivation of the estimated phytoplankton pigment from ocean color (Gordon and Morel, 1983). The average phaeophytin concentration within 100 samples from a March 1979 cruise to the Mid-Atlantic Bight, however, was only 11% of the sum of phaeophytin and chlorophyll content as determined by acetone extraction in the fluorometric method (Yentsch and Menzel, 1963). Within the inherent error of the satellite estimate of pigment (Gordon et al., 1983), such an overestimate of chlorophyll a would be small. We thus compare the satellite-derived estimate of both "chlorophyll-like" and chlorophyll pigments with the shipboard estimates of chlorophyll a from 13 cruises in March–April 1979. Recall that the pigment estimate at each CZCS pixel of 825 m resolution was navigated and averaged over a 5-nm grid (SASC, 1984), such that each satellite data point in this discussion represents the mean of about 100 color measurements.

Surface chlorophyll observations taken along cruise tracks of the R/V Albatross IV, Edgerton, and Eastward during March 16–26, 1979 (Fig. 74a) in a traditional survey suggest separate areas of high chlorophyll, and presumably high primary production, within slope waters, at the shelf-break, and at mid-shelf of the Mid-Atlantic Bight (Fig. 74b). The >8 μg chl l^{-1} surface observa-

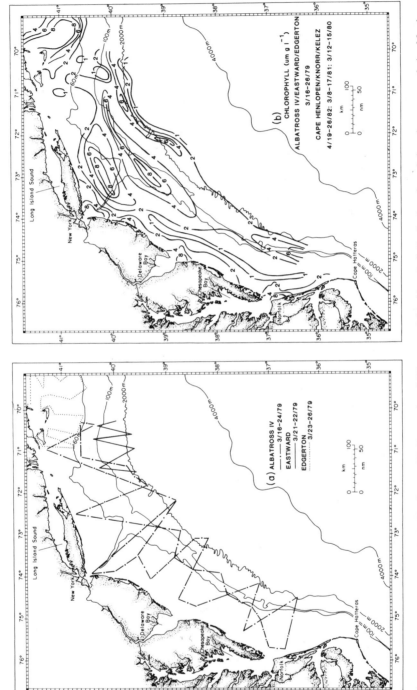

FIG. 74. (a) Cruise tracks of the R/V *Albatross IV*, *Eastward*, and *Edgerton* and (b) chlorophyll distributions from their underway maps during March 12–16, 1979, and from other cruises in March–April 1980–1982. [After Walsh *et al.* (1987a); © 1987 Pergamon Journals Ltd., with permission.]

tions (Fig. 74b) near the mouths of the Hudson, Delaware, and Chesapeake estuaries were obtained on other cruises during 1982. These latter data are presented for comparison with the CZCS observations at the mouth of the estuaries. They suggest that values of >8 μg chl l^{-1} are correct in each March image (Figs. 75a–d) and that turbid waters did not affect the accuracy of the CZCS algorithm (Viollier and Sturm, 1984) within the Mid-Atlantic Bight.

Based on previous shipboard data from the spring blooms of 1975–1977 (Walsh *et al.*, 1978), it was thought that such high chlorophyll concentrations at midshelf in 1979 (Fig. 74b) were a quasi-permanent feature of the spring bloom within this shelf ecosystem. Except for the estuarine plumes, the CZCS images for March 18, 1979 (Fig. 75a), and March 23, 1979 (Fig. 75d) instead indicated ubiquitous chlorophyll concentrations of 0.5–1.0 μg l^{-1} within shelf and slope waters. Such values were an order of magnitude less than other shipboard data collected in March 1975 and 1977 as well.

During March 20, 1979 (Fig. 75b), and March 21, 1979 (Fig. 75c), however, the CZCS images displayed the "expected" chlorophyll values of 8–10 μg chl l^{-1} at midshelf. Analysis of ship positions (Walsh *et al.*, 1987a) during these four CZCS overflights suggests that the CZCS estimates of chlorophyll variability are also correct in their depiction of an order of magnitude increase of chlorophyll over 1 day (March 20–21, 1979), and a similar decline over 2 days (March 21–23, 1979). These shipboard data thus appear to be badly aliased, even on a sampling time scale of 8 days.

At a measured daily growth rate of 0.4 day^{-1} in March 1979, that is, a population doubling time of $\sim$2 days—the phytoplankton increase of biomass from 0.5 to 8.0 μg chl l^{-1} at mid-shelf between the images of March 20 and 21 (Figs. 75b–c) cannot be due to just *in situ* primary production. Such an increment of pigment would have taken at least a week, with *no* consumption of the algal populations by herbivores or bacterioplankton. Resuspension of a chlorophyll concentration of $\sim$11 μg l^{-1} within the bottom 10 m of the water column (such as Figs. 76a and c) at the 30-m isobath and mixing of it with 0.5 μg chl l^{-1} within the upper 20 m would yield a mean concentration of 4 μg chl l^{-1}, after vertical homogenization by a wind event. A doubling of this phytoplankton abundance over 2 days would then yield a euphotic-zone biomass of $\sim$8 μg chl l^{-1}, similar to the maximum surface values detected by both the March CZCS images (Fig. 75) and the continuous vertical profiles (Fig. 76) taken by a fluorescence temperature depth (FTD) instrument on the *Albatross IV* cruise.

Zooplankton grazing rates in March are $\sim$10% of the daily primary production in the Mid-Atlantic Bight (Walsh *et al.*, 1978; Dagg and Turner, 1982; S. Smith, personal communication), such that the 10-fold decline of pigment seen by the CZCS between March 21 and 23 (Figs. 75c and d) cannot be attributed to consumption by herbivores. A comparison of vertical profiles of chlorophyll at the 50-m (Fig. 76a) and 60-m (Fig. 76c) isobaths on March 23 and 24, 1979,

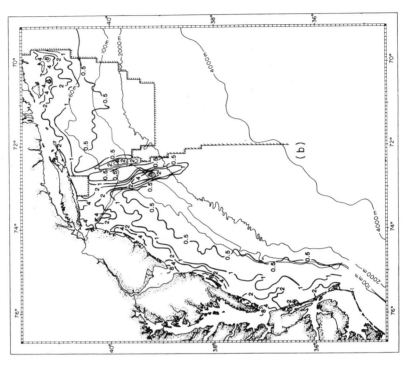

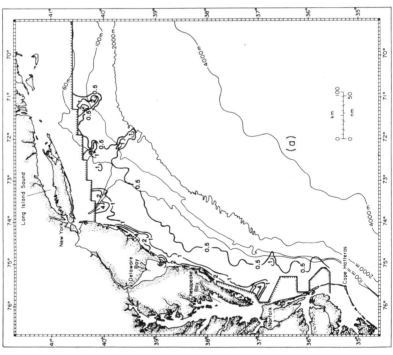

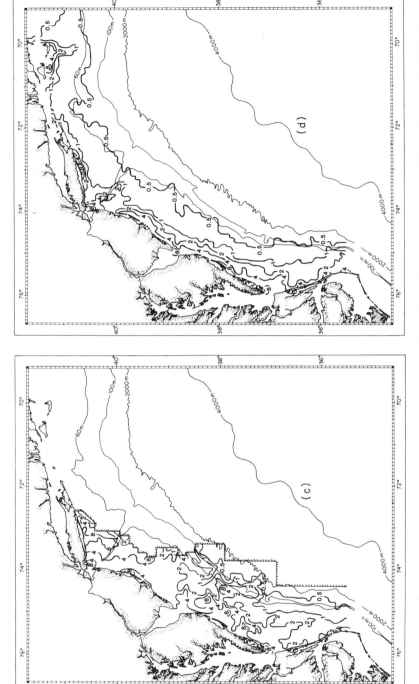

FIG. 75. The CZCS estimates of chlorophyll $\mu g\ l^{-1}$ during (a) March 18, 1979, (b) March 20, 1979, (c) March 21, 1979, and (d) March 23, 1979. [After Walsh *et al.* (1987a); © 1987 Pergamon Journals Ltd., with permission.]

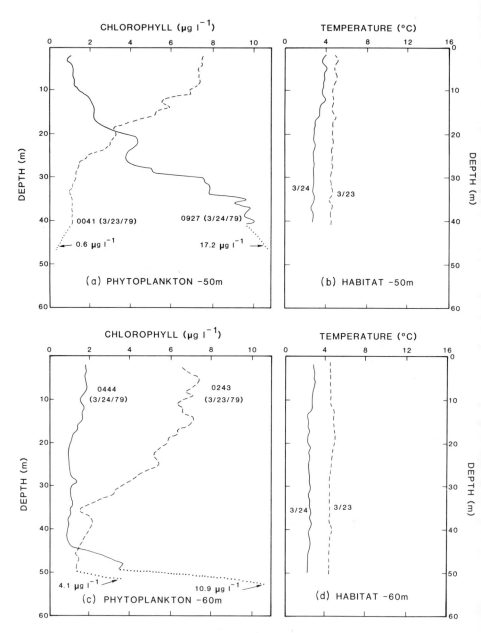

FIG. 76. The continuous vertical distribution of (a, c) chlorophyll and (b, d) temperature at the 50- to 60-m isobaths south of Long Island and Martha's Vineyard on March 23–24, 1979. [After Walsh *et al.*, 1987a); © 1987 Pergamon Journals Ltd., with permission.]

corroborate the surface decline of algal biomass measured by the CZCS. The water column was isothermal on both March 23 and 24 (Figs. 76b and c), yet surface maxima of chlorophyll may have become near-bottom maxima the following day, implying rapid sinking of diatoms and/or entrainment of the microalgae within downwelling water at a rate of at least 20 m day^{-1}.

Using a one-dimensional model (Niiler, 1975) of a wind-induced mixed layer, Wroblewski and Richman (1987) computed a vertical eddy coefficient K_z of 68 m^2 hr^{-1} over a 44-m-deep mixed layer of weak vertical stratification ($0.35\sigma_t/50$ m), after 8.5 hr of a 10-m sec^{-1} wind forcing. During February–May 1979–1982, a wind event ≥ 10 m sec^{-1} occurred about every 8 days in the Mid-Atlantic Bight (i.e., once per week), with 80% of these events originating from the northwest quadrant, 10% from the northeast, and the remainder from the south quadrants. At 27 shelf stations during March 16–24, 1979, the vertical density gradient was a mean $0.33\sigma_t/52$ m and the surface mixed layer appeared to extend at least 40–50 m down into the water column (Fig. 76). Over a 44-m surface mixed layer and an appropriate gradient of algal biomass, the equivalent vertical displacement rate, from such a K_z of 68 m^2 hr^{-1}, is 37.1 m day^{-1}, that is, the inferred downward transfer of chlorophyll in Fig. 76 is a feasible scenario.

Rapid resuspension, offshore transport, and sinking/downwelling events of phytoplankton can also be inferred from the CZCS time series in April 1979 (see Fig. 78). During April 17–19, 1979, 10 ships provided ground-truth chlorophyll measurements in the Mid-Atlantic Bight (Fig. 77a) as part of the LAMPEX experiment for calibration of aircraft and satellite overflights (Thomas, 1981). The high chlorophyll concentrations of the coastal zone (<20 m depth) and the low chlorophyll at the 60- to 100-m isobaths, southeast of Nantucket Island, measured aboard the ships on April 17–19 (Fig. 77b), matched quite well these parts of the two CZCS images on the same days (Figs. 78b and 78c). However, the major midshelf resuspension of near-bottom chlorophyll, observed south of Delaware Bay by the CZCS, on April 19 (Fig. 78c) and subsequent disappearance by April 21 (Fig. 78d) remained undetected (Fig. 77b) by the conventional shipboard surveys, since the research vessels were not present during this event.

Figure 78 represents various chlorophyll patterns within an upwelling circulation induced by northwesterly and southerly winds. The April 10, 1979 CZCS image of chlorophyll (Fig. 78a) exhibits, for example, a decline of algal biomass with distance offshore during mean northwest wind forcing (296°T) of 1.07 dyn cm^{-2} over April 5–12, 1979. Recall from Fig. 38 that the depth-averaged currents of the circulation model for this case were ≤ 5 cm sec^{-1}, with offshore flow between the 20- and 40-m isobaths and southwesterly flow between the 40- and 60-m isobaths at midshelf, south of the Hudson Canyon and north of Norfolk. In response to such northwest wind events, surface waters are pushed offshore, and the predominantly westward alongshore flow is slowed down (Beardsley and Butman, 1974).

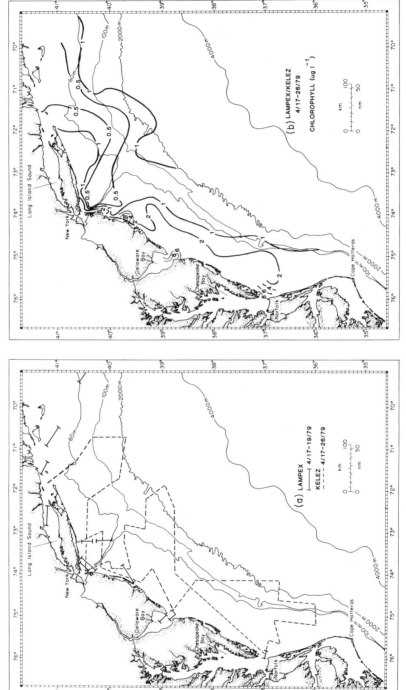

FIG. 77. (a) Cruise tracks of 10 research vessels during the LAMPEX study and (b) the chlorophyll distributions from their underway maps during April 17–26, 1979. [After Walsh *et al.* (1987a); © 1987 Pergamon Journals Ltd., with permission.]

During April 17, 1979, tongues or streamers of $1-2$ μg chl l^{-1} extended within the CZCS image (Fig. 78b) from the shelf to slope waters in the same areas of the shelf, south of Long Island, south of New Jersey, off Delaware Bay, and off Norfolk, as observed in the March CZCS time series (Fig. 75); recall the April 1984 aircraft time series (Fig. 69). In contrast to these April observations of ~ 1 μg chl l^{-1} at midshelf off Virginia, however, 2 days later as much as 16 μg chl l^{-1} was apparently detected by the CZCS on April 19, 1979. The region of high chlorophyll (≥ 8 μg chl l^{-1}) extended from off Norfolk to Cape Hatteras, 150 km south along the 40- to 60-m isobaths (Fig. 78c). At the same population growth rate of one doubling every 2 days, such an increase of algal biomass cannot be attributed solely to *in situ* growth of phytoplankton.

Resuspension of near-bottom phytoplankton is the most likely source. An accumulated, near-bottom chlorophyll concentration of $15-30$ μg chl l^{-1} within the lower 10 m of the water column at the 40-m isobath, and 0.5 μg chl l^{-1} within the upper 30 m before a resuspension event, would yield $4.13-7.88$ μg chl l^{-1} after vertical homogenization in response to such a sequence of upwelling wind events. A doubling of such phytoplankton populations after two days would then yield the extensive CZCS chlorophyll concentrations of $8-16$ μg chl l^{-1} seen on April 19, 1979 (Fig. 78c).

Within this midshelf region, the chlorophyll concentrations again declined by an order of magnitude within 2 days, that is, by April 21, 1979 (Fig. 78d). Another 10-m sec^{-1} wind event from the south on April 20, 1979 would have imparted kinetic energy to a surface mixed layer as from the northeast event on April 18, 1979, but the vertical chlorophyll gradient had been reversed by then. With more chlorophyll biomass in surface offshore waters on April 20, 1979, a downward flux of algal biomass may have occurred, as inferred for March 23, 1979 (Fig. 76). The vertical gradient of density had increased to a mean of $1.02\sigma_t/56.3$ m at 41 shelf stations by May $1-6$, 1979, however, such that K_z might have been only 23 m^2 hr^{-1} over a mixed layer depth of 18 m on April 20, 1979. An equivalent downward displacement rate of only 12 m day^{-1} would be sufficient to remove most of the algal biomass below the detection depth (<10 m) of the CZCS sensor in 1 day.

Dilution of chlorophyll concentrations of the euphotic zone has been observed after wind events in other coastal ecosystems as well. During an upwelling event off California, CZCS imagery similarly detected a 10-fold decline in algal biomass over 2 days, within 20 km of this coast (Abbott and Zion, 1985). After an upwelling event off southwest Africa, a CZCS time series suggested at least a threefold decline in algal biomass within 4 days as well (Shannon et al., 1984). After March–April wind transport events off Peru, the integrated chlorophyll biomass over the upper 40 m of the water column also decreased by $10-25\%$ in 1976 and $50-75\%$ in 1977 (Walsh et al., 1980). Since diatoms are nonmotile, the natural sinking rate of these organisms would add to their downward flux of

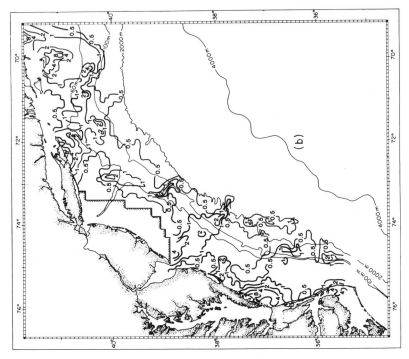

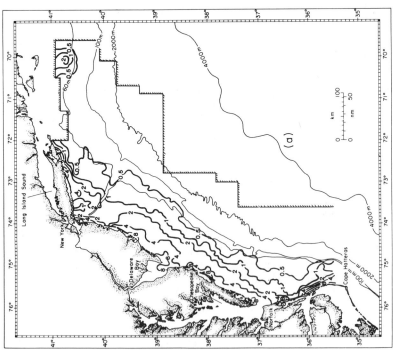

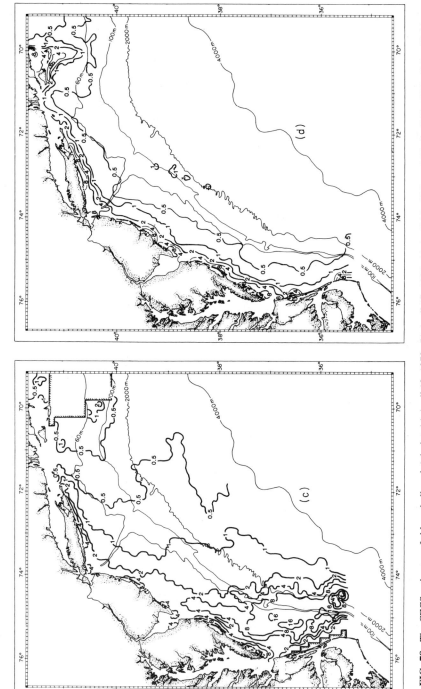

FIG. 78. The CZCS estimates of chlorophyll $\mu g\ l^{-1}$ during (a) April 10, 1979, (b) April 17, 1979, (c) April 19, 1979, and (d) April 21, 1979. [After Walsh *et al.* (1987a); © 1987 Pergamon Journals Ltd., with permission.]

biomass in these situations of dilution events, where more chlorophyll is initially found at the surface than near bottom.

The seaward tongues of chlorophyll within the aircraft and satellite data sets suggest that inferences might also be drawn from remote sensing about important scales of horizontal exchange processes in the Mid-Atlantic Bight. For example, twice as much near-surface chlorophyll was found during the April 19, 1979, resuspension event (Fig. 78c) above the 40- to 60-m isobaths, compared to the previous March 21, 1979, event above the 20- to 40-m isobaths (Fig. 75c). One possible explanation for this temporal sequence is a seasonal buildup of chlorophyll within the aphotic zone and a gradual transfer of uneaten phytodetritus seaward. A 30- to 40-km offshore migration of the algal resuspension area, from the 20- to 40-m isobaths to the 40- to 60-m isobaths, within $\sim$30 days from March 21 to April 19, 1979, suggests a mean net seaward movement of algal particles of $1.0–1.3$ km day^{-1} ($\sim 1.0–1.3$ cm sec^{-1}).

Much of the wind forcing is from the north in the Mid-Atlantic Bight during February–April, with more frequent northwest storms from off the North American continent. Not surprisingly, the average net flow of the upper 30 m of the water column past four current meter arrays, between the 45- to 105-m isobaths off Martha's Vineyard at the same time of these CZCS overflights, was 7.73 cm sec^{-1} to the west and 1.43 cm sec^{-1} offshore during April 1979, as well as in February–April 1980 (Beardsley *et al.*, 1983). This seaward movement is in agreement with the CZCS estimate of offshore advection of phytoplankton populations.

To evaluate the possible export of phytoplankton carbon to the adjacent continental slope during this time of the year, one must be able to determine a "representative" chlorophyll value, which reflects both the daily increment from *in situ* growth and the physical resuspension of uneaten algal residues, for computation of the associated flux of phytoplankton en route to the slope. Assuming an average primary production of 0.73 g C m^{-2} day^{-1}, a specific growth rate of 0.41 day^{-1}, a C/chl ratio of 45 : 1, and a homogeneous distribution of chlorophyll within a 27.2 m euphotic zone, one can calculate (Walsh *et al.*, 1987a) a 1- to 2-month mean chlorophyll for these shelf waters of 1.44 μg chl l^{-1}. Within the shelf resuspension areas during March (Fig. 75) and April (Fig. 78), the CZCS estimates of chlorophyll were as much as 10-fold higher than 1.44 μg chl l^{-1}. These high CZCS estimates were observed only within near-surface waters for 2–3 days, however, so that a mean value of 1.44 μg chl l^{-1} may be a good lower bound of the phytoplankton biomass available for export to the slope.

An offshore flow of $\sim$1.5 cm sec^{-1} during spring and a CZCS estimate of 1.44 μg chl l^{-1} within the first attenuation depth, on the landward side of the shelf break between Cape Hatteras and Martha's Vineyard, imply a net seaward transport of 2.1 ng chl cm^{-2} sec^{-1} from the Mid-Atlantic Bight. Assuming a mean C/chl ratio of 45 : 1 (Malone *et al.*, 1983), a carbon export from the spring bloom of 0.8 $\times$ 10^2 g C m^{-2} day^{-1} might occur across the shelf break. The tacit assumption in this calculation, however, is that a phytoplankton cell exits the continental shelf only at the surface of the water column.

With an alongshore flow of 8.0 cm sec^{-1} and an offshore flow of 1.5 cm sec^{-1}, an ungrazed algal cell would take 66 days to cross the 100-km-wide shelf in a 528 km along-shore trajectory, that is, half the length of the Mid-Atlantic shelf. During these 2 months a diatom frustule would certainly experience a number of downwelling or sinking events, with perhaps more chlorophyll exiting the shelf near the bottom than at the surface of the water column. Consideration of the fate of phytoplankton within the aphotic zone leads us to the subject of *in situ* time series.

3.4   Time-Dependent Chlorophyll Fields

Some of the most useful measurements of biological variables are chemical analyses in which the organisms are ground up, rather than counted. An early estimate of chlorophyll in terms of photosynthetic pigment units (PPU) within the English Channel allowed one of the first complete descriptions of the annual production cycle in 1933 (Harvey *et al.*, 1935). An absorption spectrophotometry technique developed at the same time (Zscheile, 1934) was eventually used to characterize functional groups of phytoplankton by their pigment content (Richards, 1952; Richards and Thompson, 1952), but this method required as much as 4–6 l of seawater at chlorophyll concentrations <1 μg l^{-1}. Determination of extinction values at several wavelengths was also a slow procedure and required critical alignment of the cuvettes in the light beam of the spectrophotometer.

Not all of the light energy absorbed by phytoplankton pigments during photosynthesis may be passed to electron acceptors within the cell, but some may be emitted as fluorescence, usually at longer wavelengths as pointed out by G. G. Stokes in 1852. Following Kalle's (1951) suggestion 100 yr later that chlorophyll concentrations in seawater could be quantitatively determined by excitation fluorometry, routine measurements of *in vitro* fluorescence were obtained at sea in the 1960s (Yentsch and Menzel, 1963; Holm-Hansen *et al.*, 1965). Filtered chlorophyll samples, from 1 l or less of seawater, were ground, extracted with 85–90% acetone solution, and read in Turner fluorometers, with an excitation wavelength of 430–450 nm and a maximum emission spectra of 650–675 nm. A major modification of this technique was the continuous estimate of chlorophyll concentration by *in vivo* fluorescence (Lorenzen, 1966), which led to both the underway, surface chlorophyll data from research vessels

(Figs. 59 and 62) and the development of undulating batfish, vertical FTD probes, and moored fluorometers.

3.4.1   Moored Fluorometers

To overcome the traditional sampling inadequacy of biological field measurements, which limited further testing of simulation models, a moored fluorometer program was initiated in 1976. The field efforts involved upper-ocean measurements of currents, temperature, conductivity (Walsh et al., 1978), and fluorescence (Whitledge and Wirick, 1983) in the nearshore shallow region of the continental shelf off Long Island. The original buoys (Fig. 79) were constructed from sections of "Shelton" spar with patented J-lock interconnections to minimize rotation movements. Each instrument package was mounted on a buoy in protective cages. The sensors used on each mooring for current measurements were Marsh–McBirney electromagnetic orthogonal meters to minimize wave contamination within the euphotic zone. Temperature and conductivity were also routinely measured (Fig. 80). Power was supplied by 16 lead acid battery cells, with a nominal 480 amp-hr capacity, packaged in a free-flooding housing partially filled with mineral oil. The battery package weighed nearly 500 kg in seawater and acted as the anchor for each moored array.

The first *in situ* fluorometer instrument package consisted of a model 10-005 RU Turner Designs fluorometer with a high-pressure cuvette. The fluorometer was operated at $12-16$ V supplied by the same battery package, with an excitation wavelength of 470 nm and the emitted fluorescence read at 665 nm. Water was circulated through the cuvette with a magnetically coupled pump on a drive motor, having a flow rate of 600 ml min^{-1} and a hold-up volume in the piping system of approximately 200 ml. Later modifications to conserve battery power decreased the pumping rate to 180 ml min^{-1} and a hold-up volume of about 65 ml. The cuvette was cleaned with a bottle brush and examined for fouling before and after deployment. Fouling of the cuvette window was minimized by the placement of tin butoxide antifouling agent in a section of the inlet hose. The latex tubing inlet hose was soaked in a 2% tin butoxide–toluene solution for $20-30$ min immediately before deployment. No material was found on the walls of the cuvette after it had been submerged for as much as 6 weeks. The intake of the latex tubing was connected to a bilge-pump strainer, which prevented blockage of the pump by seaweed or other flotsam.

This buoy system was used to examine the biological response to wind-driven upwelling and downwelling circulation events during the weakly stratified conditions of the early fall and late spring blooms off the Long Island coast at a water depth of 28 m. Moored *in situ* fluorometers were placed at 10.5 m depth for 11 days in September–October 1977 (Fig. 67) and at 8.6 m and 16.2 m for 14 days in April–May 1979. Nearby ($\sim$30 m), other spar buoys contained the

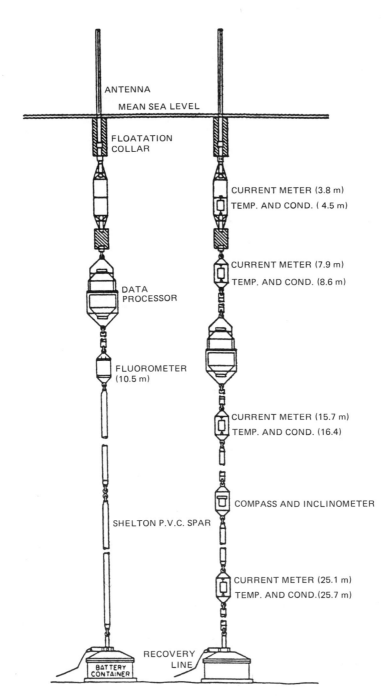

FIG. 79. Initial mooring design for an array of current meters, thermistors, conductivity and fluorescence sensors, involving line-of-sight telemetry from shallow waters off New York. [After Whitledge and Wirick (1983); © 1983 Pergamon Journals Ltd., with permission.]

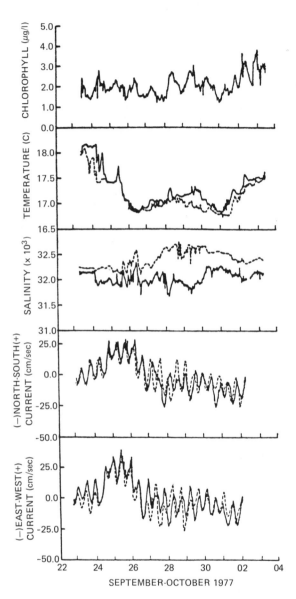

FIG. 80. A time series of chlorophyll at 10.5 m and of temperature, salinity, and current components at 8.0 and 16.0 m on the 28-m isobath, south of Long Island, from September 22 to October 4, 1977. [After Whitledge and Wirick (1983); © 1983 Pergamon Journals Ltd., with permission.]

Marsh–McBirney electromagnetic current meters located at 3.8, 7.9, 15.7, and 25.1 m depth; thermistors at 1.4, 4.5, 8.6, 15.5, 16.4, 20.4, 22.6, and 25.7 m; and conductivity sensors at 4.5, 8.6, 16.4, and 25.7 m. Fluorescence data were averaged by the data processor on the buoy, logged every 20 min, and transmitted ashore by line-of-sight telemetry during the fall deployment and every hour during the spring deployment.

A continuous monitoring of chlorophyll concentration at the depth of the fall moored in situ fluorometer (10.5 m) was undertaken, using both an FTD made from another Turner Designs fluorometer, and an attached submersible pump from which chlorophyll calibration samples were collected every 8–10 min for 4 hr, and then every 4 hr for 2 days. Data from the FTD were logged every minute for the 4-hr period. The resulting calibration of the in vivo FTD fluorescence using the in vitro pump samples had an r^2 of 0.98. Comparison of the FTD fluorescence data obtained from the ship with that from the fall-moored fluorometer showed a good correspondence with an r^2 of 0.97 for the calculated chlorophyll values.

The estimation of chlorophyll a concentration in the ocean from in vivo fluorescence requires shipboard calibration, of course, each time the fluorometers are deployed and recovered. Bench testing of the fluorometers against a laboratory fluorometer for both in vivo and in situ determinations ensures that the instruments are operating properly, but really does not produce a fully calibrated instrument. Experience from ~50 deployments shows that in situ calibration with (a) a profiling fluorometer, (b) discrete samples from a rosette attached to the profiling fluorometer for both in vivo and in situ analysis, and (c) in situ pump samples for shipboard in vivo and in situ determinations are much preferred to bench testing. All three methods have shortcomings, but the phytoplankton cells used are representative of the ambient population in terms of light and nutritional history, which is not the case with bench calibration procedures. Once these calibrations are made, the accuracy of the in situ fluorescence estimate of chlorophyll is within the range of half or double the true value (Whitledge and Wirick, 1983), while the precision of the chlorophyll estimate is ± 0.05 μg l^{-1} from fluorometers at high sensitivity (Whitledge and Wirick, 1986).

A dominant feature of the fall (Fig. 68) time series was a diel periodicity of fluorescence at time scales similar to tidal motion and daily sea breezes. The high-frequency variance of the chlorophyll record appeared to have a cyclic behavior that was not entirely determined by tidal motion, however, reflecting probably both diel changes in chlorophyll synthesis (Owens et al., 1980) and grazing stress (Whitledge and Wirick, 1983). The lower frequency variance of these records (such as Fig. 80) was clearly associated with downwelling and upwelling wind events. Wind forcing from the east and a downwelling circulation pattern (Beardsley *et al.*, 1976; Scott and Csanady, 1976; Walsh *et al.*, 1978) were observed in both the fall and spring time series; note, for example, the intense southward (offshore) and westward (alongshore) flow during September 24–25, 1977, at 7–16 m (Fig. 80) within Ekman layers of onshore flow at the surface and offshore at depth. With reversal of the winds by September 26, 1977, an upwelling flow regime occurred, with increasing salinity and declining temperature at 16 m until October 1–2, 1977, when downwelling winds again briefly prevailed. The downwelling circulation mode on this shelf is particularly important for the export of near-bottom particles toward the slope, as discussed in Section 5.1.3.

After these successful nearshore experiments, the design of the moored fluorometers was modified to allow deployment of the instruments at any bottom depth. The fluorometer package was converted to a self-contained unit, similar to conventional current meters. When mounted in its protective cage, the fluorometer's dimensions were 1.2 × 0.3 × 0.3 m, which could be accommodated within the standard taut wire moorings of current meters. A pack of "D"-cell batteries powered the instrument and its sampling life (~2–3 months) was, of course, determined by the sampling and cell cleaning frequency. Each instrument had a four-channel data acquisition system with a crystal time base. The analog channels were burst sampled at fixed intervals in time, with the digital values recorded on magnetic tape rather than transmitted ashore by telemetry. Each tape had a capacity of 50,000 16-bit words. Two data channels were used for the fluorescence sensor, with the others committed to temperature and transmissometer sensors.

Instead of a Turner Designs fluorometer, the modified instrument package used a fluorescence sensor similar to that developed by Aiken (1981). The excitation light was produced by a xenon flash tube, with a photodiode used to detect the emitted fluorescence. The energy in each xenon flash was also detected and recorded for reference. The sample cell of the second moored fluorometer was an open-ended tube, which was periodically scrubbed with a brush to prevent fouling. The temperature sensor was either a YSI or a Fenwal thermistor, while light transmission was measured by a Sea Tech transmissometer, with a path length of 25 cm, to estimate total particles for comparison with the fluorescence records.

As a result of nutrient discharge within Port Jefferson harbor on the north shore of Long Island, a large horizontal chlorophyll gradient extends between the harbor and adjacent Long Island Sound. During the test mooring of the modified second fluorometer system, at the mouth of the harbor in a water depth of 10 m from July 9 to 15, 1982, tidal excursions of the high-chlorophyll, warm-temperature harbor water past the moored fluorometer were quite evident (Fig. 81). In this experiment, a burst of 9 xenon flashes occurred every 8 min to

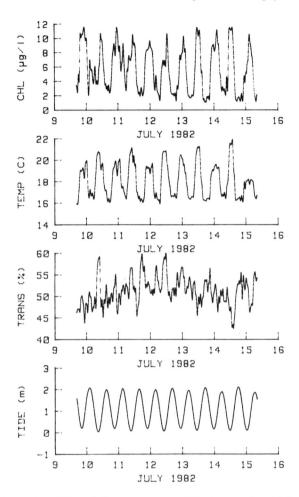

FIG. 81. A time series of chlorophyll, temperature, and light transmission at 7.0 m, in relation to the local semidiurnal tide, on the 10-m isobath in Long Island Sound, from July 9 to 15, 1982. [After Whitledge and Wirick (1986); © 1986 Springer-Verlag, with permission.]

generate 1320 data sets of *in vivo* fluorescence, with calibration samples taken during deployment and recovery of the mooring.

The data were low-pass filtered (frequency centered at 1.0 cycles hr^{-1}) to yield this brief 1982 time series, clearly depicting (Fig. 81) the ability to resolve high- and low-chlorophyll features in the sea with a second suite of *in situ* sensors. No evidence of fouling was observed during this experiment, and the new system was then deployed in a series of Eulerian studies in the Mid-Atlantic

Bight, the South Atlantic Bight, and the Bering Sea during 1984 and 1985. A total of ~2000 days of fluorometer records was obtained in conjunction with temperature, currents, and light-transmission data, of which a small subset is now described.

3.4.2 Implications

The downstream coherence, or similarity, between chlorophyll observations at one mooring to the next will depend on the strength of the flow field as well as the birth and death processes of the phytoplankton population. Analogous to the previous discussion about critical length scales at which the growth of micro-algal populations is faster than the losses of physical dispersion, sinking, and grazing, a set of Eulerian observations of chlorophyll fluctuations in time will exhibit coherent lags as a function of their separation distance. If the downstream flow is weak, the memory of algal fluctuations of the upstream mooring transmitted by the flow field—that is, once again equating distance with time—will propagate only a short distance, before biological processes interact to smear physical dislocations in response to an altered habitat, perhaps a wind event. Analysis of time-series records of temperature, chlorophyll, light transmission, and currents will provide an estimate of their coherence in time, but complete understanding of these records requires additional spatial consideration of the birth and death processes as well, that is, simulation models as presented below and in Section 5.1.3.

A measure of the coherence of a variable is the cross-correlation coefficient r_l, at different time lags l given by

$$r_l = \frac{n\sum\limits_{i=1}^{n} X_i(t)Y_i(t+l) - \left[\sum\limits_{i=1}^{n} X_i(t)\right]\left[\sum\limits_{i=1}^{n} Y_i(t+l)\right]}{\left[\left\{n\sum\limits_{i=1}^{n} X_i^2(t) - \left[\sum\limits_{i=1}^{n} X_i(t)\right]^2\right\}\left\{n\sum\limits_{i=1}^{n} Y_i^2(t+l) - \left[\sum\limits_{i=1}^{n} Y_i(t+l)\right]^2\right\}\right]^{-1/2}}$$

(189)

where X_i and Y_i are observations at two separate moorings during the time interval i. If the time lag l were 0, r_l would be the simple correlation coefficient between two variables, such that Eq. (189) is essentially sliding one data set against the other at different time lags. If $X_i = Y_i$ (i.e., at the same buoy), then Eq. (189) expresses the autocorrelation coefficient of a time series, sliding a data set against itself, and can be used to calculate the Fourier transform (Blackman and Tukey, 1958). The maximum value of an autocorrelation coefficient, 1.0, is found at zero time lag (see Fig. 88), but the maximum values of cross-correlation coefficients are usually at some finite lag (see Figs. 84 and 85). Furthermore, although Eq. (189) only provides information in the time domain, Fourier transforms can be made of the cross-covariance, leading to coherence spectra and phase spectra, which are analogous to the variance spectra discussed previously.

3.4.3 Bering Sea Time Series

Visual inspection of time series of fluorescence (Fig. 82) and temperature (Fig. 83) at 19–22 m depths from three moorings on the 44- to 49-m isobaths, be-

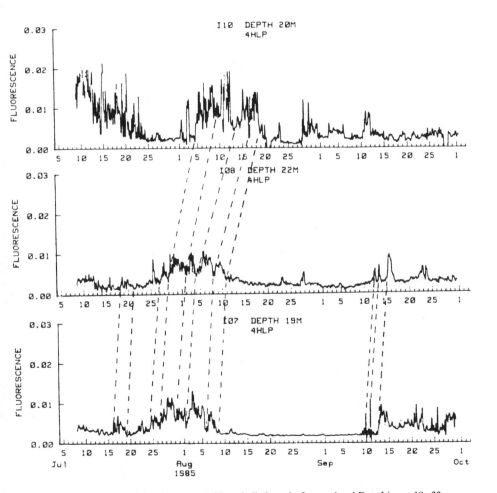

FIG. 82. Time series of 4-hr low-passed chlorophyll along the International Date Line at 19–22 m on the 44- to 49-m isobaths within Anadyr Strait (I07), Chirikov Basin (I08), and Bering Strait (I10) in the northern Bering Sea from July 7 to October 1, 1985.

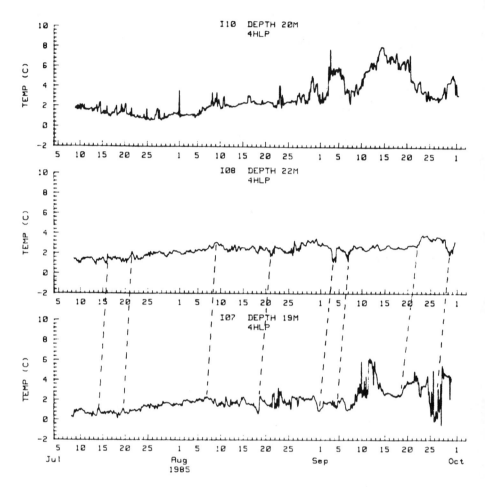

FIG. 83. Time series of 4-hr low-passed temperature along the International Date Line at 19–22 m on the 44- to 49-m isobaths within Anadyr Strait (I07), Chirikov Basin (I08), and Bering Strait (I10) in the northern Bering Sea from July 7 to October 1, 1985.

tween Anadyr and Bering Straits in the northern Bering Sea (see Fig. 86), suggests that some "events" of low chlorophyll and high temperature may have propagated downstream during July 6 to October 2, 1985 (i.e., the dashed lines). The mean flow over this time period at the Anadyr Strait mooring (I07) was northeast at 35 cm sec^{-1}, while at the intermediate mooring (I08) located 85 km downstream it was also northeast at 12 cm sec^{-1}, and at the Bering Strait mooring (I10) 144 km farther north, the flow was due north at 33 cm sec^{-1}, similar at each site to the steady-state flow field of the 0.6-Sv case in Fig. 34. Using the "upstream" velocities, a water parcel could take 67 hr (2.8 days) to travel from

Anadyr Strait to I08, and 333 hr (13.9 days) from I08 to Bering Strait, a record of which might be discernible in the time-series data of its constituents.

The highest cross correlations of 0.3–0.5 for both fluorescence (Fig. 84) and temperature (Fig. 85), between the moorings in Anadyr Strait and at I08, are at a time lag of 48 hr, reasonably close to the hypothetical travel time of a water parcel. The cross-correlation of temperature (0.45) at this lag is somewhat higher than that of fluorescence (0.33), with a small decline in algal biomass measured between Anadyr Strait and I08 (Fig. 82). Either no growth of the phytoplankton occurred during these 2 days of transit—that is, the algae were behaving as a conservative tracer like temperature—or the losses of sinking and grazing were somewhat larger than the photosynthetic gain of biomass. With algal population growth rates of ~0.6–1.0 day^{-1} observed during 1983–1984 cruises between St. Lawrence Island and Bering Strait, however, an assumption of no *in situ* growth is unrealistic. A simple steady-state model (Walsh and Dieterle, 1986) of phytoplankton growth, limited by nitrate availability, and of all losses parameterized as a sinking flux, was thus coupled to the physical submodel of Section 2.4.3 for initial analysis of these moored fluorometer data.

Upon solution of Eqs. (70)–(72), the depth-averaged, barotropic velocities u and v were entered at each grid point in the state equations for nitrate (N) in units of μg-at. NO$_3$ l^{-1}, and phytoplankton (M) in units of μg chl l^{-1},

$$\frac{\partial N}{\partial t} = \frac{\partial}{\partial x}\left(K_x \frac{\partial N}{\partial x}\right) + \frac{\partial}{\partial y}\left(K_y \frac{\partial N}{\partial y}\right) - \frac{\partial(uN)}{\partial x} - \frac{\partial(vN)}{\partial y} - \varepsilon M \quad (190)$$

$$\frac{\partial M}{\partial t} = \frac{\partial}{\partial x}\left(K_x \frac{\partial M}{\partial x}\right) + \frac{\partial}{\partial y}\left(K_y \frac{\partial M}{\partial y}\right) - \frac{\partial(uM)}{\partial x} - \frac{\partial(vM)}{\partial y} + 2\varepsilon M \quad (191)$$

$$- w_s H^{-1} M$$

where K_x and K_y were the implicit numerical diffusion coefficients derived from the finite-difference approximation of Eqs. (190) and (191), and where ε was the growth rate (hr^{-1}) of the phytoplankton, expressed by

$$\varepsilon = d(1.43 \sin 0.2618t)\left(\frac{N}{n + N}\right) \quad (192)$$

in which $d = 0.025$ hr^{-1} to allow a maximum daily growth rate of 0.43 day^{-1} over 12 hr of daylight, while the sine function was set to zero at night (i.e., $\varepsilon = 0$) then as well; and [$N/(n + N)$] was a Michaelis–Menton expression for nitrate limitation of the algal growth, with the half-saturation constant n taken to be 1.5 μg-at. NO$_3$ l^{-1}.

The other parameters of Eq. (191) were an assumed conversion ratio of chlorophyll/particulate nitrogen of 2 within the phytoplankton and w_s, their sinking

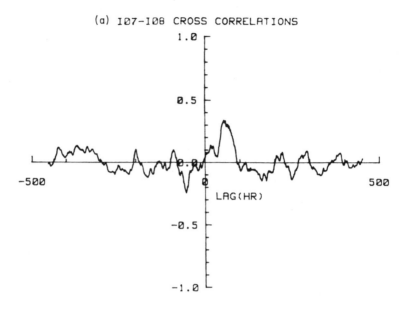

(a) I07-I08 CROSS CORRELATIONS

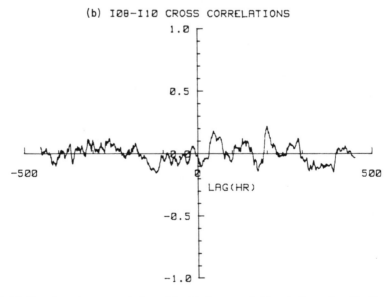

(b) IO8-I1O CROSS CORRELATIONS

FIG. 84. Time-lagged cross-correlations of the 2-hr low-passed chlorophyll records at 19–22 m on moorings (a) between Anadyr Strait (I07) and Chirikov Basin (I08), and (b) between Chirikov Basin (I08) and Bering Strait (I10) from July 10 to August 15, 1985.

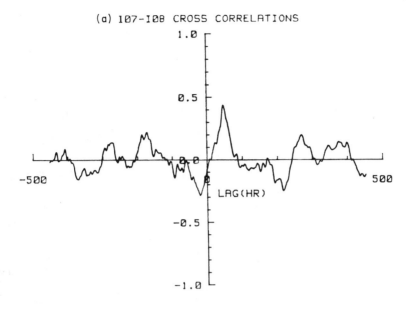

(a) I07-I08 CROSS CORRELATIONS

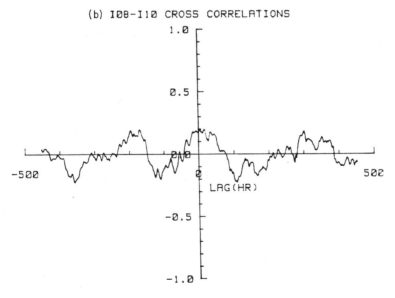

(b) I08-I10 CROSS CORRELATIONS

FIG. 85. Time-lagged cross-correlations of the 2-hr low-passed temperature records at 19–22 m on moorings (a) between Anadyr Strait (I07) and Chirikov Basin (I08), and (b) between Chirikov Basin (I08) and Bering Strait (I10) from July 10 to August 15, 1985.

velocity, of 10 m day^{-1}. Since there was no term in Eqs. (190)–(192) for light limitation of phytoplankton growth with depth, M and N were not really averages over the water column, but can be considered representative of nitrate and chlorophyll concentrations within the first attenuation depth of the water column (i.e., 1–10 m), where photosynthesis is maximal (Platt and Herman, 1983). The algal sinking rate w_s of Eq. (191) was divided by the local depth H such that the units of this loss term (μg chl l^{-1} hr^{-1}) were the same as those of the growth and advective terms. The implicit vertical balance of chlorophyll fluxes was thus taken to be the influx at the surface, which was zero, and the outflux at the bottom of the water column to the sediments, where M near the bottom was the same as M near the surface, that is, a homogeneous distribution, $\partial M/\partial z = 0$. The selection of $w_s = 10$ m day^{-1}, instead of sinking rates of $\sim$1 m day^{-1} measured in the laboratory, will be discussed in Section 5.1.

The upstream boundary conditions of nitrate and chlorophyll across Anadyr and Shpanberg Straits were taken from previous cruises in 1969–1984, during which similar longitudinal gradients of both variables were found, decreasing from west to east. The assumed values of nitrate were 10 μg-at. NO$_3$ l^{-1} in the middle of Anadyr Strait and 1 μg-at. NO$_3$ l^{-1} near the middle of Shpanberg Strait. At these boundaries, values of chlorophyll were 6 μg chl l^{-1} in the middle of Anadyr Strait and 1 μg chl l^{-1} in Shpanberg Strait. At the air and land boundaries, M and N were both taken to be zero (i.e., no influx of nutrients from either rainfall or the Yukon River). The initial conditions of nitrate and chlorophyll within the interior of the grid (Fig. 30) were $N = M = 0$ in simulation cases where growth and sinking were not considered. The initial conditions were $N = (N + M)/2$ from the previous solutions of the advective/dispersive cases, while M was set to 0.5 μg chl l^{-1}, when other cases of growth and sinking were subsequently studied. Finally, at the downstream boundary, conditions of M and N between Capes Dezhneva and Hope were not required, since Eqs. (190)–(191) were solved by upstream finite differences, which introduced the numerical diffusion implicitly represented by K_x and K_y.

A number of cases were run in the model, of course, but only the phytoplankton results at the 0.43-day^{-1} growth rate and the 10 m day^{-1} sinking rate are presented here. At an average flow of 15–20 cm sec^{-1} within the 0.6-Sv case of the model (Fig. 34), a phytoplankton cell would take $\sim$10–14 days to be moved from Anadyr Strait to about 200 km downstream towards Bering Strait, where a nitrate minimum was found after 14 days of simulated time (Walsh and Dieterle, 1986). Within the first 4 days of such a phytoplankton trajectory, the chlorophyll biomass derived from primary production south of Anadyr Strait would have sunk out of a 40-m water column at a sinking rate of 10 m day^{-1}. This might occur at 66 km downstream, that is, near the I08 fluorometer mooring where the model predicts a chlorophyll concentration of 4 μg chl l^{-1} (Fig. 86), compared to the 4.5-μg chl l^{-1} peaks observed in the time series (Fig. 82). Removal of the previously grown 6 μg chl l^{-1} of algal biomass from the boundary condition

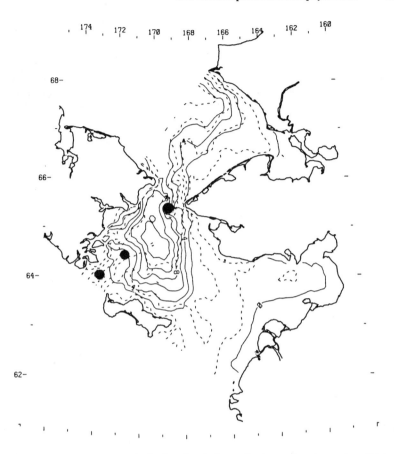

FIG. 86. Simulated chlorophyll distributions in relation to the time-series observations (●) in the Bering Sea after 14 days in a steady flow field of 0.6 Sv, with a growth rate of 0.43 day⁻¹ and a sinking rate of 10 m day⁻¹. [After Walsh and Dieterle (1986); © Elsevier Science Publishers B.V., with permission.]

would then deepen the euphotic zone, making available more light in the water column for initiation of primary production north of St. Lawrence Island.

After sinking losses were imposed for 7 days of simulated time, for example, $\leqslant 2$ μg chl l⁻¹ was found over most of the model's domain, in sharp contrast to a growth-only case. The algal population from the Anadyr Strait boundary condition had sunk out of the model in 7 days, but the local populations at the interior grid points did not have enough time to significantly increase their biomass within just 1 week. With a maximal population growth rate of only 0.43 day⁻¹, or a doubling time of ~1.5 days, it would take at least 6–7.5 days, with *no* losses, for an exponential increase of the local populations to occur from the 0.5 μg chl l⁻¹ of initial conditions to biomass levels of 8–16 μg chl l⁻¹. These high levels of chlorophyll biomass were found after 14 days of simulated time (Fig. 86), however, in the region of the nitrate minimum. Within Bering Strait, about 8 μg chl l^{-1} was simulated at the grid point adjacent to the fluorometer mooring (Fig. 86), where as much as 9 μg chl l^{-1} was found in the I10 time series. These simulation results suggest that the accumulated algal biomass, grown south of St. Lawrence Island, sinks out between this island and Bering Strait, whereas that increment of algal biomass grown north of Anadyr Strait is likely to sink out within the Chukchi Sea.

Over the larger separation distance and weaker flow regime between I08 and Bering Strait, the cross-correlation coefficients for temperature at the two moorings (Fig. 85) were reduced to 0.1–0.2 at lags of 300–400 hr (i.e., the hypothetical transit time). At these time lags, the cross-correlation coefficients of fluorescence at the two moorings (I08-I10) were actually negative (Fig. 84), in contrast to the above coincidence of temperature and fluorescence memory at 48-hr lag between Anadyr Strait and I08. Over a period of 12–16 days in the northern Bering Sea, the biological processes have become relatively uncoupled from prior physical forcing, in contrast to a 2-day biological memory of a change in habitat. In fact, the above simulation results suggest that the local algal populations of Bering Strait are not closely linked to seed populations transiting Anadyr Strait.

A tacit assumption in the above cross-correlation analysis of the downstream propagation of fluorescence and temperature events from Anadyr to Bering Strait is that the patches of high or low chlorophyll detected 48 hr later, between I08 and Anadyr Strait, are of advective origin, with the source of algal variance, at this time scale, located upstream of Anadyr Strait. The peaks of high chlorophyll in the fluorescence records of I08 and I07 (Anadyr Strait) from July 10 to August 15 (i.e., the period of the cross-correlation analysis) are spaced 3–4 days apart in time, similar to wind reversals measured at the Kotzebue airport (Fig. 87); in particular, high chlorophyll values were observed under northerly wind forcing, such as during July 16, 25, and 28, 1985 (Figs. 82 and 87). The variance spectra of temperature and fluorescence in Anadyr Strait (Fig. 88) exhibit peaks at these frequencies of 0.010–0.014 cycles hr^{-1} as well.

To simulate a wind-event response in the simple Bering Sea model, the northward transport of water through the Bering Strait was increased to a stronger flow field of 1.2 Sv. After 14 days of growth and sinking within the 1.2-Sv case, the chlorophyll isopleths in the faster flow regime were ~50 km farther north off the Siberian coast (Fig. 89) than those within the 0.6-Sv flow field (Fig. 86). The model thus predicted an algal biomass of 2 μg chl l^{-1} (rather than 4 μg chl l^{-1}) near the I07 fluorometer mooring in the 1.6-Sv case (Fig. 89), similar to the minima of time series observations (Fig. 82) during those southerly wind events (Fig. 87). Essentially, low-chlorophyll water is advected north during these stronger flow events, because the dilution rate increases, but the algal growth rate remains the same in the model.

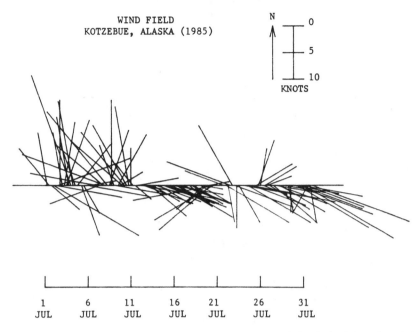

FIG. 87. Six-hour averaged wind vectors recorded at Kotzebue, Alaska, during July 1–31, 1985.

Within the shelf region directly north of St. Lawrence Island, however, there was little change in the strength of the currents between the 0.6-Sv (Fig. 34) and 1.2-Sv cases. Consequently, the 12-μg l^{-1} isopleth of chlorophyll remained in the same location after 14 days of both flow regimes (Figs. 86 and 89). It continued to be positioned between the 40-m isobaths on the Siberian and Alaskan sides of the Chirikov Basin, north of St. Lawrence Island, where the smallest daily sinking loss (0.25 day^{-1}) occurred in both flow regimes. The implication of these simulation runs is that one might expect the algal biomass in the southern region to migrate north or south in response to wind events, that is, to exhibit a strong coupling to physical perturbations, while the spatial loci of algal biomass in the northern region might remain fixed during changes in the flow field linking the Bering and Chukchi Seas (i.e., a weaker coupling as evidenced by the cross-correlation coefficients) (Figs. 84 and 85).

3.4.4   South Atlantic Time Series

A similar advective response of high- and low-chlorophyll concentrations transported past a fluorometer mooring is shown in Fig. 90 of a time series (S2) at 11 m depth on the 15-m isobath, off Savannah, Georgia, from April 1 to June 5, 1985. As in the Mid-Atlantic Bight, wind forcing from the north in the South Atlantic Bight leads to a downwelling circulation pattern and observations of

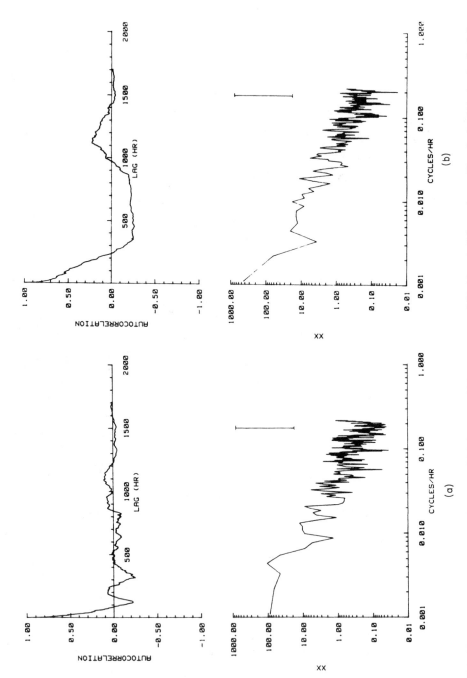

FIG. 88. Autocorrelations and variance spectra of 2-hr low-passed (a) temperature and (b) chlorophyll fluctuations at 20 m depth within Anadyr Strait from July 10 to September 29, 1985.

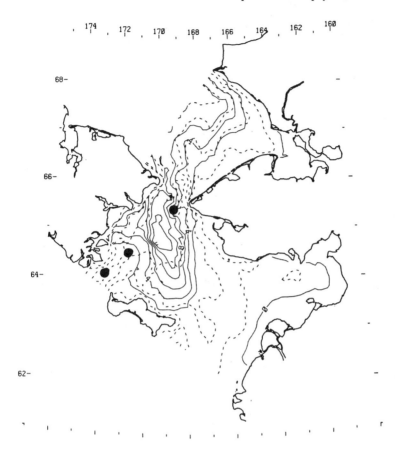

FIG. 89. Simulated chlorophyll distributions in relation to the time-series observations (●) in the Bering Sea after 14 days in a steady flow field of 1.2 Sv, with a growth rate of 0.43 day^{-1} and a sinking rate of 10 m day^{-1}. [After Walsh and Dieterle (1986); © Elsevier Science Publishers B.V., with permission.]

high chlorophyll at the fluorometer mooring on April 14 and 19 and May 7 and 31 (Fig. 90). A reversal of wind forcing from the south leads to an upwelling circulation pattern and dilution of chlorophyll at this fluorometer mooring during April 20, May 2 and 15, and June 2. As a result of nutrient outwelling from estuaries and marshes of the South Atlantic Bight (Turner *et al.*, 1979), high primary production of 285 g C m^{-2} yr^{-1} occurs on the inner shelf at depths of 0–20 m (Haines and Dunstan, 1975). A declining gradient of chlorophyll thus extends seaward (Fig. 91) past this fluorometer mooring to another one (Fig. 92), 53 km farther offshore at the 30-m isobath (S3).

The April nitrate content at midshelf in the South Atlantic Bight is less than 1 μg-at. NO$_3$ l^{-1} (recall Fig. 25), such that chlorophyll concentrations at the 30-

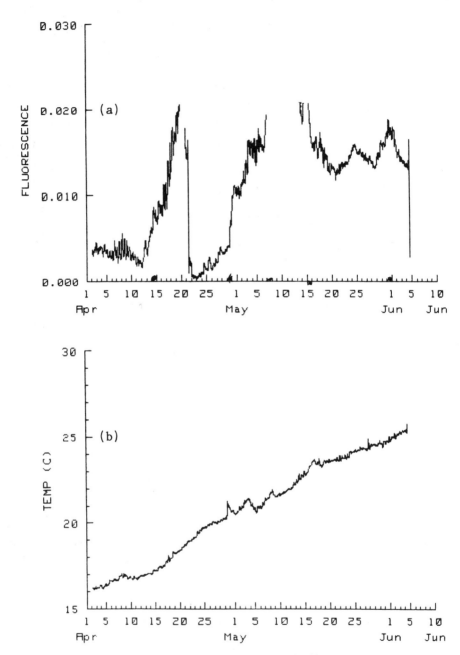

FIG. 90. A 2-hr low-passed time series of (a) chlorophyll and (b) temperature at 11 m on the 15-m isobath (S2), off Savannah, Georgia, in the South Atlantic Bight, from April 1 to June 5, 1985.

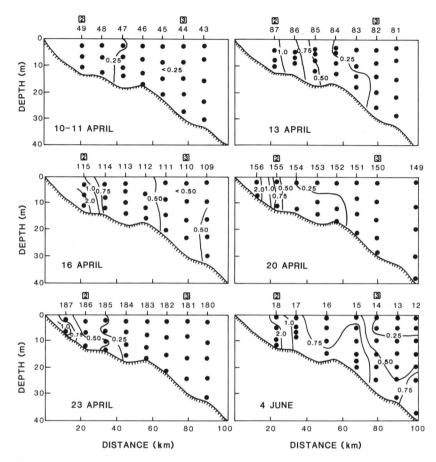

FIG. 91. A time series of chlorophyll sections taken past the S2 and S3 fluorometer moorings in the South Atlantic Bight from April 10 to June 4, 1985 (J. A. Yoder, L. P. Atkinson, and T. E. Whitledge, personal communications).]

m isobath ranged from 0.1 to 0.5 μg chl l^{-1} during April–June 1985 (Figs. 91 and 92), compared to 0.3–3.0 μg chl l^{-1} at the S2 mooring on the 15-m isobath (Fig. 90). The apparent seaward movement and shoreward retreat of the 1.0-μg chl l^{-1} isopleth past the first fluorometer at the 15-m isobath and the 0.3-μg chl l^{-1} isopleth past the second at 30 m, in response to downwelling on April 14 and upwelling on April 20, can be seen in the successive cross-shelf transects on April 13, 16, and 20 (Figs. 91b, c, and d). During the onshore intrusion of low-chlorophyll water in an upwelling circulation pattern, the cross-isobath component of flow, u, reached speeds of 10–20 cm sec^{-1} at the S3 mooring on the 30-m isobath (Fig. 93a). In this time series, a positive value of u denotes offshore

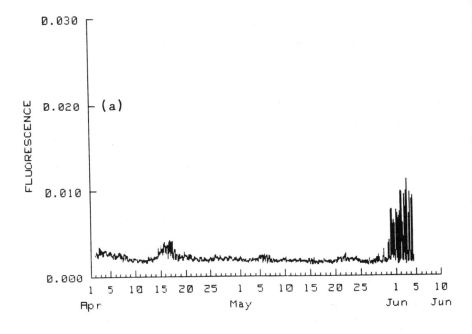

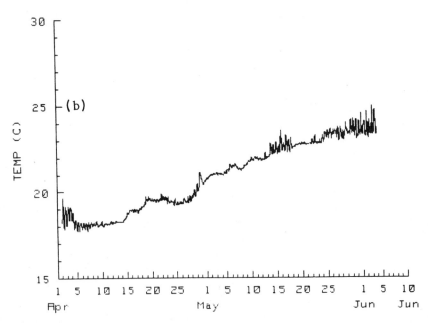

FIG. 92. A 2-hr low-passed time series of (a) chlorophyll and (b) temperature at 11 m on the 30-m isobath (S3), off Savannah, Georgia, in the South Atlantic Bight, from April 1 to June 5, 1985.

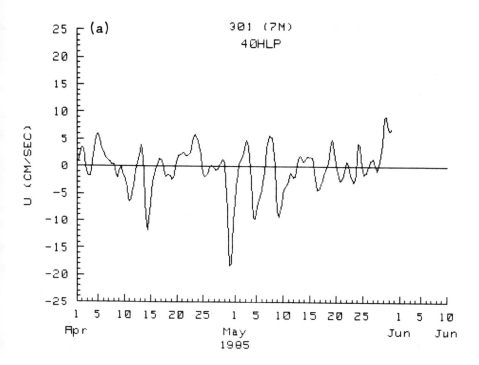

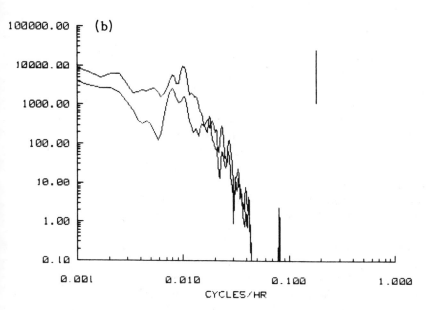

FIG. 93. A 40-hr low-passed (a) time series and (b) variance spectrum of the onshore–offshore (u) component of flow at 7 m on the 30-m isobath, off Savannah, Georgia, in the South Atlantic Bight, from April 1 to June 5, 1985 (T. N. Lee, personal communication).

flow, peaking at only 5 cm sec^{-1}; the same sign convention is used in Fig. 97 for currents in the Mid-Atlantic Bight.

Most of the wind forcing was from the south during February–June 1985 in the South Atlantic Bight, with onshore bursts of flow at 4- to 5-day intervals (Fig. 93a). The variance spectrum of the cross-isobath component of flow thus contained a significant part of the kinetic energy at wind-event time scales of 0.008–0.014 cycles hr^{-1} (Fig. 93b), similar to variance spectra of temperature and chlorophyll in the Bering Sea (Fig. 88). At onshore flow rates of 10–20 cm sec^{-1}, a water parcel of low chlorophyll content might transit the 53-km separation distance between the S3 and S2 moorings within 3–6 days, or 72–144 hr. The cross-correlation coefficient between the two fluorometer time series was, in fact, 0.4 at a time lag of ~100 hr.

From February 16 to June 29, 1985, the long-term mean of the flow at 3 m above the bottom was an onshore movement of 1.3 cm sec^{-1} at the 30-m isobath and of 2.9 cm sec^{-1} at the 15-m isobath. At flow rates of only 1–2 cm sec^{-1}, a water parcel would move shoreward between S3 and S2 within 30–60 days, or 720–1440 hr; an r_l of 0.2 occurred at time lags of 300–700 hr between these two fluorometer records as well. The discussion of *in situ* time series from moored fluorometers has thus far been restricted to shallow depths of 15–50 m off Long Island, within the Bering Sea, and in the South Atlantic Bight. On the inner shelf, resuspended chlorophyll could be detected either by satellite at the surface or by one moored instrument in the middle of the water column, but this is not the case for the outer shelf.

3.4.5   New England Time Series

A series of three moorings on the outer shelf of the Mid-Atlantic Bight (Fig. 69) also provided confirmation of advective transport of chlorophyll from February 15 to April 10, 1984, but the lateral coherence of temporal algal fluctuations within the euphotic zone was significantly reduced. The doubling of chlorophyll between April 2 and 8, 1984, detected by the aircraft (Fig. 69), was similarly measured by the moored fluorometer at ~40°30′N, 71°W (Fig. 94a). The cross-correlation coefficient of 2-hr low-passed, surface (13 m) chlorophyll data (Fig. 66a) at this mooring on the 80-m isobath south of Martha's Vineyard (E8) with those downstream at the 120-m isobath south of Long Island (W12), for example, was only 0.21 at a zero time lag. A 40-hr low-pass filter of the data, at zero time lag, removed more of the high-frequency fluctuations in these chlorophyll records (Fig. 94), such that the r_l of the data at E8 and W12 then became 0.35. The r_l of 40-hr low-passed chlorophyll records of 13 m depth at E8 with another mooring on the 80-m isobath south of Long Island (W8) was only 0.14, while that of 0.04 for W8 with W12 was even less, despite the reduction in high-frequency variance (Fig. 94), compared to the 2-hr low-passed records.

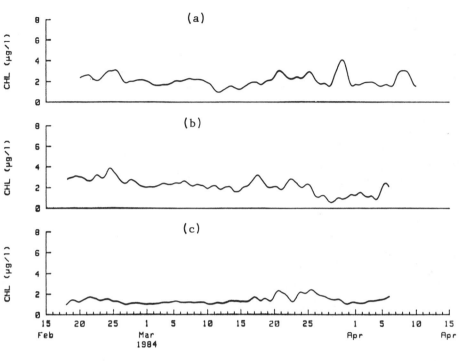

FIG. 94. Time series of 40-hr low-passed chlorophyll records at 13 m depth on (a) the 80-m isobath (E8) south of Martha's Vineyard, and (b) the 80-m (W8), and (c) 120-m (W12) isobaths south of Long Island in the Mid-Atlantic Bight from February 17 to April 11, 1984. [After Walsh *et al.* (1987b); © 1987 Pergamon Journals Ltd., with permission.]

The downstream component of the surface flow (15 m) at the W12 mooring was, at times, greater than 40 cm sec^{-1}, compared to 10-cm sec^{-1} peaks of offshore flow, such that the mean vector flow along a trajectory from E8 to W12 from February 15 to April 10 was ∼15 cm sec^{-1}. Over a separation distance of 80 km between these two moorings and an average flow of 15–20 cm sec^{-1}, a particle would take ∼4–6 days to move from E8 to W12, *if* it were retained within surface waters. The r_l of the 40-hr low-passed, surface chlorophyll data between E8 and W12 at 4- and 5-day time lags, in fact, became much smaller, respectively 0.032 and 0.020 (i.e., an order of magnitude less than at zero lag.)

The poor coherence of these surface (13-m) chlorophyll records over the presumed transit time suggests both that sinking of phytoplankton is rapid in the Mid-Atlantic Bight and that resuspension of near-bottom particles does not occur within the euphotic zone on the outer part of the shelf. The memory of algal population changes is thus not maintained here throughout the whole water column, to be later recorded, downstream of a propagation event, at a near-surface

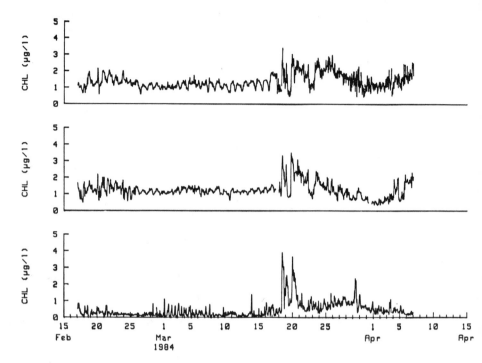

FIG. 95. Time series of 2-hr low-passed chlorophyll records at 13, 23, and 81 m depths on the 120-m isobath (W12), south of Long Island, from February 17 to April 11, 1984. [After Walsh *et al.* (1987b); © 1987 Pergamon Journals Ltd., with permission.]

depth of 13 m. This is in contrast to the South Atlantic Bight and Bering Sea, where over a similar separation distance, at bottom depths of 15–45 m, the time-lagged r_l was 0.40 between S2 and S3, as well as 0.33 between I08 and I07, compared to 0.032 between E8 and W12.

The fluorometer records of 13, 23, and 81 m depths at W12 on the 120-m isobath (Fig. 95), south of Long Island, show high cross-correlations, compared to the low spatial coherence of temporal fluctuations in algal biomass. For example, the r_l of the 40-hr low-passed records of algal biomass between 13 and 23 m at W12 was 0.62, twice that at zero lag between the 13-m records at W12 and E8. During calm wind periods (i.e., March 19–20, 1984), when diabathic horizontal motion was small, the overlying phytoplankton sank fast enough for the near-bottom signal to be recorded at the same isobath as the surface signal; more chlorophyll was then observed at 81 m than at 23 m or 13 m (Fig. 95). Over the whole time series, however, the r_l of the 13-m and 81-m chlorophyll records at W12 was somewhat less, 0.53, because the near-bottom phytoplankton were entrained within a different flow regime, moving onshore when the surface water was advected seaward and moving offshore when the surface flow was onshore.

To distinguish between settling and resuspension events of phytoplankton, one can first compute those parts of the light extinction of Eq. (114a) attributed to phytoplankton (0.0852 × chl) and coastal surface water (0.05 m^{-1}), and then subtract them from observations obtained by the moored transmissometers (Fig. 96). At 5 m above bottom on the 80-m isobath (Fig. 96c), for example, the abiotic particles were a major component of light extinction after offshore flow events on February 28, March 9–10, March 17–18, March 24, March 29–30, and April 5, that is, following northeasterly wind forcing in 1984. During northwest wind events on March 2–6 and April 1–4, however, the abiotic contribution to light extinction was negligible (Fig. 96c), similar to the transmissometer observations during most of the time at 23 m below the surface on the 120-m isobath (Fig. 96a). The same biotic sinking events after northwest forcing were

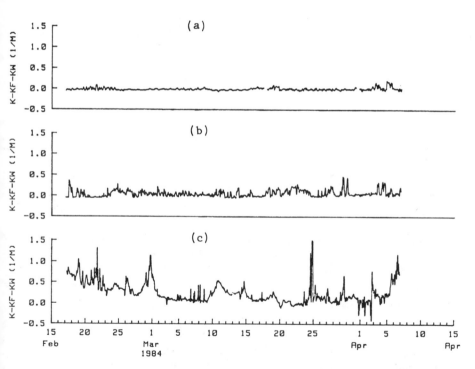

FIG. 96. Time series of 2-hr low-passed light attenuation (m^{-1}) records, attributed to abiotic particles (measured minus computed due to phytoplankton and water), at (a) 23 m depth on the 120-m isobath (W12), (b) 81 m depth on the 120-m isobath (W12), and (c) 75 m depth on the 80-m isobath (W8), south of Long Island, from February 17 to April 6, 1984. [After Walsh *et al.* (1987b); © 1987 Pergamon Journals Ltd., with permission.]

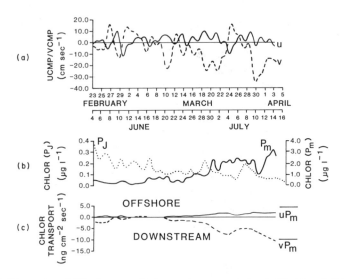

FIG. 97. Time series of 40-hr low-passed (a) current components (u, v), (b) chlorophyll (P_M), and (c) chlorophyll flux (ng cm^{-2} sec^{-1}) past a mooring 3–5 m above the 80-m isobath, south of Long Island, from February 23 to April 4, 1984, and (b) of chlorophyll (P_J) at another mooring 3 m above the 110-m isobath, south of Martha's Vineyard, from June 4 to July 16, 1983. [After Walsh *et al.* (1987a); © 1987 Pergamon Journals Ltd., with permission.]

captured by a transmissometer, moored 3 m above the bottom on the 120-m isobath.

Evidence for export of chlorophyll during these offshore flow events of near-bottom water was also provided by the deep-fluorometer (Fig. 97b) and current-meter (Fig. 97a) records at the 80-m isobath. In contrast to the surface time series (Fig. 94), the near-bottom chlorophyll record (Fig. 97b) at 80 m exhibited an order of magnitude increase in biomass (P_M) with time, reflecting the seasonal buildup, seaward export, and sinking to the bottom of the spring bloom. For example, from June 4 to July 16, 1983, a previous fluorometer mooring at 3 m above the 110-m isobath, south of Martha's Vineyard, yielded a mirror image of the 1984 spring time series (Fig. 97b). The second chlorophyll time series (P_J) declined from 0.40 μg l^{-1} in June 1983 to 0.04 μg l^{-1} in July 1983 within the aphotic zone. Note the scale change between these two time series of spring (P_M) and summer (P_J) chlorophyll observations near the shelf break.

One can estimate the near-bottom flux of chlorophyll, $\partial(uM)/\partial x$, of Eq. (191) by combining simultaneous measurements of the onshore–offshore flow and the scalar, chlorophyll fluorescence, that is, uM. The cumulative mean of this product, averaging over the cross-isobath flow events yields a net seaward chlorophyll flux uP_M of 3.0 ng chl cm^{-2} sec^{-1} (Fig. 97c), or 2.67 g chl m^{-2} day^{-1}. In terms of carbon, this flux amounts to 1.2 × 10^2 g C m^{-2} day^{-1} with a C/chl ratio of

45. Note that this cumulative 1984 offshore flux is quite similar to the previous estimate of 0.8×10^2 g C m^{-2} day^{-1} from the 1979 satellite time series.

Such a possible shelf export of $0.8-1.2 \times 10^2$ g C m^{-2} day^{-1} in 1979 and 1984 inferred from high-frequency sampling in space and time by the CZCS and moored fluorometer lacks vertical resolution from either of these data sets. The Bering Sea plankton model also lacked depth resolution. Another model of the seasonal dynamics of Mid-Atlantic Bight plankton at three depth layers, over 59 days from March 1, 1979 to April 27, 1979, was thus coupled to the physical circulation model of Section 2.6.6. The details of the time-dependent light and nutrient regulation, vertical mixing, sinking, and grazing stress at 1800 grid points, in conjunction with the changing wind forcing over this 1979 spring bloom, are discussed in Section 5.1.3. However, the algal biomass detected by the CZCS over space and the moored fluorometer over time only represents what is left in the water column by bacteria, zooplankton, and benthos. After the discussion of consumption of primary production in the next chapter, we will return to a consideration of the vertical and horizontal transport of uneaten algal residues and their ultimate fate in the sediments.

4 Consumption

In the morning when the sardine fleet has made a catch, the purse-seiners waddle heavily into the bay blowing their whistles. The deep-laden boats pull in against the coast where the canneries dip their tails into the bay. . . . Then cannery whistles scream and all over the town men and women scramble into their clothes and come running . . . to clean and cut and pack and cook and can the fish. The whole street rumbles and groans and screams and rattles while the silent rivers of fish pour in out of the boats and the boats rise higher and higher in the water until they are empty. The canneries rumble and rattle and squeak until the last fish is cleaned and cut and cooked and canned and then the whistles scream again and . . .Cannery Row becomes itself again—quiet and magical. . . . When the war came to Monterey and to Cannery Row everybody fought it more or less, in one way or another. When hostilities ceased everyone had his wounds. . . . The canneries themselves fought the war by getting the limit taken off fish and catching them all. It was done for patriotic reasons, but that didn't bring the fish back. . . . The pearl-gray canneries of corrugated iron were silent and a pacing watchman was their only life. The street that once roared with trucks was quiet and empty.

(J. Steinbeck, 1945,1954)

Our understanding of the time-dependent, nonlinear, spatial properties of marine ecosystems suffers because we have a short life span and no gills. We inherently lack a proper temporal or spatial perspective to analyze the way different scales of variability interact to produce the observed distributions of variables in the ocean. Oceanographers must be resigned to recording events rather than predicting them and their consequences as long as we act as eulerian predators located at a point on the coast and attempt to harvest the sea as it flows by.

A climatic shift (a transient that occurs on a scale of years rather than seasons, months, or weeks; recall Fig. 27) apparently induces a relocation of the ambient marine community away from one particular location, and appears as a serious change in prey availability for us. The landings of sardine off California dropped dramatically over a 15-yr period (Fig. 98) in response to the fishing pressure described by John Steinbeck. Similar fluctuations of sardine and anchovy numbers (Fig. 99) apparently occurred off California over longer time periods of 150 yr (Soutar and Isaacs, 1974), however, when presumably humans were not the major predator. There are several distinct groups of anchovy (Husby and Nelson, 1982) off Alta and Baja California (Fig. 100), moreover, whose population fluctuations must be considered at larger spatial and longer temporal scales (Fig. 55a) than previously considered in this monograph.

If we were lagrangian aquatic predators and always stayed at sea, following prey without necessarily a fixed reference to land, a geographic relocation of prey might have little effect on us unless the climatic scale of variability interacted with the event scale. The timing of a particular pulse of energy up the food chain may be altered from the normal progression of events as a result of the lagrangian shift of the community within the ocean. A disruption of the phenological sequence, such as perhaps a mismatch in prey availability and the success of first-feeding fish larvae, may then lead to a change in the structure of the ecosystem rather than its simple translation in space. Such a change may be permanent, if the dominant species lack resilience (Holling, 1973) to recover their previous levels of standing stocks after a perturbation.

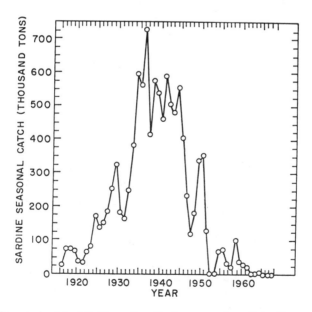

FIG. 98. The annual catch of Pacific sardine, *Sardinops caerulea,* off the Oregon–Washington, California, and Baja California coasts from 1916 to 1966. [After Longhurst (1971); © Allen and Unwin.]

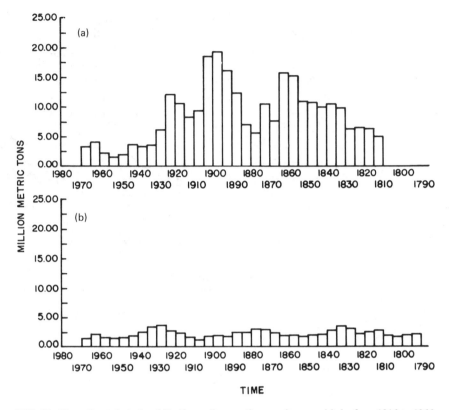

FIG. 99. The estimated stocks of Pacific sardine, northern anchovy, and hake from 1816 to 1966 off (a) California and (b) Baja California. [After Soutar and Isaacs (1974).]

For example, the North Atlantic, and in particular the North Sea, is a well-studied and yet poorly understood marine ecosystem, as evidenced by numerous speculations on causes of the recent decline of the Plymouth (Cooper, 1956) and Swedish (Devold, 1963) herring stocks. With accumulation of data over the life-span of more than one oceanographer, however, a frame of reference to interpret long-term biological changes appears to be emerging (Russell *et al.*, 1971; Cushing, 1975). The climatic temperature fluctuations off the British Isles have been linked to warm temperatures and high salinities during periods of southerly winds (Dickson, 1971) (i.e., from 1930 to 1950), and to cold temperatures and low salinities during periods of westerly winds (i.e., from 1910 to 1930). At the same time, the Plymouth and Swedish stocks of herring were high during the cold periods and the Norwegian cod and herring stocks (see Fig. 189b) increased during the warm periods.

These physical and biological changes of the North Sea and the Norwegian

Sea can both be attributed to oscillations in large-scale meteorological and oceanographic features of the North Atlantic. The shift from westerly to southerly winds over the North Sea is a function of the relative strengths of the Azores High and the Iceland Low (Bjerknes, 1964). The wind system, in turn, influences the flow of the Gulf Stream across the Atlantic (Stommel, 1960), and the transport of Atlantic waters into the North Sea and the Norwegian Sea (Helland-Hansen, 1934). An anomalous influx of warmer, more saline Atlantic water in 1929, for example, coincided with commercial catches of cod off Spitsbergen in 1931, for the first time in the preceding 50–60 yr (Mosby, 1938). Such geographical shifts of the European cod and herring stocks, ignoring effects of overfishing, could represent a spatial but not a structural response of the marine

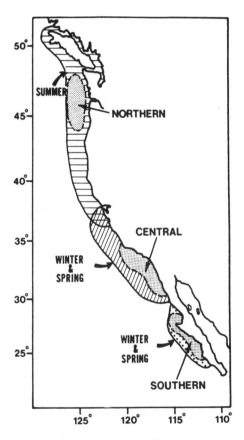

FIG. 100. The geographic ranges (hatched) and principal spawning areas (shaded) of three subpopulations of northern anchovy, *Engraulis mordax,* off the Oregon–Washington, California, and Baja California coasts. [After Husby and Nelson (1982).]

ecosystem to a transient in climate. The arrival of pilchard off the Plymouth coast during the last warm period (1930–1950) can be viewed as a latitudinal shift of the ecotone between northern and southern communities of European waters.

If long-term temperature anomalies do indeed reflect changed circulation patterns in response to climatic shifts of wind forcing, then other parts of the ocean should exhibit biological patterns of response similar to those of the North Sea and the Norwegian Sea. For example, one might expect an intensified Labrador Current and cooling of New England waters in response to a weakened Gulf Stream system during warming of European waters (Namias, 1964). The sea-surface temperature anomaly off Boothbay Harbor, Maine, during the last 30 yr (Colton, 1972) was, in fact, the inverse of that of the British Isles, and some organisms appeared to shift their geographic center of abundance in response to this latest climatic transient.

The lobster population increased off Maine during its warm period (1948–1958) when European waters were cooling. Mann (1976) has pointed out that during this warming phase of shelf waters between New York and Nova Scotia, peak catches of lobster were obtained progressively off Rhode Island (1939), New Hampshire (1943), Nova Scotia (1951), and Newfoundland (1955), the northern limit of the lobster's range. This trend was reversed during the cooling phase (1958–1968), with peak catches obtained in New Brunswick (1960), Quebec (1962), Massachusetts (1965), and Connecticut and New York (1967).

A similar shift in patterns of geographic distribution off the east coast of the United States occurred with the green crab (Welch, 1968), although some species of groundfish did not appear to significantly alter their home range during this time (Colton, 1972). Further, it has been suggested that there are even more extensive teleconnections of climate, current, and biological response between the Atlantic and the Pacific Oceans (Ishevskii, 1964; Cushing, 1975). One recent U.S. newspaper article even attributed the increase of rattlesnakes in Montana to the occurrence of the 1982 El Niño off Peru. It is usually the local synoptic space scale ($\sim 1 \times 10^5$ km^2), however, over which the response of herbivores and carnivores of a shelf ecosystem occurs to interannual transients of their habitat, not at the shorter time and smaller space scales of phytoplankton, nor at the greater ones of migratory mammals. The consumption of primary production is thus now considered in this chapter over a time domain (Figs. 54 and 55) of ~ 100 days.

4.1 Mass Balances

Adult fish and crustaceans, unlike plankton, have motility, adding a biological vector to the physical movement of organisms on the continental shelves. Routine fish migration circuits (Harden Jones, 1968) involve active movement of

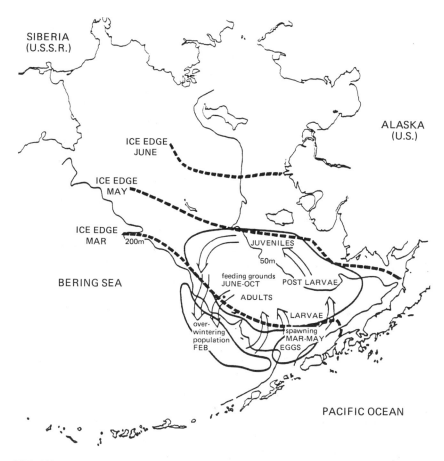

FIG. 101. A possible migratory circuit of Alaska pollock, *Theragra chalcogramma,* on the continental shelf and slope of the southeastern Bering Sea in relation to the seasonal location of the ice edge. [After Walsh and McRoy (1986); © 1986 Pergamon Journals Ltd., with permission.]

adults to the spawning grounds, and then passive transport of the larvae to nursery grounds, with eventual recruitment of the surviving juvenile fish to the adult stocks on their feeding grounds (Fig. 101). For example, interannual variations in the catch of Pacific halibut, *Hippoglossus stenolepis,* from 1963 to 1969 (Kihara, 1971) and of yellowfin sole, *Limanda aspera,* from 1956 to 1976 (Maeda, 1977) in outer Bristol Bay, Alaska, were associated with increased onshore penetration of adults within warm slope water. Over the last 20 yr, the dominant year classes of yellowfin sole (Maeda, 1977) and perhaps of Pacific halibut (McLain and Favorite, 1976) were also spawned during these periods of warm bottom temperatures in the eastern Bering Sea.

Successful year classes of pollock, *Theragra chalcogramma,* originated during years of warm temperature anomalies on the Bering Sea shelf as well, perhaps reflecting early development of phytoplankton blooms, and subsequent zooplankton cohorts as prey for pollock larvae, during warm years. The Alaska pollock lay positively buoyant eggs on or near the shelf bottom from February to June (Serobaba, 1974). Incubation time from fertilization to hatching is ~25 days at 2°C (Hamai *et al.,* 1971), and the larvae begin feeding at an age of ~15 days and a length of ~5.0 mm, a short time before complete yolk-sac absorption. During the next 20 days, until they obtain a length of ~10.0 mm, the diet of the larvae is mainly copepod nauplii, which they must obtain to avoid starvation (Kamba, 1977; Clarke, 1978). Egg and larval surveys in 1976, 1977, and 1978 indicated two major regions of spawning: (1) during March on the slope and (2) during April on either the outer or middle shelf in the southeast Bering Sea (Fig. 101). Pollock eggs, collected near the shelf break (Waldron, 1978) from March 11 to 16, 1978 (Fig. 102a), apparently grew to larvae of 6.0–6.4 mm length (T. Cooney, personal communication) by April 10–20, 1978 (Fig. 102d). Temperatures of bottom water on the middle shelf in the spring can be as much as 5°C colder than slope water, affecting the metabolism of marine organisms, (e.g., Fig. 52). In cold years, such as 1976 and 1980, for example, bottom water in June was still − 1 to 0°C on the middle shelf.

During mid-April 1978, however, the bottom temperature was >3°C at the 100-m isobath, a second spawning of pollock had occurred, and eggs were found between the 50- and 100-m isobaths north of the Aleutian Island Chain (Fig. 102b). By mid-May 1978, the pollock larvae on the slope were 9.0–9.4 mm long, and those of the second cohort, derived from the April spawning, were 5.0–5.4 mm in length on the middle shelf (Fig. 102e). Finally, by mid-June 1978 (Fig. 102f), the first cohort of larvae had obtained a median length of 11.5–11.9 mm and the second was 7.0–7.4 mm long. The daily growth increment in 1978 for these size classes was ~0.1–0.2 mm day^{-1} (T. Cooney, personal communication).

A similar spawning of pollock in slope water presumably occurred in March 1976, but in April 1976 the bottom temperature on the middle shelf was <1°C and the second set of eggs was restricted to the 100- to 200-m isobaths of the outer shelf (Fig. 102c). During 1960, bottom temperatures on the middle Bering shelf were cold (Maeda *et al.,* 1967), similar to 1976, and the 1960 year class of pollock was poor (Ishida, 1967). Conversely, in 1963, bottom temperatures were warm and the 1963 year class of pollock was strong (Chang, 1974). To explore the consequences of spawning on the outer Bering shelf during a cold year, compared to the middle shelf in a warm year, we will examine a series of numerical experiments using wind and spawning conditions from 1976 to 1978 in Section 4.5.1. Before proceeding with an analysis of the timing of some crucial

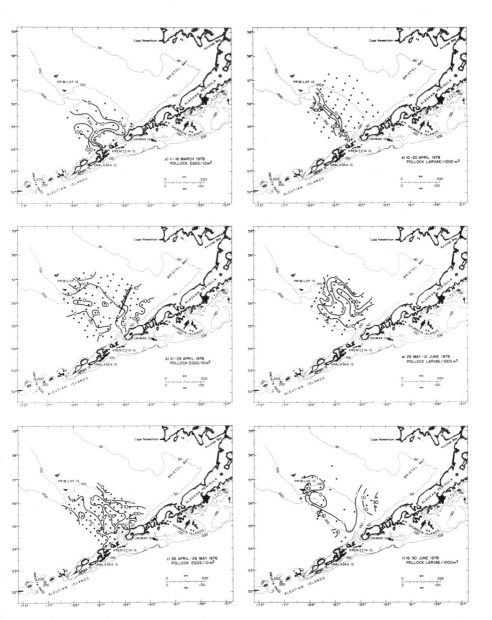

FIG. 102. The distribution of Alaska pollock eggs and larvae during a warm (1978) and a cold (1976) year in the southeastern Bering Sea. (a) March 11–16, 1978, eggs/10 m²; (b) April 21–29, 1978, eggs/10 m²; (c) April 26–May 28, 1976, eggs/10 m². [After Walsh and McRoy (1986); © 1986 Pergamon Journals Ltd., with permission.]

events in seasonal primary and secondary production, however, one must consider the transfer of energy to all higher trophic levels.

Alaska pollock was selected as a biological tracer because it was the major fishery species in the Bering Sea and considerable data on its abundance and life history were available. Also, during their early life history, pollock larvae are not ecologically very different from the larvae of other fishes and crustaceans, such as the Tanner crabs (Incze, 1983) and copepods (Smith and Vidal, 1986). In all cases, these planktonic creatures must be where their food is when they start eating, according to the general patchiness panacea for theories of food limitation in the sea (Walsh, 1983). Adult fish and crustaceans can search out their prey, however, and their predatory activities impose a significant source of mortality as well. An assessment of the causes of interannual variation in abundance of ground fish, crustaceans, or pelagic foragers thus first requires a quantitative description of the food web in which they are all embedded. Specifically, we will first consider annual, rather than daily, carbon and nitrogen budgets of shelf ecosystems at 60, 45, and 30°N latitude.

4.2 Bering Sea Budget

The only exchange of water between the Pacific and Arctic Oceans occurs within the 85-km-wide, shallow (<50 m) Bering Strait, linking the broad continental shelves (500–800 km) of the Bering and Chukchi Seas (Fig. 31). Some 250 km south of the Bering Strait, St. Lawrence Island is bordered by two other passages, the 75-km-wide Anadyr Strait to the west and the 190-km-wide Shpanberg Strait to the east. Mean currents within these straits are about 25 cm sec^{-1} in the Bering Strait, 15 cm sec^{-1} within Anadyr Strait, and 5 cm sec^{-1} in Shpanberg Strait (Aagaard *et al.*, 1985), just south of the mouth of the Yukon River—recall Fig. 34. Farther south, over most of the southeastern Bering shelf, the residual flow is weak, of the order of 1–3 cm sec^{-1} on the inner shelf (<50 m depths), and <1 cm sec^{-1} on the middle shelf (50- to 100-m isobaths), with tides constituting >90% of the kinetic energy here, in contrast to <10% north of St. Lawrence Island.

Along the Bering Sea shelf break (150 m) and slope, the Bering Slope Current instead flows northwest at 10–25 cm sec^{-1} (Kinder *et al.*, 1975). At Cape Navarin, Siberia, this Bering Slope Current bifurcates, with part of the flow entrained as a strong, northwest barotropic current, following the 60- to 70-m isobaths around the Gulf of Anadyr to Anadyr Strait and the western part of Shpanberg Strait (Coachman *et al.*, 1975). More than 60% of the mean, northward transport (~1–2 Sv) of water through Bering Strait is derived from this Anadyr Water; the remaining two water types are a mixture of Yukon River and the original southeastern Bering Shelf Water, with the resultant water type termed Alaska Coastal Water.

Within the strongly advective physical regime on the northern Bering–Chukchi shelves, the influence on primary production of the addition of fluvial nitrate from the Yukon River to Alaska Coastal Water is an insignificant 3% of that supplied from slope water. In contrast, the major nutrient supply is contributed from the adjacent water masses, Anadyr Water and Bering Shelf Water, particularly the former, which originates at depth along the shelf break of the Bering Sea. Anadyr Water retains the nitrate signature of the Bering Slope Current and still contains $15-25$ μg-at. NO_3 1^{-1} north of St. Lawrence Island. This nutrient source is primarily responsible for the prodigious phytoplankton production during summer in Bering Strait (Sambrotto et al., 1984). In comparison, primary production within Alaska Coastal Water follows a typical pattern of a polar spring bloom after the breakup of the sea ice (Fig. 101), with relatively little algal production thereafter.

Without significant horizontal advection, or river discharge on the southeastern Bering shelf, the supply of nitrate for this spring bloom depends instead on an apparent eddy "diffusion" (Csanady, 1976) process for a shoreward flux of nutrients over the 500-km distance from the slope water. The daily onshore input of nitrate, below the euphotic zone at the Bering Sea shelf break, is similar to that introduced by upwelling at the edge of the shelf off Peru (Walsh, 1975). A combination of seasonal wind mixing and the 20- to 50-cm sec^{-1} tidal regime on the much wider Bering Sea shelf leads to an effective vertical velocity of only 10^{-3} cm sec^{-1} from either diffusion (Coachman and Walsh, 1981) or advection (Stigebrandt, 1981) considerations, however, in contrast to 10^{-2} cm sec^{-1} off the Peru coast, or 10^{-1} cm sec^{-1} on Georges Bank. As a result of the difference in local vertical flux of nutrients, both the Peru shelf and Georges Bank remain eutrophic year-round, while the surface waters of the southeastern Bering Sea eventually become depleted of nitrogen by summer. Specifically, with a seasonal decrease in wind forcing and an increase in thermal stratification of the water column, the average depth of the surface mixed layer shoals from $\sim$45 m in March to 10 m by June. As a consequence, the rate of nutrient resupply diminishes at a time when uptake by phytoplankton increases.

Surface waters are thus stripped of nutrients by the end of May in all parts of the Bering Sea (Whitledge et al., 1986), except in the shallow western waters between Anadyr and Bering Straits. Without continual, physical resuspension of phytoplankton cells, as from the early spring wind events, the dominant diatom populations sink out of the summer euphotic zone on the middle and outer regions of the southeastern Bering Sea. Like the differences in nitrate and annual productivity between Anadyr Water ($\sim$20 μg-at. NO_3 1^{-1}, $\sim$300 g C m^{-2} yr^{-1}) and Alaska Coastal Water ($<$1 μg-at. NO_3 1^{-1}, $\sim$50 g C m^{-2} yr^{-1}), floristic

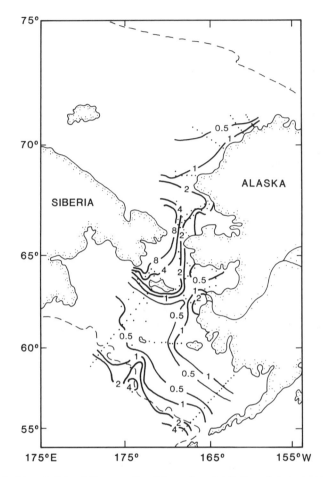

FIG. 103. A chlorophyll (μg l^{-1}) composite of the surface distribution of phytoplankton biomass within the Bering/Chukchi Seas during June–August 1978–1984 ($N = 180$). [After Walsh *et al.* (1986a); © E. J. Brill, with permission.]

compositions of the two water types are also distinct in the northern Bering/ Chukchi Seas. During the summer of 1985, for example, diatom populations constituted the high algal biomass (>10 μg chl l^{-1}) of Anadyr Water, while flagellates dominated the low standing stocks (<1 μg chl l^{-1}) of Alaska Coastal Water (P. McRoy, personal communication).

The summer surface chlorophyll distributions during June–August 1978–1984 within the Bering/Chukchi Seas (Fig. 103) accordingly reflect this pattern of nitrate supply. North of St. Lawrence Island, for example, an order of magnitude more chlorophyll (8–16 μg chl l^{-1}) was found on the western side of Bering/Chukchi Seas than on any of their eastern sides (0.5–1.5 μg chl l^{-1}),

from the Aleutian Island Chain to the Arctic Ocean. During a joint U.S.–
U.S.S.R. cruise on the *Akademik Korolev* through Anadyr Strait in 1984, 10–20
μg chl l^{-1} were observed within Anadyr Water at 15- to 30-m depths, as well,
similar to the chlorophyll plumes (Fig. 59) found off the Peru coast (Walsh *et
al.*, 1980). This is not surprising, since based on the measured current speeds in
Anadyr Strait (Coachman and Aagaard, 1981), the distance from the shelf break,
and a nutrient source depth of 150–200 m, a vertical motion of ~8 m day^{-1} is
implied during transit of slope water to the Bering Strait (Sambrotto *et al.*,
1984). In contrast, an effective upward velocity of ~1 m day^{-1} was calculated
for the southeastern Bering Sea (Coachman and Walsh, 1981), where the annual
primary production is 165 g C m^{-2} yr^{-1}, instead of 285 g C m^{-2} yr^{-1} within
Bering Strait, despite the longer growing season farther south.

4.2.1 Outer Shelf

The fate of this Bering Sea primary production is considered in annual carbon
budgets of the outer, middle, and inner shelf domains, which support large
pelagic and demersal fisheries, as well as extensive populations of sea birds and
mammals. Within these regions, algal production is apparently transferred to
higher trophic levels through two distinct pathways. On the outer shelf, primary
production supports an essentially pelagic food web, a suite of species that are
grazers or predators in the water column, leading to the pelagic fishery for the
Alaskan pollock (Fig. 104). By the mid-1970s, for example, pollock had replaced the Peru anchovy as the object of the world's largest single species fishery,
with total landings of 5 × 10^6 tons yr^{-1} during 1974–1978, of which 20–50%
was derived from the southeastern Bering Sea (Fig. 189). The benthos on the
outer shelf is a relatively small proportion of the present yield.

The situation is reversed on the middle and inner shelves, however, where
less of the primary production is grazed by zooplankton but settles instead to the
bottom, nurturing a rich benthic food web (Figs. 105 and 106). This spatial
distinction of pelagic and benthic food webs within the Bering Sea annual carbon
budgets is possible because of the broad width and distinct flow fields of these
shelves, allowing separation of both oceanographic and ecological processes, as
indicated by the distribution and abundance of numerous shelf species whose
feeding habits are known. The pelagic, oceanic animals—ranging from the ontogenetic copepods and large fin whales, seals and birds, such as the red-legged
kittiwake, *Risa brevirostris*—are confined to the outer shelf domain of the southeastern Bering Sea, the realm of the pelagic food web of the Alaska pollock
(Iverson *et al.*, 1979), and to Anadyr Water in the northwestern Bering Sea.

On the southeastern central shelf the macrobenthos reaches its greater abundance (Haflinger, 1981), and it is in this region that benthic predators congre-
gate. Walruses are abundant here, and there used to be intensive fisheries for the

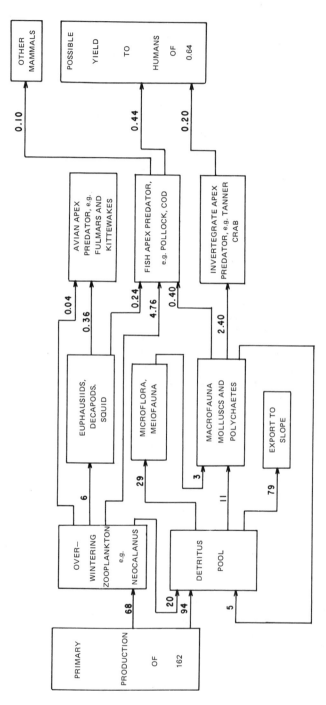

FIG. 104. An annual carbon budget (g C m^{-2} yr^{-1}) of the food web on the outer shelf of the southeastern Bering Sea. [After Walsh and McRoy (1986); © 1986 Pergamon Journals Ltd., with permission.]

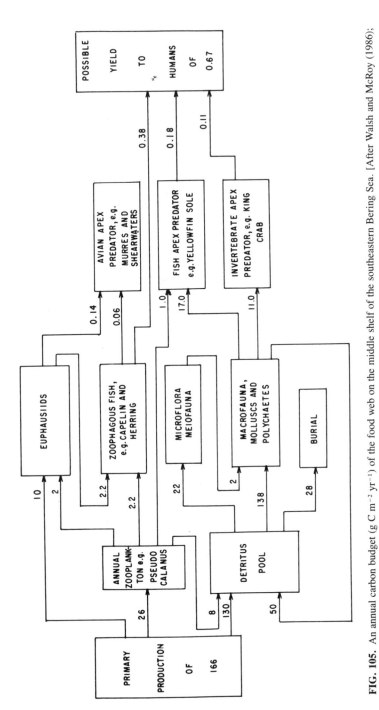

FIG. 105. An annual carbon budget (g C m^{-2} yr^{-1}) of the food web on the middle shelf of the southeastern Bering Sea. [After Walsh and McRoy (1986); © 1986 Pergamon Journals Ltd., with permission.]

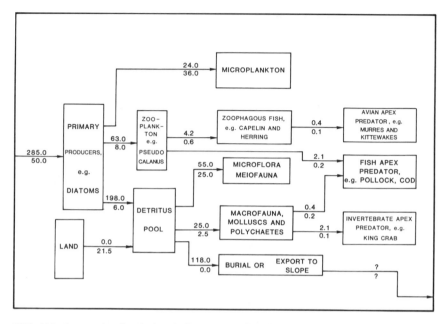

FIG. 106. An annual carbon budget (g C m⁻² yr⁻¹) of the food webs within Anadyr Water (upper values) and Alaska Coastal Water (lower values) on the inner shelves of the northern Bering and Chukchi Seas. [After Walsh *et al.* (1986a); © E. J. Brill, with permission.]

yellowfin sole and king crab, *Paralithodes camtschatica* (Pereyra *et al.*, 1976). The stocks of one of these apex predators of the middle shelf, red king crab, collapsed in 1982, however, and the yield of Tanner crab from the outer shelf was 165% greater that year (L. Incze, personal communication).

The Bering Sea zooplankton community (Ikeda and Motoda, 1978; Iverson *et al.*, 1979; Cooney and Coyle, 1982) is similarly partitioned into two size classes of different life histories. The first consists of short-lived, small organisms such as *Pseudocalanus* spp., *Oithona* spp., *Oncaea* spp., and *Acartia longiremus*, whose cohorts develop after the spring bloom (Heinrich, 1962b) and extend mainly across the inner and middle shelves. The larger, long-lived species, such as *Neocalanus cristatus, N. plumchrus,* and euphausiids, occupy mostly the outer shelf in the southeastern Bering Sea and Anadyr Water in the northern Bering/Chukchi Seas. They originate from slope water, where an overwintering population of the larger zooplankton introduces immature stages into surface water each spring as a result of ontogenetic migration (Heinrich, 1962a; Minoda, 1972), similar to other areas of the North Pacific (Fulton, 1973; McAllister *et al.*, 1960)—recall Section 1.5.2.

These different reproductive strategies of zooplankton within the shallow and deep parts of the Bering Sea and the eastern and western sides of the Chukchi Sea partially determine the timing of grazing stress exerted on the spring bloom, that is, the amount of phytoplankton carbon, either diverted directly to the pelagic food web or left behind in the water column to eventually sink to the demersal food web. Based on copepod ingestion (Dagg *et al.*, 1982), respiration, and growth rates (Vidal and Smith, 1986), for example, the small zooplankton prey of the larval pollock *and* the larger, ontogenetic migrators together remove on the outer-shelf only 68 g C m^{-2} yr^{-1} (Fig. 104). The rest of the outer shelf primary production evidently sinks out of the water column as phytodetritus, either to be consumed on the bottom or to be advected seaward (Walsh, 1983).

The macrobenthos on the outer shelf has a relatively small biomass with an estimated annual production of 2.8 g C m^{-2} yr^{-1} (K. Haflinger and H. Feder, personal communication). The amount of carbon input to the micromeiobenthos, based on bottom respiration measurements in the Bering Sea during June 1981, is estimated to be 22–29 g C m^{-2} yr^{-1} (W. Phoel, personal communication). These data indicate that the carbon demands of the benthic organisms are similar to those at 4°C in the Mid-Atlantic Bight (Florek and Rowe, 1983). In this budget, 20% transfer efficiencies were assumed for the zooplankton and benthic macrofauna, while for most of the other food web components a 10% efficiency was used (Fig. 104).

A few additional assumptions were necessary to deal with higher trophic levels which can span several year classes, that is, over a longer period than the present calculations (Figs. 104–106). For the demersal fish and invertebrate apex predators of the middle shelf, for example, a production : biomass (P : B) ratio of 0.1 was assumed, since bottom temperatures can be as low as 1°C, in contrast to surface temperatures >10°C on the outer shelf. To allow for these temperature differences, the major fish apex predator in the pelagic system of the outer shelf (pollock) was assumed to have a P : B of 1.0 (Fig. 104). These different P : B ratios were also based on fishery statistics showing that, although pollock can live up to 15 yr, about 70% of the annual pelagic catch are 2- to 3-yr-old fish (Francis and Bailey, 1983); in contrast, the yellowfin sole can also live to 15 yr, but 70% of the demersal catch was 5–10 yr old in the 1970s (Pereyra *et al.*, 1976), and 10–15 yr old in the 1980s (Bakkala and Low, 1983).

With a P : B ratio of 1 for pollock, a steady-state yield of 0.44 g C m^{-2} yr^{-1} to humans (Fig. 104) implies a pollock biomass of 0.44 g C m^{-2}, or 7.3 g wet weight (ww) m^{-2}, on the outer shelf. This is, in fact, the average density observed by G. B. Smith (1981). The peak yields of pollock from 1971 to 1974 were 1.7 × 10^6 tons yr^{-1}, or 7.7 tons km^{-2} yr^{-1} (0.46 g C m^{-2} yr^{-1}) over the 2.2 × 10^5-km^2 area of the outer shelf. Other pelagic animals—fin whales, seals, and surface-feeding birds—were assumed to be in the outer-shelf domain and to remove carbon as well (Fig. 104). With these generous assumptions, however, about half of the primary production of the outer shelf is still unaccounted for and must be buried or exported from the shelf (Walsh *et al.*, 1981). To check on the assumptions and implications of this carbon budget, a second, independent budget for the middle shelf was constructed.

4.2.2 Middle Shelf

In contrast to the outer shelf, the food web of the central shelf is primarily benthic, because of the larger amount of algal production that apparently sinks to the sea floor. Only 26 g C m^{-2} yr^{-1} is taken annually by zooplankton (Fig. 105). Macrobenthic infaunal biomass at midshelf instead varies from <4 to >24 g C m^{-2}, 10-fold that of the outer shelf (Haflinger, 1981). A maximum secondary production of 28 g C m^{-2} yr^{-1} was assumed (Fig. 105) for the infauna and epifauna in this budget (H. Feder and K. Haflinger, personal communication). This is also 10-fold that of the estimated production of the outer-shelf macrobenthos.

Because of the unusually large infaunal biomass on the middle shelf, the annual respiration demands of this component of the benthos were independently estimated. Using a respiration rate for *Mya* at 3°C of 0.3 μl O$_2$ mg dry weight^{-1} (dw) hr^{-1} (Kennedy and Mihursky, 1972), the maximum observed biomass of middle-shelf infauna, and a 0.5 g C g dw^{-1} conversion, we obtain a respiration estimate of 14.4 ml O$_2$ m^{-2} hr^{-1}. Such a respiration rate is approximately threefold that of the middle-shelf micromeiobenthos, as measured by shipboard incubation of sediment cores in June 1981. On an annual basis, the carbon equivalent of this respiration would be ~62 g C m^{-2} yr^{-1}, or a sum of 84 g C m^{-2} yr^{-1} for total release of CO$_2$ from the three size classes of the benthos on the middle shelf (Fig. 105).

Assuming that a 20% macrobenthic growth efficiency yields the 28-g C m^{-2} yr^{-1} secondary production, which is all consumed by predators within an annual steady state, an annual ingestion demand of 140 g C m^{-2} yr^{-1} is required for the macrobenthos populations on the middle shelf (Fig. 105). Their respiration flux of 62 g C m^{-2} yr^{-1}, with a combined growth and reproductive increment of 28 g C m^{-2} yr^{-1}, implies that the assimilation efficiency of the macrobenthos is 64%—that is, a fecal flux of 50 g C m^{-2} yr^{-1} is available for possible reingestion (Fig. 105). The assimilation efficiency of a filter feeder such as the Pacific oyster, *Crassostrea gigas*, is only ~40% (Bernard, 1974), however. The efficiency of infaunal deposit feeders may be even less, if the phytodetritus is of poor food quality (i.e., with a high C : N ratio) as a result of prior remineralization within the water column or sediments.

The C : N content of surface sediments on the middle Bering shelf is <6, however, similar to the average for live phytoplankton, while the C : N ratio in these sediments increases seaward toward the shelf break (Walsh *et al.*, 1985).

As the result of weak residual currents on the middle shelf, low pelagic grazing stress, and direct phytoplankton input to the bottom, the macrobenthos may be unusually efficient herbivores here. These herbivores account for a disproportionate share of the metabolic fluxes, compared to the smaller size classes within usual benthic communities, such as the outer Bering shelf (Fig. 104), or the Mid-Atlantic Bight. Furthermore, they constitute the basis of an unusually productive benthic food web, for example, the past yields of Alaska king crab.

With respect to fish apex predators, annual yields of $4-5 \times 10^5$ tons yr^{-1} of yellowfin sole during the early 1960s reduced the biomass of exploitable fish (>6 yr old) to $\sim$40% of the virgin stocks by 1963 (Bakkala, 1981). A 10-fold decline of their harvest in the mid-1970s, to $0.4-0.5 \times 10^5$ tons yr^{-1}, however, then led to an increase in stocks by 1978 to approximately 70% of virgin biomass. These interannual changes of sole biomass, presumably in response to varying harvest rates, suggest that the actual secondary production of yellowfin sole on the middle shelf (4×10^5 km^2) is <2 tons km^{-2} yr^{-1}, but >0.2 tons km^{-2} yr^{-1}. The second budget estimate (Fig. 105) of the potential midshelf fish harvest thus includes a yield of 1.4 tons km^{-2} yr^{-1} (0.08 g C m^{-2} yr^{-1}) for the second-most-abundant groundfish in the Bering Sea (after pollock). A yield of 0.10 g C m^{-2} yr^{-1} is also estimated for all other demersal fish (i.e., halibut, Alaska plaice, arrowtooth flounder, flathead sole, and rock sole). Furthermore, a yield of 0.10 g C m^{-2} yr^{-1} is estimated for pollock recruited from this part of the shelf during warm years. Finally, this second carbon budget also suggests that comparatively small herbivore production in the midshelf water column might nourish a large stock of zoophagous fish, such as herring or capelin (Wespestad and Barton, 1981).

After satisfying the annual carbon demands of all these higher trophic levels, only $\sim$17% of the midshelf primary production is left for additional consumption and burial; that is, within the myriad of uncertainties inherent in such an exercise, the middle-shelf carbon budget is probably balanced. The role of bacterioplankton is considered in the next budgets of the inner shelf. An independent check is also available from measurements of dissolved CO_2 in the midshelf water column (Codispoti et al., 1986a). From May 15 to July 15 in 1980 and 1981, the average respiration estimate ($N = 11$) was 0.20 g C m^{-2} day^{-1}, a net increase of CO_2 in the water after previous depletion by the spring bloom. Recall that the annual carbon budget for the middle shelf (Fig. 105) suggested a total respiration for the meiomicro- and macrobenthos of 84 g C m^{-2} yr^{-1}, or 0.23 g C m^{-2} day^{-1}.

At a seaward flow of only 1 cm sec^{-1} on the middle shelf of 200 km width and of 5 cm sec^{-1} on the outer shelf of 150 km width, a particle would take 267 days to exit the shelf break from the 50-m isobath. Wind forcing for such off-shore transport usually ends by May, however, $\sim$30 days after initiation of the spring bloom, or 10% of such a transit time. Therefore, it is not surprising that all of the phytoplankton production might be consumed by zooplankton, fish, and benthos on the middle shelf rather than transferred to slope depocenters. From the 100-m isobath, however, the transit time for shelf export would be only ~35 days. An analysis of 9- to 18-m sec^{-1} wind events during spring of 1979–1981 in the Bering Sea (Walsh, 1983) indicated, furthermore, that 50% were favorable for offshore transport of water and phytoplankton in a surface Ekman layer, as indicated by changes in wind direction and algal biomass on the outer and middle shelves. In some years, such northerly wind events persist throughout the summer (McLain and Favorite, 1976). Within 100–150 km of the shelf break, resuspension and seaward transport to the Bering slope may be the fate of outer shelf phytodetritus. Of the 94 g C m^{-2} yr^{-1} of primary production left behind by outer shelf zooplankton (Fig. 104), for example, as much as 84% may be exported to the adjacent slope (Walsh et al., 1985). This hypothesis is tested with construction of two more carbon budgets (Fig. 106) for inner shelf regions of very different primary production, 285 g C m^{-2} yr^{-1} within Anadyr Water and 50 g C m^{-2} yr^{-1} within Alaska Coastal Water.

4.2.3 Inner Shelf

Within shallow waters, the importance of the benthic community in terms of both carbon metabolism and nitrogen remineralization increases. For example, the flux of regenerated inorganic nitrogen from the benthos in Anadyr Water could account for 6–10% of the estimated daily nitrogen assimilation by phytoplankton here. Within a 750-km transect along the International Date Line on the inner shelf, the near-bottom ammonium concentration increased downstream from 1.5 μg-at. NH$_4$ l^{-1} in Anadyr Strait to >6.0 μg-at. NH$_4$ l^{-1} in the Chukchi Sea (T. Whitledge, personal communication). Similarly, shipboard ^{15}N incubations suggested that "new," (i.e., based on nitrate) production of the phytoplankton was 86% within Anadyr Water off St. Lawrence Island, compared to 32% in Alaska Coastal Water near Cape Lisburne, during the summer of 1985 (J. Goering, personal communication).

Over this length scale of >6° latitude, the benthic infaunal biomass of the northern Bering/Chukchi Seas changes by an order of magnitude (Stoker, 1981), from ~2.5 g C m^{-2} at 60°N to ~25.0 g C m^{-2} at 66°N, again 10-fold that of the outer shelf. An annual P : B ratio of 0.1 for these organisms at 0°C, and a 10% food chain efficiency (Walsh and McRoy, 1986), imply respective annual phytodetrital carbon demands of at least 2.5–25.0 g C m^{-2} yr^{-1}. They would consume only 1–10% of the annual primary production of Anadyr Water, however, leaving the rest either to be consumed by other heterotrophs or to be exported northward.

During August 1983 and July 1984, bottom metabolism measurements of oxygen consumption, sulfate reduction, and nitrification were made from St.

Lawrence Island to the Bering Strait, in sediments of relatively low organic carbon (J. Grebmeier, personal communication). Carbon consumption by the small benthic organisms beneath the chlorophyll-rich Anadyr Water was two- to threefold higher than within sediments beneath Alaska Coastal Water, ranging respectively from $0.14-0.25$ g C m^{-2} day^{-1} to $0.07-0.11$ g C m^{-2} day^{-1} (T. Blackburn and K. Henriksen, personal communication). Assuming that the winter (November–May) carbon consumption of these smaller benthos here is 60% that of ice-free conditions during June–October, their annual mean food requirement might be 55.0 g C m^{-2} yr^{-1} for those benthos under Anadyr Water and 25.0 g C m^{-2} yr^{-1} for residents of the Alaska Coastal Water (Fig. 106). These estimates are respectively 19% and 50% of an annual primary production of 285 g C m^{-2} yr^{-1} within Anadyr Water and 50 g C m^{-2} yr^{-1} within Alaska Coastal Water.

If it is assumed that the carbon demands of the infaunal biomass were not included in the shipboard measurements of oxygen consumption and ^{15}N, ^{35}S turnover within cores, the input of carbon to the total benthos must be greater. The macrobenthic biomass was only 58 g ww m^{-2} under Alaska Coastal Water in August 1983, compared to 545 g ww m^{-2} within Anadyr Water, such that most of the above estimated infaunal metabolic demands would occur beneath Anadyr Water—we assume it is 10-fold that of Alaska Coastal Water (Fig. 106). The estimates of the total benthic carbon demand then become respectively 80.0 g C m^{-2} yr^{-1} and 27.5 g C m^{-2} yr^{-1}, or 28% and 55% of the annual primary production (Fig. 106).

Rates of benthic nitrification within Anadyr Water and Alaska Coastal Water were respectively 6.3 mg NO$_3$ m^{-2} day^{-1} and 4.2 mg NO$_3$ m^{-2} day^{-1}, or 2.5% and 42% of the phytoplankton uptake rates of ^{15}NO$_3$. These estimates were in the same range measured within Danish waters at similar seasonal temperatures and depths (Henriksen et al., 1981; Blackburn and Henriksen, 1983). Over 150 days of ice-free conditions, 0.95 g NO$_3$ m^{-2} might be released to a column of Anadyr Water, about 15% of the depth-integrated inventory of nitrate in June. At these rates of in situ production of nitrate on the Bering/Chukchi shelves, the source of "new" production in Anadyr Water clearly depends on "preformed" nitrate, advected northward from source waters of the Bering continental slope.

Assuming that the new production is 80% in Anadyr Water and only 30% in Alaska Coastal Water, about 0.37 g C m^{-2} day^{-1} of primary production in the former and 0.23 g C m^{-2} day^{-1} in the latter must be supported by a daily flux of recycled nitrogen. Daily fluxes of recycled nitrogen, in the form of NH$_4^+$, from the micro- and meiobenthos were actually somewhat smaller than those measured for nitrate, respectively 2.1 mg NH$_4^+$ m^{-2} day^{-1} and 2.8 mg NH$_4^+$ m^{-2} day^{-1} within the Anadyr Water and Alaska Coastal Water. Shipboard experiments (T. Blackburn and K. Henriksen, personal communication) with the benthic infaunal species of the Bering/Chukchi shelves indicated, however, that the flux of NH_4^+ from the sediments increased considerably in the presence of these zoobenthos by about 150–350%.

Estimates of infaunal abundance and these shipboard rate experiments suggest that an additional ammonium flux of 9.8 mg NH_4^+ m^{-2} day^{-1} might occur beneath Anadyr Water and 2.8 mg NH_4^+ m^{-2} day^{-1} within Alaska Coastal Water. If mixed into the euphotic zone, a total ammonium flux of 11.9 mg NH_4^+ m^{-2} day^{-1} and 5.6 mg NH_4^+ m^{-2} day^{-1}, with a C/N assimilation ratio of 6 : 1, would support respective carbon fixations of 0.07 g C m^{-2} day^{-1} and 0.03 g C m^{-2} day^{-1} for phytoplankton populations on the western and eastern sides of the Bering Strait. The remainder of the carbon productivity supported by recycled nitrogen, 0.31 g C m^{-2} day^{-1} and 0.20 g C m^{-2} day^{-1}, must be supplied from excretory products of zooplankton and bacterioplankton within the water column, since nutrient inputs from higher trophic levels would be minimal.

Estimates of bacterial production in the northern Bering/Chukchi Seas were obtained both with ^{15}N techniques and from estimates of the grazing rates by their predators, heterotrophic microflagellates (P. Andersen, personal communication). The mean bacterioplankton biomass in Alaska Coastal Water (13.5 mg C m^{-3}) was somewhat larger than that of Anadyr Water (8.4 mg C m^{-3}), while the average abundance of microflagellates exhibited a similar trend of respectively 50.6 ml^{-1} and 41.4 ml^{-1}. Bacterial production rates of 0.4–0.6 mg C m^{-3} day^{-1} in the Bering Sea were similar to those found in Antarctic waters (Fuhrmann and Azam, 1980). Over a 40-m water column, however, a bacterial secondary production rate of 16–24 mg C m^{-2} day^{-1} is only about 1% of the daily primary production of Anadyr Water, similar to observations during the spring bloom in the New York Bight (Walsh et al., 1986b).

Assuming a 90% respiratory loss of bacterioplankton carbon over the 150 days of summer–fall, annual carbon inputs to these organisms might be at least 24.0 g C m^{-2} yr^{-1} within Anadyr Water and 36.0 g C m^{-2} yr^{-1} in Alaska Coastal Water (Fig. 106). The source of this carbon need not be uneaten diatom cells, however. Zooplankton fecal pellets and dissolved organic matter are other food supplies, and are not considered in these budgets; they are discussed in Sections 4.3.4 and 4.3.6. In any case, if bacterioplankton, in the process of remineralization of detritus for their own body nitrogen, were to also incorporate only 10% of the nitrogen in a C/N ratio of 6 : 1, with the remainder lost as dissolved amino acids, urea, and ammonium to the water column, as much as 24–36 mg N m^{-2} day^{-1} might be added to the respective recycled nitrogen pools of Anadyr Water and Alaska Coastal Water. In this situation, the recycled nitrogen requirement of algal carbon production in Alaska Coastal Water can be met by excretory products of the bacterioplankton and benthos.

However, at least 28 mg N m^{-2} day^{-1} must be supplied by zooplankton excretion within Anadyr Water to support the remaining 0.17 g C m^{-2} day^{-1} derived from recycled nitrogen. The same swift current, which carries high nitrate concentrations from the shelf break of the northwestern Bering Sea to Anadyr Strait, imports copepods of the outer shelf domain here as well. *Neocalanus plumchrus* is routinely found within these inner shelf waters (Springer and Roseneau, 1987), and rarely seen on the inner shelf of the sluggish southeastern Bering Sea (Cooney and Coyle, 1982). With an assimilation efficiency of 60%, a C/N ratio of 6, and this required macrozooplankton excretion rate, the offshore zooplankton of the Anadyr Water must then ingest 70 mg N m^{-2} day^{-1}, or 0.42 g C m^{-2} day^{-1} (i.e., 22% of the daily primary production).

The same offshore zooplankton species on the outer shelf in the southeastern Bering Sea (Fig. 104) consumed 42% of the annual production (Walsh and McRoy, 1986). These grazing losses are compared to only 16% on the southeastern Bering midshelf (Fig. 105) by similar inshore zooplankton species found both there and in Alaska Coastal Water. Since a smaller number of offshore zooplankton species would be expected to transit the longer distance over both the outer and middle shelves of the northwestern Bering Sea, an annual zooplankton grazing rate of 22% of the primary production of Anadyr Water might be a reasonable conclusion of these budgets; perhaps 16% is also consumed by zooplankton in Alaska Coastal Water (Fig. 106).

Losses of the oceanic zooplankton biomass during transit of the northern Bering Sea, from Anadyr to Bering Strait, appeared to be negligible in 1985, with a range in input of zooplankton fluxes of 2.4–7.4 g C day^{-1} to the Chukchi Sea (A. Springer, personal communication). Planktivorous birds on the Diomede Islands removed <2% of this zooplankton supply, with their dietary patterns clearly demarcated by δ ^{15}N signatures. Piscivorous birds were enriched in ^{15}N by 4‰ more than the avian predators of zooplankton (P. Parker, personal communication), for example, as would be expected for terminal consumers of a longer food web (Fig. 106).

To summarize, a carbon consumption of 63.0 g C m^{-2} yr^{-1} by zooplankton, 24.0 g C m^{-2} yr^{-1} by bacterioplankton, and 80.0 g C m^{-2} yr^{-1} by the benthos still leaves 118.0 g C m^{-2} yr^{-1} of algal production in Anadyr Water to be either exported to the Arctic Ocean or buried in the Chukchi sediments (Fig. 106). This amounts to 41% of the "new" production, derived mainly from Pacific slope waters. In contrast, a total carbon ingestion by the consumer communities of Alaska Coastal Water of 71.5 g C m^{-2} yr^{-1} requires an additional input of at least 21.5 g C m^{-2} yr^{-1} from terrestrial sources (Fig. 106), that is, the Yukon River, assuming that *no* detrital carbon exits the Chukchi Sea within this water type.

In fact, sediments of the Bering/Chukchi shelves mainly have an organic ^{13}C signal of plankton (-19.8 to -21.9), except for those sediments with a terrestrial marker (-23.1 to -27.5) off the mouth of the Yukon River and within the adjacent Norton Sound (P. Parker, personal communication). The significance of these isotope tracers will be discussed in Sections 5.2.1 and 5.3.1, but they clearly support this last budget's suggestion of an import of terrestrial carbon within Alaska Coastal Water.

With specification of the amount of phytoplankton consumed annually by both pelagic and benthic herbivores and saprovores, reasonable carbon budgets (Figs. 104–106) appear to reflect cross-shelf differences in the physical habitat of the Bering/Chukchi Seas. At an inner-shelf primary production level of only 50 g C m^{-2} yr^{-1} within a 10- to 50-cm sec^{-1} flow regime, little algal carbon apparently escapes biological utilization; a terrestrial subsidy of organic debris may, in fact, be required here. Similarly, with an annual phytoplankton fixation of 165 g C m^{-2} yr^{-1} within a 1- to 5-cm sec^{-1} flow regime, a large benthic community apparently consumes most of the production of the overlying water column on the middle shelf. Phytodetritus of the outer-shelf food webs, supported by an input of 165–285 g C m^{-2} yr^{-1}, however, is evidently removed to the adjacent slopes within both the 1- to 5-cm sec^{-1} flow field of the southeastern Bering Sea and within the 10- to 50-cm sec^{-1} flow field of the northern Chukchi Sea.

With export of excess food from the outer-shelf food web as a regular phenomenon, most pelagic organisms may not be food limited in the Bering/Chukchi Seas. In this context, for example, the body size of *Neocalanus* CV during summer on the outer Bering shelf (Vidal and Smith, 1986) is almost threefold that of its counterparts in the open North Pacific Ocean (Frost *et al.*, 1983), suggesting little food limitation of this species in the coastal area. The annual landward excursions of pollock to the middle shelf and of these large zooplankton to the outer shelf may be viewed as an evolutionary foray from the food-limited open ocean during atypical periods of this Ice Age, such as the presence of continental shelves. On shelves of narrower width, such as the Mid-Atlantic Bight, where a benthic community is not isolated within a weak flow field, export should instead occur over the whole shelf.

4.3   Mid-Atlantic Budget

The impressive fish yield of the New England shelf was described as early as 1616 by Captain John Smith in his "Generall historie of Virginia, New England, and the Summer Isles." Voluminous information on prey preferences and metabolic requirements of this fish community on the mid-Atlantic shelf (Table VIII) was also compressed into annual estimates of their carbon demands (Fig. 107). Confrontation of these data with estimates of primary production, which increased from 170 g C m^{-2} yr^{-1} (Ryther and Yentsch, 1958) to 300 g C m^{-2} yr^{-1} (Walsh *et al.*, 1978; O'Reilly and Busch, 1984), raised questions about the fate of primary production in this and other shelf ecosystems (Walsh *et al.*, 1981). Unlike the Bering Sea, however, primary production occurs year-round

Table VIII.

Dominant Commercial Fish Biomass, Excluding Menhaden, and Their Annual Food Consumption
(Metric Tons $\times$ 10^3) on the Northeast Continental Shelf[a]

	1963–1965		1968–1969		1972–1974	
	Biomass	Food	Biomass	Food	Biomass	Food
Zooplankton feeders						
Herring	3566	26,026	1137	8,297	310	2,261
Redfish	165	1,335	287	2,356	195	1,601
Butterfish	148	1,409	168	1,600	161	1,529
Mackerel	472	4,065	3266	28,159	1418	12,225
Alewife	102	767	43	323	29	218
Subtotal	4453	33,602	4901	40,735	2113	17,834
Benthos feeders						
Red hake	279	1837	186	1220	84	553
Haddock	783	5221	254	1692	161	2120
Thorny skate	272	4690	289	4990	237	4083
Little skate	146	1540	43	455	97	1022
Winter skate	129	1226	40	385	83	787
Winter flounder	71	504	36	257	38	269
Yellowtail flounder	77	610	94	746	41	325
Ocean pout	105	362	36	124	27	93
American plaice	64	312	39	192	28	138
Scup	68	364	38	201	79	422
Subtotal	1994	16,666	1055	10,262	875	9812
Fish feeders						
Silver hake	999	11,487	414	4,763	523	6008
Pollock	216	1,567	213	1,546	245	1776
Dog fish	1002	16,431	1024	16,973	362	5938
Cod	156	1,335	106	2,356	141	1601
Barn door skate	116	997	10	84	11	96
Goose fish	74	1,115	37	564	41	712
Subtotal	2563	32,932	1804	26,286	1323	16,131
Total	9,010	83,200	7760	77,283	4311	43,777
All commercial fish	9,720	89,783	8665	83,308	5350	51,215

[a] After Edwards and Bowman (1979).

in the Mid-Atlantic Bight (Fig. 108), with only the spring bloom escaping consumption by higher trophic levels. We thus must consider both annual and spring bloom budgets of carbon and nitrogen.

The annual nutrient cycle and its qualitative transformation into higher trophic levels within the Mid-Atlantic Bight have been discussed in Section 1.3. The classic diatom–copepod–herring food web was portrayed, with scant dis-

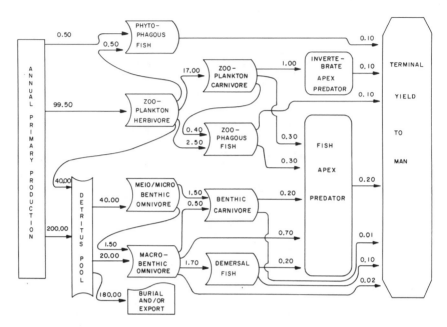

FIG. 107. An annual carbon budget (g C m^{-2} yr^{-1}) of the food web on the New York shelf. [After Walsh (1981a); © Academic Press.]

cussion of energy utilization by bacteria, protozoans, and gelatinous zooplankton (Pomeroy, 1974). A quantitative analysis, in this section, of data from 113 cruises in the Mid-Atlantic Bight will generate such an updated budget of the nitrogen cycle, inherent in the original carbon budget of Fig. 107. As in the case of the inner Bering/Chukchi shelves, the role of the bacterioplankton–protozoan food chain, as well as dissolved organic matter, is now explicitly considered.

We will find, however, that ~41% of the nitrogen demand of the annual primary production, off either New York or on Georges Bank, must still be supplied by physical fluxes of NO$_3$ as slope input (Table IX) at the shelf boundaries. This is not significantly different from the 47% "new production," estimated on the basis of 30 earlier cruises taken in the New York Bight during 1974–1976 (Walsh *et al.*, 1978). Furthermore, an independent seasonal analysis of the nutrient supply to Georges Bank (Klein, 1986) similarly suggests that recycled nitrogen contributes to only 28% of the winter production and 60% of the summer primary production. Another recent study of the nitrate supply to Georges Bank (Loder *et al.*, 1982) concludes, as well, that about 50% of the nitrogen demand during spring, summer, and fall is supplied from external sources, allowing a potential export of 50% of the primary production as either fish or detrital carbon.

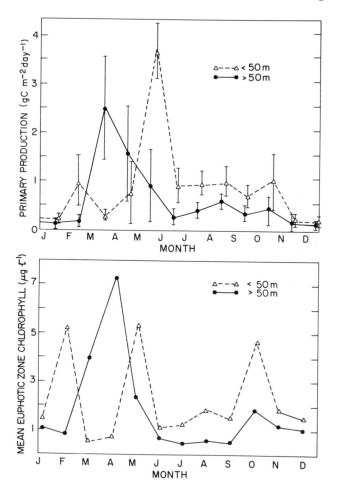

FIG. 108. Seasonal changes of primary production (g C m^{-2} day^{-1}) and chlorophyll biomass (μg l^{-1}) within nearshore (<50 m) and offshore (>50 m) waters of the New York shelf. [After P. G. Falkowski, personal communication.]

4.3.1 Microbial Nitrification

The source of nitrate on the mid-Atlantic shelf is, of course, bacterial nitrification within either slope or shelf waters, with lateral exchange from the first and vertical exchange in the second. Direct estimates of nitrification and ammonification on European shelves at similar latitudes range from 4–20 mg NO$_3$ m^{-2} day^{-1} and 37–50 mg NH$_4$ m^{-2} day^{-1} off the Danish coast (Henriksen *et al.*, 1981; Blackburn and Henriksen, 1983) to 22–48 mg NO$_3$ m^{-2} day^{-1} and 37–74

Table IX.

Contribution of Recycled Nitrogen to the Annual Nitrogen Demand
of Primary Productivity on Georges Bank
(80 g N m^{-2} yr^{-1}) and the New York Shelf (60 g N m^{-2} yr^{-1})[a]

Nitrogen demand	Georges Bank (g N m^{-2} yr^{-1})	New York Shelf (g N m^{-2} yr^{-1})
Recycled sources		
Microbial nitrification	1.5	3.0
Benthic ammonification	0.6	1.2
Sediment organic nitrogen	4.8	7.4
Zooplankton excretion	5.3	6.8
Bacterioplankton and protozoan remineralization	28.7	14.0
Phytoplankton release	7.5	1.5
Total recycling	48.4	33.9
Required slope input	31.6	26.1
Advective input: from 20 storms during 180 days of unstratified water column	34	34
Annual diffusive input Seasonal total over 180 days of unstratified water column	36.7	30.6
Seasonal total over 180 days of stratified water column	6.1	0
System losses		
Phytoplankton export	48.0	31.0
Fishing	0.1	0.1

[a] [After Walsh et al. (1986b); © MIT Press.]

mg NH$_4$ m^{-2} day^{-1} in shallow waters off the Belgian coast (Billen, 1978). As much as 50% of the NO$_3$ produced in these shelf sediments may be consumed during denitrification (Henriksen et al., 1981), however, such that only 10–20 mg NO$_3$ m^{-2} day^{-1} may be released to the overlying, shallow-water columns. At this rate on the New York shelf for 150 days of strongly stratified conditions of the water column (June–October), 1.5–3.0 g NO$_3$ m^{-2} yr^{-1} would be the contribution from in situ nitrification. This would be only 2.5–5.0% of the 60 g N m^{-2} yr^{-1} nitrogen demand (Table IX), with a C/N ratio of 5 : 1 and an annual primary production of 300 g C m^{-2} yr^{-1} (Fig. 107).

If bacterial nitrification occurs only in the lower 30 m of the water column, below the seasonal pycnocline but not in the sediments, then the resupply of the NO$_3$ to the euphotic zone by summer nitrification from the New York water column is estimated to be 0.05–0.10 g NO$_3$ m^{-3} yr^{-1}, similar to water column rates of 0.02–0.35 g NO$_3$ m^{-3} yr^{-1} off California and other regions (Ward et al., 1982). For the purposes of this analysis, an annual input is assumed of 3.0

g NO_3 m^{-2} yr^{-1} from nitrification in the sediments and water column off New York, and 1.5 g NO_3 m^{-2} yr^{-1} from a water column of 400 g C m^{-2} yr^{-1} production on Georges Bank (Table IX), in contrast to ~1 g NO_3 m^{-2} yr^{-1} within the colder Bering Sea. The precursor of this nitrogen resupply process on the shelf, ammonification, involves a diverse group of organisms, zooplankton, benthos, and microplankton. Their relative importance, as well as the ability of phytoplankton to compete for ammonium with the nitrifiers, both depend on the amount of vertical stratification of the water column.

4.3.2 Zooplankton Regeneration

Because of the traditional importance assigned to zooplankton as major consumers of phytoplankton blooms on the continental shelves (Walsh, 1983) and on Georges Bank (Schlitz and Cohen, 1981) in particular, specific grazing, respiration, excretion, and abundance studies of zooplankton were conducted from 1974 to 1978 in the New York shelf and Georges Bank regions (Dagg and Grill, 1980; Judkins et al., 1980; Vidal and Whitledge, 1982). The annual species succession and pulses of copepod biomass off New York and on Georges Bank reflect seasonal responses to both temperature cycles and the timing of phytoplankton blooms (recall Fig. 12). *Pseudocalanus minutus* and *Centropages typicus* are the dominant shelf copepods off New York and Georges Bank (Bigelow and Sears, 1939; Clarke, 1940; Sears and Clarke, 1940; Grice and Hart, 1962; Sherman, 1978; Judkins et al., 1980), while *Calanus finmarchicus* is mainly abundant offshore in slope waters and within the Gulf of Maine (Bigelow, 1926; Redfield, 1941; Riley, 1947). The offshore (>50 m) peak of zooplankton biomass occurs during May–June, as the deeper waters warm about 1 month after the March–April phytoplankton bloom on the outer shelf region. Inshore zooplankton (<50 m) become abundant during the even warmer period of July–August, during the lower phytoplankton biomass periods of May–June.

Assuming that the abundant copepods, *Calanus finmarchicus, Centropages typicus, Pseudocalanus minutus, Paracalanus parvus, Temora longicornis, Acartia* spp., *Penilia avirostris,* and *Oithona similis,* are all herbivores, their weight-specific grazing rates (Mullin and Brooks, 1976) and seasonal changes in biomass were used (Dagg and Turner, 1982) to calculate grazing stress in relation to seasonal food abundance during 1978 in the New York Bight and on Georges Bank. After a time lag in development of the offshore zooplankton cohorts during late spring, 30% of the 1.0 g C m^{-2} day^{-1} primary production of the declining spring bloom was removed by zooplankton on Georges Bank and 28% in the New York Bight (Fig. 108). Before this time, the grazing stress was a seasonal minimum of 1–7% of the late winter bloom production of 2–3 g C m^{-2} day^{-1} on Georges Bank (Riley and Bumpus, 1946; O'Reilly et al., 1986) and at midshelf off Long Island (Walsh et al., 1978).

With intense stratification of the water column in early summer, areas of $1.5-2.5$ g C m^{-2} day^{-1} primary production are confined to shallow waters, either receiving river discharge or subject to coastal upwelling within the New York Bight, where as much as 28–53% of the daily production can be consumed by zooplankton (Walsh *et al.*, 1978; Dagg and Turner, 1982). But only 5–10% of the $0.8-1.6$ g C m^{-2} day^{-1} summer production is consumed on Georges Bank, where the tidal stirring allows continual input of recycled nutrients to this system (recall Fig. 11). Before fall overturn and physical resupply of nutrients, the New York shelf zooplankton community consumed 118% of the late summer ~ 0.3 g C m^{-2} day^{-1} primary production off Long Island (Dagg and Turner, 1982) and 206% of the same production at the 65-m isobath (Falkowski *et al.*, 1983). In contrast, only 38% of the 0.8 g C m^{-2} day^{-1} fall production on Georges Bank was consumed by the zooplankton community (Dagg and Turner, 1982).

Compiling these seasonal estimates of zooplankton ingestion, we estimate that 83 g C m^{-2} yr^{-1}, or 21% of the annual primary production, is consumed by pelagic herbivores on Georges Bank, compared to 103 g C m^{-2} yr^{-1}, or 34% of the primary production, on the New York shelf. Assuming an excretion/ingestion ratio of $0.33 : 1.00$ (Walsh, 1975) and a C/N ratio of $5 : 1$, we obtain an annual macrozooplankton (>103 μm) excretory input of 5.5 g N m^{-2} yr^{-1} on Georges Bank and 6.8 g N m^{-2} yr^{-1} on the New York shelf (Table IX), about 10% of the annual nitrogen demands. Using biomass estimates of zooplankton caught with larger 333- and 165-μm nets during 1978 and 1979 on Georges Bank, and respective excretion estimates of 0.38 and 0.10 mg NH$_4$ m^{-3} day^{-1} for two size classes, Schlitz and Cohen (1981) suggested that the total copepod input of recycled nitrogen to Georges Bank might be 13.8 g NH$_4$ m^{-2} yr^{-1}. Their estimates, however, neglected seasonal temperature effects on plankton metabolism (e.g. Fig. 51).

Measured zooplankton excretion rates in August–September on the Nova Scotian (Fournier *et al.*, 1977) and New York (Harrison *et al.*, 1983) shelves are quite similar—0.35 and 0.41 mg NH$_4$ m^{-3} day^{-1}. Reflecting seasonal changes in temperature and zooplankton biomass, however, excretion rates on the Nova Scotian shelf in March, May, and November were, respectively, 0.03, 0.22, and 0.15 mg NH$_4$ m^{-3} day^{-1} (Fournier *et al.*, 1977). Applying an annual mean of only ~ 0.20 mg NH$_4$ m^{-3} day^{-1} for zooplankton return of nitrogen to the New York shelf, or Georges Bank, at the 60-m isobath, we obtain an independent estimate of 4.3 g N m^{-2} yr^{-1}. This is in reasonable agreement with the calculated nitrogen input from the more numerous, seasonal ingestion measurements, which also reflect ambient temperature changes.

Although the slope source waters for Georges Bank contain less nitrate than coastal upwelling water at the same latitude off Oregon (Fig. 9), the relatively continuous tidal flux of recycled nutrients and the seasonal nitrate input from slope water together lead to an annual primary production at the 60-m isobath of ~400 g C m^{-2} yr^{-1}, in contrast to $\sim200-300$ g C m^{-2} yr^{-1} on either the Oregon or the New York shelves. Yet grazing and excretion by copepods on Georges Bank consume only $\sim21\%$ of the annual primary production and supply only $\sim10\%$ of its nitrogen demand (Table IX), allowing a possible detrital input of ~64 g N m^{-2} yr^{-1} to the benthos. Similarly, in the nitrogen budget for the New York shelf (Table IX), the meager utilization of the primary production by the herbivorous zooplankton leads to an annual secondary production of 4 g N m^{-2} yr^{-1}, available for consumption by the pelagic food web, as well as a zooplankton fecal pellet flux of 8 g N m^{-2} yr^{-1} to the demersal food web. The rest of the annual plant production (~40 g N m^{-2} yr^{-1}) is presumably transferred to the shelf sediments, for either consumption and remineralization of this particulate nitrogen, or its resuspension and export to the continental slope.

4.3.3 Benthos Regeneration

The number of meiobenthic organisms in some areas of the Mid-Atlantic Bight, south of Martha's Vineyard ($0.1-1.0 \times 10^6$ animals m^{-2}; Wigley and McIntyre, 1964), for example, far outnumbers the macrobenthos (1.4×10^3 animals m^{-2}; Wigley and Theroux, 1979). They appear to be responsible, along with the bacteria and protozoa, for remineralization of material on the bottom, rather than serving as much of a pathway to higher trophic levels (Fenchel, 1969; Tenore, 1977). The detritus flux from the water-column production is thus assumed to be either buried, exported, or mainly remineralized by both the meiobenthos and the microbenthos in our analysis, since attempts to partition carbon flow through benthic communities in the past have shown that only a small portion is used by larger macrobenthic organisms (Smith, 1973; Smith et al., 1973), with the exception of the Bering Sea benthos.

Some phytodetritus and fecal matter must be used as an energy source by the two smaller size categories of benthos, and several studies have estimated their community metabolism off New York; for example, Thomas et al. (1979) measured nearshore bottom oxygen utilization by incubating cores aboard ship at in situ temperatures. Their results of 360 ml O$_2$ m^{-2} day^{-1} [or 54 g C m^{-2} yr^{-1} with a Respiratory Quotient (RQ) of 0.75] in shallow waters off New York were similar to seasonal oxygen demands of the inshore bottom biota (Smith et al., 1974) measured with in situ bell-jar incubations. During 1977–1979, similar core incubations of the oxygen demand were made over a much broader area of the shelf of the Mid-Atlantic Bight, which suggested the average level of meio- and microbenthic utilization of organic carbon at midshelf off New York might be ~43 g C m^{-2} yr^{-1}, in contrast to ~27 g C m^{-2} yr^{-1} on Georges Bank (Florek and Rowe, 1983). Since little refractory organic matter accumulates on the mid-Atlantic shelf, with a high C/N ratio as a result of selective release of dissolved nitrogen from the sediments (Walsh *et al.*, 1985), a C/N ratio of 5 : 1 is assumed for benthic metabolic processes. This suggests that a total of about 5.4–8.6 g N m^{-2} yr^{-1} of the annual nitrogen demand might be attributed to recycling of either ammonium or organic nitrogen by the benthic organisms, respectively, on Georges Bank and off New York (Table IX).

Direct rates of ammonium and nitrate flux from sediments of the Mid-Atlantic Bight to the overlying water column were also estimated (Walsh *et al.*, 1986b) from ~175 cores collected during two cruises in August and September 1980 to quantify the potential role that the seabed plays as a source of nutrients (Rowe *et al.*, 1975). Rates (averages of 2–4 seabed cores) of ammonium flux ranged from a maximum release of ~166 mg NH_4 m^{-2} day^{-1} in the nearshore, sewage sludge dumping area off New York to ~0.5 mg NH_4 m^{-2} day^{-1} at the edge of the shelf. Excluding special areas such as the estuaries, shelf valleys, and dump sites, ammonium release rates were higher on the shelf between Cape Cod and Delaware (3.4 mg NH_4 m^{-2} day^{-1}), than on regions of the shelf to the north on Georges Bank (1.7 mg NH_4 m^{-2} day^{-1}), or south to Virginia (1.0 mg NH_4 m^{-2} day^{-1}), where nitrification may have been dominant as indicated by a flux of ammonium into the sediments.

Similarly, the shelf macrobenthic biomass declines in a north–south gradient, from the Great South Channel off southern New England to Cape Hatteras within the Chesapeake Bight (Wigley and Theroux, 1979). In fact, there appears to be an order of magnitude less benthic biomass on most of Georges Bank and the shelf south of Cape May, compared with the benthos found off Martha's Vineyard (Emery and Uchupi, 1972). Several factors influence the rate of ammonium release from the sediment, such as the amount and nature of oxidizable organic matter, the temperature and ambient oxygen concentrations, the amount of macrobenthic and meiobenthic fauna, and the activity of bacteria present in the sediments. All affect the type of nitrogen conversions that take place and the rates at which they proceed—that is, whether dissolved organic nitrogen (DON), ammonium, nitrite, nitrate, or gaseous nitrogen and nitrous oxide make up the major product formed from decomposition of the particulate organic nitrogen.

These estimates of midshelf ammonification are an order of magnitude less than NH_4 releases observed in shallow waters off the Danish and Belgian coasts (Billen, 1978; Henriksen *et al.*, 1981; Blackburn and Henriksen, 1983). The measurements in Buzzards Bay (18 mg NH_4 m^{-2} day^{-1}), in New York Harbor (77 mg NH_4 m^{-2} day^{-1}), and in the Hudson Shelf Trough (94 mg NH_4 m^{-2} day^{-1}), however, exhibited a similar range to those encountered in European waters, such that the estimates of ammonium release from the 60-m isobath may be reasonable. It is possible that these shelf values represent a relatively high state of ammonium flux in the seasonal cycle as well, since bottom temperatures are approaching the annual maximum by September (recall Fig. 8). One would expect, for example, winter ammonification to be at its annual minimum, when detrital fallout has just begun, fluxes of particulate nitrogen off the shelf may be highest, and bottom water temperatures are lowest.

Assuming that the September 1980 ammonification rates are applicable to benthic fluxes throughout the year, we obtain an annual estimate of 1.2 g NH_4 m^{-2} yr^{-1} for benthic ammonification off New York and 0.6 g NH_4 m^{-2} yr^{-1} on Georges Bank (Table IX). This regional difference in ammonium released from bottom metabolism (8.6 g N m^{-2} yr^{-1} off New York, 5.4 g N m^{-2} yr^{-1} on Georges Bank) is similar to that obtained from the core measurements of oxygen consumption, but thus far only inorganic forms of dissolved nitrogen have been considered. Release of dissolved organic nitrogen from Buzzards Bay sediments in winter (98 mg DON m^{-2} day^{-1}) can be 10-fold the inorganic nitrogen flux (Christensen et al., 1983), while the summer release of organic nitrogen from Narragansett Bay (34–136 mg DON m^{-2} day^{-1}) may be of the same order as the inorganic flux (Nixon et al., 1976).

We have assumed that the remaining benthic nitrogen releases, estimated from the O_2 measurements, are in the form of DON, 4.8 g N m^{-2} yr^{-1} on Georges Bank and 7.4 g N m^{-2} yr^{-1} off New York. It is equally possible that the oxidation of organic matter within the incubated cores partly reflects nitrification rather than decomposition of DON; we may thus be overestimating the importance of the recycled nitrogen sources by adding the additional organic nitrogen term in the budgets. Before fall overturn, however, such DON input from the benthos at the 60-m isobath would reach only the euphotic zone of the tidally mixed Georges Bank in significant daily amounts. Previous respiration studies of the waters on the New York shelf and Georges Bank (Thomas et al., 1978, 1979) have, in fact, pointed out that remineralization of organic matter in the upper water column by bacterioplankton, above the seasonal thermocline, may instead represent a major source of recycled nutrients.

4.3.4 Microplankton Regeneration

Based on total water column and seabed respiration measurements on Georges Bank in March and July 1977 (Thomas et al., 1978), an RQ of 1 suggested that 47% of the nitrogen demand of the primary productivity in March and 70% in July could be met by nitrogen remineralization in the water column, compared to a benthic inorganic nitrogen input here of 7% in March and 3% in July. As in the case of the Bering Sea studies, we have attempted to identify how much of the water column remineralization of nitrogen can actually be attributed to the microplankton (<100 μm), such as protozoa, and the bacterioplankton (<2 μm) by size-fractionation studies involving uptake of ^{14}C- and ^3H-labeled amino and nucleic acids (Walsh et al., 1978; Ducklow et al., 1982), as well as dilution of ^{15}N-labeled ammonium sulfate (Harrison et al., 1983). Assuming that dissolved amino acids were the only substrate for ammonifying bacteria, we initially estimated a minimal regeneration rate of 0.21 mg NH_4 m^{-3} day^{-1} in the New York water column during May–September 1976 (Walsh et al., 1978).

Subsequent studies of the ammonium released from just microplankton remineralization of all dissolved and particulate organic substrates suggested a "potential" regeneration rate of 4.8 mg NH_4 m^{-3} day^{-1}, with 74% of the flux attributed to bacteria in August 1980 (Harrison et al., 1983). At this time on the New York shelf, the zooplankton were estimated to contribute 30% of the total recycled nitrogen in the water column, the benthos only 8%, the protozoa 16%, and the bacterioplankton 46%. A previous oxygen budget of summer conditions on the New York shelf (Falkowski et al., 1980) similarly indicated that bacterial consumption of dissolved organic matter amounted to 50% of the respiration demands above the pycnocline, compared to 25% by the other plankton. During March 1981, however, the bacterial productivity on the New York shelf was as little as 1% of the phytoplankton productivity (Ducklow et al., 1982), similar to the secondary production of the zooplankton at this season (Walsh et al., 1978), representing an annual minimum of metabolic activity within both the saprovore and herbivore components of the food web.

Based on bacterial uptake of [14]C- and ^{3}H-labeled organic compounds, more protozoan grazing estimates, additional dissolved organic carbon (DOC) release rates of phytoplankton, and biomass estimates of the bacteria and protozoa, the seasonal maximum of ammonium production by these organisms on Georges Bank is estimated to be 10.2 mg NH_4 m^{-3} day^{-1} in August–September, with the protozoa contributing 71% of the nitrogen flux at this time (Hobbie et al., 1986). Higher rates, of course, have been observed in nearshore summer waters, such as 29 mg N m^{-3} day^{-1} in August within Vineyard Sound (Glibert, 1982). When bacteria on Georges Bank are not abundant enough prey for the protozoa, however, the minimum input of nitrogen from the microplankton might be 0.1 mg NH_4 m^{-3} day^{-1} during other seasons (Hobbie et al., 1986). If we assume that the protozoan and bacterial remineralization fluxes parallel seasonal changes in their biomass (Hobbie et al., 1986), as well as the metabolic fluxes of the phytoplankton and zooplankton, a seasonally adjusted estimate each quarter of 4.8, 0.1, 0.1, 0.1 mg NH_4 m^{-3} day^{-1} and 10.2, 0.1, 0.1, 0.1 mg NH_4 m^{-3} day^{-1} over the upper 30 m off New York and on Georges Bank would yield annual recycled nitrogen inputs of, respectively, 14.0 and 28.7 g NH_4 m^{-2} yr^{-1} from these microbial components of the pelagic food web (Table IX).

A bacterial share of 74% of this ammonium flux off New York, ~10.4 g NH_4 m^{-2} yr^{-1}, and 29% of the input on Georges Bank, 8.3 g NH_4 m^{-2} yr^{-1}, implies consumption of a similar amount of the precursor, DON, since we have assumed that less than 10% of the excreted ammonium would be retained as body nitrogen of the bacterioplankton. They could intercept either some of the dissolved organic matter released from the sediments, or the extracellular loss of phytoplankton DON during photosynthesis. Based on 838 measurements of primary production in the Mid-Atlantic Bight (Walsh *et al.*, 1986b), we estimate that ~16.5% of the total carbon photoassimilated by phytoplankton is excreted as DOC, or 11.9 g DON m^{-2} yr^{-1} on the New York shelf and 15.8 g DON m^{-2} yr^{-1} on Georges Bank, with a C/N ratio of 5 : 1 and the respective differences in annual particulate primary production. The metabolic demands of the bacteria imply that most of the photosynthetic exudate of dissolved organic matter could be consumed heterotrophically on the New York shelf, leaving only 1.5 g N m^{-2} yr^{-1} for autotrophic utilization, compared to 7.5 g N m^{-2} yr^{-1} on the tidally mixed Georges Bank (Table IX).

Approximately 57–61% of the annual nitrogen demand of primary production on the New York shelf and Georges Bank may be supplied by regenerated nitrogen from marine organisms during mainly the summer and fall, with the rest (~30 g N m^{-2} yr^{-1}) in the form of nitrate, required from offshore sources. The outer shelf members of the Mid-Atlantic Bight food web receive less of their nitrogen supply from these biological cycling processes as the depth of the water column increases (Harrison, 1980) (i.e., recall the Bering Sea budgets) and thus require more from the cross-shelf boundary exchange. For example, it is estimated that the amount of nitrogen that enters the Gulf of Maine through deep slope water in the Northeast Channel represents about 40% of the total amount needed for primary production on Georges Bank (Ramp *et al.*, 1980). However, Ramp *et al.* (1980) emphasize that the flow field inside the Gulf of Maine is not yet fully understood and that the nutrients entering through Northeast Channel cannot all be used to support primary production on just Georges Bank, thereby requiring additional inputs across the shelf break front from Atlantic Slope Water.

4.3.5 Slope-Shelf Exchange

In terms of carbon, the revised nitrogen budget suggests that 110 g C m^{-2} yr^{-1} of the Georges Bank primary production may be consumed by pelagic and demersal herbivores (Table IX), compared to 145 g C m^{-2} yr^{-1} off New York. The bacterial–protozoan food chain could instead be supported by phytoplankton exudates, such that 290 g C m^{-2} yr^{-1} may be available as phytodetritus on Georges Bank and 155 g C m^{-2} yr^{-1} off New York. Modern marine carbon may be accumulating at a rate of 0.36 cm yr^{-1} within 2% (dw) carbon deposits at 174 m near the edge of Georges Bank (Hathaway *et al.*, 1979), or ~50 g C m^{-2} yr^{-1} with a specific gravity of 2.6 and a porosity of 0.73 for these sediments. Little phytodetritus is thought to be now accumulating on most of the New York shelf (Walsh *et al.*, 1985), however, and such depocenters are the subject of Chapter 5. Assuming that the shelf carbon storage occurs over the same area as the primary production on Georges Bank, as much as 240 g C m^{-2} yr^{-1} from this ecosystem and 155 g C m^{-2} yr^{-1} off New York may exit the shelf as particulate export (Walsh, 1983). With a C/N ratio of 5 : 1 for living phytoplankton, these export fluxes reflect shelf nitrogen losses of ~48 g N m^{-2} yr^{-1} from Georges Bank and ~31 g N m^{-2} yr^{-1} from the New York system (Table IX).

An annual export of 30–50 g N m^{-2} yr^{-1} of phytoplankton particulate nitrogen, from either the New York shelf or Georges Bank, to the adjacent continental slope requires an influx of at least the same amount of dissolved nitrogen from deeper slope waters to maintain an annual mass balance of shelf nitrogen from one year to the next. The dissolved nitrate pool of upper slope waters is large (~130 g N m^{-2}), compared to winter concentrations of ~6 g N m^{-2} on the shelf, and appears to be seasonally invariant at 500 m, the maximum depth of the winter mixed layer (Leetma, 1977). Nitrate might thus be supplied to the shelf from offshore intermediate depths, while particulate nitrogen is lost to slope sediments, with a time delay of years to decades before replenishment of the slope water nitrate stocks is required from remineralization of nitrogen during early diagenesis in slope sediments. The nitrogen buried on Georges Bank, in contrast, is subject to wind mixing, tidal resuspension, and bioturbation (Fanning *et al.*, 1982; Walsh *et al.*, 1985) and may be returned to the water column on a much shorter time scale.

Such an input of nitrate from the slope waters is apparently either "advected" onshore at the event time scale (Walsh, 1976) in response to upwelling favorable winds, or "diffused" onshore at longer time scales by turbulent motion. Wind-induced variations of the cross-shelf nitrate distribution have been observed over relatively short time periods during winter–spring (Walsh *et al.*, 1978). Over a 2-month period in April–May 1976, for example, the 1 μg-at. l^{-1} isopleth of nitrate appeared to move shoreward across 50 km of the New York shelf. It then evidently retreated seaward nearly 70 km, however, raising questions about how much nitrogen actually remained on the shelf, as in the case of eddy-induced nitrate intrusions in the South Atlantic Bight (Bishop *et al.*, 1980).

Surface winds and nearshore currents observed at these times show close coupling, with an upwelling circulation and increases of midshelf nitrate concentrations, and suggest a nitrogen input of 1.7 g N m^{-2} storm^{-1}, *when* weak stratification allows mixing of this influx over the euphotic zone (Walsh *et al.*, 1978). A series of March–April 1979 upwelling events and the phytoplankton response are simulated in Section 5.1.3. The peak cyclone activity within the Mid-Atlantic Bight is during January (5–6 storms month^{-1}), and a total of 20 wind events, favorable to upwelling between November and May, could possibly supply the rest of the annual nitrogen demand from the slope boundary off New York (Table IX). In fact, between just February and June, an average of 15 cyclones were encountered each year from 1950 to 1970 within the Mid-Atlantic Bight (Reitan, 1974). These same wind events, which might induce nutrient imports from slope water, also favor phytoplankton exports from shelf water (Walsh, 1983), moreover.

Alternative attempts to estimate the longer-term exchange of nutrients, salinity, and other dissolved compounds across the shelf break have averaged over the above wind events by relying on the usual eddy flux parameterization of horizontal mixing. Recent (Stommel and Leetma, 1972) and earlier (Ketchum and Keen, 1955) salinity budgets of the freshwater input to the Mid-Atlantic Bight, for example, derived a lateral eddy coefficient K_y of ~3 × 10⁶ cm² sec⁻¹ on the shelf. This value is what Riley (1967a) essentially used to estimate horizontal nutrient fluxes across the New York shelf. Additional considerations of drogue dispersion within surface waters (EG&G, 1979), of temperature patterns (Loder et al., 1982), and of vertical shear of tidal currents (Klein, 1986), all on Georges Bank, suggest respective ranges in K_y of 1.0–5.5 × 10⁶ cm² sec⁻¹ as well.

During unstratified water column conditions in either November or April (recall Fig. 8), a maximum cross-isobath gradient of 0.4 μg-at. NO₃ l⁻¹ km⁻¹ (0.4 × 10⁻⁸ μg-at. NO₃ cm⁻⁴) is found within the near-bottom 60-m layer of the outer shelf/slope waters of the Mid-Atlantic Bight; during February–March, however, a minimum horizontal, seasonal gradient of 0.1 μg-at. NO₃ l⁻¹ km⁻¹ is observed. With a mean of the horizontal eddy coefficients K_y of 3 × 10⁶ cm² sec⁻¹ and an average seasonal gradient of 0.2 μg-at. NO₃ l⁻¹ km⁻¹, an onshore flux of ~36 μg-at. NO₃ cm⁻¹ sec⁻¹ might occur over the 60- to 120-m depth interval at the shelf break.

Assuming a well-mixed water column from November through April (i.e., a vertical eddy coefficient K_z ≥5 cm² sec⁻¹), compared to 0.05–0.50 cm² sec⁻¹ in summer off New York (Falkowski et al., 1980; Harrison et al., 1983), about 0.2 μg-at. NO₃ l⁻¹ day⁻¹ would be added to the upper 60 m of the shelf water column, within 25 km of the shelf break off New York or Georges Bank. At the 60-m isobath in both regions, the integrated daily resupply flux from Atlantic slope water might be 0.17 g NO₃ m⁻² day⁻¹, or 30.6 g N m⁻² yr⁻¹ over 180 days (November to April) of destratification (Table IX). Since Georges Bank is vertically mixed year-round by the tidal currents, affecting the horizontal dispersion as well, nitrate could also be introduced into the euphotic zone during the other 180 days of the year, unlike off New York.

With development of seasonal stratification of the slope water column and of thermal fronts (Beardsley and Flagg, 1976; Voorhis et al., 1976; Horne, 1978; Herman and Denman, 1979) at the edge of Georges Bank along the 60- to 75-m isobaths, however, the apparent K_y across the slope/shelf boundary is reduced to ~3 × 10⁵ cm² sec⁻¹ (Klein, 1986). Using the maximum cross-isobath gradient of 0.4 × 10⁻⁵ μg-at. NO₃ cm⁻⁴, reflecting impoverished nutrient conditions on the shelf, and the seasonally reduced K_y of 3 × 10⁵ cm² sec⁻¹, we obtain an estimated flux of 7.2 μg-at. NO_3 cm^{-1} sec^{-1} over the 60–120 m depth of the summer shelf break front off Georges Bank. At such a reduced onshore flux, about 0.04 μg-at. NO_3 l^{-1} day^{-1} might be introduced into the well-mixed water column at the 60-m isobath through May–October on Georges Bank, or an additional 6.1 g NO_3 m^{-2} yr^{-1} (Table IX).

Finally, in a third approach to estimation of cross-shelf fluxes of nitrate, Houghton *et al.* (1978) used regressions of nitrate, oxygen, temperature, and salinity, and used a current-meter record of velocity, temperature, and salinity from December 1975 to April 1976 at the Nova Scotian shelf break, that is, during destratified water column conditions and presumably including the period of high nitrogen demands of the spring bloom. Their mean horizontal eddy fluxes, excluding tidal and inertial scales of motion, were an onshore turbulent transfer of 38 μg-at. NO_3 m^{-2} sec^{-1} at 50 m, an offshore transfer of 10.7 μg-at. NO_3 m^{-2} sec^{-1} at 150 m, and zero transfer at 230 m. If their flux estimate is applied over the 50- to 110-m depth interval of the Scotian shelf water column, where evidently most of the onshore flux of NO_3 occurred (Houghton *et al.*, 1978), an onshore eddy transfer of 23 μg-at. NO_3 cm^{-1} sec^{-1} is obtained, similar to our already discussed average of 36 μg-at. NO_3 cm^{-1} sec^{-1} over the same 60-m depth and time interval at the Mid-Atlantic shelf/slope boundary. At either time scale, we thus apparently can obtain an estimate of the onshore flux of nitrate each year, which could balance the phytodetrital export and is more than 300-fold that of the fish harvested from this ecosystem (Table IX).

4.3.6   Spring Bloom

Annual nitrogen budgets suggest that ~40% of the primary production of the Mid-Atlantic Bight might be available for export to slope waters, but analyses of moored fluorometer records, during June–July 1983 and February–April 1984 at the shelf break off New England (Fig. 97) suggest an annual export of perhaps 2% instead (Walsh *et al.*, 1987a). Using a mean C/chl *a* ratio of 45 : 1 for live phytoplankton within the spring bloom of the Mid-Atlantic Bight (Malone *et al.*, 1983) during February–April 1984, this near-bottom chlorophyll flux of 2.67 g chl m^{-2} day^{-1} implied a net daily carbon export of 1.2 $\times$ 10^2 g C m^{-2} day^{-1} past the 80-m isobath. This seaward export reflects, of course, the previous accumulation on the inner shelf of phytoplankton carbon within a water parcel, during its trajectory from the coast to the slope. The alongshore flux of carbon at this fluorometer mooring was much larger, ~5.0 $\times$ 10^2 g C m^{-2} day^{-1} (Fig. 97), and we shall see in Section 5.1.3 that most of the algal export from the Mid-Atlantic Bight probably takes place south of Virginia, not off Long Island.

At such a local export off New England, however, and a mean primary production rate of 1 g C m^{-2} day^{-1} during March–April 1984 (Falkowski *et al.*,

1987) with no grazing losses, an accumulation interval, or a residence time, of 120 days on the mid-Atlantic shelf is derived for the winter–spring bloom of phytoplankton. Within a mean offshore flow regime of 1 cm sec^{-1} (0.86 km day^{-1}), successive daughter cells of a diatom population could, in fact, transit the 100-km-wide mid-Atlantic shelf after 116 days, departing Georges Bank in January and arriving at the Long Island moorings in April. Imposition of a seasonal grazing stress on the primary producers would both increase the presumed residence time of this algal population on the shelf, of course, and introduce detrital carbon particles to the spring ecosystem.

Use of a larger C/chl a ratio, reflecting such phytodetritus, for conversion of the fluorescence records would decrease the required residence time of all particles, however, since more total carbon would be exported per unit chlorophyll over the same time period. To place the daily estimates of grazing and sinking losses of phytoplankton made during the shelf edge exchange processes (SEEP) experiment in perspective, the local daily rate of change of chlorophyll, $\partial(uP)/\partial x$, in the aphotic zone is computed, over a separation distance of 12 km, instead of the synoptic scale of 1×10^5 km^2 for the whole Mid-Atlantic Bight. The cumulative mean transports, $\overline{uP}$, at 75 m on the 80-m isobath (Fig. 97a) and at 81 m on the 120-m isobath (Fig. 95c) were used for this calculation.

The 40-hr low-passed chlorophyll records of zero lag at these depths had an r^2 of 0.32, as high as the vertical coherence between the surface and the deepest fluorometers at the two isobaths. Recall that the r^2 was only 0.002 for the W8 and W12 chlorophyll records at 13 m (Fig. 94). This hundredfold difference in horizontal coherence between the surface and near-bottom chlorophyll time series at W8 and W12 underscores the importance of vertical transfer processes on this shelf—that is, a particle can no longer sink when at the bottom boundary.

The mean diabathic flow past the current meter at 84 m on the 120-m isobath was weakly onshore, however, while this flow past the meter at 78 m on the 80-m isobath was offshore. These current meter records suggest that a convergence of water at 78–84-m depth might have occurred somewhere over the ~12-km separation distance between W8 and W12, to satisfy the bottom Ekman layer constraints on the predominantly downstream parabathic flow. The mean diabathic, near-bottom (118 m) flow at W12 was also offshore at 1.0 cm sec^{-1} during February–April 1984, similar to an offshore flow of 1.4 cm sec^{-1} near the bottom on the outer shelf during February–May 1979–1980. Downwelling of water, in addition to passive sinking of phytoplankton, may thus be involved in particle transport on the mid-Atlantic shelf (Walsh *et al.*, 1987c).

Without additional vertical resolution of the flow field, however, we can only make an estimate of the mean horizontal flux over time at 75–81 m depths in finite difference form, with $(uP\text{W8} - uP\text{W12})/\Delta x$—recall Eq. (158). We obtain 0.26 μg chl l^{-1} day^{-1} from this expression, the current-meter data, and Figs. 95c and 97a, or 11.9 mg C m^{-3} day^{-1} with the C/chl a ratio of 45 : 1. Since the primary production, daily grazing losses, and organic matter caught with sediment traps are usually reported as depth integrals, a depth range of possible carbon export must now be specified.

The simulation model of the 1979 spring bloom in Section 5.1.3 suggests that as much as 90% of the algal export from the Mid-Atlantic Bight might occur within the lower third of the water column. Over a 30- to 40-m bottom layer at the 120-m isobath, we thus further estimate that the daily horizontal algal export might have been 0.35–0.47 g C m^{-2} day^{-1} from the mid-Atlantic shelf during February–April 1984. This estimate is also based on the moored temperature, current meter, fluorometer, and transmissometer data at the 80- and 120-m isobaths, south of Long Island.

Such possible carbon fluxes of algal export constitute 78–100% of the 1984 March mean primary production, 0.45 g C m^{-2} day^{-1}, and 23–30% of that during April 1984, 1.55 g C m^{-2} day^{-1} (Falkowski et al., 1987). During this time period, the biomass of the dominant zooplankton herbivore, Calanus finmarchicus, was observed to increase eightfold over a 33-day period of the 1984 spring bloom, ingesting a daily mean phytoplankton ration of 0.18 g C m^{-2} day^{-1} in March and 0.50 g C m^{-2} day^{-1} in April (Smith and Lane, 1987). Assuming that the seaward horizontal export of phytoplankton occurred over only the lower 30 m of the water column at the shelf break, that is, 0.35 g C m^{-2} day^{-1}, the physical and biological losses of algal carbon (export and pelagic grazing) would have consumed 118% of the March primary production and 55% of the April fixation of carbon.

Sediment traps, moored 10 m and 70 m off the bottom on the outer shelf at W12 during March–April 1984 (G. T. Rowe, personal communication), caught respectively 0.16 and 0.10 g C m^{-2} day^{-1}, about half that estimated from the moored instruments within the lower part of the water column. This sediment-trap flux represents presumably a combination of the daily vertical fallout of both fecal-pellet carbon and phytodetrital carbon, as well as the particulate matter resuspended off the bottom from previous inputs. At a herbivore assimilation efficiency of 60%, as much as 0.20 g C m^{-2} day^{-1} of the April grazing flux might sink as fecal pellets, at rates of 150 m day^{-1}, to be caught in these near-bottom sediment traps. The slower 1–20 m day^{-1} sinking flux of phytoplankton might instead have been advected past these sediment traps on the shelf.

Earlier estimates of benthic metabolism during August–October on the mid-Atlantic shelf (Florek and Rowe, 1983) suggested that, as the bottom of the water column warms from its seasonal minimum in March, the benthos might consume an average of 0.16 g C m^{-2} day^{-1} (Walsh, 1983). This would be 25% of the April 1984 primary production not lost to horizontal export or pelagic grazers, if ingested directly by the benthos. A more recent analysis (Rowe et al., 1987) of benthic metabolism during the SEEP experiment suggests that a similar amount of 0.15 g C m^{-2} day^{-1} would be required by the benthos, that is, less than the potential fecal-pellet flux and equal to that caught by the near-bottom trap at the 120-m isobath.

Finally, bacterioplankton productivity is less than 2% of the spring algal production in the Mid-Atlantic Bight (Ducklow *et al.*, 1982), decomposing perhaps 20% of the organic carbon caught in sediment traps (Ducklow *et al.*, 1985). Bacterioplankton may be a carbon sink in coastal food webs (Ducklow *et al.*, 1986), however, with little carbon or nitrogen passed to other organisms in the form of food for higher trophic levels. The previous nitrogen budget suggested, furthermore, that the secondary production of bacterioplankton can be maintained from dissolved organic matter excreted each day by phytoplankton in the Mid-Atlantic Bight. These heterotrophs may thus be a sink for dissolved organic carbon, but not the particulate algal carbon. Little direct remineralization of algal biomass may occur by attached bacteria on living phytoplankton cells.

The local fate of algal carbon within the SEEP arrays at the shelf break may thus have been both consumption and a short transit to offshore waters, with ~50% of this March–April 1984 primary production quickly exported to the adjacent slope to be caught in sediment traps on the 500- and 1250-m isobaths (Biscaye *et al.*, 1987). Annual estimates of ^{210}Pb and ^{14}C mixing rates (described in Chapter 5), combined with vertical carbon gradients in the surface sediments (Walsh *et al.*, 1985), suggest that the average daily carbon accumulation rate in slope sediments off the Mid-Atlantic Bight might be only 0.03 g C m^{-2} day^{-1}, however (i.e., <10% of the export estimate). Some of the slope import of carbon must be remineralized by the local benthos, although the benthic biomass and metabolism are lower on the slope than on the shelf (i.e., 0.005 g C m^{-2} day^{-1} at 1800 m) (Rowe *et al.*, 1987).

Marked seasonal fluxes of particulate matter were observed with the sediment traps moored on the slope during this SEEP experiment (Biscaye *et al.*, 1987)—see Section 5.1.2. During April 1984, for example, 0.08–0.26 g C m^{-2} day^{-1} was found respectively within sediment traps 50 m above the bottom at the 1250- and 500-m isobaths, compared to 10-fold less from August 1983 to November 1984. Little import of phytoplankton carbon may thus occur on the mid-Atlantic slope during summer and fall, with most of the annual carbon loading as the result of the spring bloom. If the annual accumulation rate of carbon, estimated by ^{14}C and ^{210}Pb mixing rates on this continental slope, were actually to occur over only 100 days, the daily sediment accumulation rate on the slope during the spring bloom would then be 0.11 g C m^{-2} day^{-1}, that is, 31% of the export estimate and similar to that measured in the April 1984 sediment trap samples on the slope.

A set of shipboard, aircraft, and satellite observations, sparsely sampled in time, and a set of moored data, sparsely sampled in space, provide corroboration of initial hypotheses of shelf export of phytoplankton from the Mid-Atlantic Bight (Malone *et al.*, 1983; Walsh *et al.*, 1981, 1985, 1987a). Within the verti-

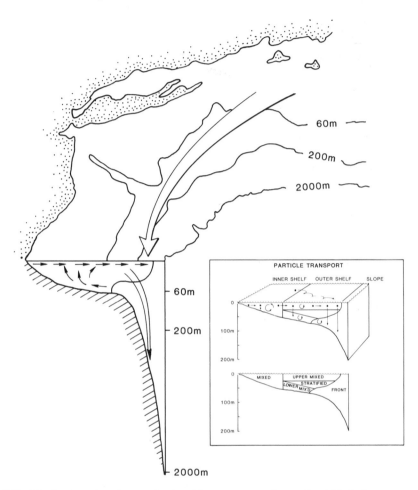

FIG. 109. A schematic trajectory of surface Ekman transport of a particle within the homogeneous water column of the inner shelf and of bottom Ekman transport within the stratified water column of the outer shelf during the spring bloom in the Mid-Atlantic Bight.

cally homogeneous water column of the inner shelf (<60-m isobath), algal biomass of the spring bloom that has previously sunk out is evidently resuspended and transported seaward within a surface Ekman layer in response to northwest wind events (recall Figs. 69, 75, and 78). Within the vertically stratified waters of the outer shelf (60- to 100-m isobaths), where the shelf break front impedes vertical exchange in the middle of the water column (Fig. 109), near-bottom phytoplankton are evidently resuspended and instead transported seaward within a bottom Ekman layer, in response to northeast wind events (Fig. 97).

I have sketched this process in Fig. 109, where the shelf export of phytoplankton carbon is estimated to be a surface process at midshelf and a subsurface phenomenon at the shelf break during a longshore trajectory of 2- to 3-month duration, similar to the transit time of a water parcel from the shelf break of the northwestern Bering Sea to a Chukchi Sea depocenter. Algal populations may not survive a similar hegira on tropical shelves, however, because the turnover time of organic matter by bacterioplankton decreases from 18 days at $-0.2°C$, to 11 days at $4°C$, and to 2 days at $25°C$ (Pomeroy and Deibel, 1986). We thus consider a third set of budgets for shelf communities under varying nutrient loading of subtropical waters at 25–30°N latitudes.

4.4 Gulf of Mexico Budget

The annual discharge of the Mississippi River is fivefold that of the total freshwater influx of the Mid-Atlantic Bight (Bue, 1970) and three times that of the Yukon River flow (Lerman, 1981), with a nitrogen content of as much as 150 μg-at. NO_3 1^{-1} during spring flood (Everett, 1971). After dilution and phytoplankton uptake on the otherwise oligotrophic Louisiana shelf, a seasonal maximum of 30 μg-at. NO_3 1^{-1} (Fig. 110) is still found ~50 km downplume from the mouth of the Mississippi River (Fucik, 1974). As a result of this nutrient injection, annual primary production of the Mississippi River Delta is ~250–350 g C m^{-2} yr^{-1} (Thomas and Simmons, 1960; Fucik, 1974), in contrast to ~25 g C m^{-2} yr^{-1} for the open Gulf of Mexico and parts of the west Florida and Texas shelves (El-Sayed, 1972). The annual shrimp yield of the Gulf of Mexico has, in fact, been related to the daily freshwater discharge of the Mississippi River (Griffin et al., 1976).

The C/N ratio of suspended particulate matter in the Mississippi River is >10, in contrast to <6 within the euphotic zone of the Louisiana shelf (Armstrong, 1974). In the surface sediments, the C/N ratio reflects the above source and deposition patterns with >10 ratios found near the coast (Fig. 111) and <6 observations found under the Mississippi River plume and on the upper slope (Trask, 1953; Armstrong, 1974). Similarly, the δ ^{13}C value of suspended particulate matter is -25.6 to -25.9 in the Mississippi River (Hoffman, 1974), and -24.1 within the passes (Eadie, 1972), indicative of terrestrial carbon, whereas 10–20 km off the mouth, particulate values of -19.6 to -20.0 are found in the water column; the latter are representative of warm-water plankton (Sackett et al., 1965). The δ ^{13}C values of the surface sediments of the Mississippi River, of adjoining bays, and of the salt marshes are similarly -22.4 to -23.6, in contrast to -19.0 to -20.8 of sediments of the outer continental shelf (50–100 m) in the eastern Gulf of Mexico (Sackett and Thompson, 1963) and between the Mississippi River Delta and the Rio Grande (Hedges and Parker, 1976).

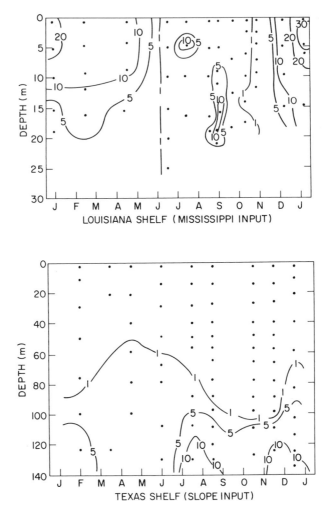

FIG. 110. Time series of nitrogen input (μg-at. NO_3 l^{-1}) to the Texas–Louisiana shelves from the coastal and open boundaries of the Gulf of Mexico. [After Walsh (1983); © 1983 Pergamon Journals Ltd., with permission.]

Like the other shelf ecosystems, there may be seaward export of phytoplankton carbon, derived from the Mississippi River effluent, to the upper slope of the Gulf of Mexico, where 1–2% organic carbon muds (Trask, 1953; Flint and Rabelais, 1981) are located in areas of low oxygen content (Richards and Redfield, 1954). Near the shelf break at 100 m depth, about 30 km off the mouth of the river, independent [14]C (Parker, 1977), [210]Pb (Shokes, 1976), and [239,240]Pu (Scott et al., 1983) analyses suggest very high sedimentation rates of ~1.0 cm yr^{-1}.

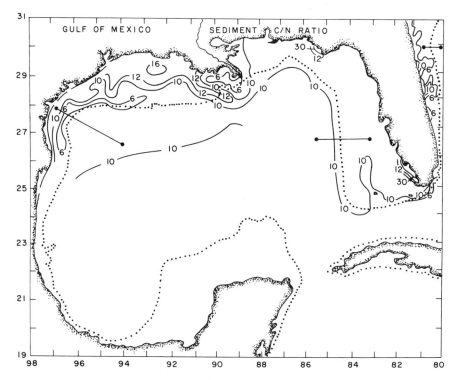

FIG. 111. The distribution of the organic C/N ratio within surficial sediments of the Gulf of Mexico. [After Walsh (1983); © 1983 Pergamon Journals Ltd., with permission.]

The adjacent slope sedimentation rates range from 0.3 to 0.8 cm yr^{-1}, with evidence of slumping (Booth, 1979; Scott *et al.*, 1983) as was found in the southeastern Bering Sea and the Mid-Atlantic Bight (see Section 5.2.2). In contrast to the 30–40% $CaCO_3$ sediments of the open gulf, the slope sediments of the northern Gulf of Mexico contain only 1–7% carbonate (Scott *et al.*, 1983), implying that these slope sediments may be of diatomaceous origin as a result of shelf export from the Mississippi River plume.

During the 1930s, Alfred Redfield was one of the first to suggest that the low oxygen content of such slope waters, adjacent to western boundary currents, was a tracer of upwelled, nutrient-rich water (Redfield, 1936). Away from the region of major river input of nitrogen in the Gulf of Mexico, for example, onshore fluxes of nitrate occur at the shelf break off the west Texas coast. The presence and seasonal intensification of a surface western boundary flow, the Mexican Current, off the Mexican–Texas shelf have been correlated with seasonal changes in the wind curl (Sturges and Blaha, 1976) that might lead to eddy-induced upwelling of nutrients off this coast (Fig. 110). In contrast to coastal upwelling at the same latitude off Baja California (Walsh *et al.*, 1974), however, high nitrate concentrations are apparently confined to the shelf edge off Texas, rather than penetrating 100 km to the coast.

An annual time series of nitrate content at this shelf break (Fig. 110) suggests that eddy-induced upwelling might be intensified in the western Gulf of Mexico during winter and summer, in phase with changes in the wind curl forcing (Sturges and Blaha, 1976). Increases of chlorophyll on the Texas outer shelf during February 1977 were, in fact, attributed to such an upwelling event (Flint and Rabelais, 1981). These transients in nutrient supply are restricted to shelf areas, however, for an extensive survey of the open Gulf of Mexico during February 1980 found consistently low chlorophyll (<0.1 μg chl l^{-1}) seaward of the shelf break (S. El-Sayed, personal communication); after a fivefold increase in primary production over a 115-m euphotic zone during February 1981, only $0.1-0.2$ μg chl l^{-1} was found offshore in the southeastern Gulf of Mexico (Ortner *et al.*, 1984).

An annual primary production of as much as 103 g C m^{-2} yr^{-1} may occur nearshore in this region of the Gulf of Mexico (Flint and Rabelais, 1981), as a result of nutrient injection and further recycling within shallow depths ($\sim$30 m of water), compared to the 250–300 g C m^{-2} yr^{-1} derived from the Mississippi River effluent. The only low C/N ratios of the west Texas shelf sediments are found at the shelf break (Fig. 111), moreover, suggesting that the impact of eddy-induced upwelling is mainly restricted to this part of the shelf ecosystem. Using the same approach as in the other budgets of the Bering/Chukchi Seas and the Mid-Atlantic Bight, the annual flux of carbon was estimated through the food web of the 1.3 $\times$ 10^5 km^2 Louisiana–Texas shelves (Fig. 112), downstream of the Mississippi River effluent.

4.4.1   Texas–Louisiana

Unlike the rich data sets of the previous two shelf ecosystems, however, the data base for this exercise is poor at best. Estimates of annual primary production vary from $\sim$300 g C m^{-2} yr^{-1} above regions of sediment C/N ratio <6 at the mouth of the Mississippi River to 20 g C m^{-2} yr^{-1} above areas of sediment C/N ratio >10 on the west Texas shelf (El-Sayed, 1972). Furthermore, most of the harvest of the gulf menhaden, *Brevoortia patronus*, is taken on the Louisiana–Mississippi shelves (Gunter, 1967), while much more brown shrimp, *Penaeus aztecus*, are taken on the east Texas shelf (Heald, 1969). In an attempt to address these regional differences, the primary production was estimated to be a minimum of $\sim$100 g C m^{-2} yr^{-1} off Louisiana–Texas (Fig. 112).

A P/B ratio of 7 for copepods (Steele, 1974) and 10 for protozoans, together with their combined biomass, yields a pelagic secondary production of $\sim$4 g C

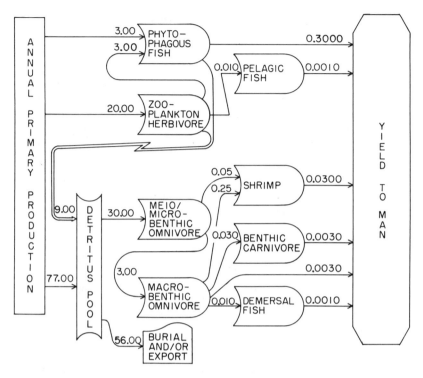

FIG. 112. An annual carbon budget (g C m^{-2} yr^{-1}) of the food web on the Texas–Louisiana shelves. [After Walsh *et al.* (1981); © 1981 Macmillan Journals Ltd., with permission.]

m^{-2} yr^{-1} on the west Texas shelf (Flint and Rabelais, 1981). A 20% growth efficiency, as in the other carbon budgets, implies an annual grazing loss of 20 g C m^{-2} yr^{-1} to these planktonic herbivores (Fig. 112). The annual yield (0.3 g C m^{-2} yr^{-1}) of gulf menhaden is derived from 1- to 3-yr-old fish, with the majority of the catch from the first two year classes (Arenholz, 1981). This multiyear secondary production is simulated with an apparent annual growth efficiency of 5%, and thus an ingestion demand of 6 g C m^{-2} yr^{-1} (Fig. 112), composed of phytoplankton food for adults and zooplankton crustaceans for juveniles. The production (~1 g C m^{-2} yr^{-1}) of the protozoan herbivores is assumed to be eaten by other pelagic fish with an additional step in this part of the food web.

Herbivore fecal pellets and ungrazed phytodetritus suggest an average shelf input of at least 86 g C m^{-2} yr^{-1} to the Texas–Louisiana benthic community (Fig. 112). Estimates of benthic oxygen consumption in very shallow (15 m) waters, ~10 km off the west Texas shelf, imply an annual carbon respiration of ~200 g C m^{-2} yr^{-1} (R. Flint, personal communication), similar to nearshore estimates of 160 g C m^{-2} yr^{-1} within the Mid-Atlantic Bight (Rowe and Smith, 1977). At a nearshore carbon export rate of ~183 g C m^{-2} yr^{-1} for the Gulf of Mexico (Happ et al., 1977), however, most of this inshore benthic carbon demand is probably met by macrophyte detritus (Fry, 1981). Furthermore, the densities of meiobenthos and macrobenthos (Flint and Rabelais, 1981) both decline by more than an order of magnitude across the rest of the 100-km-wide Texas shelf (30–180 m), suggesting that a mean consumption rate of 30 g C m^{-2} yr^{-1} is a reasonable estimate for the annual carbon budget (Fig. 112).

Macrobenthic biomass at the 30-m isobath ranges from 0.02 g C m^{-2} (Flint and Rabelais, 1981) to 0.25 g C m^{-2} (Rowe et al., 1974) in the northwest Gulf of Mexico. At an annual temperature range of ~10–30°C within nearshore waters, a P/B of 1 for the Gulf of Mexico macrobenthos yields a maximum annual secondary production of ~0.3 g C m^{-2} yr^{-1} (Fig. 112). Similarly, the yield of Texas–Louisiana shrimp (Heald, 1969; Flint and Rabelais, 1981) is estimated to be ~0.03 g C m^{-2} yr^{-1} with a P/B ratio of 1 as well. Demersal fish yields (Heald, 1969) suggest very small secondary production from other apex predators on the Texas–Louisiana shelves.

With these estimates (Fig. 112), over 50% of the annual primary production is apparently not consumed by this hypothetical shelf food web off Texas and Louisiana. We have not included the carbon shunt of a bacteria–protozoan–gelatinous zooplankton food chain, however, and addition of these components is considered in the next section; it is nevertheless doubtful that more carbon can be removed in this present budget by the diatom–copepod–fish food chain. The yield of menhaden from the Gulf of Mexico appears to have leveled off from 1 × 10^5 tons yr^{-1} in 1936 (Gunter, 1967) to 5–6 × 10^5 tons yr^{-1} in 1965 (Heald, 1969) and 1975 (Hoss and Hettler, 1979), with annual maxima as high as 8 × 10^5 tons yr^{-1} (Arenholz, 1981); that is, the proposed carbon demands of the higher trophic levels may be an accurate estimate of their present ingestion fluxes (Fig. 112).

During this time period, the nitrogen content of the Mississippi River more than doubled (Walsh, 1980; Walsh et al., 1981), from concentrations of ~40 μg-at. N l^{-1} in 1905 (Gunter, 1967) and in 1935 (Riley, 1937) to as much as ~130 μg-at. NO$_3$ l^{-1} during February 1972 and May 1973 (Ho and Barrett, 1977). Similarly, the phosphate content at the mouth of the Mississippi River increased almost 10-fold from ~0.5 μg-at. PO$_4$ l^{-1} in 1935 (Riley, 1937) to 3–4 μg-at. PO$_4$ l^{-1} 30 yr later (Alberts, 1970; Ho and Barrett, 1977). Since the yields of the phytophagous menhaden stocks only increased by 25% from 1962 to 1975, primarily as a result of expansion of the fishery (Lindall and Saloman, 1977), the adult gulf menhaden populations are presumably not food limited. The increased algal shelf production of 300 g C m^{-2} yr^{-1}, associated with this eutrophication may have thus resulted in more carbon export rather than an increase of fish stocks.

4.4.2 Florida–Georgia

Nitrate sections across the Loop Current at the edge of the west Florida shelf (O'Brien and Wroblewski, 1972; Paluszkiewicz et al., 1983) suggest shelf-break upwelling, and spin-off eddies have been observed on this shelf as well (Maul, 1977). The upward slope of the nutrient isopleths in the offshore region of the eastern Gulf of Mexico (Morrison and Nowlin, 1977) delineates the anticyclonic Loop Current as it penetrates farther into the Gulf from the Yucatan Strait. During 1973–1977, however, the Loop Current was found due north of 28°N only 10% of the time, and against the shelf break less than 30% of the time north of 26°N (Vukovitch et al., 1979), where the annual primary production is ~30 g C $m^{-2} yr^{-1}$ (R. Iverson, personal communication). Without frequent generation of positive vorticity from frictional drag of the Loop Current against the west Florida shelf, less eddy-induced upwelling would occur in response to this western boundary current, compared to that from the Florida or Mexican Currents (Figs. 25 and 110).

The only area of low C/N ratio in the west Florida shelf sediments (Hathaway, 1971; Folger, 1972) is, in fact, found at the Key West shelf break, where the Loop Current continually passes through the Straits of Florida (Fig. 111) to become part of the Florida Current. Marine biogenic deposits have apparently not overlain the mangrove detritus of the nearshore zone, nor the relict deposits on the rest of the west Florida shelf, while the annual fish yield here is ~20% that of the Mid-Atlantic Bight (Heald, 1969). In contrast, the sediments at the edge of the east Florida and Georgia shelves, adjacent to the Florida Current, have a C/N ratio of <6 from Miami to Savannah (Fig. 111), but again the fish yield of the South Atlantic Bight is about 10% that of the Mid-Atlantic Bight.

During studies of a frontal eddy of the Loop Current as it propagated south on the west Florida Shelf at 26°N during April and September 1982, the nutrient input, chlorophyll biomass, and primary production were found to be much less than within another eddy of the Florida Current at the same time of year at 30–32°N on the Georgia shelf (Yoder et al., 1983, 1985, 1987). They found that (1) the depth of the nutricline was a mean of 21 m on the Georgia shelf, compared to 54 m on the west Florida shelf; (2) the algal biomass averaged 30 mg chl m^{-2} within the euphotic zone at the Georgia shelf break, compared to 13 mg chl m^{-2} off Florida; and (3) the daily primary production was a mean of 2.0 g C $m^{-2} day^{-1}$ in this part of the South Atlantic Bight, compared to 0.5 g C m^{-2} day^{-1} on the southwest Florida shelf. Based on estimates of eddy frequency in the two regions, the west Florida shelf south of 25°N might have an annual primary production of 90 g C $m^{-2} yr^{-1}$ (Yoder et al., 1987), compared to 360 g C $m^{-2} yr^{-1}$ on the outer Georgia shelf (Yoder et al., 1985).

Using an estimate of net particulate production of 137 g C $m^{-2} yr^{-1}$, from an earlier carbon budget (Pomeroy, 1979), in a series of hypothetical stranded

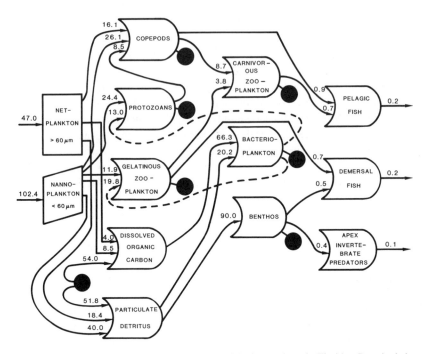

FIG. 113. An annual carbon budget (g C m^{-2} yr^{-1}) of the food web on the Florida–Georgia shelves. [After Pace *et al.* (1984).]

eddies within a simulation model of this shelf type, Pace *et al.* (1984) essentially produced an analog (Fig. 113) of this less-productive west Florida eddy ecosystem. Addition of the bacterioplankton shunt in the second model (Fig. 113) led to consumption of all of the primary production, both particulate and dissolved, in contrast to the first model (Fig. 112), which contained protozoa, but not the picoheterotrophs. Consequently, the secondary production of protozoa and herbivorous copepods increased from 4.0 g C m^{-2} yr^{-1} (Fig. 112) to 29.2 g C m^{-2} yr^{-1} (Fig. 113), while that of the macrobenthos rose from 0.3 g C m^{-2} yr^{-1} in the former to 4.6 g C m^{-2} yr^{-1} in the latter.

A more realistic approximation of the shelf food web within a spin-off eddy at the edge of the Georgia shelf was obtained when the primary production of this second model (Fig. 113) was increased to 213 g C m^{-2} yr^{-1} and an advective loss term was added (Pace *et al.,* 1984). In this additional simulation, 36% of the annual primary production from the spin-off eddies was then exported from a hypothetical South Atlantic Bight, similar to the results of another Mid-Atlantic Bight model in Section 5.1.3. The secondary production of the macrobenthos in this other South Atlantic Bight case was 3.6 g C m^{-2} yr^{-1}, somewhat less than that of the closed system (Fig. 113) but equivalent to that of the Mid-Atlantic Bight (Fig. 107).

The biomass of the macrobenthos on the Georgia shelf (Tenore, 1985) is higher than that of the northern Gulf of Mexico, ranging from a seasonal mean of 3.7 g C m^{-2} on the outer shelf to 10.6 g C m^{-2} on the inner and 22.5 g C m^{-2} on the middle shelves, assuming the previous 0.5 g C g dw^{-1} conversion. Although Pace *et al.* (1984) thought that the simulated production of macrobenthos might be too large, the observed macrobenthos biomass on the Georgia outer shelf and their estimates of secondary production of 3.6–4.6 g C m^{-2} yr^{-1} yield a reasonable P/B ratio of 0.8–1.0, that is, the same as that employed in the Texas–Louisiana model (Fig. 112). If the commercial yield of fish is an accurate estimate of their secondary production on the west Florida or Georgia shelves, however, the second model's output of apex predators (Fig. 113) is overly optimistic.

Landings of menhaden on the east coast constitute the second largest fishery of the United States, surpassed only by that of the gulf menhaden (Fig. 112). The northerly migration of Atlantic menhaden, *Brevoortia tyrannus,* into the Mid-Atlantic Bight consists of grazing adults and juveniles (Fig. 107), however, with most of their yield taken north of Cape Hatteras during the summer fishery of May to September within the 0- to 40-m isobaths (Roithmayr, 1963); during 1952–1972, less than 30% of the total menhaden landings of the east coast was taken within the South Atlantic Bight (Nelson *et al.*, 1977). Spawning of mature 3-yr-old menhaden mainly occurs instead on the outer shelf of the South Atlantic Bight during the winter (Nelson *et al.*, 1977; Nicholson, 1978), with recruitment of the surviving larvae within the adjacent estuaries. A suite of other fish species also spawn on the outer shelf of the South Atlantic Bight in winter, such as spot, pinfish, mullet, croaker, flounder (Weinstein, 1981), and bluefish (Kendall and Walford, 1979), resulting in larval fish abundances of ~100 m^{-2} here, compared to ~10 m^{-2} on the inner shelf (Yoder, 1983).

A second peak spawning of anchovy and goby occurs in the summer (Weinstein, 1981), resulting in larval fish evenly distributed across the southeastern U.S. shelf in concentrations of ~150 m^{-2} (Yoder, 1983), perhaps twice the larval fish abundance of the Mid-Atlantic Bight (Fig. 12). Estimates of the required daily rations of the larvae of three subtropical fish species (Houde, 1978) and the reproductive rates (Checkley, 1980) of copepods during summer in the South Atlantic Bight suggest, however, that the supply of copepod nauplii would be insufficient to allow more than 10% larval survival within a spin-off eddy (Paffenhofer, 1985).

If displaced onshore toward the nursery grounds of the South Atlantic Bight estuaries (Thayer *et al.,* 1974; Shenker and Dean, 1979), more fish larvae might survive, while offshore movement to the Gulf Stream would result in starvation of this summer larval fish community (Paffenhofer, 1985). Similarly, winter storms would both dilute the microzooplankton prey of larval menhaden (Govoni *et al.*, 1983; Stoecker and Govoni, 1984) and disperse these fish to impoverished offshore waters (Yoder, 1983; Reish *et al.*, 1985). We now return to this problem of larval survival in an analysis of variations of year-class recruitment of Alaska pollock within the southeastern Bering Sea.

4.5 Food Limitation

The critical period in the early life history of fish is when the strength of a year class may be determined (Hjort, 1914; Marr, 1956; May, 1974). The availability of food and the causes of mortality during this period are thus important in assessing and managing a fishery. For many marine fish, high rates of mortality during pelagic larval stages suggest that the critical phase is during early development. However, larval mortality is difficult to understand and estimate, because it results from the combination and interaction of many biological and physical processes. Despite intrinsic difficulties, some aspects of mortality can be studied when particular care is taken to isolate variables. This approach had been used successfully in laboratory studies of larval growth (Lasker *et al.*, 1970; Laurence, 1977), larval susceptibility to starvation (Blaxter and Staines, 1971), the effect of prey density on larval searching and feeding (Hunter, 1972), and the suitability and availability of prey in the habitat (Lasker, 1975; Walsh *et al.*, 1980).

The critical-period hypothesis assumes that many larvae die from starvation, or from other causes such as vulnerability to predation. Neither source of mortality has been adequately tested (Lasker, 1985), but starving fish larvae have been found in the sea (O'Connell, 1980). In a weakened state, they may succumb to either a lack of food or diminished predator avoidance (Bailey and Yen, 1983). In any case, overall larval survivorship is low such that the survivors are select individuals who have encountered favorable paths through the resources and hazards of their habitat. Extrapolating this hypothesis to real populations is difficult because larval metabolism, the distribution of resources and hazards, and the exact manner in which larvae encountered them must all be understood. Simulation models appear to be the most appropriate way to make this extrapolation to field studies at annual, seasonal, and event scales of habitat variability (Fig. 27).

Over a number of years, the numerical codes were developed to simulate the trajectory of such individual particles within the sea, from larval fish (Walsh and McRoy, 1986) to phytoplankton (Falkowski and Wirick, 1981), or waste materials (Hopkins and Dieterle, 1986). The first application of this Lagrangian approach involved dispersal and survival of larval fish within a simulated flow field

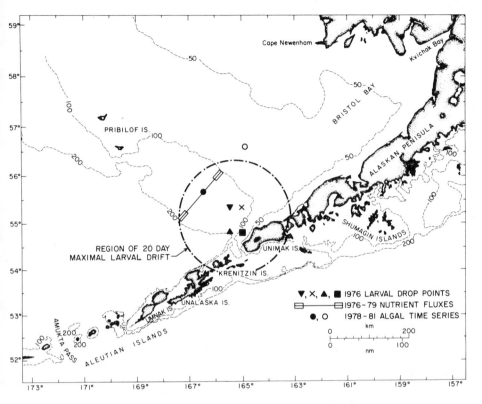

FIG. 114. The locations of larval drop points and maximal larval drift after 20 days of simulated survival of Alaska pollock within the southeastern Bering Sea. [After Walsh and McRoy (1986); © 1986 Pergamon Journals Ltd., with permission.]

of the southeastern Bering Sea. Using the techniques described in Section 3.1.5, a model simulated (1) the drift of larvae spawned at particular sites; (2) the growth of these larvae; (3) the mortality processes acting upon them; and (4) their distribution, length frequency, etc., after 20 days (Fig. 114). Correspondingly, the simulation model consisted of (1) a circulation submodel; (2) a submodel to simulate the distribution of larvae within the flow field; and (3) a biological processes submodel for individual pollock larvae.

4.5.1 Larval Survival

Because wind events are an important cause of habitat variability on continental shelves (Walsh, 1976; Wroblewski, 1984), diluting the prey both of clupeid larvae off California (Lasker, 1975) and Peru (Walsh *et al.*, 1980), and of gadoid larvae on Georges Bank (Lough, 1984) and off the Norwegian coast (Tilseth and

Ellertsen, 1984), we concentrate on the results of interannual changes in wind-induced larval transport of Alaska pollock. The opportunistic emergence of larvae of another demersal Arctic spawner, capelin (*Mallotus villosus*), for example, is evidently stimulated by onshore winds, leading to increased feeding success and reduced mortality from predation (Frank and Leggett, 1983; Leggett *et al.*, 1984).

Over the southeast Bering Sea, the mean wind speed is $\sim$10 m sec^{-1} during February to March and declines to $\sim$5 m sec^{-1} by June to July. Wind events also change direction, from predominantly northerly forcing in winter to southerly in summer, as indicated by a 1977 time series of wind speed and direction at St. Paul Island (Walsh and McRoy, 1986). Over a $\sim$25-yr interval (1946 to 1969), the transition period between the different wind forcings, favorable for offshore or onshore surface Ekman transport, was generally in May (McLain and Favorite, 1976), during the middle of the spring bloom. On the southeastern Bering shelf, low-frequency flow of water at these event time scales of 2–10 days accounts for up to 20% of the kinetic energy and correlates with winds $>$10 m sec^{-1} (Kinder and Schumacher, 1981).

The computed currents were derived from the same linearized, single-vertical-layer, barotropic, free-surface, hydrodynamic model of the northern Bering/Chukchi Seas (Walsh and Dieterle, 1986). In the southeastern Bering Sea case, a two-dimensional horizontal grid of $\sim$20 km spacing was used, and details of the numerical methods are given in Section 2.4.2. The subgrid scale diffusion was treated by a particle-in-cell method (Sklarew *et al.*, 1971; Lange, 1973), which first solved the advection and biological state variable equations by the usual finite difference approximations in Cartesian coordinates as in the hydrodynamic model (Tingle *et al.*, 1979). Larval fish were then represented by lagrangian marker particles within the eulerian grid and moved at each time step by advection and Fickian diffusion. Each particle, up to as many as $\sim$2000, had an assigned vector containing present position and past history.

The hydrodynamic submodel incorporated local bottom topography of the southeastern Bering Sea, friction, Coriolis force, the geostrophic pressure gradient, and a time-dependent wind stress derived from the 12-hourly 950-mbar winds measured at St. Paul Island—recall that the previous model of the Bering/Chukchi Seas was not forced by the wind. Omitting tidal motion of $\sim$50 cm sec^{-1} (Hastings, 1976), the calculated daily, depth-averaged currents in May 1978 (Fig. 115) matched quite well the mean residual, parabathic circulation of 5–10 cm sec^{-1} on the slope, 1–5 cm sec^{-1} on the outer shelf, and $<$1 cm sec^{-1} on the middle shelf as indicated by drogues (Coachman and Charnell, 1979) and current meter records (Kinder and Schumacher, 1981). In response to 10 m sec^{-1} winds from the north, diabathic offshore flows of 1–5 cm sec^{-1} were also calculated (Fig. 115a). Southerly wind forcing caused onshore flow of similar magnitude (Fig. 115b), with cross-shelf currents weaker on the middle than

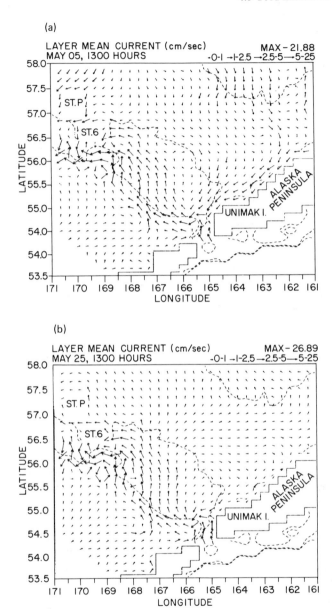

FIG. 115. The depth-averaged, simulated flow on the southeastern Bering shelf under (a) northerly and (b) southerly wind forcing during May 1978. [After Walsh and McRoy (1986); © 1986 Pergamon Journals Ltd., with permission.]

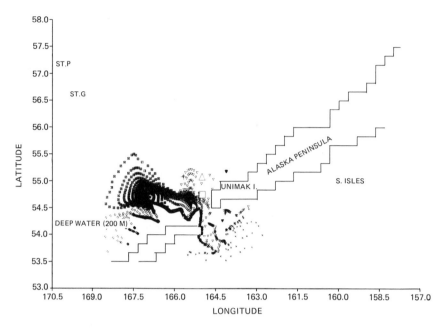

FIG. 116. The distribution of 20-day-old pollock larvae spawned on May 1, 1976. [After Walsh and McRoy (1986); © 1986 Pergamon Journals Ltd., with permission.]

outer shelf in both cases. Contrast these results with the much stronger flow field simulated north of St. Lawrence Island, in Fig. 34.

In the biological submodel, each pollock larva experienced a feeding, a growth, and perhaps a mortality event each day. The results of the feeding and mortality events were stochastic—that is, the local food concentration about a larva was a random deviate, drawn from an assumed distribution with a spatially changing mean and variance. Ingestion was proportional to the local food concentration and the larva's weight, while growth in weight was a fraction of the ingestion. The probability that a larva fell prey was inversely proportional to its physiological condition and proportional to the local predator density. The local predator density was also stochastic, but the mean density did not vary spatially. In summary, the outcome of a biological event was governed by probabilistic relationships between the larva's food, its predators, and its biological properties (that is, age, length, and weight).

To explore the consequences of pollock spawning on the outer Bering shelf in a cold year, compared to the middle shelf in a warm year as described in Section 4.1, a series of numerical experiments was performed using wind and spawning conditions encountered from 1976 to 1978. The results are shown from the 1976 simulation runs of a cold year, in which 501 pollock larvae were placed within the model's grid at four drop points on the outer Bering shelf during May 1, 6,

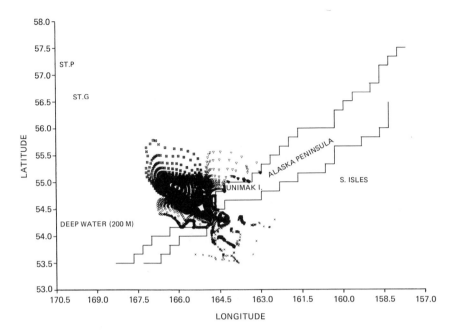

FIG. 117. The distribution of 20-day-old pollock larvae spawned on May 6, 1976. [After Walsh and McRoy (1986); © 1986 Pergamon Journals Ltd., with permission.]

and 11, 1976; none were released on the middle shelf. The location of each drop point is indicated in Fig. 114, while the 20-day-old larvae originating from these points are denoted in Figs. 116–118.

The model was used to determine whether interannual variation in spawning location and/or food abundance was more important: could the larval fish drift landward from the outer to the middle shelf before starvation? A total of 2004 yolk-sac larvae were tracked over 20 days of the first-feeding, larval survival period (growth from 5 to 10 mm length). In each of these simulations, every larva was assumed to begin feeding on copepod nauplii 3 days after entering the model. To isolate food availability from larval drift as an important factor, it was assumed in these results (Figs. 116–118) that the abundance of copepod nauplii was proportional to the total zooplankton biomass in mainly warm years. The zooplankton gradient of only a twofold spatial change in mean abundance was taken from Japanese survey cruises of 1954 to 1970 (Motoda and Minoda, 1974). As we shall see, however, under the same flow regime, the transition from warm to cold-year food conditions actually involved a 10-fold reduction of mean prey abundance of copepod nauplii on the middle shelf.

The May 21, 1976, distribution of larvae developed from the 2001 fish launched on May 1, 1976, under food conditions of a warm year is shown in Fig. 116. Recall that individual larva are represented by symbols that also indi-

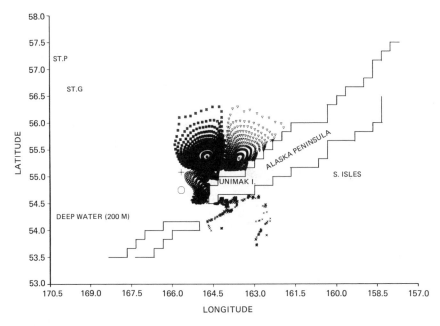

FIG. 118. The distribution of 20-day-old pollock larvae spawned on May 11,1976. [After Walsh and McRoy (1986); © 1986 Pergamon Journals Ltd., with permission.]

cate their original drop point (Fig. 114). Within the first 10 days, net drift had been to the southwest as a result of wind-induced advection from the north and subgrid scale dispersion. Some larvae passed through Unimak Pass into the Gulf of Alaska, and many more were in the surf along the northern shores of the Fox Islands. Ten days later, on May 21, 1976, the distribution of surviving larvae was somewhat different as a result of wind-driven flow from the south. Fig. 116 illustrates that after 20 days these simulated larvae were dispersed more to the west and north, like their May 6 and 11, 1976, counterparts.

For comparison, distributions (after 20 days of dispersal) of larvae dropped on May 6 and 11 under food conditions of a warm year show that the net dispersal of the later-hatched larvae is more towards the northeast, that is, on the middle shelf (Figs. 117 and 118). Zooplankton nauplii were assumed to be twofold more abundant northeast of Unimak Island, and the survivorship of the later-spawned fish larvae was thus greater. In fact, since no spatial inhomogeneity of predation was assumed, the mortality estimate for larvae dropped on May 1 was two times greater than that of the larvae modeled on May 11. Figure 119 is a frequency distribution of the surviving larvae in terms of their individual biomass (x 10^2 μg dw) after each experiment.

Of the three launch dates considered above, May 11 was the best for larval survival, and for a given date those released at the most northeast drop point had

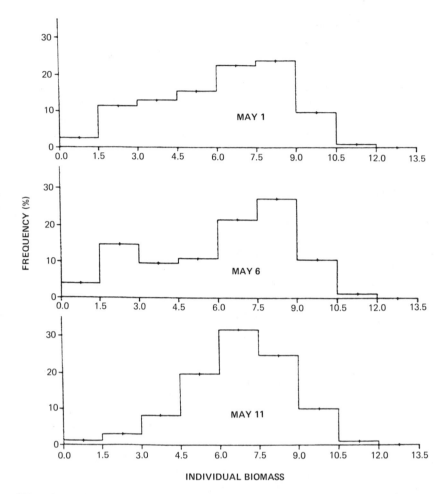

FIG. 119. The frequency distribution of individual larval biomass ($\times$ 10^2 μg dry weight) of 20-day-old pollock hatched on May 1, 6, and 11, 1976. [After Walsh and McRoy (1986); © 1986 Pergamon Journals Ltd., with permission.]

Results of additional simulations led to the conclusion that under 1976, 1977, or 1978 conditions of wind forcing, larval pollock hatched on the outer shelf the highest survival, that is, on the middle shelf rather than the outer shelf. The daily weight increment of 6-mm pollock larva dropped on May 11 was a mean of 12%, in good agreement with observed daily growth of 8.5–12.0% for larval haddock, *Melanogrammus aeglefinus* (Jones, 1973), and cod, *Gadus morhua* (Laurence, 1978). Independent calculations (Incze *et al.*, 1986) suggest a body weight increment of 12% day^{-1} for Alaska pollock larvae <7 mm and 5.8% day^{-1} for larvae >7 mm length.

could drift to the middle shelf within 20 days (Fig. 114). If the abundance of nauplii food on the middle shelf were reduced in the model by an order of magnitude under the same flow conditions, however, to mimic cold-year food availability, the larval survival of pollock would be 10-fold less in this region. Such a calculation was thus based on an assumed difference in prey availability for first-feeding larvae, not on advective changes between a warm or cold year. The following section examines the naupliar abundance of copepods on the outer and middle shelves during cold and warm years in relation to the available primary production.

4.5.2 Interannual Variation

Mesoscale, synoptic-scale, and seasonal scale processes (Table I) constitute the "weather of the sea," while longer-term scale phenomena affect both the oceanic and atmospheric climates. Although year-to-year variability is a climatic effect, one of its main consequences is a modification of typical sea-weather patterns. This may result from changes in the general ocean circulation and marine energy transports, responding to similar changes in the atmosphere, with noted differences in typical atmospheric-weather patterns, such as El Niño, over the marine ecosystem under investigation. Thus, in some exceptional years, important modifications will be observed in mesoscale and synoptic-scale processes, characteristic of a particular time of year, which will entail changes in the flow field and the many chemical and biological processes that depend on physical transport and diffusion, such as photosynthesis, timing of zooplankton cohorts, survival of larval fish, and sinking of particulate and dissolved forms of carbon.

During the 1982 El Niño (Fig. 1), for example, loci of high primary production both shifted from the west to the east side of the Galapagos Islands (Feldman et al., 1984) and contracted toward the coast off Peru. There was perhaps a three- to fourfold reduction in offshore extent of the productive coastal upwelling habitat (Feldman, 1986). North of Iceland, where one might expect to see teleconnections of biological and chemical signals associated with interannual modulation of water transport between the Bering and Fram Straits, vertical mixing, nutrient input, and primary production of surface waters, during May–June, all change by an order of magnitude from one year to the next (Stefansson, 1985).

Interannual variation in the northward transport of water through the Bering Strait is thought to be controlled by year-to-year changes of the intensity of northerly wind forcing (Aagaard et al., 1985). Intensified wind stress from the north during winter causes a set-down of sea level south of Bering Strait, a diminished northward transport of water, increased ice cover in the Bering and Chukchi Seas during spring, and low temperatures at Cape Lisburne during mid-July (Fig. 120). Consequent reproductive failure of sand lance, capelin, and cod within Alaska Coastal Water, as well as of their avian predators, then apparently

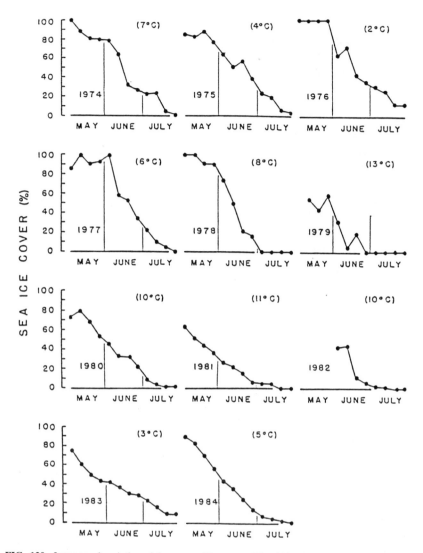

FIG. 120. Interannual variation of the seasonal ice cover (%) within the Bering/Chukchi Seas and of the mid-July (∓3 days) surface temperature (°C) at Cape Lisburne from 1974 to 1984. [After Walsh *et al.* (1986a); © E. J. Brill, with permission.]

occurs during cold years at Cape Lisburne (Springer *et al.*, 1984). During 1976, a year of very cold temperature (Fig. 120) and weak northward transport, apparent food shortages also led to a poor breeding season of sea birds on St. Lawrence Island, that is, those feeding on the food web of Anadyr Water (Springer and Roseneau, 1987).

Within the southeastern Bering Sea, the onset and amount of the spring bloom of phytoplankton do not evidently change from a cold to a warm year (Walsh and McRoy, 1986), but like the bacterioplankton off the Newfoundland coast (Pomeroy and Deibel, 1986), the zooplankton are severely impacted by colder temperatures. In mid-April of 1980, bottom water temperatures (-0.8 to $1.3°C$) were $\sim 2°C$ lower on the southeastern middle Bering shelf than during mid-April 1981, suggesting that 1980 might have been analogous to 1976. Within the upper water column, the middle-shelf temperatures during mid-May 1980 were still $\sim 4°C$ colder than those of the outer shelf in mid-May of 1978, 1979, and 1981. Although the 1980 phytoplankton bloom started on the middle shelf about the same time as in mid-April 1978 and 1979, the initial 1980 reproductive success at midshelf of small zooplankton species (e.g., *Pseudocalanus*, was much less than in 1979 and 1981.

Female and juvenile (CV) *Pseudocalanus* first appeared in warmer waters of the outer shelf in March to April 1980 and were not abundant on the middle shelf until the end of May (Smith and Vidal, 1986). Using a Belehradek function of temperature and *Pseudocalanus* growth (McLaren, 1978), the generation time, from hatching to mature female, would be ~ 40 days longer on the midshelf at $0°C$ than at $4°C$, that is, the temperature difference between 1979 and 1980. Tenfold fewer *Pseudocalanus* females were, in fact, found on the midshelf before May 15, 1980, than before May 15, 1979 (Dagg *et al.*, 1982), and fourfold fewer than before May 30, 1981 (Dagg *et al.*, 1984).

We would also expect smaller production of nauplii during cold years, and, in fact, only $5-8$ nauplii l^{-1} were found at midshelf on May 16, 1980 (M. Clarke, personal communication), fourfold to 10-fold fewer than in 1981 and 1979. Larval pollock spawned and hatched in this region would probably not survive at these low prey concentrations; those drifting from the outer to the middle shelf would also not survive the critical period of first feeding. During April to May 1980, pollock larvae were found at only 20% of the stations. An order of magnitude more larvae were caught on the outer shelf ($\leq 50/10^3$ m^3) than the middle ($\leq 5/10^3$ m^3), similar to the cross-shelf gradient of larvae in April to May 1976 (T. Nishiyama, personal communication).

In contrast, the abundance of *Pseudocalanus* CV on the outer shelf (Smith and Vidal, 1986) in May 1980 (29 m^{-3}) was only threefold less than in May 1979 (72 m^{-3}) and about the same as in May 1981 (25 m^{-3}). This reflected outer-shelf water temperatures in 1980, which were just $1-2°C$ colder than in 1978, 1979, or 1981. The mean number of *Pseudocalanus* CII and CIII on the outer shelf between April 20 and June 7, 1980 (39 m^{-3}), was actually more than during the same period in 1981 (32 m^{-3}). As many as 11 nauplii l^{-1} were also found in slope waters on May 18, 1980, compared to $15-20$ nauplii l^{-1} earlier in the spring on April 20, 1979, suggesting that avoidance of starvation during late spring by larval pollock on the outer shelf and slope may be similar during warm and cold years.

First-feeding pollock larvae of 5–10 mm length are visual feeders after yolk absorption, ingesting eggs, nauplii, and copepodites of small calanoid copepods ranging in size from 80 to 340 μm (Clarke, 1978). Survival of the larvae is thus not coupled to the population dynamics of the larger copepods (e.g., *Neocalanus*) on the outer shelf. Feeding studies of these larvae in the laboratory from 1980 to 1981 (R. Lasker, personal communication), and their respiratory demands (T. Nishiyama, personal communication), suggest that each larva of this size class might ingest a maximum of 10 nauplii day^{-1} at prey concentrations between 12 and 24 nauplii l^{-1}. This is well below food densities required by other gadoids and flatfish (Wyatt, 1972; Laurence, 1974).

During the warm years of 1978 and 1979, only 1–10 nauplii l^{-1} were found on the outer and middle shelves during April (T. Cooney, personal communication), which is an insufficient prey density for larval survival of the pollock. However, 10–20 nauplii l^{-1} were found at this time in 1979 in slope waters, where pollock eggs and larvae occurred in March and April 1978 (Figs. 102a and d). By the end of May in 1979, more than 120 nauplii l^{-1} were found in subsurface maxima on the middle shelf, and 20–40 nauplii l^{-1} on the outer shelf, reflecting cross-shelf gradients in the biomass of adult *Pseudocalanus* and *Oithona* spp. (Cooney and Coyle, 1982). These prey densities are sufficient to support larvae spawned during April (Figs. 102b and c).

Two years later during the warm year of 1981, a sufficient prey density, 15–20 nauplii l^{-1}, was also found between May and June in subsurface maxima on the middle shelf. The mean number of pollock larvae (100/10^3 m^3) was then the same as in May to June 1978 (Fig. 102e). They consumed <1% of the daily naupliar production of adult *Pseudocalanus* and *Oithona* spp. in 1981 (Dagg *et al.*, 1984), however, implying that larval fish on the middle and outer shelves were not food-limited in May of 1978, 1979, or 1981.

During cold years when larval pollock are mainly restricted to the outer shelf, additional mortality may be induced here by avian predators, such as kittiwakes (G. B. Smith, 1981). These surface feeders remove threefold more carbon from the outer shelf than from the middle regime (Schneider and Hunt, 1982). During a cold (1980) year, more carbon may also be taken by avian predators on the outer shelf than during a warm (1981) year (Schneider *et al.*, 1986), possibly reflecting increases of prey available to the birds. The low reproductive success of kittiwakes and murres on St. George Island, during the subsequent warm year of 1981, in fact, may be related to a poor year class of pollock spawned in 1980. During warm years, pollock larvae spawned on the middle shelf may suffer less predation from birds and encounter sufficient naupliar concentrations to avoid starvation, enabling recruitment of a larger year class.

Individual larva may find sufficient food only on the outer shelf in cold years, and in both regions during warm years, within the short drift from their hatching location (Fig. 114). Onshore movement of adult pollock to spawn during warm years thus allows higher larval survival in terms of both food and predation. We conclude from the results of the simulation analyses that years of decreased zooplankton biomass and less larval survival of fish on the middle shelf may be associated with cold sea-surface temperature anomalies from severe winters (McLain and Favorite, 1976). For example, after the cold year of 1960, the herring class of that year had lower growth rates than those of 1957, 1962, and 1967 (i.e., from warm years) (Rumyantsev and Darda, 1972). In terms of year-class strength of Alaska pollock spawned in the southeastern Bering Sea between 1972 and 1979 (Francis and Bailey, 1983), the 1978 year class from a warm year was the strongest and the 1976 year class from a cold year was one of the worst, with a sevenfold variation in recruitment.

We have been able to document the circumstances under which a successful year class of pollock might emerge from the shelf of the southeastern Bering Sea during a warm year. Later spawning of copepods and fish may occur on the middle shelf during cold than warm years, but the major phytoplankton bloom here is terminated by nutrient depletion before the end of May (Fig. 103), thereby curtailing the amount of energy that may be passed up the food web that year. Despite the occurrence of late-summer blooms of phytoplankton during warm and cold years in the Bering Sea (Rumyantsev and Darda, 1972), successful cohorts of pollock, yellowfin sole, halibut, or herring are apparently determined only by larval interaction with copepod grazers of the spring bloom. Once past this critical period, the juvenile and adult fish evidently have sufficient motility to obtain their daily rations, moving over large distances in response to either natural cues or disruption of their habitat.

4.5.3   Predator Control

El Niño (Stevenson and Wicks, 1975) is mainly signaled by the penetration south along the Peru coast of anomalously warm and low-salinity water beyond the usual summer limits, as occurred in 1982. Similar northward influxes of warm water may occur off the California coast (Walsh, 1978). Local winds may not weaken significantly during this phenomenon (Wyrtki et al., 1976), and it is thought that the hydrographic anomalies reflect a much larger atmospheric and oceanic coupling (Bjerknes, 1966). Weakening of the westerlies in the southern hemisphere (Schell, 1965) during March–November was first thought to lead to warm-temperature anomalies off Peru in the subsequent months of December–February. Strong trade winds in the equatorial region also occur when the westerlies are weak. The mechanism for El Niño is now attributed to varying strength of the Equatorial Countercurrent, 8 months before the El Niño, in response to a sequence of events within hemispheric wind systems (Namias, 1973). Prediction of El Niño (Quinn, 1974) presently involves large-scale differences in atmospheric pressure between Australia and Easter Island, as well as the Juan Fernandez Islands.

Water is believed to pile up in the western Pacific Ocean during the weak westerlies because the strong trade winds build up an east–west slope of the sea level through the North and South Equatorial Currents. If the westerlies then intensify (i.e., the trades weaken), the piled-up water returns across the Pacific Ocean as a Kelvin wave (Wyrtki et al., 1976). The southerly transport of warm water into the Peru region and northerly transport into California coastal regions from the Equatorial Countercurrent may occur as additional poleward Kelvin waves (Gill and Clarke, 1974). Upwelling of cold water then resumes in these eastern boundary currents as the westerlies continue to remain strong, feeding the southeasterly trades that tend to oppose the Equatorial Countercurrent.

The 1972 El Niño could be sensed (Wooster and Guillen, 1974) as early as March 1971 in a warm-temperature anomaly of equatorial water between 80 and 110°W, for example, and consisted of perhaps three periods off Peru. The first major warm anomaly occurred along the coast during February–June 1972 as far as 14°S. A period of smaller anomaly was observed during July–November 1972 (i.e., the austral winter). A second major anomaly of shorter duration lasted from December 1972 to February 1973 as far as 17°S. By March 1973 the coastal sea-surface temperatures returned to normal. A biological signal of this perturbation off Peru in 1972 was a southerly transport of tropical organisms such as the pelagic crab, Euphylax dovii (Vildoso, 1976), in a similar manner to northerly transport of the red galatheid crab off San Diego, California, in 1972 (Walsh et al., 1977).

The long-term atmospheric temperature anomaly (Hubbs and Roden, 1964) off San Diego over the last century (Fig. 121b) shows a very similar pattern to secular changes of an upwelling index (Bakun, 1973) over 25 yr off the west coast of the United States. The atmospheric warming phases off San Diego during 1850–1870, 1890–1905, and 1930–1950 are also correlated with intrusions north to at least Monterey of the range of subtropical fish (Hubbs, 1948). Peaks of clupeids off Santa Barbara (Soutar and Isaacs, 1974) during these times (Fig. 121c) may also reflect a northward translation of the larval drift of these fish. Finally, the northward extension of the red crab, Pleuroncodes planipes, from its home range off Baja California during 1859, 1926, 1941, 1958, and 1972–1973 (Fig. 121c) may reflect both the long-term climatic process and an interaction with the El Niño scale of variability.

The available sea-surface temperature data (Ramage, 1975) off Peru (Fig. 121b) appear to be in phase with the longer atmospheric record off San Diego. The biological response to this decadal scale of habitat variability may be the same in both systems. In response to long-term changes of wind, current, and

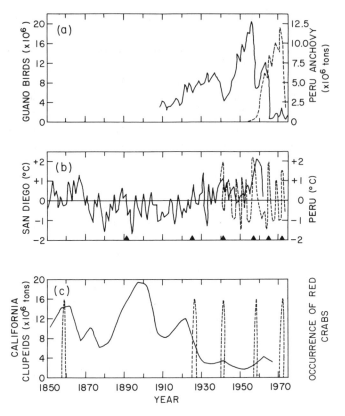

FIG. 121. The long-term oscillations of (a) Peru anchovy (– – –), and guano birds (————), and (c) California clupeids (————) and red crabs (– – –) in relation to (b) air (————) and sea-surface (– – –) temperature anomalies off San Diego and Peru. The triangles indicate a major El Niño. [After Hubbs (1948), Hubbs and Roden (1964), Longhurst (1967b), Soutar and Isaacs (1974), Ramage (1975), Santander (1976), and Walsh (1978); © International Council for Exploration of the Sea, with permission.]

subsequent sea-surface temperature, Peru anchovy may move south (Sears, 1954). The smaller number of guano birds off Peru during the warming phase of 1920–1940, in contrast to larger populations of birds during a colder period of 1940–1960 (Fig. 121a), could thus reflect such latitudinal excursions of the Peru anchovy stocks.

The anchovy fishery was not initiated until 1955, and the previous oscillations of the guano birds may actually represent such an alteration in their food supply. When humans replaced the guano birds as the major predator of anchovy off Peru, however, the bird populations were decimated by the time the 1965 El Niño occurred. Before this time, the bird populations had recovered from their losses during the 1941 and 1957 El Niños (Fig. 121a).

The normal clupeid response to the El Niño scale of variability off Peru may be a southward movement of the anchovy stocks to a different habitat (Walsh *et al.*, 1980). If the guano birds had then remained close to their nesting sites in the north, they would have starved after their prey emigrated. The Peru anchovy, *Engraulis ringens,* may have evolved a natural resilience to this scale of variability and changing habitat until the additional predation pressure, introduced by humans in 1955, culminated in 1972 with a dramatic drop in their abundance as well (Fig. 121a). During this same year, El Niño had occurred again. The arrival of El Niño off Peru has been associated with biological disasters (Brongersma-Sanders, 1957), but these are not quantitative observations. With the exception of the guano birds, a biological disaster off Peru in response to El Niño may only have been an impression rather than a fact until overfishing occurred. overfishing occurred.

For example, spawning of the Peru anchovy was already low in 1971, as it had been in 1969 and 1970 (Fig. 122). The fat content of the adult clupeids was also high, indicating disruption of reproduction (Valdivia, 1976). Recruitment was thus low before the onset of the 1972 El Niño, for very little of the 1971

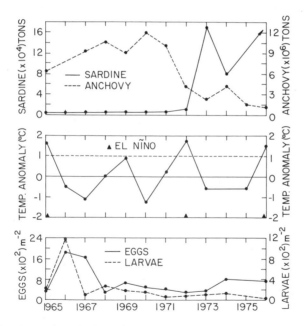

FIG. 122. The short-term oscillations of Peru anchovy ($\times$ 10^6 tons yr^{-1}) and Chilean sardine ($\times$ 10^4 tons yr^{-1}) landings along the Peru coast in relation to the January sea-surface temperature anomaly (°C) at Lima, Peru, and of the mean August–September spawning level ($\times$ 10^2 eggs m^{-2} station^{-1}) and larval abundance ($\times$ 10^2 larvae m^{-2} station^{-1}) of anchovy between 4 and 18°S. [After Walsh *et al.* (1980); © 1980 Pergamon Journals Ltd., with permission.]

year class was found in 1973. It is possible that, irrespective of the 1972 El Niño, the clupeid yield off Peru might have collapsed as a result of overfishing. During 10 cruises off Peru in 1972–1973, there was no evidence that the oceanographic disturbances had induced mass mortality of the remaining adult Peruvian anchoveta (Valdivia, 1976).

The response of the stock appeared, instead, to be a shift south in the location of its geographic center of abundance. During March 1972, at the height of the warm-temperature anomaly, the anchoveta were concentrated inshore as the coastal divergence weakened and warm offshore water moved near the coast. Most of this catch was taken below 10°S (Valdivia, 1976). In further response to the warm-water intrusion from the north, the anchoveta were mainly found south of first 12°S and then 14°S during April 1972. After March 1973, the stocks were again found along the whole Peru coast, but the yield was still low in both 1973 and 1974 (Fig. 122). Thus, the guano birds have remained eulerian predators off Peru, while humans, in our shift from dugout canoe to factory ship, have become more of a lagrangian predator like the anchovy.

The catch of the Peru anchovy fishery declined precipitously from the peak years of 1969 to 1971 (Fig. 121). As a result of overfishing and the 1972 El Niño, the estimated biomass of anchovy was only $2-4 \times 10^6$ metric tons during 1972, in comparison with $15-20 \times 10^6$ tons during 1969 (Walsh, 1975). Fishing was permitted in only 18 of the 48 months between June 1972 and June 1976, with the result that anchoveta stocks may have returned to $10-11 \times 10^6$ tons by January, 1976. Resumption of fishing, natural mortality, and the 1976 El Niño again reduced the 1977 anchovy stocks to the levels of 1972, $\sim 3 \times 10^6$ tons by March 1977. Because of the poor recruitment after this 1976 El Niño (Santander and de Castillo, 1977), fishing was postponed in March 1977, with only 0.5×10^6 tons of anchovy caught by April 1977. The fishing effort was then switched to sardine, with a total clupeid landing of $\sim 2 \times 10^6$ tons in 1977 (Fig. 1). The 1978–1982 catches of anchovy and sardine remained at this level, until the 1982 El Niño led to further reduction in the 1983 landings (Fig. 1).

The austral sardine, *Sardinops sagax,* used to support a small fishery of 1×10^4 tons yr^{-1} off northern Chile (Longhurst, 1971), in contrast to peak Chilean catches of 1×10^5 tons yr^{-1} of *E. ringens* in 1966 (Brandhorst et al., 1968). Few sardine eggs were observed off Peru from 1964 to 1971 (Santander and de Castillo, 1977). Only 1×10^3 tons yr^{-1} of *S. sagax* was landed in Peru until 1972, when 1×10^4 tons was caught (Fig. 122). A large number of sardine eggs was found earlier than normal in July 1972, and similar to oscillations of sardine and anchovy populations off California, 1.7×10^5 tons of *S. sagax* was taken off Peru by March and April, 1973, after the recruitment failure of the *E. ringens* stock in 1972 (Murphy, 1973). The 1973 sardine yield was then 8% of the annual clupeid catch off Peru (Fig. 122), however, similar to catch ratios off Chile during 1966. By 1983, a decade later, the sardine catch was more than

90% of the clupeid yield off Peru (Fig. 1) and similar to peak landings of *S. caerulea* off California during 1937 (Fig. 98).

An analysis of the decadal trends of the Peru ecosystem suggests that the apparent clupeid species succession is the result of both altered survival of larval anchovy during the 1965, 1972, 1976, and 1982 El Niños and the effects of long-term fishing mortality on the adults. The spawning area at 10°S represents a more favorable habitat for survival of larval anchovy off Peru with a larger number of larvae found there in 1964 than at 15°S (Walsh *et al.*, 1980). A possible south-ward displacement of the spawning stock in 1965 and the occurrence of an order of magnitude fewer anchovy larvae during the 1965 El Niño (Santander and de Castillo, 1969) did not appear, however, to seriously affect the landings of the next year (Fig. 122). The fishery had not peaked by 1966, and the guano birds were no longer important predators on anchovy at that time (Walsh, 1978). The highest spawning levels of anchovy between 1964 and 1976 (Santander and de Castillo, 1977) were in 1966 and 1967 (Fig. 122). During 1968 and 1969 the abundance of eggs decreased to about one-quarter of the maximum (1967), but they were still more abundant (Jordan, 1971) than in later years. The production of anchovy spawn continued to decrease in 1970 and 1971, apparently not as a result of adverse oceanographic conditions but possibly because of the continuing fishing pressure.

Over these long migration distances off Peru, adult anchovy would encounter sufficient heterogeneity of prey to obtain their rations each day (Wirick, 1981), similar to the daily vertical excursions of adult copepods within California waters (Mullin and Brooks, 1976). Adult predators of pelagic food webs, unlike their larval stages, may thus not be food-limited (Gulland, 1971; Cushing, 1975), exerting little control on their prey, in contrast to terrestrial (Hairston *et al.*, 1960), fresh-water (Smith, 1968; Zaret and Paine, 1973), or littoral (Paine, 1969; Simenstad *et al.*, 1978; Wharton and Mann, 1981) food webs.

As a consequence, there appears to be little relationship between the abundance of adult stocks and the amount of annual recruitment within the anchovy (Fig. 122), salmon (Peterman, 1980), cod (Garrod and Clayden, 1972), plaice (Bannister, 1978), herring (Saville, 1978), mackerel (Hamre, 1978), halibut (Southward, 1967), and haddock (Cushing, 1968) populations. There is the haunting possibility that nekton are poorly coupled to the lower trophic levels in the sense that, once their stocks are fished below their carrying capacity, the release of predator stress on their food resource is not a sufficiently strong feedback to allow rapid increase of prey and thus rapid recovery of the predator; we will return to this subject in Section 6.3.

The budgets and simulation models of this chapter, except for Section 4.5.1 and one case of Pace *et al.* (1984), have been deterministic with respect to either the forcing functions or the fluxes between state variables. Espousing Albert Einstein's philosophy that "God does not roll dice," it has been assumed that all the rules, or laws, of the universe are knowable, with the included first-order principles of sufficient importance to accurately simulate, or describe, the cycles of elements within continental shelf ecosystems. Ignoring the nonlinear terms of the Navier–Stokes equations, tidal forces, departures from mean wind stress at shorter time scales, species composition and succession, as well as a host of other biological and chemical variables, constitutes, however, a rather poor description of the second-order processes.

In aggregate, this lack of definition of nonlinear feedback between simple state variables and forcings of a model can lead to a poor homolog of the modeled system. Furthermore, apparent random behavior can be observed from inadequate sampling of the "real world." Deviation of a model's predictions from such real-world observations presumably can be remedied by either a more detailed definition in the model of these second-order interactions, or a better parameterization of the apparent randomness of the observations with models of stochastic processes. A simple, deterministic stock-recruitment model (May and Oster, 1976) can lead, however, to three different solutions of stable equilibria, periodic oscillations, and apparently random fluctuations, without the additional complexity of multispecies interactions (Andersen and Ursin, 1977).

We could proceed farther into the morass of complex predatory interactions of unstructured pelagic food webs (Longhurst, 1985a; Isaacs, 1973), and their stochastic description by either Markov chains (Leslie, 1945) or earlier random walk models (Pearson, 1906). We will instead now focus on their implications of export, however, since significant mathematical reduction of the complexity of these food webs has little impact on the amount of simulated detrital loss (B. Frost, personal communication). It would appear that among some shelf food webs, supported by an annual primary production of ≥ 200 g C m^{-2} yr^{-1} at 30, 45, and 60°N latitudes, all of the fixed carbon of algal photosynthesis is not consumed by herbivores, saprovores, or carnivores. Presumably, evidence of this export remains stored in the sediments of these shelves or adjacent continental slopes.

5 Storage

There had been such floods before . . . a thing that happens only once in a hundred years . . . the farmers were awakened by the plaintive bellowing of their animals. Swinging their feet out of bed, in the dark, they put them down in a foot of cold, muddy water. It was salt . . . the North Sea had come to visit them . . . trees and bushes were growing in a moving gray ground and thick yellow foam was washing over the stretches of their ripening corn.

(Dinesen, 1934)

Now these floods did not occur each year . . . the flood of 1543 . . . was forty days in reaching its crest. . . . And it was a most magnificent spectacle to behold. That which previously had been forests and fields was now converted into a sea . . . and nothing was visible except the pine needles and branches of the highest trees.

(de la Vega, 1605)

The boundaries of land and sea are transient interfaces, migrating back and forth across the continental shelves at diurnal, seasonal, decadal, millenial, and glacial time scales. Each year, for example, most of the human deaths from hurricanes, or typhoons, are the result of drowning. Six years after publication of de la Vega's account of a Mississippi River flood, about 3000 people perished in a 25-m-high tidal wave off the east coast of Japan on December 2, 1611 (Imamura, 1934). Similarly, another tsunami of 29-m amplitude killed at least 2986 people in the same region on March 3, 1933 (Imamura, 1934), while Isak Dinesen was writing the description of a hundred-year flood from the North Sea. On January 31, 1953, it occurred with 1853 people drowned, when the North Sea last came to visit the Netherlands. Over a 1064-yr-long time series, however, only three other major seismic sea waves arrived off Japan, in 869, 1640, and 1896, such that the most significant perturbations of sea level may occur at glacial time

283

scales, in terms of storage of organic carbon and the depth over which a falling particle must escape consumption.

Present shelf regions, for example, are an anomalous marine habitat of the relatively rare interglacial periods over the last 2 million years. Since the beginning of the Pleistocene, the Earth's climate has been colder than it is today for about 90% of this time. Threefold more of the Earth's area was thus covered by ice sheets during the predominantly glacial periods, resulting in about a 3% decrease of the ocean's volume and a land boundary most of the time at about the 100-m isobath of today's shallow seas. Ancient rivers then discharged sediments at the shelf break rather than within the estuaries of today, where terrestrial carbon is now mainly trapped. Rapid rise in sea level ceased ~5000 years ago (see Fig. 164), but the sediments of the continental shelves are still relict (Emery, 1968) today, with little deposition of marine organic matter in these regions.

Past the "mud line" at the shelf break (Murray and Renard, 1891), however, the organic carbon contents of slope and upper-rise sediments off the Mid-Atlantic Bight (Fig. 123) and the Bering Sea (Fig. 124), for example, are 1–2% of dry weight (dw) sediment, an order of magnitude greater than most of the adjacent shelf sediments. The Hudson Shelf Valley, the "mud hole" south of Martha's Vineyard, and the shelf regions west of St. Lawrence Island and north of Bering Strait appear to be local depocenters; note that the organic carbon in slope sediments also increases downstream from the Aleutian Island Chain to Siberia and from New England to Virginia (Figs. 123 and 124). Furthermore, on the Peru slope, as much as 10–20% dw carbon sediments are found (Fig. 125), suggesting that all continental slopes may be depocenters of organic carbon (Premuzic et al., 1982), as indicated by other properties as well.

The grain size of surface sediments in the Mid-Atlantic Bight generally decreases with increasing bottom depth of the sample (Fig. 126a). Sediment porosity is inversely related to grain size, with values of 53–58% found in the coarse, upper-slope sediments and 70–80% in the fine, lower-slope sediments off the Mid-Atlantic Bight (Keller et al., 1979). Similarly, the porosity of sediments at 390 m on the Bering slope is 49%, compared to 80–90% on both the lower Bering slope (Walsh et al., 1985) and the shelf off Peru (Henrichs and Farrington, 1984). With the exception of stations taken in the "mud patch" or the "mud hole," indicated by the 1% bottom carbon isopleth (Fig. 123) south of Martha's Vineyard, the mid-Atlantic shelf sediments are dominated by sand-sized material, typical of the Holocene sand sheets of this region (Swift et al., 1972). From the shelf break to a water depth of 1100 m on the upper slope, the sediments exhibit a bimodal grain size distribution, with common occurrence of both fine sand and fine silt and clay. Samples from the lower slope contain primarily fine silt and clay, similar to surface deposits of the mid-Atlantic rise as deep as 4400 m (Bulfinch et al., 1982; L. Doyle, personal communication).

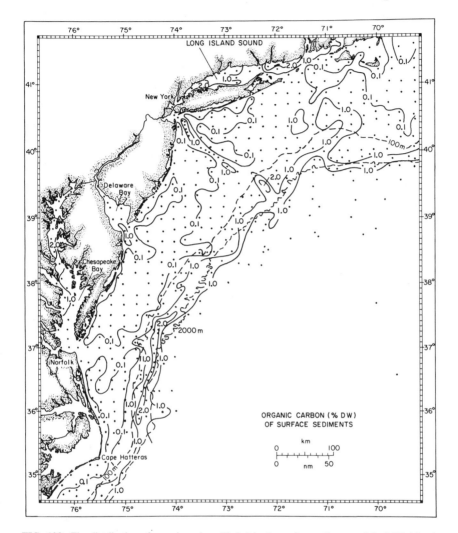

FIG. 123. The distribution of organic carbon (% dw) in the surface sediments of the Mid-Atlantic Bight from 1962 to 1982. [After Hathaway (1971), Walsh *et al.* (1985); © 1985 Pergamon Journals Ltd., with permission.]

The amount of calcium carbonate on the mid-Atlantic shelf is somewhat variable (Fig. 126b), but the majority of sediment samples contain less than 2% dw; in contrast, $CaCO_3$ is undetectable within the Peru shelf sediments (Henrichs and Farrington, 1984), and recall that the open Gulf of Mexico sediments contain 30–40% dw $CaCO_3$ (Scott *et al.*, 1983). A nearly linear increase in carbonate occurs from shelf values of <2% to as much as 13% dw at a depth of ~1500 m

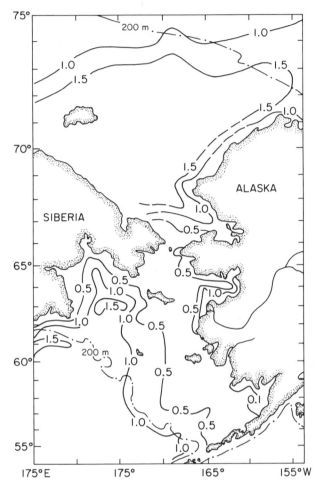

FIG. 124. The distribution of organic carbon (% dw) in the surface sediments of the Bering Sea. [After Gershanovitch (1962), Creager and McManus (1966), Lisitzin (1966), Sharma (1974, 1979), Romankevich (1977), and Walsh *et al.* (1985); © 1985 Pergamon Journals Ltd., with permission.]

on the lower mid-Atlantic slope. A decline in carbonate, organic carbon (Fig. 126c), and amorphous silica (Fig. 126d), as well as coarsening grain size, and older surface sediment (Fig. 126h) are found at deeper stations of 1600–1800 m bottom depths. At the lower boundary of the slope, the increasing trend of these properties with depth reappears; as much as 35% dw carbonate can be found in surface sediments at 2500–3000 m on the upper rise of the Mid-Atlantic Bight (Balsam, 1982), reflecting a coccolithophore-based food web rather than a diatomaceous one.

Amorphous SiO_2 content of the mid-Atlantic shelf, for example, is 0.5–1.0%

dw sediment, similar to that of deep ocean sediments in the Equatorial Atlantic, and in contrast to ~60.0% dw of diatomaceous oozes in parts of the Antarctic Ocean (Eggimann *et al.*, 1980). In going from the shelf to the mid-Atlantic slope, however, the biogenic silica of surficial sediments increases 10-fold (Fig. 126d) and exhibits distributional patterns similar to those of organic carbon and fine silt (Figs. 126a and 126c). Within the high-latitude ecosystem of diatomaceous sediments in the Bering Sea, the amorphous SiO_2 and carbon (Fig. 124) contents of surface sediment on the slope are similarly 10-fold those of the shelf, with as much as 23.9–30.6% dw SiO_2 found at 1773–1965 m (Kotenev, 1972).

The spatial trends of sediment grain size and carbonate content in the Mid-Atlantic Bight suggest that below a bottom depth of 1400 m the influence of present terrigenous sedimentation is substantially reduced, although pollen grains can be found at least as far offshore as 4000 m bottom depth (Balsam and Heusser, 1976). The mean diameter of ragweed pollen, *Ambrosia artemisiifolia,* is ~20 μm, similar to those of some diatom frustules, such that the recent input of pollen to slope sediments could result from both atmospheric and surface Ekman transport, although pollen is introduced to the marine environment mainly by rivers (Balsam and Heusser, 1976). On the lower slope, the increasing influence of Holocene pelagic deposition is also supported by the relative abundance of planktonic foraminifera in sediment samples from the deeper stations.

The source of the organic carbon in these sediments can be surmised from the C/N content (Fig. 126e), the $^{13}C/^{12}C$ (Fig. 126f) and $^{15}N/^{14}N$ isotope ratios, and the presence of lignin (Fig. 126g). Away from terrestrial carbon residues of C/N ratio >10 on the shelf, the mean C/N ratio is 8.0 on the mid-Atlantic slope and has not changed significantly over the last 20 yr. The average δ ^{13}C of organic matter is −20.7‰ on this slope and is considered to be of marine origin (Hunt,

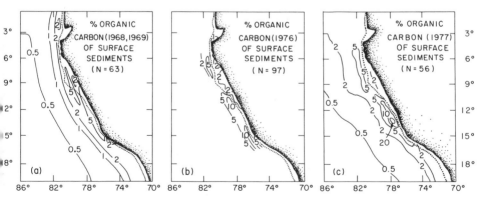

FIG. 125. The distribution of organic carbon (% dw) in the surface sediments along the Peru coast in (a) 1968–1969, $N = 63$; (b) 1976, $N = 97$; (c) 1977, $N = 56$. [After Walsh (1981); © 1981 Macmillan Journals Ltd., with permission.]

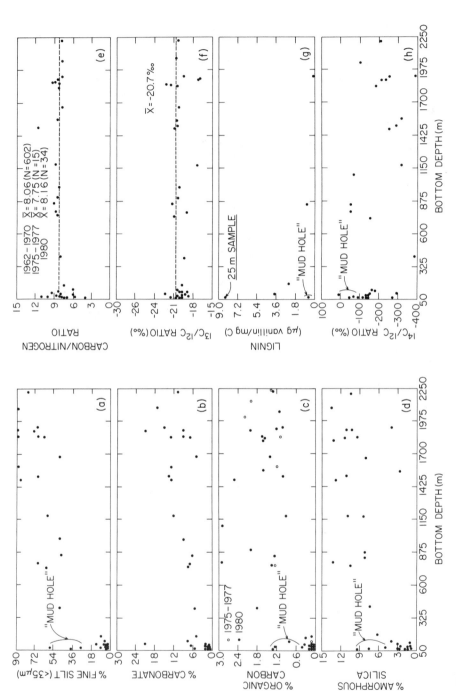

FIG. 126. The surface distribution of (a) fine silt and clay, (b) calcium carbonate, (c) organic carbon, (d) biogenic silica, (e) C/N ratio, (f) $^{13}C/^{12}C$ ratio, (g) lignin, and (h) $^{14}C/^{12}C$ ratio from depths of 50–2250 m within the Mid-Atlantic Bight. [After Walsh *et al.* (1985); © 1985 Pergamon Journals Ltd., with permission.]

1966)—see Section 5.3.1 for its definition. The mean δ ^{15}N of 12 slope sediment samples from the Mid-Atlantic Bight is $+5.4\%o$ (Walsh *et al.*, 1985), similar to that of $+6.4\%o$ either for open-ocean plankton (Saino and Hattori, 1980) or for sediments from Wilkinson Basin in the Gulf of Maine (Macko, 1981), and is also considered to be of marine origin (Peters *et al.*, 1978). Finally, the lignin content from terrestrial plants on the mid-Atlantic slope south of New York is less than 10% of that found in nearshore sediments at a depth of 25 m (Fig. 126g), similar to other patterns observed in slope sediments south of Martha's Vineyard (I. Kaplan, personal communication).

Off the Peru coast, where little river run off occurs, the δ ^{13}C ratio of organic matter in sediments (Fig. 125) at 9°S is similarly $-20.9\%o$, the C/N ratio is ~7.6, and the *n*-alkane fraction of organic carbon is of predominantly marine origin (Walsh, 1983). Within anoxic 16.9% dw carbon deposits, near the edge of the Peru shelf at 12°S, the fatty-acid component of the organic carbon is also of diatom origin (D. J. Smith *et al.*, 1983). In terms of the sedimentary nitrogen pool, hydrolyzable amino acids within other anoxic sediments at 15°S constitute 40–70% of the total nitrogen (Henrichs *et al.*, 1984), similar to that of 50–80% of the nitrogen within marine plankton (Mopper and Degens, 1972).

Surface δ ^{13}C values of the outer Bering shelf sediment indicate marine origin, -21.0 to $-21.5\%o$ (Peters *et al.*, 1978), as well. Similar to other shelves (Walsh, 1983), the gradient of sediment carbon in the Bering Sea (Fig. 124) is from <0.1% dw sediment near shore (Lisitzin, 1966; Sharma, 1974) to >1.0% dw on the lower slope (Gershanovitch, 1962). On the outer shelf (126–161 m) during June 1982, the organic-carbon content of surface sediments at six stations ranged from 0.47% dw to 1.12% dw, with little variation in C/N (6.7–7.8) and δ ^{13}C (-20.5 to $-21.3\%o$) values. The *n*-alkane hydrocarbon fraction of these shelf sediments, unlike the δ ^{13}C data, suggests a mixed terrestrial source, however (Venkatesan *et al.*, 1981). The origin of any terrigenous sediments, 250–500 km from the shore, may be either recent and related to seasonal ice formation, rafting, and breakup, or it may reflect fossil deposits (Emery, 1968), which raises the perplexing problem of age and sediment mixing.

Some of the carbon deposited within the surface sediments off Peru is less than 1 yr old, as indicated by the ^{14}C label of atmospheric bomb tests begun in the 1950s (Smith *et al.*, 1983). Since 1970, the marine food web of surface waters has been labeled with δ^{14}C values of approximately $+200\%o$ (Walsh, 1983), as discussed in Section 5.2.1. Comparison of ^{14}C activity in the tops of cores taken near the edges of these three shelves at depths of 161–182 m (Walsh *et al.*, 1985) indicates, however, that the organic carbon instead has a δ^{14}C value of $-70\%o$ off Peru, $-159\%o$ off New York, and $-133\%o$ off the southeastern Bering Sea. The ^{14}C age of this surface sediment off New York is ~1500 yr old, but other techniques suggest it may be recent sediment, deposited since the onset of the Industrial Revolution about 1850.

Palynological analysis of eight other core tops (0–4 cm) taken between bottom depths of 2191 and 4395 m in the Mid-Atlantic Bight (L. Heusser, unpublished data), for example, indicates that (1) ragweed pollen was abundant compared to oak (*Quercus* spp.) and (2) the top sediment was thus deposited here since ~1840. An increase in ragweed of adjacent fields and a decrease in the ratio of oak to ragweed pollen within coastal sediments has been correlated with species succession of *Ambrosia* during increased land clearance of the nineteenth century "agricultural revival," well under way within the mid-Atlantic coastal region by 1840 (Smith and Walton, 1980; Brush *et al.*, 1982). Presumably, the age of recent surface sediments has been diluted by mixing with older, subsurface sediments.

It is important to distinguish between such mixing and sedimentation of carbon, because oscillation of burial sites of organic debris within estuaries, shelves, and slopes is thought to have impacted both nutrient cycles of the open ocean and atmospheric CO_2 pools during transitions from glacial to inter-glacial periods (Broecker, 1982a, b; McElroy, 1983). The uses and limitations of geochemical tracers, such as radioactive and stable isotopes, for inferring the input and fate of phytodetritus, will thus be discussed in this chapter. An analysis of the fate of both present and past carbon deposits at the ocean's margins will also provide perspective on possible future sequestering of elements in response to human acceleration of fluxes between their atmospheric, terrestrial, and oceanic reservoirs.

5.1 Sinking Losses

Within a fluid at rest, the trajectory of a spherical particle can be described (Stokes, 1856) by its radius (r_p) and mass, or density (ρ_p), the acceleration of gravity (g), as well as the viscosity (η) and density (ρ) of the surrounding water, that is,

$$w_s = \frac{g r_p^2 (\rho_p - \rho)}{18 \eta \rho} \tag{193}$$

Since most biogenic particles are only slightly more dense than seawater (McCave, 1975; Hawley, 1982) (i.e., $[(\rho_p - \rho)/\rho \approx 0.1]$, and the viscosity of seawater is $\sim 10^{-6}$ m^2 sec^{-1} (Csanady, 1986), Eq. (193) becomes

$$w_s = (2 \times 10^5 \quad \text{m}^{-1} \text{ sec}^{-1})(r_p^2) \tag{194}$$

For biogenic particles of 10- and 100-μm radii (i.e., respective diameters of 20 and 200 μm), the Stokes settling velocities w_s would be 1.7 and 172.8 m day^{-1}.

A picoplankton cell of 2 μm diameter at the sea surface would only sink 6.2 m during a year, effectively never leaving the euphotic zone. In contrast, fecal matter, of 200 μm diameter under microbial attack, might sink to ~1400 m in about 15 days, with a reduction in size to 68 μm diameter (Csanady, 1986); such a partly decomposed particle would only sink another 34 m before complete dissolution occurred. Without decomposition, the same particle would arrive at a depth of 1434 m within only 8.3 days. To arrive intact within slope sediments (Figs. 123–125), individual algal cells of 20 μm diameter must thus increase either their density (Anderson and Sweeney, 1978), and/or their diameter through aggregation processes to form both chains and larger marine snow (Alldredge, 1979; Shanks and Trent, 1980; Silver and Alldredge, 1981; Knauer *et al.*, 1982); alternatively, they might encounter sinking water of course.

Macroaggregates of >1000-μm length contain diatom frustules, other phytoplankton, and fecal pellets, as well as detritus, and are ubiquitous in coastal waters (Riley, 1970). Within oligotrophic waters of the Gulf of California, the marine snow appeared to be of zooplankton origin, that is, appendicularians and mucous-feeding gastropods (Alldredge, 1979). Within rich coastal upwelling waters off Santa Barbara, California, snow of zooplankton origin is rare (Alldredge, 1979), however, and may represent gelatinous self-aggregations of diatoms (Smetacek, 1985) as a result of mucous secretion (Degens and Ittekot, 1984). Using a modified form of Eq. (193), Alldredge (1979) calculated a sinking velocity of 91 m day^{-1} for marine snow, similar to experimental observations of 43–95 m day^{-1} (Shanks and Trent, 1980). It is of interest to note that the C/N ratio of marine snow in California coastal waters was 9.5 : 1, while that of total particulate matter was 6.3 : 1 (Alldredge, 1979); the C/N content of slope sediments of the Mid-Atlantic Bight (Fig. 126e) is 8.0 : 1, suggesting perhaps an equal input of fast- and slowly-sinking particles to the sediments.

Such settling velocities of marine snow are equivalent to those of 15–376 m day^{-1} measured for fecal pellets (Smayda, 1969; Turner, 1977), the usual candidate for rapid transport of organic matter to the sea bottom (Steele, 1974; Bishop *et al.*, 1977; Honjo and Roman, 1978). Recall in Section 1.5.2, however, that a bulk settling velocity of 150 m day^{-1} for phytodetritus was estimated from a time series of sediment traps at Station P (Honjo, 1984) during summer when zooplankton grazing was estimated to be minimal. Similarly, abrupt deposition of phytodetritus after the spring bloom has been observed visually off the Irish coast over a number of years at bottom depths of 2000 (Billet *et al.*, 1983) to 4000 m (Lampitt, 1985), with an inferred sinking rate of 100–150 m day^{-1}. Rapid mass sinking of phytoplankton cells has also been observed during the spring bloom in the Baltic (Bodungen *et al.*, 1981) and North (Davis and Payne, 1984) Seas.

In contrast, copepod fecal pellets constitute only a few percent of the vertical flux of organic matter in a number of sediment-trap deployments (Knauer et al., 1979; Rowe and Gardner, 1979; Urrere and Knauer, 1981; Honjo et al., 1982; Bodungen et al., 1986). These fecal pellets may instead remain in suspension on the shelf (Krause, 1981), with most of them consumed within the water column (Hofmann et al., 1981). Instead of sinking at $1-2$ m day^{-1} (Smayda, 1970), algal populations of the continental shelves may, at times, settle at rates of $100-200$ m day^{-1}. In the process of scavenging individual phytoplankton cells, marine snow and smaller detritus may also sweep coastal waters clear of particle-reactive radionuclides.

Continental shelf waters, for example, are strongly depleted in ^{210}Pb, relative to its parent ^{226}Ra, within the New York Bight (Li et al., 1979), off the Washington coast (Schell, 1977), and in the North Sea (Spencer et al., 1980). Shelf residence times are only a few months for this radionuclide, compared to 54 yr in the deep sea (Craig et al., 1973). A winter–spring deployment of sediment traps on the slope in the Santa Barbara Basin (Moore et al., 1981) obtained sinking fluxes of ^{210}Pb that were sevenfold those predicted from both atmospheric input of ^{210}Pb and its water-column production from decay of ^{226}Ra. Offshore export of fine-grain and phytodetrital particles has been invoked as a mechanism for large inputs of this tracer, as well as for bomb-derived ^{239,240}Pu and ^{55}Fe to California slope sediments (Krishnaswami et al., 1973; Koide et al., 1975; Moore et al., 1981). Section 5.2.2 discusses the use of this naturally occurring radioisotope of lead, in conjunction with bomb-produced ^{14}C, to estimate the mixing and sedimentation rates of particles within the underlying sediments. We must first consider, however, the trajectory of such falling particles within the turbulent water column of the continental shelf.

5.1.1   Particle Trajectories

The net vertical displacement of nonmotile plankton in the water column is a balance of the upward movement imparted by kinetic energy of the winds, or tides (James, 1977), and downward movement as a result of gravity (w_s), downwelling (w), and turbulent mixing (K_z). Parameterization of the vertical turbulent kinetic energy by K_z suggests that transfer of dissolved or particulate matter from this process across the main thermocline of the ocean is small, that is, $K_z \approx 0.1-1.0$ cm^2 sec^{-1}. Above the thermocline in the surface mixed layer of the water column, however, values of K_z may be as high as $100-200$ cm^2 sec^{-1} for perhaps $0.5-1.0$ day after a wind event. Depending on the nature of the vertical gradient of plankton biomass (i.e., a larger amount either in surface water or near the bottom), a net downward or upward flux of particles could occur at a turbulent rate, equivalent to $\sim 20-40$ m day^{-1} displacement over a mixed-layer depth of ~ 45 m.

Ignoring stratification processes of surface heating and freshwater discharge on the shelf, the transient response of the depth of the mixed layer h_m to an impulsive wind stress τ_s, can be empirically described (Niiler, 1975) by

$$h_m \approx \left[\frac{\tau_s}{\rho N^2}\right]^{1/2} \left[(12tFN)^{1/3} + \frac{tN}{30F}\right] \tag{195}$$

where τ_s/ρ is the turbulent kinetic energy of the wind, with τ_s derived from Eq. (17); F is the fraction of this energy available for mixing; N^2 is the Brunt–Väisälä frequency [recall Eq. (16)]; and t is time. After a half-pendulum day of π/f (i.e., about 8.5 hr), the initial rapid response of a deepening mixed layer ceases and, despite continued wind stress, approaches a final value of h_m asymptotically (Niiler, 1975).

Based on mixing-length hypotheses (Mellor and Yamada, 1974; Kundu, 1980), the eddy diffusivity within this mixed layer can be calculated from

$$K_z = L \left(\frac{\tau_s}{\rho}\right)^{1/2} \text{Ri}' \tag{196}$$

where L is the length scale, approximated by $0.1h_m$; and Ri' is a stability factor, involving the Richardson number of Eq. (15). When the wind stress ceases, the kinetic energy dissipates as a function of time (Niiler, 1975) by

$$\frac{d(\tau_s/\rho)^{1/2}}{dt} = -\frac{(\tau_s/\rho)^{3/2}}{25L} \tag{197}$$

such that K_z decays with time as well, of the order of 24 hr after cessation of the wind impulse. A simple description of this process (Fig. 127) was provided by Wroblewski and Richman (1987), with F assumed to be 0.5 and N to be 30.2 hr^{-1} (i.e., a temperature gradient of ~4°C over a 100-m water column).

Applying various wind speeds over a 24-hr period (Fig. 128), estimates of the maximum depth and eddy diffusivity of the mixed layer of such a weakly stratified water column range from an h_m of 0 to 44 m, and for K_z from 1 to 190 $cm^2 \, sec^{-1}$ (Wroblewski and Richman, 1987). Such high vertical diffusivity cases are typical of the spring bloom period, whereas the low values of calm wind periods are typical of stratified, summer conditions of the shelf water column. To contrast the effects of various vertical mixing and settling rates on particle trajectories within the Mid-Atlantic Bight, a time-invariant K_z of 5 $cm^2 \, sec^{-1}$ and a range in w_s of 0–3 m day^{-1} are first used for a model of the August shelf ecosystem (Hopkins and Dieterle, 1986) within this section of the book.

Time-dependent values of K_z and settling velocities of 1–20 m day^{-1} are then used in another model (Walsh et al., 1987c) of the March–April shelf ecosystem in Section 5.1.3. Since we wish to distinguish between downwelling (w), vertical

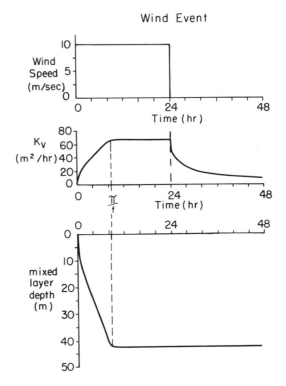

FIG. 127. The simulated response of the vertical mixing coefficient ($m^2\ hr^{-1}$) and mixed layer depth (m) of a weakly stratified ocean to a wind pulse of 10 m sec^{-1}, lasting for 24 hr. [After Wroblewski and Richman (1987); © IRL Press, with permission.]

mixing (K_z), and sinking of phytoplankton (w_s), estimates of K_z during spring wind events were computed (Csanady, 1976) instead from

$$K_z = \frac{\tau \rho^{-1}}{200f} \tag{198}$$

This expression yields smaller values than those from Eqs. (195) and (196), where τ is now the mean wind stress, ρ is of course the density of water, and f is the Coriolis parameter. Values of K_z ranged in this spring bloom model from 17 to 80 $cm^2\ sec^{-1}$, at most about half of Wroblewski and Richman's (1987) estimates, except for one simulation run when all values of K_z were doubled (see Table X).

The diagnostic circulation model of Section 2.7 was used to compute an August flow field over the apex of the New York Bight (Fig. 41) under a southwest wind forcing (225°T) of 0.44 dyn cm^{-2}. An upwelling circulation results (Fig. 129), with offshore flow at the surface. The geostrophic flow field at a depth of

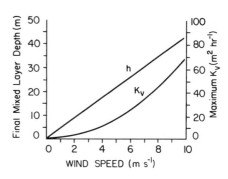

FIG. 128. The mixed layer depth (m) and associated vertical mixing coefficient (m^2 hr^{-1}), under varying wind speeds (m sec^{-1}), of a weakly stratified ocean. [After Wroblewski and Richman (1987); © IRL Press, with permission.]

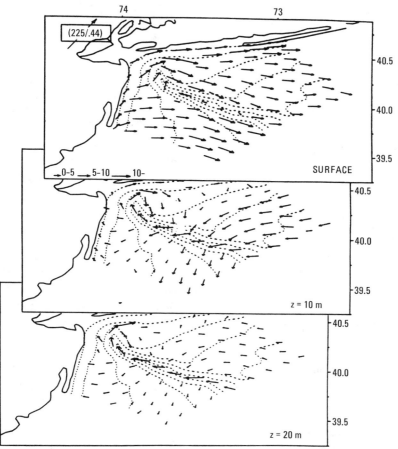

FIG. 129. The simulated flow field within the apex of the New York Bight at the surface and at depths of 10 and 20 m under an August 1978 wind forcing of 0.44 dyn cm^{-2} from the southwest. [After Hopkins and Dieterle, 1986); © International Council for Exploration of the Sea, with permission.]

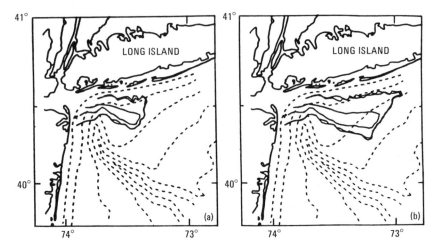

FIG. 130. The particle distribution (0.033, 0.33, 33.0, 330.0 particles m^{-3}) of a neutrally buoyant, inert effluent discharged from the Hudson River estuary, within the 0- to 5-m surface layer of a model under (a) 4 days and (b) 8 days of wind forcing from the southwest. [After Hopkins and Dieterle (1986); © International Council for Exploration of the Sea, with permission.]

10 m is less influenced by the wind, however, and reflects the barotropic component of Eq. (112). In contrast, the flow at 20 m is below the pycnocline of this diagnostic model and displays the result of baroclinic shear on the barotropic flow field (Hopkins and Dieterle, 1986).

As in the case of the larval fish model in Section 4.5.1, 2700 lagrangian, but biologically inert, particles were released every 12 hr over the upper 2 m of this model at a grid point, adjacent to the Hudson River estuary (Fig. 41), for 8 days. These simulated particles did not grow, become prey, or decompose. They also were not diffused over the horizontal dimension in the cases presented here, with settling velocities of either 0 (Fig. 130) or only 3 m day^{-1} (Fig. 131). The consequences of an estuarine source of biogenic particles in the Mid-Atlantic Bight are further considered, however, in both Sections 5.1.3 and 6.4.

The trajectories of neutrally buoyant particles (i.e., $w_s = 0$ m day^{-1}), within the 0- to 5-m depth layer of the model after 4 (Fig. 130a) and 8 (Fig. 130b) days, reflect only the surface flow field (Fig. 129). A horizontal plume of particles grows eastward away from the mouth of the Hudson River and north of the Hudson Shelf Valley. Since the vertical circulation, induced by the southwest wind, is upward over most of the apex, particles do not enter the subsurface layers of the model by downwelling. Since K_z is also relatively small, little vertical flux, or subsequent transport of particulate matter to the sea bottom, occurs in this situation.

A settling velocity of 3 m day^{-1} was larger than the model's upwelling rates,

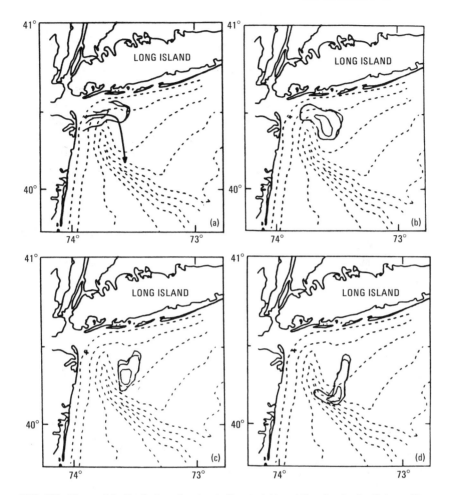

FIG. 131. The particle distribution of an inert effluent, sinking at 3 m day^{-1}, after 8 days of input from the Hudson River estuary under a southwest wind case within (a) the 0- to 5-m layer, (b) the 5- to 10-m layer, (c) the 10- to 15-m layer, and (d) the 20- to 25-m layer, at the same concentration levels as Fig. 130. [After Hopkins and Dieterle (1986); © International Council for Exploration of the Sea, with permission.]

however, such that the particles no longer remained within just the surface layer (Fig. 131a). The trajectory of the centroid of particle mass (drawn in Fig. 131a) instead became three-dimensional, with successive southerly displacement of sinking particles (Figs. 131b–d) within the deeper flow fields of the model (Fig. 129). The distribution of particles within the 20- to 25-m near-bottom layer after 8 days of the second case (Fig. 131d) is quite different from the 0- to 5-m layer of the first case (Fig. 130b). It is similar, moreover, to the observed carbon gradients within the surface sediments of this region (Fig. 123), suggesting that some biogenic particles do sink out of the water column above the Hudson Shelf Valley.

Anoxic conditions of the lower layer of the water column, which initiated in this region of the New York Bight during August 1976, were perhaps the consequence of a rain of phytodetritus here (Falkowski *et al.*, 1980; Stoddard and Walsh, 1986). At simulated sinking rates of 1–3 m day^{-1} for phytoplankton off Peru (W. O. Smith *et al.*, 1983), however, the algal cells were instead advected offshore within the upper 30 m of the more vigorous upwelling circulation at 15°S. The residence time of an algal cell within the shelf water column, and its concomitant susceptibility to grazing or decomposition processes, will determine whether shelf or slope depocenters are the more likely refuge for seeding the next bloom of phytoplankton (Smith *et al.*, 1983; Smetacek, 1985). Since the New York shelf is 100 km wide, compared to the narrow 10-km shelf off Peru at 15°S, a particle trajectory over 8 days is probably too short a time scale to resolve this question. A long-term time series of sediment traps at the edge of the New England shelf provides insight into the appropriate scale of phytodetritus transport and deposition within the Mid-Atlantic Bight.

5.1.2   Sediment Traps

Like moored fluorometers and CZCS sensors, sediment traps are a relatively recent technique for data acquisition in the sea (Bloesch and Burns, 1980; Blomquist and Hakanson, 1981). As in the case of G. Schott's color maps of the South Atlantic Ocean (Hentschel, 1933–1936) and G. Harvey's time series of algal biomass in the English Channel (Harvey *et al.*, 1935), however, today's particle samplers had their antecedents as well, in earlier studies of the 1930s (Scott and Miner, 1936). Since then, many shapes and sizes of drifting and moored sediment traps have been deployed in such diverse oceanic regions as the Panama Basin (Honjo, 1982), the Sargasso Sea (Deuser, 1986), the North Pacific (Honjo, 1984), the Scotia Sea (Wefer *et al.*, 1982), the Black Sea (Izdar *et al.*, 1984), the Equatorial Pacific (Betzer *et al.*, 1984), and in coastal waters off Japan (Iseki *et al.*, 1980), Alaska (Walsh, 1983), Hawaii (Lorenzen *et al.*, 1983b), California (Knauer *et al.*, 1979), Peru (Staresinic *et al.*, 1983), Antarctica (Bodungen *et al.*, 1986), the Bahamas (Wiebe *et al.*, 1976), New York (Biscaye *et al.*, 1987), Canada (Hargrave and Burns, 1979), Norway (Lutter, 1984), Germany (Zeitzschel *et al.*, 1978), and Great Britain (Davis and Payne, 1984).

Various poisons, such as sodium azide (Spencer *et al.*, 1978), mercuric chloride (Hartwig, 1976), phenol (Matsuyama, 1973), formalin (Lawacz, 1969), and chloroform (Zeitzschel *et al.*, 1978), have been used to preserve the trapped organic matter. Great care has also been taken to remove from the sediment traps organisms that presumably swam rather than fell into these containers (G. Knauer, personal communication). Observations of the episodic nature of particle fluxes have also led to varying lengths of trap, or multiple collector, deployment, from usually 1- to 3-month sampling intervals in the deep sea (Figs. 22–23) to 1- to 3-week duration near the coast (Figs. 132–135). The results of two such time series in the Kiel Bight (Fig. 132) and the Mid-Atlantic Bight (Fig. 134) are of particular interest, because they represent year-long data sets that bracket the spring bloom.

During January–December 1976 (Fig. 132) (Zeitzschel et al., 1978) and February–June 1980 (Fig. 133) (Peinert et al., 1982), sediment traps were moored 2 m above the 20-m isobath within the Kiel Bight (Smetacek, 1984). The winter peaks of settling detritus, with high C/chl ratios of 200 : 1 in January–February (Figs. 132–133), are the result of resuspension of surface sediments. The sedimentation events during the spring bloom consist instead of fresh phytodetritus, with C/chl ratios of <45 : 1, which blanket the sea bottom and lead to maximum growth, oxygen demand, and carbohydrate/lipid storage of clams, Macoma baltica (Ankar, 1980; Graf et al., 1982, 1983).

A carbon input of 1–2 g C m^{-2} day^{-1} to these shallow sediments at 20 m lasts for 1–2 weeks in March (Figs. 132–133); it is ~100-fold more than the seasonal peak of carbon flux at 1000 m in the Pacific (Fig. 22), and ~1000-fold that at 3200 m in the Atlantic (Fig. 23). With seasonal development of zooplankton and bacterioplankton cohorts, the amount of summer algal carbon that escapes the water column is an annual minimum of 0.1–0.2 g C m^{-2} day^{-1}, with again high C/chl ratios of >200 : 1. A fall bloom in October–November adds more fresh phytodetritus to the Kiel Bight sediments, such that two-thirds of the organic matter settling out of the water column on an annual basis is intact phytoplankton cells, or phytodetritus (Smetacek, 1984).

During the SEEP experiment from September 1983 to October 1984, 20 sediment traps, with multiple samplers, were moored at the 500-m, 1250-m, 2300-m, and 2750-m isobaths across the New England slope (Biscaye et al., 1987). The total particle flux (mg m^{-2} day^{-1}) within each sample over time (Fig. 134) and the April 1984 carbon flux over depth (Fig. 135) are shown here; the particulate organic carbon rain was about 10% of the total particle flux (P. Biscaye, personal communication). Winter resuspension events of fine-grain sediments (Churchill et al., 1987) can be seen in the sediment trap samples (Fig. 134) from the upper slope (500–1250 m), as in the case of the Kiel Bight time series.

Although redeployment of the instruments led to a hiatus in the Mid-Atlantic Bight time series, the impact of the spring bloom is clearly seen at moorings 4–5 in April and 6–7 in May (Fig. 134). As mentioned in Section 4.3.6, an April 1984 sedimentation flux of 0.26 g C m^{-2} day^{-1}, on mooring 4 at 50 m above the 500-m isobath (Fig. 135), represents 58% of the mean March primary production and 17% of the April carbon fixation on the adjacent shelf. Farther

POC g m^{-2} day^{-1} POC / Chl g equiv.

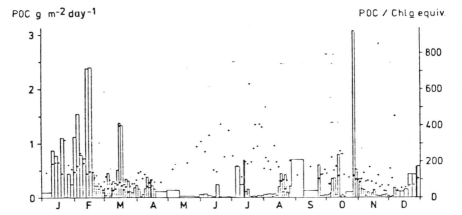

FIG. 132. Sedimentation rates (g m^{-2} day^{-1}) of particulate organic carbon (POC) monitored with a multisample sediment trap, deployed 2 m above bottom in a 20-m water column of the Hausgarten, Kiel Bight during 1976. The dots are ratios POC/chlorophyll *a* equivalent (= chlorophyll *a* + pheopigments) of the collected material. [After Smetacek (1984); © Plenum Press.]

POC g m^{-2} day^{-1} POC / Chl. g equiv.

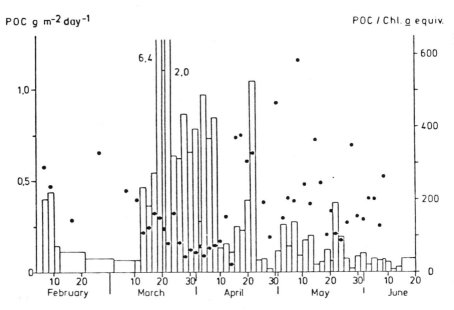

FIG. 133. Sedimentation rates (g m^{-2} day^{-1}) of particulate organic carbon (POC) and POC/chlorophyll *a* equivalent of the material collected in 1980, from the same depth and site as in Fig. 132. [After Smetacek (1984); © Plenum Press.]

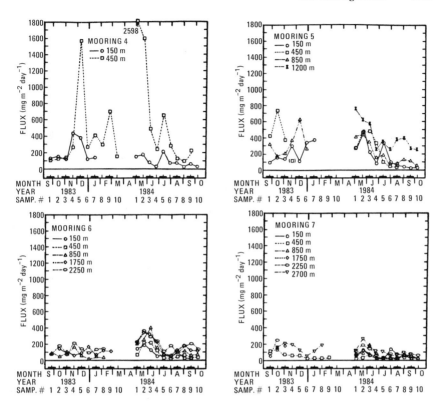

FIG. 134. Time series of total particulate fluxes (mg m^{-2} day^{-1}) at four sediment trap moorings, perpendicular to New England continental slope, from the SEEP-I experiment at water depths of 500 (#4), 1250 (#5), 2300 (#6), and 2750 (#7) m along 70. 2°W. [After P. Biscaye et al. (1987); © 1987 Pergamon Journals Ltd., with permission.]

offshore and deeper in the water column, April carbon fluxes of 0.08 g C m^{-2} day^{-1} at 1200 m on mooring 5 and of 0.02 g C m^{-2} day^{-1} at 2250 m on mooring 6 represent only 18% and 4% of the March primary production. A diatom chain of 100-μm diameter, which left the surface in March at a sinking rate of 43.2 m day^{-1}, would have taken respectively 27.8 and 52.1 days, with no decomposition losses, to arrive at these near-bottom traps at the 1250-m and 2300-m isobaths.

The May 1984 sediment flux at the near-bottom trap on mooring 6 (2300-m isobath) was actually larger than in April (Fig. 134), and represented 4% of the March shelf production. Similarly, the annual maximum of carbon input also occurred at the 2700-m trap on mooring 7 in May, amounting to 3% of the March fixation of carbon. By June, most of the sediment trap fluxes were at an annual

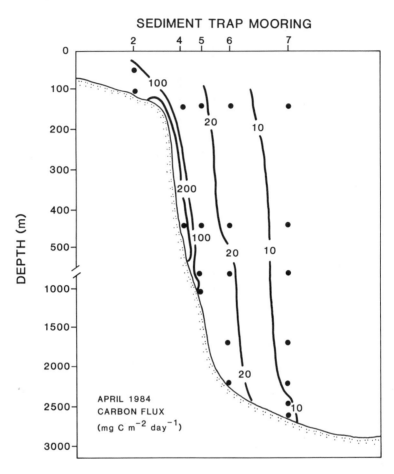

FIG. 135. The daily carbon flux (mg C m⁻² day⁻¹) caught in sediment traps moored on the continental shelf, slope, and rise of the Mid-Atlantic Bight during April 1984. [After P. Biscaye and G. Rowe (personal communication).]

minimum, when the Mid-Atlantic Bight shelf phytoplankton community, like that of Kiel Bight, underwent a species succession and was subjected to increased grazing losses. Sporadic pulses of sinking organic matter were observed in the fall trap samples at the 1250-m isobath, perhaps in response to the fall algal blooms; they were not detected at the other moorings.

The seasonal pulses of primary production in shallow waters of the Kiel Bight and in the outer-shelf and slope waters of the Mid-Atlantic Bight are both captured in these time series. Tenfold more organic carbon was caught in April at a sediment trap depth of 18 m in the former, however, compared to the particle flux at 450 m in the latter. Since the March–April primary production is similar ($0.5–1.5$ g C m^{-2} day^{-1}), most of the difference in carbon input to these shallow shelf and upper slope sediments can be attributed to the longer residence time of particles within the water column of the Mid-Atlantic Bight. Recall from Section 4.3.6 that it might take a diatom population ~116 days to cross the 100-km-wide distance between the 20-m and 500-m isobaths. It might take perhaps another 7 days for the algal cells to fall from the edge of the shelf at the 150-m isobath to the near-bottom sediment trap at Mooring 4 (Fig. 135). We shall explore this possible 3-month trajectory in the next section.

As we move farther towards the deep sea, the seasonal peak of carbon flux at 1200 m on the New England slope (Fig. 134, mooring 5) is approximately three-fold that at 1000 m in the North Pacific Gyre (Fig. 22), while the maximum input of carbon at 2700 m on the New England rise (Fig. 134, mooring 7) is approximately fivefold that at 3200 m in the Sargasso Sea (Fig. 23). On a daily basis, the mean input of sinking carbon at the near-bottom traps on slope moorings 4–6 is ~40 mg C m^{-2} day^{-1}, compared to a combined mean of ~8 mg C m^{-2} day^{-1} at 1000 m at Station P and 3200 m at Station S, the same as that throughout the water column at mooring 7 (Fig. 135).

Since the combined surface area of the continental shelves and slopes is ~20% of that of the open ocean (Table II), however, the total particle input that eventually arrives at 2000 m, within the continental margins and the open ocean, may be equivalent. Their fate thereafter is different, with shelf and slope particles intercepting the sea bottom, and open-ocean particles falling another 2000–3000 m. At a mean daily rate of 40 mg C m^{-2} day^{-1} at 50 m above the bottom of the slope on moorings 4–6, between the 500- and 2300-m isobaths, the annual carbon input to these sediments of the Mid-Atlantic Bight might be ~15 g C m^{-2} yr^{-1}. Section 5.2.2 considers the possible accumulation rate of organic matter within the slope sediments in relation to such an annual carbon flux.

5.1.3   Shelf Export

Within Sections 3.3.3 and 3.4.5, the satellite time series of the 1979 spring bloom within the Mid-Atlantic Bight (Walsh et al., 1987a) and the moored fluorometer time series of the 1984 spring bloom in the same region (Walsh et al., 1987b) suggested that the daily export of algal biomass to slope waters might be $1.8–2.7$ g chl m^{-2} day^{-1} across the shelf break. Such time series lacked depth resolution, however, since the CZCS sampled, at most, the upper 10 m of the water column (Figs. 75 and 78) and the in situ instruments sampled perhaps the lower 5 m at the 80-m and 110-m isobaths (Fig. 97). Furthermore, the algal biomass detected by the CZCS over space and by the moored fluorometer over time only represented what was left behind in the water column by an unknown

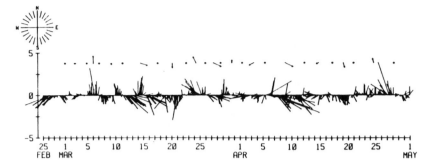

FIG. 136. The daily and mean wind stress (dyn cm^{-2}) over 13 periods as measured at the John F. Kennedy airport between February 25 and May 1, 1979. [After Walsh *et al.* (1987c); © 1987 Pergamon Journals Ltd., with permission.]

balance of primary production, physical losses, and those due to consumption by the bacteria, zooplankton, and benthos of this shelf ecosystem.

To place these daily estimates of export in the context of seasonal production, consumption, and transport of the spring bloom, a simulation model of the plankton dynamics, over 6000 grid points in the horizontal plane (Fig. 37) and three depth layers in the vertical, was run for 58 days, from February 28 to April 27, 1979, in response to wind-induced changes of the barotropic circulation field (Fig. 136). The details (Walsh *et al.*, 1987c) of the time-dependent terms for the physical circulation, for the light and nutrient regulation of photosynthesis, for the vertical mixing, and for the algal sinking at $\sim$18,000 grid points have been discussed previously in this monograph.

Equations (190) and (191) of the Bering Sea model were expanded to include the vertical dimension and an explicit grazing loss as

$$\frac{\partial N}{\partial t} = -\frac{\partial(uN)}{\partial x} - \frac{\partial(vN)}{\partial y} - \frac{\partial(wN)}{\partial z} + K_z \frac{\partial^2 N}{\partial z^2} - \varepsilon M \qquad (199)$$

$$\frac{\partial M}{\partial t} = -\frac{\partial(uM)}{\partial x} - \frac{\partial(vM)}{\partial y} - \frac{\partial(wM)}{\partial z} + K_z \frac{\partial^2 M}{\partial z^2} + 2\varepsilon M$$

$$- w_s \frac{\partial M}{\partial z} - gM \qquad (200)$$

Where u and v are the horizontal velocities, first derived as a continuous function of depth from the circulation model in Section 2.6.6 [i.e., Eqs. (94), (96), and (48) and (49)]. For solution of N and M in Eqs. (199) and (200), u and v were then entered as the mean of their depth integral over each of three vertical layers, where the depth Z of a layer varied across the shelf as a function of the bottom depth H (i.e., $Z = H/3$).

The vertical velocity w is obtained from the continuity equation in Section

2.1.6. For each of the 13 wind cases (Fig. 136), a time-dependent K_z is first computed from Eq. (198), and then inserted both in Eqs. (48) and (49) to calculate u and v from $\partial e/\partial x$ and $\partial e/\partial y$ of Eqs. (94) and (96), and in Eqs. (199) and (200) for an estimate of the vertical mixing of nitrate and phytoplankton during a particular wind event. The horizontal eddy coefficients K_y and K_x are implicitly included in Eqs. (199) and (200) as well, since an upstream finite-difference expression was used for their numerical integration (i.e., recall Section 3.1.5).

In contrast to Eq. (192) of the simpler Bering Sea model, however, the specific growth rate (hr^{-1}) of the phytoplankton, ε, is now a function of ambient light (I) and nitrate (N) in the more complex Mid-Atlantic Bight model as

$$\varepsilon = \varepsilon_m [I/I_s \, e^{(1 \, - \, I/I_s)}] \left(\frac{N}{n + N} \right) \tag{201}$$

where ε_m is a maximum growth rate of $0.026 \, hr^{-1}$, assuming a mean temperature of 5°C and a phytoplankton assemblage of 50% netplankton (Malone, 1982). The saturation light intensity I_s of Steele's (1962) inhibition expression of Eq. (122) is taken to be 5 g-cal cm^{-2} hr^{-1} (Malone, 1977b). The Michaelis half-saturation constant n of the nutrient limitation term (Caperon, 1967), at which $\varepsilon = \varepsilon_m/2$, is taken to be 1 μg-at. NO_3 1^{-1} for coastal phytoplankton species (Carpenter and Guillard, 1971). The depth integral of ambient light over each of the three layers is actually a time-dependent function of season, day length, diel periodicity, and algal biomass, as discussed in Section 3.1.1.

Such a complex light field is described by a depth-integrated form of Eq. (119) as

$$I(z, d, t,) = \delta I_m \sin^3 [\pi(t \, - \, a)/i(d)]$$

$$e^{[-(k_w \, + \, k_s)z \, - \, k_p \int_0^z M \, dz]} \tag{202}$$

where all the terms have been previously defined. The specific attenuation coefficient for plant pigments in this model is assumed to be 0.020 m^2 mg^{-1} chl a (Bannister, 1974b; Jamart et al., 1977). The water and detrital contribution to water clarity, $k_w + k_s = 0.13 \, m^{-1}$, is derived as the average residual from the observed extinction coefficients and pigment concentrations during March 15–24, 1979 (Walsh et al., 1987a).

Since Z is a function of the bottom depth, the upper layer of the model is 10 m deep at the 30-m isobath and 30 m deep at the 90-m isobath as a result of entering the known bottom topography H every ~3 km in the model. The observed depth of the euphotic zone in the Mid-Atlantic Bight during March ranges, however, from about 10 m in the Hudson River plume (Malone et al., 1983) to ~30 m at the shelf break (Walsh et al., 1987a). Thus a surface layer of

Table X.

Parameter Evaluation of the 1979 Spring-Bloom Simulation[a]

Experiment	Sinking rate[b]	Mixing rate	Grazing rate	Estuarine boundary	Upstream boundary	Primary production[c]	Grazing loss[c]	Shelf export[d]	Slope export[d]
a	1	1×	Exp	1×	1×	10.4	6.9	21.5	1.6
b	10	1×	Exp	1×	1×	8.2	5.1	20.0	1.1
c	20	1×	Exp	1×	1×	6.8	3.9	18.9	0.9
d	1	2×	Exp	1×	1×	10.2	6.8	21.6	1.5
e	10	1×	None	1×	1×	10.0	0	39.5	2.7
f	10	1×	Linear	1×	1×	4.7	5.1	7.5	0.2
g	20	1×	Exp	10×	1×	7.2	4.1	19.2	0.9
h	20	1×	Exp	10×	0.17×	6.3	3.8	19.2	0.8
i	20	1×	Exp	1×	0.17×	6.4	3.5	18.9	0.8

[a] After Walsh et al. (1987c); © 1987 Pergamon Journals Ltd., with permission.
[b] In m day^{-1}.
[c] In mg chl m^{-2} day^{-1}.
[d] In g chl m^{-2} day^{-1}.

variable depth (as well as the middle and bottom layers) in the model mimics fairly well the spatial changes of biological processes across the shelf.

In this simulation analysis, we are concerned with the amount of phytoplankton export from the mid-Atlantic shelf, which depends on the amount of "new" dissolved nitrogen added from the slope and estuarine boundaries (Eppley and Peterson, 1980; Walsh, 1981), not the recycled production based on ammonium or urea regenerated within the shelf ecosystem. Thus the growth of phytoplankton in Eq. (200) from excretory products (Walsh, 1975), implicit within the grazing-loss term gM is not considered. The input of nitrate from the coastal boundary is instead assumed to be 10.0 μg-at. NO_3 l^{-1} from the Hudson River, 1.3 μg-at. NO_3 l^{-1} from Chesapeake Bay, 1.0 μg-at. NO_3 l^{-1} from Delaware Bay, and 1.0 μg-at. NO_3 l^{-1} from Long Island Sound. Most of the model's cases (Table X) assumed that the nitrate boundary condition of slope water and of upstream shelf water was 6.0 μg-at. NO_3 l^{-1} throughout the water column. The estuarine boundary conditions of nitrate were later increased 10-fold for the eutrophication case, while in two simulation runs the upstream nitrate content was set instead to 1.0 μg-at. NO_3 l^{-1}.

Initial conditions of the spatial chlorophyll field of Eqs. (199), (200), and (202) are obtained from a CZCS image of the Mid-Atlantic Bight on February 28, 1979 (Fig. 71), at about the same spatial resolution as the model's grid (Fig. 37). The 1.0 μg l^{-1} surface isopleth of chlorophyll is located near the 20-m isobath, and the 0.5 μg l^{-1} isopleth at the 60-m isobath, with the assumption of uniform pigment distribution within the three vertical layers of the model. The estuarine boundary conditions of algal biomass are respectively 7.5, 5.0, 4.0, and 3.0 μg chl l^{-1} from the Hudson River, Delaware Bay, Long Island Sound, and Chesapeake Bay. The chlorophyll contents of inflowing shelf and slope waters are instead set to 0.5 and 0.2 μg chl l^{-1} at the upstream and offshore boundaries.

With respect to the biological loss term of Eq. (200), zooplankton grazing rates in the Mid-Atlantic Bight during March are about 10% of the daily primary production (Walsh et al., 1978; Dagg and Turner, 1982; Smith and Lane, 1987), with an increase to ~40% by April (Dagg and Turner, 1982; Smith and Lane, 1987), and more than 100% by October (Dagg and Turner, 1982). From March to April 1984, the dominant herbivore, *Calanus finmarchicus*, increased eightfold in abundance over a 33-day period (Smith and Lane, 1987), suggesting an exponential increase of this herbivore population, with a doubling time of about 10 days.

In this model, an exponential increase of the grazing stress gM is thus mainly employed from February 28 to April 27 by using

$$g = \frac{\ln(1 - G)}{[24 - i(d)]} \tag{203}$$

where G is $0.03e^{-0.023(59-d)}$ for $H > 50$ m, while G is a constant, 0.06, for $H < 50$ m; that is, there is a spatial gradient of grazing pressure, like the larval fish model of Section 4.5.1. To simulate diurnal migration of the copepod grazers, this grazing stress is only imposed on the phytoplankton of the upper and middle layers of the model at night. In terms of primary production, the exponential grazing stress leads to a loss of about 10% of the daily growth increment in February and 100% at the end of April; losses to benthos and bacterioplankton are then represented by this term as well. In one experiment of this model (Table X), no grazing stress was assumed, while in another experiment, a linear increase in grazing stress from March to April of

$$G = 0.1 + 0.0078(d - 69) \qquad \text{for} \quad d > 69 \qquad (204)$$

was instead applied.

With respect to the physical loss term of sinking in Eq. (200), sinking rates of $1-20$ m day^{-1} are employed within the model, as discussed in Section 5.1.1. Specifically, nine separate cases of this simulation model were run (Table X), with, for example, sinking rates of (a) 1 m day^{-1}, (b) 10 m day^{-1}, and (c) 20 m day^{-1}, at an exponential grazing rate, at the normal mixing rate, and at the normal estuarine or upstream boundary conditions. In another run (d), the mixing rate was doubled, with a sinking rate of 1 m day^{-1}, while the rest of the parameters remained the same. In a third experiment, the zooplankton grazing rate was assumed (e) to be zero, or (f) to vary linearly with time rather than as an exponential function, at an intermediate sinking rate of 10 m day^{-1}, with the normal mixing rate and boundary conditions. A fourth experiment increased 10-fold the estuarine input of dissolved nitrogen, at upstream boundary conditions of (g) 6 μg-at. NO$_3$ l^{-1} and (h) 1 μg-at. NO$_3$ l^{-1}, with a sinking rate of 20 m day^{-1}, an exponential grazing rate, and the normal mixing rate. Finally, the last experiment (i) considered an upstream boundary condition of only 1 μg-at. NO$_3$ l^{-1}, at the normal estuarine flux of nitrogen, the usual mixing rate, a sinking rate of 20 m day^{-1}, and an exponential grazing rate.

The response of this circulation model to changes in wind forcing during April 1979 has been discussed in Section 2.6.6, while the migration of near-bottom nitrate isopleths back and forth across the shelf during April 1976 has been described in Section 4.3.5. As displayed in the cumulative cross-isobath flux of chlorophyll during April 1984 in Section 3.4.5, however, we are mainly concerned here with the net transport of algal biomass over the spring bloom and whether it accumulates in shelf or slope depocenters. Using a constant sinking velocity of 20 m day^{-1} and the time-dependent growth, mixing, and grazing rates of case (c) in Table X, we will thus explore the seasonal change in biological response to three downwelling cases on March 3, April 4, and April 16, and three upwelling cases on March 16, April 11, and April 20, 1979.

At the beginning of March, for example, the incident radiation in the Mid-

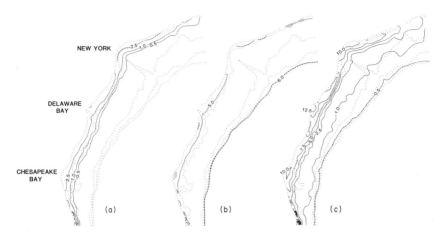

FIG. 137. The simulated chlorophyll (μg l^{-1}) of text case (c) within the (a) upper and (c) lower layers, as well as (b) nitrate (μg-at. l^{-1}) in the upper layer, on March 3, 1979. [After Walsh *et al.* (1987c); © 1987 Pergamon Journals Ltd., with permission.]

Atlantic Bight is <250 g-cal cm^{-2} day^{-1} (Fig. 47), which, over a well-mixed water column, would yield a mean *in situ* light intensity of respectively 64 and 32 g-cal cm^{-2} day^{-1} within the surface layers of the model at the 30- and 60-m isobaths. Below an *in situ* light intensity of 40 g-cal cm^{-2} day^{-1}, the spring bloom does not occur at these latitudes (Riley, 1967b; Hitchcock and Smayda, 1977). Consequently, under a weak northeast wind forcing of 0.21 dyn cm^{-2} from 088°T through March 4, 1979 (Fig. 136), the 0.5- and 1.0-μg l^{-1} isopleths of simulated chlorophyll in the surface layer are then located along the 20-m isobath (Fig. 137a).

The dashed lines of Fig. 137 and subsequent plots are respectively the 10-m, 20-m, 60-m, 100-m, and 200-m isobaths. After 4 days of nearshore growth at this time in the simulation, the 5.5 μg-at. l^{-1} isopleth of nitrate of the upper layer is also coincident with the 20-m isobath (Fig. 137b). With a sinking velocity of 20 m day^{-1}, and a maximum nearshore (<30 m depth) downwelling velocity of 11.3 m day^{-1}, 5.0 μg chl l^{-1} is computed in the lower layer near the 20-m isobath (Fig. 137c). The 1 μg l^{-1} isopleth of chlorophyll in the lower layer is instead coincident with the 60-m isobath, while the 0.5-μg l^{-1} isopleth is found at the 100-m isobath, ~50–100 km farther seaward than those of the upper layer. At a grazing loss of only about 10% of the daily primary production, most of this nearshore algal biomass of the model is not consumed in early March, but advected downstream and seaward.

By the northwest wind forcing of 0.90 dyn cm^{-2} from 287°T on March 16, 1979 (Fig. 136), the 2.5-μg l^{-1} isopleth of chlorophyll in the lower layer begins to impinge on the 60-m isobath as well (Fig. 138c). More than 10 μg chl l^{-1} is

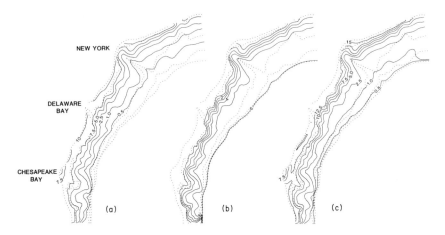

FIG. 138. The simulated chlorophyll ($\mu g \ l^{-1}$) of text case (c) within the (a) upper and (c) lower layers, as well as (b) nitrate (μg-at l^{-1}) in the upper layer, on March 16, 1979. [After Walsh *et al.* (1987c); © 1987 Pergamon Journals Ltd., with permission.]

then found in the lower layer of the model at the 20-m isobath (Fig. 138c), similar to shipboard observations during March 24, 1979 (Fig. 76). With this upwelling circulation of 10–14 m day^{-1} and a strong mixing coefficient of 74.2 cm^2 sec^{-1}, the 0.5-$\mu g \ l^{-1}$ isopleth of surface chlorophyll occurs at the 60-m isobath (Fig. 138a), similar to the spatial pattern detected on March 18, 1979, during CZCS orbit 2010 (Fig. 75a). After 2 weeks of growth in the model, the nearshore dissolved nitrogen stocks become depleted, with the 1-μg-at. l^{-1} isopleth of surface nitrate found at the 20-m isobath, and the 5.5-μg-at. l^{-1} isopleth at the 60-m isobath (Fig. 138b). Subsequent generations of phytoplankton within shallow waters (<20 m depth) will now grow in the model under nutrient limitation, since the ambient nitrate concentrations here have become less than the half-saturation constant—recall Eq. (201).

After a month of increasing solar radiation (Fig. 47), the *in situ* light intensity within the upper 30 m at the 90-m isobath becomes >40 g-cal cm^{-2} in early April, while the nitrate concentration is still 5.5 μg-at. NO$_3$ l^{-1} here (Fig. 139b). The daily grazing stress is now about 20% of the outer-shelf production from Eq. (203), however, and as much as 50% of the inner-shelf production. Within another downwelling circulation on April 4, analogous to March 3, a uniform algal biomass of >3 μg chl l^{-1} occurs in surface waters, landward of the 50-m isobath (Fig. 139a).

Within the lower layer of the model, which reflects the history of the surface primary production, a midshelf maximum of >10 μg chl l^{-1} instead occurs between the 20- and 60-m isobaths on April 4 (Fig. 139c). With less than 0.5 μg-at. NO$_3$ l^{-1} found landward of the 20-m isobath, except for the estuarine

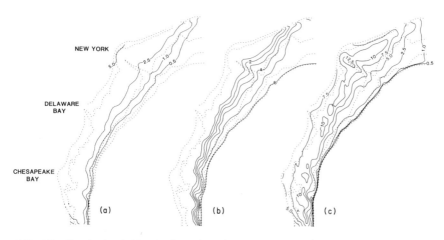

FIG. 139. The simulated chlorophyll (μg l^{-1}) of text case (c) within the (a) upper and (c) lower layers, as well as (b) nitrate (μg-at. l^{-1}) in the upper layer, on April 4, 1979. [After Walsh *et al.* (1987c); © 1987 Pergamon Journals Ltd., with permission.]

discharge, the dissolved source of particulate nitrogen, sinking out of the surface to nearshore bottom waters, has been curtailed by this time in the simulated spring bloom. The midshelf maximum of the model on April 4, 1979, is also the result of prior offshore advection of uneaten phytoplankton from shallow waters. A near-bottom maximum of >20 μg chl l^{-1} was found at midshelf on the 50- to 60-m isobaths during an April 1–5, 1984, cruise (Walsh *et al.,* 1987b).

A strong upwelling circulation of 9–14 m day^{-1} by April 11, 1979, in response to another northwest wind forcing (Fig. 136) leads to a simulated resuspension of chlorophyll from this bottom layer (Fig. 140c) to the top layer (Fig. 140a) at midshelf off New Jersey to Virginia, and immediately south of Long Island. The lower layer of the model is depleted of chlorophyll, while the upper layer is enhanced. Thus *in situ* growth within the euphotic zone may not have accounted for all of the surface increment of chlorophyll, calculated at midshelf between April 4 and 11 (Figs. 139a and 140a).

Similar midshelf maxima of surface algal biomass (Fig. 140a) had been detected aboard ship in April 1975 (Walsh *et al.,* 1978) and during orbit 2452 of the CZCS on April 19, 1979 (Fig. 78c). Although nutrient-rich water is advected shoreward within the aphotic zone in response to the upwelling favorable winds (Fig. 136), the nitrate content of the upper layer at midshelf is actually reduced by ~1 μg-at. NO$_3$ l^{-1} between April 4 (Fig. 139b) and 11 (Fig. 140b). At a nitrate/chlorophyll ratio of 0.5, about 40–50% of the simulated primary production (i.e., biomass increment plus grazing loss) is the result of particulate incorporation of dissolved nitrate.

Another reversal of the winds from the northeast (Fig. 136) yields the third

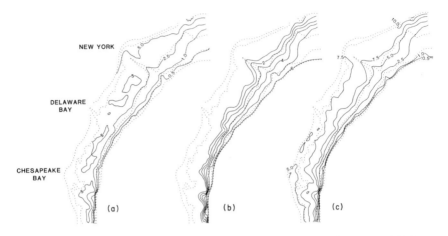

FIG. 140. The simulated chlorophyll (μg l^{-1}) of text case (c) within the (a) upper and (c) lower layers, as well as (b) nitrate (μg-at. l^{-1}) in the upper layer, on April 11, 1979. [After Walsh et al. (1987c); © 1987 Pergamon Journals Ltd., with permission.]

downwelling case of the model by April 16, 1979, with a K_z of only 31.2 cm^2 sec^{-1}, but with a maximum downward entrainment velocity of 22–30 m day^{-1}. High chlorophyll (>10 μg l^{-1}) is again found within the bottom layer south of Long Island and at midshelf, from New Jersey to Delaware (Fig. 141c). In contrast to the second downwelling case of April 4 (Fig. 139a), however, a midshelf maximum of 2.5 μg chl l^{-1} is simulated by April 16 in the surface layer of the model (Fig. 141a), with plumes of chlorophyll extending off the Hudson and Delaware estuaries. These were also observed by the CZCS during orbit 2425 on April 17, 1979 (Fig. 78b). Perhaps as a result of an apparent lack of surface export of chlorophyll from the Chesapeake estuary (Fig. 141a), the chlorophyll content of the model's lower layer at midshelf off Virginia on April 16 (Fig. 141c) is less than that on April 4 (Fig. 139c).

The third upwelling case, of 0.58 dyn cm^{-2} on April 20, 1979 from 341°T, is the weakest northwest wind forcing (Fig. 136), with maximum positive vertical velocities in the model of only 6–13 m day^{-1}, and a K_z of 35 cm^2 sec^{-1}. At the 44-m isobath, such an upwelling rate of 6 m day^{-1} and an effective mixing rate of only 7 m day^{-1} would actually result in a net downward algal transfer of 7 m day^{-1}, with the assumed sinking rate of 20 m day^{-1}. The simulated surface nitrate pattern on April 20 (Fig. 142b) is quite similar to both the previous upwelling event on April 11 (Fig. 140b) and the preceding downwelling case on April 16 (Fig. 141b), indicating little flux of nitrate to the euphotic zone.

A chlorophyll plume of >2.5 μg l^{-1}, derived from the Hudson River export, is found in the upper layer of the model on April 20 off New York (Fig. 142a), with local maxima of the same amount of algal biomass found at midshelf, far-

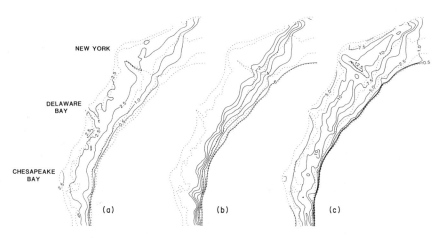

FIG. 141. The simulated chlorophyll (μg l^{-1}) of text case (c) within the (a) upper and (c) lower layers, as well as (b) nitrate (μg-at. l^{-1}) in the upper layer, on April 16, 1979. [After Walsh *et al.* (1987c); © 1987 Pergamon Journals Ltd., with permission.]

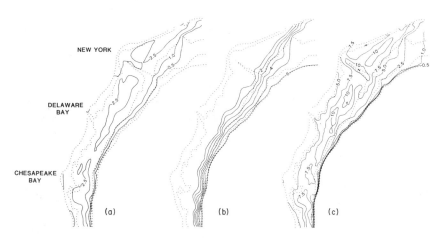

FIG. 142. The simulated chlorophyll (μg l^{-1}) of text case (c) within the (a) upper and (c) lower layers, as well as (b) nitrate (μg-at. l^{-1}) in the upper layer, on April 20, 1979. [After Walsh *et al.* (1987c); © 1987 Pergamon Journals Ltd., with permission.]

ther to the south. The higher biomass simulated previously by the model in the second upwelling case on April 11 (Fig. 140a) and observed by the CZCS during orbit 2452 on April 19, 1979 (Fig. 78c), is not reproduced by this case (c) of the model. Part of the chlorophyll increase detected by the CZCS on April 19 may have been the result of surface export of macrophytes from coastal marshes, rather than just resuspension of near-bottom phytoplankton residues. A lower

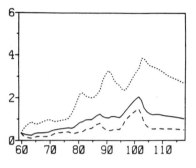

FIG. 143. The simulated chlorophyll content (μg l^{-1}) at noon of each Julian day from March 1 to April 27, 1979, within the upper (– – –), middle (———), and lower (•••) layers of the 80-m water column at a grid point, adjacent to the 1984 fluorometer moorings, south of Long Island. [After Walsh *et al.* (1987c); © 1987 Pergamon Journals Ltd., with permission.]

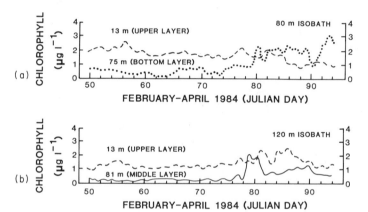

FIG. 144. The observed chlorophyll content (μg l^{-1}), after application of a 40-hr low-pass filter, from moored fluorometers both at (a) 13 m and 75 m on the 80-m isobath, and at (b) 13 m and 81 m on the 120-m isobath, south of Long Island, from February 19 to April 4, 1984. [After Walsh *et al.* (1987c); © 1987 Pergamon Journals Ltd., with permission.]

algal sinking rate of 1 m day^{-1} in the model (Table X) yields a bloom, of course, in this region of the shelf. Over the 58-day period of this case (c) simulation of the 1979 spring bloom, the phytoplankton export was a mean, over 325-km-long segments at the shelf break, of 1.30 g chl m^{-2} day^{-1} between Long Island and New Jersey, 2.25 g chl m^{-2} day^{-1} between New Jersey and Maryland, and 3.64 g chl m^{-2} day^{-1} between Maryland and Cape Hatteras (Fig. 37). More than 90% of this export occurred in the lower layer of the model, as indicated by the time history of simulated algal biomass at the 80-m isobath, south of Long Island (Fig. 143). This grid point of the model was adjacent to 1984 fluorometer moorings at the 80- and 120-m isobaths (Walsh *et al.*, 1987b), which exhibited similar records of chlorophyll fluctuations at depths of 13 m and 75–81 m (Fig. 144). Note that near-bottom chlorophyll concentrations of >2 μg chl l^{-1} were found after Julian day 80 in both the simulated (Fig. 143) and observed (Fig. 144) time series.

The ability of case (c) of the simulation model to replicate both spatial patterns of chlorophyll detected by the CZCS (Figs. 75a and 78b) and temporal patterns detected by moored fluorometers (Fig. 144) suggests that a simulated export of 2.40 g chl m^{-2} day^{-1} may be a reasonable time-averaged mean for March–April in the Mid-Atlantic Bight. If we assume that this export occurred at the 120-m isobath, with 90% in the lower third of the water column, the depth-averaged mean over the whole water column would then be 0.9 g chl m^{-2} day^{-1} (Table X). The possible range and implication of such an estimate are now discussed in relation to an evaluation of the parameter space of the model and predicted inputs to slope depocenters.

With a surface area of the model of 8.52 × 10^{10} m^2, an upstream shelf boundary area of 8.95 × 10^6 m^2, a downstream shelf boundary area of 2.10 × 10^6 m^2, a shelf-break boundary area of 8.85 × 10^7 m^2, and an estuarine boundary area of 2.54 × 10^6 m^2, the units of Table X can be converted to total mass fluxes. For example, a mean of 8.9 × 10^8 g chl day^{-1} was produced over all of the mid-Atlantic shelf during March–April 1979 for case (a) with a sinking rate of 1 m day^{-1}. An average of 66% of this synthesized chlorophyll was lost each day to grazing, while 21% was lost to export, either downshelf or within slope waters. The remaining 13% had not yet been removed from the shelf water column by the end of this simulation (Table X). In case (e) of no grazing loss (Table X), 38% of the March–April daily chlorophyll production was instead lost as physical export past the shelf and slope boundaries (Fig. 37) over the same time period.

Using a C/chl ratio of 45 : 1, the mean daily fixation of carbon, based just on nitrate, would have been 0.47 g C m^{-2} day^{-1} in case (a) and 0.31 g C m^{-2} day^{-1} in case (c) of the model (Table X). Averaging over upwelling and downwelling wind events, a mean primary production of 0.57 g C m^{-2} day^{-1} was actually measured in the Mid-Atlantic Bight (O'Reilly and Busch, 1984) during March–April 1977–1980 at 66 stations, excluding the observations from the mouths of the estuaries (Walsh et al., 1987a). Approximately 50% of the spring bloom's nitrogen demand is met by nitrate off New York (Conway and Whitledge, 1979). The total carbon fixation predicted by the model would thus have been 0.94 g C m^{-2} day^{-1} in case (a) and 0.62 g C m^{-2} day^{-1} in case (c), if uptake of recycled nitrogen had been added to Eq. (200).

The productivity difference between these two cases of the model was the result of increased light limitation, induced by changing the sinking rate from 1 m day^{-1} to 20 m day^{-1} (Table X), thereby decreasing the residence time of an algal cell within the euphotic zone. In contrast, a doubling of K_z in case (d) had little impact on the production or export of chlorophyll, compared to case (a) in Table X. The vertical structure of chlorophyll in case (a) or (d) was quite different, however, from case (c), either over time at the 30-m (Fig. 145) and 50-m (Fig. 146) isobaths, south of Delaware, or within specific spatial patterns in the downwelling (Fig. 147) and upwelling (Fig. 148) circulation modes.

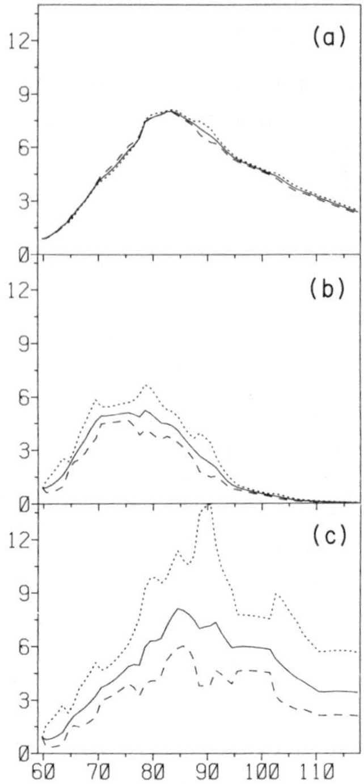

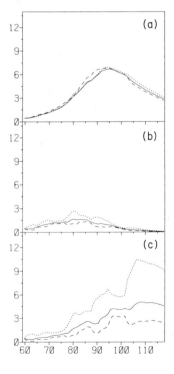

FIG. 145. The simulated chlorophyll content (μg l^{-1}) at noon of each Julian day from March 1 to April 27, 1979 within the upper (– – –), middle (———), and lower ($\cdots$) layers of the 30-m water column at a grid point, east of Delaware Bay, of (a) text case (a), (b) text case (f), and (c) text case (g). [After Walsh *et al.* (1987c); © 1987 Pergamon Journals Ltd., with permission.]

FIG. 146. The simulated chlorophyll content (μg l^{-1}) at noon of each Julian day from March 1 to April 27, 1979, within the upper (– – –), middle (———), and lower ($\cdots$) layers of the 48-m water column at a grid point, adjacent to the Philadelphia dump site east of Delaware Bay, of (a) text case (a), (b) text case (f), and (c) text case (g). [After Walsh *et al.* (1987c); © 1987 Pergamon Journals Ltd., with permission.]

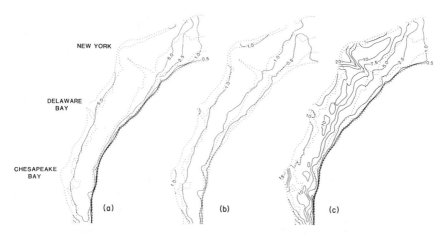

FIG. 147. The simulated chlorophyll (μg l^{-1}) within the lower layer of (a) text case (a), (b) text case (f), and (c) text case (g) on April 4, 1979. [After Walsh *et al.* (1987c); © 1987 Pergamon Journals Ltd., with permission.]

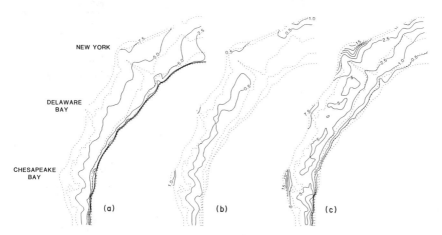

FIG. 148. The simulated chlorophyll (μg l^{-1}) within the upper layer of (a) text case (a), (b) text case (f), and (c) text case (g) on April 11, 1979. [After Walsh *et al.* (1987c); © 1987 Pergamon Journals Ltd., with permission.]

A sinking rate of 1 m day^{-1}, for example, allowed no vertical structure of chlorophyll to develop at the 30-m (Fig. 145a) and 50-m (Fig. 146a) isobaths, adjacent to the Philadelphia dump site at 38°37'N, 74°22'W, east of the mouth of Delaware Bay. A uniform 5–6 μg chl l^{-1} was instead found throughout the whole midshelf region within either the lower layer of a downwelling circulation on April 4, 1979 (Fig. 147a), or in the surface layer during an upwelling event on April 11, 1979 (Fig. 148a), that is, Julian days 94–106 of Fig. 146a. In case

(a) or (d), the seasonal peak of vertically homogeneous chlorophyll on day 95 at the 50-m isobath (Fig. 146a) followed about 2 weeks after the maximum algal biomass on day 82 at the 30-m isobath (Fig. 145a), a seaward progression of $\sim$2 km day^{-1}.

In case (a) or (d), a peak of 6 μg chl l^{-1} throughout the water column was subsequently observed 1 week later on Julian day 102 at the 80-m isobath, south of Long Island, as well. This was in contrast to about 4 μg chl l^{-1} in the lower layer and 1 μg chl l^{-1} in the upper layer of case (c)—recall Fig. 143. Since high values of 6 μg chl l^{-1} were not found during April from this region, either in the 1984 moored fluorometer records (Fig. 144) or within a previous 9-yr data base of $\sim$5000 stations (Fig. 149), the loss terms of sinking and grazing within cases (a) and (d) were probably too small.

An increased sinking rate of 10 m day^{-1} and no grazing loss of case (e) resulted (Table X) in the same primary production as in cases (a) and (d). The average April chlorophyll of the lower layer in case (e) was 11.42 μg l^{-1} (Table XI), however, about threefold that of cases (a) or (d). The average daily algal export to the slope of the lower layer of the model past the location of the fluorometer moorings in case (e) also increased to 4.0 g chl m^{-2} day^{-1}, compared to an observed 1984 mean of 2.7 g chl m^{-2} day^{-1} (Walsh et al., 1987b). The computed slope import of case (e) would actually have been 8.0 g chl m^{-2} day^{-1}, if recycled nitrogen had been included as a nutrient source of this model.

A linear grazing stress that was larger than the exponential formulation and the same 10 m day^{-1} sinking rate of case (f) instead consumed all of the March–April primary production on a mean basis (Table X). The algal biomass above the 30-m (Fig. 145) and 50-m (Fig. 146b) isobaths was reduced to zero by Julian day 115 of case (f), with only 0.5–1.0 μg chl l^{-1} found previously on April 4 (Fig. 147b) and April 11 (Fig. 148b), 1979. The average 1979 March–April export at the 80-m isobath of this third grazing experiment in case (f) was only 0.3 g chl m^{-2} day^{-1}, an order of magnitude less than the observed flux in 1984.

The exponential grazing stress and the 20-m day^{-1} sinking rate of case (c) thus appeared to be the best combination of loss terms of this simulation model to approximate prior satellite (Figs. 75a and 78b), moored fluorometer (Fig. 144), and shipboard (Fig. 149) estimates of chlorophyll. One could increase the sinking rates to 100 m day^{-1}, approximating those of macroaggregates, but such measurements (Shanks and Trent, 1980) were made in still water, not on the turbulent shelf. These parameters of the model are accordingly used to evaluate changes of the boundary fluxes of nutrients within cases (g), (h), and (i) in anticipation of Section 6.2. The lingering 5-μg-at. l^{-1} concentration of nitrate across the shelf, south of Martha's Vineyard, in Fig. 139b, for example, was an artifact of the time-invariant upstream boundary condition of 6 μg-at. NO$_3$ l^{-1} in case (c). A reduction of the upstream boundary condition to 1 μg-at. NO$_3$ l^{-1} (Table X) in cases (h) and (i) did not significantly alter the primary produc-

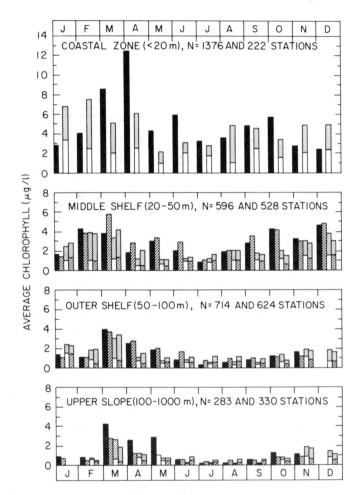

FIG. 149. Monthly distribution of mean chlorophyll (μg l^{-1}) within the upper 20 m and the lower 20–50 m, or 20–75 m, of the coastal zone, the middle shelf, the outer shelf, and the slope in the New York Bight (first and second bar graph of each month) and the Mid-Atlantic Bight (third and fourth graphs) during 1974–1982 from 4673 stations. [After Walsh *et al.* (1987c); © 1987 Pergamon Journals Ltd., with permission.]

tion, grazing loss, or export, however, compared to case (c), and we discuss instead the results of the coastal eutrophication case (g).

A 10-fold increment of the estuarine nutrient loading to the Mid-Atlantic Bight in case (g) significantly increased the nearshore chlorophyll concentrations of the upper (Fig. 148c) and lower (Fig. 147c) layers between Cape May and Montauk Point (i.e., the New York Bight), as well as off Delaware Bay and off Chesapeake Bay, compared to the previous results of case (c) in Figs. 139c and

Table XI.

April 1979 Algal Biomass (μg chl l^{-1}) Scenarios[a]

Experiment	Upper layer	Middle layer	Lower layer	Water column
a	4.01	3.78	3.97	3.92
b	2.64	3.40	4.87	3.64
c	1.81	2.90	5.20	3.30
d	3.95	3.80	3.90	3.88
e	6.16	8.07	11.42	8.55
f	0.25	0.33	0.48	0.35
g	1.97	3.12	5.54	3.55
h	1.75	2.77	4.85	3.12
i	1.59	2.54	4.50	2.88

[a]After Walsh *et al.* (1987c); © 1987 Pergamon Journals Ltd., with permission.

140a. The algal biomass of case (g) did not increase much at the 30-m (Fig. 145c) and 50-m (Fig. 146c) isobaths off Delaware, however, or within the whole spatial domain of the model (Table XI). The primary production of the entire Mid-Atlantic Bight (Table X) only increased by 5% in case (g).

Since the export of estuarine nitrogen to the shelf in case (g) was still only 10% of that from the upstream boundary, this result is not surprising. It does raise the point, however, that compression of 4×10^7 data points into one single spatiotemporal mean for each entry of Table X obscures the spatial and temporal complexity of both this model and the "real world." Table X, for example, does not indicate for case (c) either that 90% of the simulated slope import occurs in the lower layer of the model, or that the computed slope import off New York is one-third that off Virginia, where the slope carbon content at least doubles (Fig. 123).

Furthermore, the shelf export has a time dependency associated with the estuarine source function. Extension of this calculation for case (g) to May 20, 1979, for example, leads to a 25% increase of the shelf export from the lower layer, since not all of the estuarine-derived particulate matter had reached the shelf break by April 20, 1979, in the model. This effect is demonstrated by the first and second bar graphs of each month in Fig. 149, which are the mean chlorophyll concentrations from 1974 to 1982 in the upper 20 m and the lower 20–50 m, or 20–75 m, of the New York Bight, while the third and fourth bar graphs are other chlorophyll means of an independent data set, taken over the larger area of the Mid-Atlantic Bight. During March, April, and May, the chlorophyll concentrations of the New York Bight, from coastal to slope waters, are at least twice that of the whole Mid-Atlantic Bight, reflecting the increased nutrient loading of the Hudson River, that is, simulated in case (g) as Figs. 147c and 148c.

With a decline in nutrient loading from the Hudson River by June and a seasonal increase of the grazing stress, there are no differences in mean algal biomass between the data sets of the New York and Mid-Atlantic Bights for the months of June to February (Fig. 149). If a grazing stress is not imposed during a simulated 10-fold increase of nutrient export from the Hudson River during summer, anoxic conditions, in fact, develop within the New York Bight—see Section 6.4.4. Temporal and spatial changes of physical and biochemical processes within the Mid-Atlantic Bight are thus both important in determining the fate of primary production within this shelf ecosystem.

The algal biomass left behind at any one moment in the water column as a net result of the birth and death processes of shelf phytoplankton populations is, by itself, a poor index of the fate of primary production. Three alternative scenarios of light limitation, nutrient regulation, and grazing stress are each capable of inducing a twofold change in algal biomass within the Mid-Atlantic Bight, for example (Table XI). In concert, moreover, these processes can lead to as much as a 10-fold variation in downstream and offshore shelf export, ranging from 8% to 38% of the mean March–April primary production (Table X).

If recycled nitrogen had been included as a state variable, case (c) would have predicted a mean March–April primary production of 0.62 g C m^{-2} day^{-1} over the whole model domain, or an export of ~0.05–0.25 g C m^{-2} day^{-1} from such changes of the model's parameters. Such biological and physical fluxes of organic matter have indeed been measured in the Mid-Atlantic Bight (Walsh et al., 1987a,b). If algal carbon were only imported to the adjacent slope waters during a spring bloom period of 90 days, these simulation results would predict a range of 4.5–22.5 g C m^{-2} yr^{-1} as export. Additional sediment data in the next section provide a test for the robustness of such simulation models, which might eventually allow us to predict, rather than hindcast, the coastal zone's response to continuing perturbations of eutrophication, overfishing, and climate modification, the subjects of Chapter 6.

5.2 Present Depocenters

Seasonal variations in carbon losses from the shelf may appear as altered lateral sinking fluxes to the slope bottom as a result of marked changes of grazing fluxes; for example, diatomaceous varved sediments have been observed in the anoxic Santa Barbara Basin (Heath et al., 1977), and recall Fig. 134 of the sediment trap time series on the New England slope. With even a 90% grazing stress on a world shelf (2.6 × 10^{13} m^2) production of 5.2 × 10^9 tons C yr^{-1} (200 g C m^{-2} yr^{-1}), resulting from the fast growth rate of diatoms, the annual shelf export to the slope bottom may now be 0.5 × 10^9 tons C yr^{-1}. If the same primary production (Brown et al., 1985), or at least half that of the shelf (Table II), and a similar loss of diatom cells and macroaggregates were to occur in slope waters, as much as 0.2–0.5 × 10^9 tons C yr^{-1} might be added from this ecosystem as well. In contrast, the particle export to the deep sea, the other 80% of the ocean (Menard and Smith, 1966), is estimated to be only 0.2 × 10^9 tons C

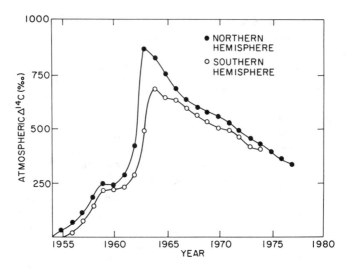

FIG. 150. A time series of bomb input of radiocarbon (Δ ^{14}C‰) to the atmosphere from 1955 to 1977. [After Nydal *et al.* (1979), and Walsh (1983); © 1983 Pergamon Journals Ltd., with permission.]

yr^{-1} from the oligotrophic oceanic ecosystems (Table II) of heavily grazed microflagellates, coccolithophores, and dinoflagellates (Walsh *et al.*, 1981). To first examine the possible fate of 10–50% of the primary production of ocean margins at an intermediate time scale of a 30-yr transient, the bomb production of ^{14}C is used as a tracer of modern carbon flow, within both shelf and oceanic food webs, to the underlying sediments.

5.2.1 Anthropogenic Radionuclides

Cosmic ray penetration within the Earth's atmosphere was the natural source of radiocarbon, ^{14}C, before the release of radionuclides from major bomb tests of 1954–1963 (Fig. 150). Before these atmospheric bomb tests, a means of dating organic carbon by the radioactive decay of this naturally occurring isotope had been developed (Libby *et al.*, 1949). Small amounts of artificial ^{14}C are also now created from neutron activation of nitrogen impurities in the fuel and air of nuclear reactors, with subsequent release at the reactor site, or during fuel processing (Osterberg, 1975). At the beginning of the "bomb" period, the atmospheric ratio of ^{14}C to the major stable isotope, ^{12}C, was ~1.2 × 10^{-12} (Machta, 1973). Between 1850 and 1950, the ^{14}C/^{12}C ratio in the atmosphere had decreased by as much as 2.5% (Suess, 1955), as a result of burning fossil fuel with no ^{14}C content, that is, ^{14}C has a Libby half-life of only 5570 ± 30 yr.

The heavier stable carbon isotope, ^{13}C, represents ~1.1% of atmospheric CO_2 and is fractionated by plants during photosynthesis, with preferential uptake of the lighter ^{12}C isotope (O'Leary, 1981). Because of the small absolute amounts of ^{14}C and ^{13}C, their abundance with respect to ^{12}C of a sample is usually reported as δ ^{14}C and δ ^{13}C where

$$\delta \ ^{14}C(\%_0) = [(^{14}C/^{12}C)/R_s - 1] \ 10^3 \tag{205}$$

$$\delta \ ^{13}C(\%_0) = [(^{13}C/^{12}C)/R_s - 1] \ 10^3 \tag{206}$$

and R_s is the respective isotope ratio of standards (1.2×10^{-12} for $^{14}C/^{12}C$ in oxalic acid and 1.1×10^{-2} for $^{13}C/^{12}C$ in Pee Dee Belemnite, PDB).

Because of the plant discrimination against ^{13}C (Craig, 1954), organic matter always has less ^{13}C than the PDB standard, with negative δ ^{13}C values of approximately -25 to $-27\%_0$ for trees; -10 to $-11\%_0$ for some land plants such as corn (Lowdon, 1969); -10 to $-12\%_0$ for marine macrophytes, such as *Zostera* (McConnaughey and McRoy, 1979) and *Spartina* (Hackney and Haines, 1980); and -19 to $-23\%_0$ for marine phytoplankton (Table XII). After recent combustion of fossil detritus, coal, and petroleum with δ ^{13}C values near -25 to $-30\%_0$ (Table XII), the δ ^{13}C value of atmospheric CO_2 ($-7.0\%_0$; Keeling, 1961) has,

Table XII.

Percent Cellular Composition and Theoretical Stable Isotope ($\delta^{13}C$) Values (%) of Both Modern Phytoplankton Groups and Fossil Remains[a]

| Group | Percent of cell dry weight | | | | | |
	Lipid (L)	Carbohydrate (C)	Protein (P)	Ash	Total	Organic
Diatoms	4	17	35	41	97	56
Chlorophytes	15	32	34	11	93	82
Dinoflagellates	19	38	27	9	93	84
Haptophytes	30	25	38	20	113	83
Chrysophytes	40	29	39	6	114	108

| | Percent of organic pool | | | $\delta^{13}C$ of cell | $\delta^{13}C$ of |
	L	C	P	component	modern organism
Diatoms	7	30	63		-19.28
Chlorophytes	19	40	41	L -30.0	-21.08
Dinoflagellates	23	45	32	C -20.5	-21.73
Haptophytes	32	28	40	P -17.5	-22.34
Chrysophytes	37	27	36		-22.94

	Diagenetic elimination of C and P	$\delta^{13}C$ of crude oil
Diatoms	$\Delta 6$	-25.28
Chlorophytes	$\Delta 6$	-27.08
Dinoflagellates	$\Delta 6$	-27.73
Haptophytes	$\Delta 6$	-28.34
Chrysophytes	$\Delta 6$	-28.94

[a] After Walsh (1983); © 1983 Pergamon Journals Ltd., with permission.

in fact, decreased almost 10% between 1956 and 1978 (O'Leary, 1981). Similar fractionation of ^{14}C during photosynthesis (Rafter, 1955) would lead to different radiocarbon ages of plants of the same "true" age. To allow for such isotope fractionation effects in measurements of the $^{14}C/^{12}C$ specific activity of various carbon compounds (Broecker and Olson, 1961), a more exact formulation involves ^{13}C as a correction of ^{14}C fractionation by

$$\Delta\ ^{14}C(\%_o) = \delta\ ^{14}C - [2(\delta\ ^{13}C + 25)](1 + 10^{-3}\ \delta\ ^{14}C) \quad (207)$$

with oak wood ($\delta\ ^{13}C = -25\%_o$) as the reference. The age (yr BP) is then determined by $8033\ \ln[1/(1 + \delta\ ^{14}C)]$, using the Libby half-life (Stuiver and Polach, 1977).

The introduction of organic fossil ^{12}C, with zero ^{14}C content, to the atmosphere and thence to the oceans (the "Suess effect") had decreased the $^{14}C/^{12}C$ specific activity of the surface ocean by only 1–2% over one century, in contrast to a 10–20% increase of this ratio in the ocean from addition of bomb ^{14}C over one decade. The prebomb level of atmospheric $\delta\ ^{14}C$ increased (Fig. 150), in fact, from $0\%_o$ in 1955 to $+750\%_o$ by 1963 (Nydal et al., 1979), or 100% above natural ^{14}C tropospheric concentrations in the northern hemisphere by this time (Harkness and Walton, 1969). With a declining trend in $\delta\ ^{14}C$ of $\sim25\%_o\ yr^{-1}$ of atmospheric CO_2 by 1972, tropospheric bomb ^{14}C activities should decrease below the natural level by the year 2000 (Stenhouse and Baxter, 1976). Unlike "bomb" tritium (^3H), which was washed out of the atmosphere with rainfall during 1962–1966 (Broecker, 1979), "bomb" ^{14}C has been more slowly transferred to the surface layers of the ocean (Fig. 151) at a characteristic adjustment time of ~7 yr.

Surface $\delta\ ^{14}C$ values of inorganic carbon increased from prebomb values of $-50\%_o$ at midlatitudes in the Pacific Ocean (Bien et al., 1960), the Atlantic Ocean (Broecker et al., 1960), and the Gulf of Mexico (Mathews et al., 1973) to values of $+150$ to $+200\%_o$ by 1967–1968 in these areas; the $\delta\ ^{14}C$ values were at similar levels by 1973 as well (Mathews et al., 1973; Broecker et al., 1980; Linick, 1980). Within a decade, bomb ^{14}C and ^3H had penetrated some parts of the ocean as deeply as 1000 m, however, confounding simple vertical models (Pritchard et al., 1971; Oeschger et al., 1975; Quay and Stuiver, 1980) that ignore horizontal or biological processes (Walsh et al., 1981). Below the oxygen minimum layer, where few biogenic particles survive to further release the bomb label, most deep ocean waters at 2000–5000 m still have prebomb $\delta\ ^{14}C$ values of -180 to $-220\%_o$ (Ostlund and Stuiver, 1980; Stuiver and Ostlund, 1980), suggesting an apparent age of ~2000 yr (Bien et al., 1965; Williams et al., 1978).

With penetration of bomb ^{14}C to a depth of at least 300 m by 1966 (Pritchard et al., 1971), the shallow, seasonally mixed shelf waters and their inhabitants would have been exposed, at this time, to total inorganic carbon (CO_2, HCO_3^-,

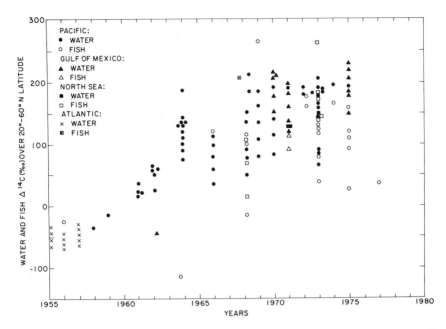

FIG. 151. A time series of radiocarbon (Δ ^{14}C‰) within surface seawater and pelagic, demersal, and bathypelagic fish from 1955 to 1977. [After Bien *et al.* (1960), Broecker *et al.* (1960), Emery (1960), Hubbs and Bien (1967), Williams *et al.* (1970), Bien and Pandolfi (1972), Harkness and Walton (1972), Erlenkeuser and Willkomm (1973), Mathews *et al.* (1973), Linick (1977, 1979, 1980), Ostlund and Stuiver (1980), Stuiver and Ostlund (1980), and Walsh (1983); © 1983 Pergamon Journals Ltd., with permission.]

CO_3^{2-}), and presumably phytoplankton, δ ^{14}C values of $+25$ to $+50$‰ at the equator, $+150$ to $+200$‰ at 30°N latitude, and $+75$ to $+100$‰ at 60°N (Linick, 1980). Phytoplankton had a δ ^{14}C content similar to ambient surface inorganic carbon, for example, $+210$ to $+240$‰ at 30°N during 1971 (Williams *et al.*, 1978). Shelf zooplankton populations off California (Williams *et al.*, 1970) and in the North Sea (Harkness and Walton, 1972) during 1967–1969 did, in fact, have respective mean δ ^{14}C values of $+133$ and $+83$‰. Because of the increasing bomb effect with time, zooplankton within Long Island Sound had an apparent age of 1000 yr in the future, that is, a δ ^{14}C of ~ 120‰, by 1975 (Benoit *et al.*, 1979). Similarly, shrimp populations in the Gulf of Mexico (Mathews *et al.*, 1973) had a δ ^{14}C of $+216$‰ by 1971, reflecting ingestion of the bomb-labeled detritus within this food chain.

Longer-lived pelagic predators of these shelf ecosystems—cod, shark, and whiting—had lower δ ^{14}C values from $+68$ to $+102$‰ in 1968–1971, however, while predators within the longer benthic food web, such as skate and flounder, initially ranged from as little as $+13$ to $+124$‰ (Fig. 151). The δ ^{14}C content of macrobenthos on the California shelf increased by as much as 200‰

Table XIII.

Radiocarbon Labeling ($\Delta^{14}C\%$) of the Shelf Food
Web off California Since the Industrial Revolution[a]

Date	Water	Plankton	Benthos
1878			-72^b
1911			-120
1921			-67
1932			-95
1935			-120
1938			-82
1939			-74
1953			-85^b
1957		-82	
1958		-60	
1959	-88		-37^b
1965	$+63$		
1966	$+71$		$+53^b$
1967		$+133$	
1975			$+37^b$ to $+116^b$

[a] After Berger *et al.* (1966), Berger and Libby
(1966, 1967), Hubbs and Bien (1967), Williams
et al. (1970), Linick (1977), and Walsh (1983); ©
1983 Pergamon Journals Ltd., with permission.

[b] *Mytilus californianus*

between 1953 and 1975, similar to changes in the specific activity of water and plankton (Table XIII). By 1973, pelagic carnivores in the North Sea, such as herring, whiting, and cod, also had a $\delta^{14}C$ content of $+142\%o$ and demersal fish, such as plaice, had values as high as $+265\%o$. In contrast, human blood protein, with a $\delta^{13}C$ of $-29.0\%o$, increased after the bomb tests from a $\delta^{14}C$ value of $-38\%o$ during 1952–1957 to about the same value of $+647\%o$ each year during 1965–1968 (Harkness and Walton, 1969), similar to that of the atmosphere (Fig. 150) and of $+542\%o$ for $\delta^{14}C$ in lamb, beef, milk, cheese, and potatoes during 1967–1972 (Harkness and Walton, 1972; Stenhouse and Baxter, 1976), reflecting the major sources of food in the human diet.

Bathypelagic predators in the open ocean during 1967–1971 had lower $\delta^{14}C$ values of only $+50$ to $+90\%o$ (Williams *et al.*, 1970; Mathews *et al.*, 1973), reflecting their trophic positions both as diurnal migrators in the surface food web, and consumers, at depth, of organisms derived from prebomb food of $-50\%o$. Before 1960, a bathypelagic fish at 640 m in the San Pedro Basin had a $\delta^{14}C$ of less than $-25\%o$ (Emery, 1960), for example, while a coelacanth at 150–200 m in the Western Indian Ocean still contained a $\delta^{14}C$ of $-37\%o$ in 1966 (Williams *et al.*, 1970), compared to $-115\%o$ for bathypelagic fish at 400–3000 m in the North Pacific during 1964 (Hubbs and Bien, 1967). By

1972–1975, bathypelagic fish caught at 1350–1450 m both within the central Pacific gyre and on the slope off Baja California had δ ^{14}C values of $+29$ to $+177\%o$ (Linick, 1979), indicating that both the shelf and oceanic food webs have contained the same bomb label as seawater over the last decade (Fig. 151).

Similar to the time lag in appearance of the bomb label between shelf and bathypelagic fish, slope macrobenthos initially had smaller δ ^{14}C values than shelf benthos (Table XIII). Before 1960, slope deposit feeders, such as echinoids and worms in the Santa Catalina Basin, had a δ ^{14}C value of $-187\%o$ (Emery, 1960), while after the bomb tests benthic organisms, such as holothurians in the Gulf of Mexico, had increased δ ^{14}C values of $-50\%o$ by 1971 (Mathews et al., 1973). In contrast, deep-sea benthos at 5000 m in the Sargasso Sea had δ ^{14}C values of $+8$ to $+17\%o$ in 1968 (Williams et al., 1970), and those of rat-tail fish at 5000 m in the Pacific gyre were $+37\%o$ by 1977 (P. Williams, personal communication), suggesting input of bomb labeled food with high sinking rates of >100 m day^{-1} (Suess, 1980), similar to those observed off Hawaii (Lorenzen et al., 1983b).

Particulate carbon at a depth of 2000 m within the Pacific Ocean contained, in fact, only 10% less bomb carbon (δ ^{14}C $= +182\%o$) than surface phytoplankton in 1971 (Williams et al., 1978), reflecting the bomb label of fast-sinking detritus. Most of the organic material, sinking at slower rates of $1-10$ m day^{-1}, is oxidized above this depth to form the oxygen minimum layer, however. Of the 10% primary production that escapes the open ocean euphotic zone as detritus (Walsh et al., 1981), a much smaller fraction of 1% finally reaches oceanic sediments at 4000–5000 m (Honjo, 1980). Bottom sediments at 6000 m depth, for example, still had δ ^{14}C organic carbon values of approximately $-700\%o$ in 1977 and an age of $\sim10,000$ yr, compared to a 2000-yr age of the dissolved inorganic carbon at 5700 m (Williams et al., 1978); little plankton with the bomb label had survived the descent to the deep sea.

Because of the shallow depths of the shelf, the much larger flux of detritus from the coastal ecosystem encounters the bottom, with partial decomposition by the benthos there, rather than being oxidized within the water column as it is in the open ocean. The ^{13}C content of dissolved inorganic carbon within nearshore bottom waters, for example, is reduced as a result of the oxidation of ^{13}C-deficient particulate organic matter in these sediments (Sackett and Moore, 1966; Spiker, 1980). Little phytoplankton carbon is thus retained within this system, because of such decomposition, resuspension events, and seaward transport discussed in the previous section. Radiocarbon estimates of the age of organic matter in shelf sediments, after the bomb tests, range therefore from 850 yr old at 28 m in the Baltic (Erlenkeuser et al., 1974) and 1300 yr old at 65–79 m on the New England shelf (Bothner et al., 1981), to 2000–3000 yr old at 14 m in Long Island Sound (Benoit et al., 1979) and 2000–4000 yr old at 36–62 m within the New York Bight (K. Turekian, personal communication).

Variation in the apparent age of nearshore surface sediments is a function of the amount of the different sources of carbon: terrestrial, sewage, refractory prebomb marine carbon, and recent additions of bomb ^{14}C-enriched organisms, mixed together with residues of ^{14}C-free fossil fuel. Within nearshore areas, subject to river run off and sewage effluents, the sediment organic carbon has similar δ ^{13}C values of approximately -24 to $-26‰$, derived from terrestrial (Hunt, 1966; Parker et al., 1972; Hedges and Parker, 1976; Tan and Strain, 1979), sewage (Burnett and Schaeffer, 1980; Rau et al., 1981), and fossil fuel (O'Leary, 1981) sources. The ^{14}C contents of these carbon sources in shallow water are quite different, however, with an age of ~2300 yr for terrestrial carbon (Benoit et al., 1979), compared to 3000 yr in the future for that of sewage (K. Turekian, personal communication), that is, close to the bomb label of human protein, while fossil fuel has a zero age, since δ [14]C ≈ 0 after about 35,000–50,000 yr.

Assuming that marine photosynthesis is the major origin of organic carbon on the mid-Atlantic slope (Fig. 123) and that source water for coastal productivity could be derived from offshore waters as deep as 1000 m by horizontal transport along isopycnal surfaces (Iselin, 1939), the maximum age of sediment carbon at the slope surface in 1955 should have been ~750 yr (Broecker et al., 1960). With bomb ^{14}C contamination of the mid-Atlantic shelf food web, the age of the sediment organic carbon on the slope surface should have become even younger by 1980. For example, by 1975, the zooplankton in Long Island Sound had an apparent ^{14}C age of ~1000 yr in the future (Benoit et al., 1979). Yet the average age of organic carbon collected in 1980 within surface sediments of the upper slope was 1900 yr old, compared to 2300 yr on the lower slope and ~1300 yr on the shelf (Walsh et al., 1985).

Farther offshore, in areas of marine sediments (δ ^{13}C of -20 to $-21‰$) a nonzero, older age at the surface on the outer shelf and slope (Fig. 126) can be attributed to (1) downward mixing and dilution of fresh "bomb" organic detritus with older, subsurface carbon by bioturbation, or mixing, of the sediments; (2) no net deposition of ^{14}C-enriched plankton at the sediment–water interface within the last 30 yr; or (3) the older age of inorganic carbon in upwelled water, taken up by the shelf organisms at the surface (Berger et al., 1966; Killingley and Berger, 1979). In 1965, for example, the δ ^{14}C of upwelled water off California (Table XIII) was as much as 100‰ lower than that of offshore waters in 1964 (Fig. 151); similarly, during 1971, water upwelled to the surface on Campeche Bank had a δ ^{14}C of $+68‰$, compared to $+173‰$ in slope surface water of the Gulf of Mexico (Mathews et al., 1973). Recall, however, from the introduction to this chapter that the age of sediments at the shelf edge off Peru was younger than of those off New York or Alaska.

In this respect, on the upper slope at 370–540 m, the organic carbon of surface sediments is still younger off Peru (δ ^{14}C of $-150‰$) than off New York or the southeastern Bering Sea (δ ^{14}C of -255 to $-589‰$). On the lower slope

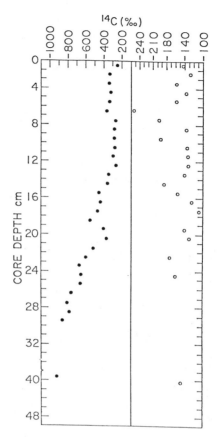

FIG. 152. The down-core distribution of ¹⁴C activity (Δ ¹⁴C‰) within two cores on the upper (●, 390 m) and lower (○, 1500 m) continental slope of the southeastern Bering Sea. [After Walsh *et al.* (1985); © 1983 Pergamon Journals Ltd., with permission.]

at 1500 m off both New York and the Bering Sea, younger organic carbon (− 147 to − 275‰) is found in surface sediments, similar to the age of sediments of the upper slope off Peru. Because of the wider shelves of the Mid-Atlantic Bight (100 km) and the Bering Sea (500 km), bomb labeled plankton and phytodetritus cannot transit the shallow waters to the shelf break within one wind-event cycle, as they can on the narrow 10-km Peru shelf at 15°S. Older carbon of >30 yr age, without a bomb label, could also be resuspended off the shelf bottom and be advected offshore as well, diluting export of the bomb-labeled plankton. In addition, bioturbation, or physical slumping, of sediments on the slope could vertically mix "younger" surface carbon with "older" subsurface deposits.

Analysis of downcore ¹⁴C activity from two Bering Sea slope cores (Fig. 152), for example, shows that at 390 m there is a rapid decline of δ ¹⁴C values, approaching − 900‰. One thus finds very old carbon (19,000 yr at 40 cm) on this upper slope, compared to a well-mixed, relatively young marine carbon on the lower slope (1500 m), with an average δ ^{14}C of $-155‰$ (1400 yr). These data suggest downslope slumping of organic matter here, a process which also may be occurring off the Mid-Atlantic Bight and Peru. The ^{14}C activity of organic carbon in the surface sediments from 700–1000 m and at 2000 m, for example, is 100–200 greater (that is, younger) than that at midslope between 1200 and 1600 m in the Mid-Atlantic Bight (Fig. 126h). In each case, the "young," predominantly marine organic carbon may have slumped onto the lower slope, leaving older carbon behind on the upper slope. The extent of such lateral loss of organic carbon to the lower slope significantly influences the distribution of remaining "young" and "old" carbon, as sampled with box cores, and hence the mixing processes discussed next.

5.2.2   Bioturbation/Slumping

Removal of ^{210}Pb from the water column by falling particles (Moore et al., 1981) results in an excess sediment activity of this natural radionuclide (half-life of 22.3 yr), with respect to that derived from crustal decay of the intermediate parent, ^{226}Ra with a longer half-life of 1622 yr (Koide et al., 1972). Measurement of ^{210}Pb distribution within a sediment core, in conjunction with ^{14}C data (Officer, 1982) and other bomb tracers, such as ^{239,240}Pu (Koide et al., 1975), can provide an independent estimate of mixing and sedimentation rates. Lead-210 may also be removed at the sediment–water interface within the benthic boundary layer by chemical coprecipitation with hydrous iron and manganese oxides in carbon-rich sediments (Bacon et al., 1976; Carpenter et al., 1981), but these processes would not affect the sedimentation rate calculation from the ^{226}Ra and ^{210}Pb activities (Koide et al., 1972).

The average ($N = 156$) ^{226}Ra activity within the upper 50 cm of sediments on the shelf, slope, and rise off the Mid-Atlantic Bight is 1.4 dpm g^{-1} (Santschi et al., 1980; Benninger and Krishnaswami, 1981; Schell, 1982). This information can be used to estimate the excess, unsupported ^{210}Pb activity from total ^{210}Pb activity measured within cores taken in shelf or slope sediments. In shelf sediments of low organic content (~0.1% C dw) off the Virginia, Maryland, and Massachusetts coasts (Fig. 123), for example, the total ^{210}Pb activity of 0.1–1.5 dpm g^{-1} is equal to, or supported by, the ^{226}Ra activity of the sediment (Walsh et al., 1985), such that there is no excess ^{210}Pb, and probably no net accumulation of phytodetritus.

Within both the Hudson Shelf Valley and the "mud patch" region, southwest of Martha's Vineyard between the 50- and 100-m isobaths, 1–2% carbon muds are found on the mid-Atlantic shelf (Fig. 123). These areas do contain excess ^{210}Pb activities (Santschi et al., 1980; Benninger and Krishnaswami, 1981; Bothner et al., 1981) both within a homogeneous, surface mixed layer of 8 cm and at deeper levels within cores taken there. An apparent sedimentation rate of ~ 0.3 cm yr^{-1} can be calculated (Walsh et al., 1985) within the lower 17 cm to a depth of ~ 25 cm, where the decay rate of ^{226}Ra then supports the ^{210}Pb activity. At a sedimentation rate of 0.3 cm yr^{-1} over the last 30 yr, however, another anthropogenic radionuclide, bomb plutonium, should have penetrated these shelf sediments to a depth of only ~ 10 cm. In fact, ^{239,240}Pu is found as deep as 35 cm in the "mud patch" (Santschi et al., 1980), off Martha's Vineyard (Livingston and Bowen, 1979), within Long Island Sound (Benninger et al., 1979), and in the Hudson Shelf Valley (Benninger and Krishnaswami, 1981).

Such deep penetration of an anthropogenic radionuclide is attributed mainly to mixing rather than to sedimentation, as a result either of bioturbation by macrofauna (Benninger et al., 1979; Cochran and Aller, 1979) or of wind stirring, with little real accumulation of detritus on this part of the shelf as well. In the absence of most benthic organisms within the anoxic shelf sediments off Peru, for example, the penetration of bomb plutonium within these sediments is only ~ 10 cm, in agreement with ^{210}Pb estimates of a sedimentation rate of 0.30 cm yr^{-1} at 183–194 m (Koide and Goldberg, 1982). Other ^{210}Pb estimates of sedimentation on the anoxic Peru shelf range from 0.16 cm yr^{-1} at 186 m (DeMaster, 1979) to 1.30 cm yr^{-1} at 92 m (Henrichs and Farrington, 1984).

The numbers and metabolism of macrobenthic animals off New York and New England decline by an order of magnitude from the shelf to the continental rise (Rowe et al., 1974, 1987). Biogenic reworking of sediments is thought to be restricted to the upper 2–3 cm on the Atlantic rise (Yingst and Aller, 1982), for example, compared to 20–30 cm on the shelf. Therefore, the ^{210}Pb data on the Atlantic upper slope (Fig. 153) might suggest an actual sedimentation rate of 0.1–0.3 cm yr^{-1}, unlike those of the shelf sediments, since presumably bioturbation on the mid-Atlantic slope is less in the presence of a small macrobenthic population, similar in size to that of the Peru shelf. Moreover, in contrast to the mid-Atlantic shelf sediments, the ^{239,240}Pu data from the oxic mid-Atlantic slope (Livingston and Bowen, 1979; Santschi et al., 1980) also suggest a sedimentation rate of 0.3 cm yr^{-1}, with a penetration depth of only ~ 10 cm since the 1954 bomb tests (Fig. 154).

Furthermore, ^{239,240}Pu data from anoxic basins off California at similar depths (Koide et al., 1980) exhibit a similar penetration depth (Fig. 154), where the ^{210}Pb sedimentation rates are 0.3–0.4 cm yr^{-1} in these slope habitats of negligible bioturbation (Koide et al., 1972). An apparent sedimentation rate of 0.1–0.3 cm yr^{-1} on the Mid-Atlantic upper slope involves the tacit assumption, however, that the decays of both ^{210}Pb and ^{239,240}Pu with depth are associated with particles accumulating in the sediment, rather than with vertical mixing (bioturbation and slumping) or mobilization of these radionuclides in a liquid phase (Livingston and Bowen, 1979). This assumption is invalid.

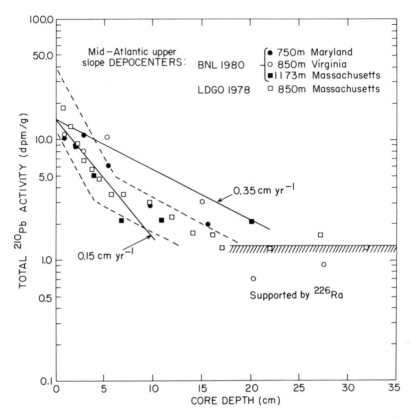

FIG. 153. The distribution of total ^{210}Pb activity (dpm g^{-1}) within four cores from 750 to 1173 m on the upper Mid-Atlantic slope. [After Santschi *et al.* (1980) and Walsh *et al.* (1985); © 1985 Pergamon Journals Ltd., with permission.]

Comparison of the vertical penetration of both ^{14}C (Fig. 152) and ^{210}Pb (Fig. 155) within the sediments of the upper Bering slope (390 m depth), for example, suggests that bioturbation and/or slumping is a major factor on this continental slope. The ^{210}Pb data on the upper Bering slope suggest an apparent sedimentation rate of ~0.1 cm yr^{-1} within the first 10 cm at the 390-m isobath, but the ^{14}C activity does not decline with depth over the upper 12 cm! The vertical profile of ^{14}C activity instead suggests a mixed surface layer of ~12 cm at 390 m and a mixed subsurface layer of ~10 cm extent within the 1500-m core as well (Fig. 152), similar to the deep penetration of bomb plutonium on the Atlantic shelf (Walsh *et al.*, 1985).

The ^{210}Pb profiles from the lower slope off Virginia (Fig. 156) to Massachusetts (Santschi *et al.*, 1980), and even farther north off Newfoundland (Smith and Schafer, 1979), exhibit similar subsurface maxima at 6–10 cm within these cores, consistent with other such slumping events. The ^{14}C age of this depth interval (Fig. 157) is younger by ~400 yr than that of the overlying surface sediment and may reflect past slumping of sediment to the lower slope, with subsequent burial by older, upper slope sediments, exposed by the previous erosion. Other bomb tracers, ^{55}Fe (Labeyrie et al., 1976) and $^{239,240}Pu$ (Santschi et al., 1980), exhibit a subsurface maximum here as well in their depth profiles. Analytical or sampling errors are unlikely, since these data are derived from four separate field programs and research groups between 1974 and 1980. Furthermore, subsurface maxima of $^{239,240}Pu$, trace metals, and ^{137}Cs on the Louisiana slope at 320 m have also been used to delineate slumping features in the Gulf of Mexico (Scott et al., 1983).

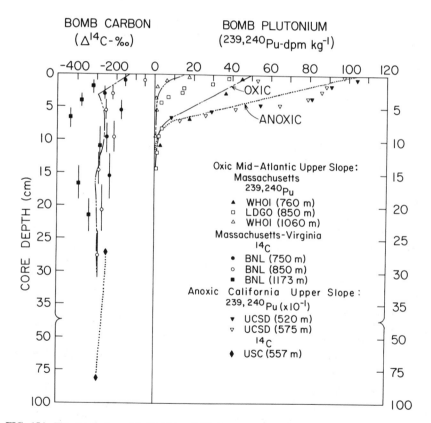

FIG. 154. The distribution of Δ ^{14}C (‰) and $^{239,240}Pu$ (dpm kg^{-1}) within six cores from 750 to 1173 m on the oxic mid-Atlantic upper slope and three cores from 520 to 575 m on the anoxic California upper slope. [After Emery and Bray (1962), Koide et al. (1980), Santschi et al. (1980), Bowen and Livingston (1981), and Walsh et al. (1985); © 1985 Pergamon Journals Ltd., with permission.]

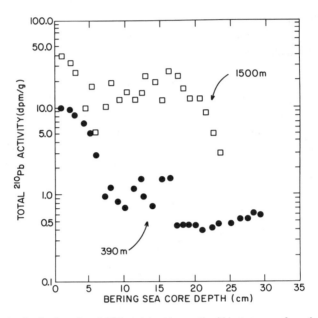

FIG. 155. The distribution of total ^{210}Pb activity (dpm g^{-1}) within two cores from the upper (390 m) and lower (1500 m) continental slope of the southeastern Bering Sea. [After Walsh *et al*. (1985); © 1985 Pergamon Journals Ltd., with permission.]

The ^{210}Pb profile at 1500 m on the lower Bering slope (Fig. 155) exhibits the same subsurface maximum found in the Atlantic lower slope cores (Fig. 156). The ^{14}C age of this subsurface ^{210}Pb maximum at 16–18 cm is younger by ~1200 yr than the sediment at 6–7 cm depth on the Bering slope. A decline of water content with depth in this Bering Sea core suggests that compaction has occurred, with again burial of the younger sediments at 16–25 cm ($\overline{X}$ = 1118 yr) by older sediments at 0–7 cm ($\overline{X}$ = 1400 yr). The intervening sediment of similar ^{210}Pb activity from 7 to 16 cm (Fig. 155) has the same mean ^{14}C age of 1400 yr found in the surficial sediment, but exhibits no vertical gradient of either isotope. Within anoxic sediments at 485 m on the Peru upper slope, a subsurface maximum of ^{210}Pb was also found (Henrichs and Farrington, 1984); bioturbation is presumably minimal here, and such a ^{210}Pb distribution may reflect slumping on this slope as well (Henrichs and Farrington, 1984).

Within the cores on the lower slope off the Mid-Atlantic Bight (Fig. 156) and the southeastern Bering Sea (Fig. 155), the ^{210}Pb distributions within the first 5 cm again suggest apparent sedimentation rates of 0.1–0.2 cm yr^{-1}. These rates are similar to other ^{210}Pb estimates from continental slopes off Peru (Koide and Goldberg, 1982), California (Koide *et al.,* 1972), Washington (Carpenter *et al.,* 1982), Southwest Africa (DeMaster, 1979), Baja California (Koide *et al.,* 1973), and within the Gulf of Mexico (Booth, 1979). Such ^{210}Pb estimates of sedimentation rates over the last century on the Mid-Atlantic and Bering slopes, however, are at least 10-fold larger than those estimated by other methods either during the whole Pleistocene ice age (Hathaway *et al.*, 1979) or since the last Wisconsin glaciation (MacIlvane, 1973; Doyle *et al.*, 1979; Balsam, 1982; Sancetta, 1983; Prior *et al.*, 1984). It is clear that these apparent ^{210}Pb sedimentation rates cannot be translated into long-term accumulation rates, but they may reflect short-term storage of sediment in transit on the slope, once mixing corrections are applied to these estimates.

Within basins on other slopes, for example, where the accumulated sediment is trapped and cannot be transported to the continental rise by slumping, the long-term sedimentation rates, based on ^{14}C ages of core intervals, are much closer to those derived from short-term estimates of ^{210}Pb, or bomb-derived ^{239,240}Pu. Radiocarbon ages suggest a sedimentation rate of 0.12 cm yr^{-1} within

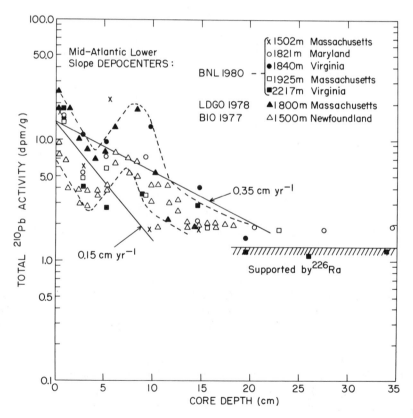

FIG. 156. The distribution of total ^{210}Pb activity (dpm g^{-1}) within seven cores from 1500 to 2217 m on the lower mid-Atlantic slope and off Newfoundland. [After Smith and Schafer (1979), Santschi *et al.*, 1980), and Walsh *et al.* (1985); © 1985 Pergamon Journals Ltd., with permission.]

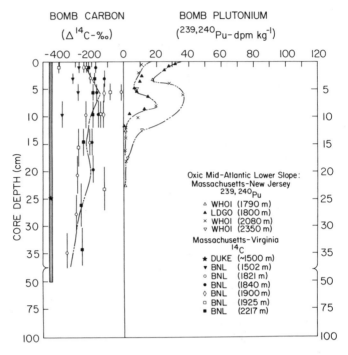

FIG. 157. The distribution of Δ ¹⁴C (‰) and ²³⁹,²⁴⁰Pu (dpm kg⁻¹) within 11 cores from 1500 m to 2350 m on the oxic Mid-Atlantic lower slope. [After Doyle *et al.* (1979), Santschi *et al.* (1980), Bowen and Livingston (1981), and Walsh *et al.* (1985); © 1985 Pergamon Journals Ltd., with permission.]

Stellwagen Basin in the Gulf of Maine (Silva and Hollister, 1973), 0.36 cm yr⁻¹ on the north slope of Georges Bank (Hathaway *et al.*, 1979), and 0.25 cm yr⁻¹ within the surface sediments of the Santa Barbara Basin off California (Emery and Bray, 1962), compared to ²¹⁰Pb and seismic estimates of 0.10–0.40 cm yr⁻¹ in these regions (Emery, 1969; Koide *et al.*, 1972).

Within shelf areas of very rapid deposition, such as the Mississippi River plume, ²¹⁰Pb (Shokes, 1976) and ¹⁴C (Parker, 1977) estimates are also similar, suggesting a sedimentation rate of ∼1.0 cm yr⁻¹ at 100 m depth, about 25 km off the mouth of this river. Near the mouth of the Columbia River, the ²¹⁰Pb shelf sedimentation rates of ∼1.0 cm yr⁻¹ off the Washington coast also agree with long-term accumulation rates inferred from seismic measurements of the thickness of Holocene sediments (Nittrouer *et al.*, 1979). Furthermore, within areas of the Potomac and Saguenay estuaries, ²¹⁰Pb sedimentation rates of ∼1.0 cm yr⁻¹ are supported by independent time horizons from pollen records (Smith and Walton, 1980; Brush *et al.*, 1982).

Within continental margins subject to resuspension and downslope losses, however, the estimated ^{14}C sedimentation rate is at least an order of magnitude less than those derived from ^{210}Pb or ^{239,240}Pu analyses. Despite high primary production at the edge of the shelf off Peru and Southwest Africa, for example, radiocarbon ages indicate sedimentation rates of $0.03-0.10$ cm yr^{-1} there (DeMaster, 1979), compared to ^{210}Pb estimates of $0.30-1.10$ cm yr^{-1} (De-Master, 1979; Koide and Goldberg, 1982; Henrichs and Farrington, 1984) at the same isobaths of these anoxic sediments. Radiocarbon dating of samples within a third upwelling ecosystem off the Oregon coast (Kulm and Scheidegger, 1979) suggests sedimentation rates as low as 0.01 cm yr^{-1} on the slope. These estimates are an order of magnitude less than the rates determined by ^{210}Pb for the Washington slope (Carpenter *et al.*, 1982), similar to the discrepancy between ^{14}C and ^{210}Pb estimates of sedimentation on the Bering and mid-Atlantic slopes.

5.2.3 Accumulation

In coastal and oceanic regions where bioturbation is thought to cause a major vertical flux within surface sediments, the discrepancy between ^{14}C- and ^{210}Pb-derived sedimentation rates is treated as a "true" sedimentation rate (^{14}C) and a "mixing" rate (^{210}Pb). Using a ^{14}C sedimentation rate of 0.075 cm yr^{-1} in Long Island Sound (Benoit *et al.*, 1979) and the depth penetration of both ^{210}Pb and ^{239,240}Pu (Benninger *et al.*, 1979), for example, a mixing coefficient of 0.63 cm^2 yr^{-1} was established. Similarly, on the mid-Atlantic ridge, a ^{14}C sedimentation rate of ~0.003 cm yr^{-1} and an apparent ^{210}Pb sedimentation rate of 0.081 cm yr^{-1} led to a mixing coefficient of ~0.19 cm^2 yr^{-1} (Nozaki *et al.*, 1977). This procedure is formalized (Officer, 1982) by

$$K_z' = \frac{(S_a - S_t)S_a}{\lambda_{\text{Pb}}} \qquad (208)$$

where K_z' is the vertical mixing coefficient in the sediments, S_a is the apparent sedimentation rate derived from the ^{210}Pb data, S_t is the "true" sedimentation rate derived from the ^{14}C data, and λ_{Pb} is the ^{210}Pb decay constant (0.0311 yr^{-1}). Within anoxic regions off the Peru coast, it is difficult to ascribe the discrepancy between ^{14}C and ^{210}Pb sedimentation rates to be the result of bioturbation, embodied in such a mixing coefficient, but we include the effects of both slumping and bioturbation in such a calculation.

Assuming a range in particle sedimentation rates of 0.10, 0.30, and 1.00 cm yr^{-1} from ^{210}Pb data on the slope and corresponding ^{14}C-derived sedimentation rates of 0.01, 0.03, and 0.10 cm yr^{-1}, the respective mixing coefficients K_z' from Eq. (208) are 0.29, 2.60, and 28.94 cm^2 yr^{-1}. In contrast, the exit of dissolved total CO_2 from these sediments is parameterized with sediment eddy coefficients of the order of 200 cm^2 yr^{-1} (Henrichs and Farrington, 1984). With this parameterization of vertical mixing and the "true" sedimentation rate, we can then estimate the net accumulation rate of particulate carbon within the slope depocenters.

Assuming that carbon only penetrates the sediment as a result of sedimentation, without bioturbation or slumping, the accumulation rate A_c is expressed over depth (Muller and Suess, 1979) by

$$A_c = S_t b \qquad (209)$$

where S_t has already been defined and b is the product of the specific gravity of the sediment, the percent carbon dry weight, and the percent volume of dry sediment, that is, $(1 - \text{porosity})$. Assuming that all slopes contain $\sim 2\%$ carbon dw (Premuzic et al., 1982), that the specific gravity of sediment is 2.60, that the porosity of this sediment is 0.80, and an intermediate ^{210}Pb sedimentation rate of 0.30 cm yr^{-1}, we obtain an accumulation rate from Eq. (209) of 31.2 g C m^{-2} yr^{-1}, that is, $(0.30)(0.020)(2.60)(0.20) \times 10^4$. Over a world slope area of 3.2×10^{13} m^2 (Table II), such a flux would amount to 99.84×10^{13} g C yr^{-1}, or $\sim 1.0 \times 10^9$ tons C yr^{-1}. An intermediate ^{14}C sedimentation rate of 0.03 cm yr^{-1}, of course, would suggest an accumulation rate of only 0.1×10^9 tons C yr^{-1}, without consideration of a mixing term.

Since K_z' is considered analogous to an "eddy diffusivity" (Officer, 1982), a diffusive flux term can be added to Eq. (209) in the depth-integrated form as

$$A_c = S_t b + K_z' \frac{\partial C}{\partial z} \qquad (210)$$

where K_z' is obtained from Eq. (208) and $\partial C/\partial z$ is the gradient in carbon over the depth of bioturbation, or of slumping. If the vertical carbon gradient in the sediment over such a depth is not zero (Fig. 158), the diffusive flux would be a parameterization of the positive carbon loading to the slope sediments by the hypothesized mixing processes. Assuming the same porosity and specific gravity, an intermediate ^{14}C rate of 0.03 cm yr^{-1}, an intermediate K_z' of 2.60 cm^2 yr^{-1} (i.e., ^{210}Pb rate of 0.30 cm yr^{-1}), and a carbon gradient of $\sim 0.5\%$ dw C/10 cm, such as the slope off the Mid-Atlantic Bight (Fig. 158), we obtain

$$3.12 \quad \text{g C m}^{-2}\text{ yr}^{-1} = S_t b \quad \text{and} \quad 6.76 \quad \text{g C m}^{-2}\text{ yr}^{-1} = K_z' \frac{\partial C}{\partial z}$$

for a combined total of

$$9.88 \text{ g C m}^{-2}\text{ yr}^{-1} = S_t b + K_z' \frac{\partial C}{\partial z}$$

Over world slopes, such an intermediate flux suggests an annual loading of 31.62×10^{13} g C yr^{-1}, or $\sim 0.3 \times 10^9$ tons C yr^{-1}.

Within an ocean of vertical sediment gradients of carbon such as those of the Bering Sea cores (Fig. 158), there would be no diffusive loading to the sedi-

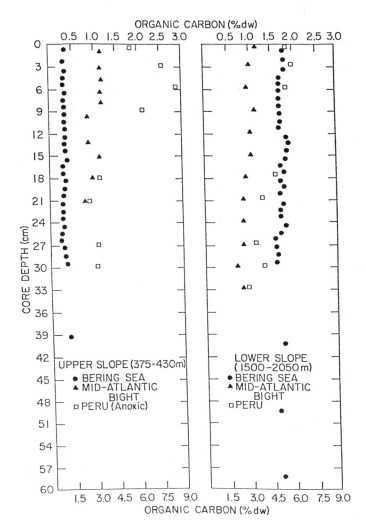

FIG. 158. The vertical gradient of organic carbon (% dw) on the continental slopes off the Mid-Atlantic Bight, the southeastern Bering Sea, and the Peru coast; the bottom scale is only for the Peru data [After Walsh *et al.* (1985); © 1985 Pergamon Journals Ltd., with permission.]

ments, since there is no change in carbon with depth; a total slope storage of 0.1×10^9 tons C yr^{-1} would occur. Within an ocean of vertical carbon gradients of ~4.5% dw C/10 cm on the Peru anoxic upper slope (Fig. 158), a porosity of 0.90, a maximum ^{14}C rate of 0.10 cm yr^{-1}, and a maximum K'_z of 28.94 cm^2 yr^{-1} (i.e., ^{210}Pb rate of 1.00 cm yr^{-1}), a slope storage of 11.24×10^9 ton C yr^{-1} would result. Most of the continental slopes do not have vertical carbon gradients like the upper Peru slope, but we do not know how many slopes are like the Bering Sea example (Fig. 158). The near-bottom water temperatures at 1500 m on the lower slopes off the Mid-Atlantic Bight, the southeastern Bering Sea, and the Peru coast are respectively 3.8, 2.3, and 3.1°C, suggesting that regional temperature effects on bottom metabolism are minimal.

It appears instead that increased carbon loading to slope sediments may have occurred off Peru in response to overfishing of the major herbivore (Walsh, 1981) and off the Mid-Atlantic Bight in response to eutrophication (Walsh *et al.*, 1981), creating transient gradients in the vertical profiles of sediment carbon (Fig. 158). The processing of carbon within the southeastern Bering Sea food web, on the other hand, appears to have remained in steady state since the beginning of the Holocene. Although the populations of Alaska king crab, *Paralithodes camtschatica,* have collapsed on the middle shelf of the Bering Sea, the rest of the food web does not yet appear to have been impacted by human fishing (Walsh and McRoy, 1986). The Yukon and Kuskokwim Rivers have not suffered eutrophication as well. Without a disturbance at the bottom or top of the Bering food web—that is, eutrophication or overfishing—the amount of carbon in the upper slope sediments has apparently remained the same from ~3000 BP to 18,500 BP and in the lower slope sediments from ~1000 BP to 2500 BP.

Because the Peru shelf is as narrow as ~10 km at 15°S, and the surface waters are subject to strong offshore Ekman transport, the diatom plumes of 20–40 μg chl l^{-1} overlie the slope (Fig. 59). It is thus not surprising that little change in carbon loading to the shelf sediments is found here (Henrichs and Farrington, 1984). In contrast, within seven cores taken on the Peru slope, the near-surface carbon was 7.13% dw, compared to 3.47% dw at 30 cm. Within another 13 cores on this slope, the organic carbon averaged 5.27% dw at the surface, and 3.47% at 50 cm. Similar to slope cores taken in 1976 and 1977 (Walsh, 1981), moreover, organic carbon at 506 m on the Peru upper slope during 1978 was two- to threefold less at 30 cm than within the surficial sediments (Henrichs and Farrington, 1984).

Such a vertical gradient of carbon within the Peru sediments is mainly attributed to increased loading of uneaten phytoplankton to the bottom, as a transient response to relaxation of grazing pressure (Walsh, 1981)—see Section 6.3. Decomposition of phytodetritus at depth within the cores is considered to be a minor factor, since ~90% of the amino acid pool may be remineralized in the water column above 50 m (Lee and Cronin, 1982), since little change in sediment C/N ratio occurs with depth in the upper 50 cm, and since only 20–30% of the sediment organic carbon is undergoing rapid diagenesis (Henrichs and Farrington, 1984). The recent departure from steady-state cycling of carbon, in response to removal of a herbivorous fish off Peru, is an anomaly, however, since most shelf food webs are more complex and upwelling ecosystems comprise less than 2% of the total shelf area (Table III).

Eutrophication of coastal ecosystems, with subsequent increases in primary production and possible shelf export, is a more ubiquitous phenomenon as discussed in Section 6.2; the carbon loading on the mid-Atlantic slope and rise is perhaps typical (Fig. 158). Recent calculations (Walsh, 1984) suggest that 6×10^7 tons N yr^{-1} may be released to shelf ecosystems from river discharge, sewage outfalls, and waste dumping. Such a present nutrient loading would be 10-fold more than the pristine situation before 1850. Unlike particulate matter trapped in estuarine depocenters, this dissolved nitrogen and phosphorus exit the estuaries to form phytoplankton blooms, with a C/N ratio of ~6, on the shelf.

After export of shelf production to the adjacent slope and selective return of nitrogen to the water column, a C/N ratio of ~8 is found in the slope sediments off the Atlantic, Bering, and Peru shelves. Such a sediment C/N ratio and the estimated anthropogenic nitrogen loading would allow an input of 0.48×10^9 tons C yr^{-1} to slope depocenters, as a non-steady-state response to increased nutrient availability since the Industrial Revolution. Recall that the intermediate estimate of present carbon loading to the slope was 0.30×10^9 ton C yr^{-1}. A departure of this magnitude from the past steady-state carbon cycle, established with the reappearance of shelves at the beginning of the Holocene, should be detected with the ^{14}C bomb label of plankton over the last 30–40 yr.

The expected age of the surface mixed layer can be calculated (Nozaki et al., 1977) from

$$\text{Age} = (1/\lambda_C) \ln(1 + \lambda_C/S_t H_m) \tag{211}$$

where λ_C is now the ^{14}C decay constant (0.000121 yr^{-1}), S_t was previously defined in Eq. (208) as the ^{14}C sedimentation rate, and H_m is the depth of the sediment mixed layer. Assuming a ^{14}C sedimentation rate of 0.0015 cm yr^{-1} with a mixed layer H_m of 10 cm on the Atlantic and Bering slopes, we obtain a predicted age of 4888 yr, or a δ ^{14}C of approximately −460‰. This is about the age of sediment (−472‰) beneath the mixed layer at 15–20 cm on the upper Bering slope (Fig. 152), as well as at the surface on the upper (540 m) and lower (1975 m) slopes off the Mid-Atlantic Bight, where older sediments are exposed (Walsh et al., 1985). However, the average surface age of sediments at 19 stations taken from the mid-Atlantic, Bering, and Peru slopes is ~2000 yr old (−220‰). Furthermore, the average age of the upper 10 cm of 11 cores taken on the Atlantic (Figs. 154–157) and Bering (Fig. 152) slopes is ~2100 yr (−232).

With a slightly larger intermediate sediment mixing coefficient of 2.88 cm² yr^{-1}, derived from a ^{210}Pb rate of 0.30 cm yr^{-1} and the ^{14}C rate of 0.0015 cm yr^{-1}, bomb-labeled plankton could have penetrated the upper 10 cm of sediment within ~35 yr, the period between the first introduction of anthropogenic ^{14}C to the atmosphere (Fig. 150) and the collection date of the above samples (Walsh et al., 1985). Using the ^{14}C activity of −460‰ predicted from Eq. (211) for carbon in the surface sediment mixed layer, of $+200‰$ for bomb-labeled plankton and of $-230‰$ for the observed value, about one-third of the carbon in the upper 10 cm of the slope sediments might be bomb-labeled phytodetritus; one could use other sedimentation rates and carbon contents (Anderson et al., 1987), of course. The rest of the organic carbon would then be ~5000 yr old, similar to the age of the sediment carbon of a homogenized sample, taken over the upper 50 cm on the Virginia slope (Fig. 157) and at 75 cm on the New England shelf (Bothner et al., 1981).

A possible prebomb age of 5000 yr for organic carbon on the Atlantic and Bering slopes implies that little net marine carbon has been stored on these slopes since the fast rise in sea level ended 5000–6000 yr ago (see Fig. 164). Presumably the shelf–slope exchanges of nutrients and plankton, described in Section 4.3.5, were then in steady state and remained so, until the recent transient of nutrient loading commenced after 1850. A 10-fold increase of nutrient loading to the coastal zone since 1850 further implies that only 0.6×10^7 tons N yr^{-1} was added from the land to the slope before 1850, that is, 0.048×10^9 ton C yr^{-1}—a rather small amount. Unlike energy, elements are mainly recycled on the planet, not dissipated, such that a transient response, involving redistribution among the steady-state pools of an element, such as C or N, requires a transient forcing.

During the initial flooding of continental shelves ~14,000 yr ago, carbon storage may have occurred (Broecker, 1982a,b) on the slope, but most of it must have been subsequently oxidized during the early Holocene to avoid even larger increases of CO_2 in the atmosphere and of nutrient depletion in the sea (Walsh, 1984). Within eight piston cores taken across the mid-Atlantic slope off Virginia, for example, an order of magnitude less carbon is found in early Holocene and late Wisconsin sediments (Doyle et al., 1979), compared to that found today. With establishment of a shelf "steady state" 5000 yr ago, remineralized nutrient input from slope water presumably balanced plankton output from shelf water, allowing little net accumulation of organic matter on the slope. Like the last deglaciation, human alteration of terrestrial and atmospheric fluxes of C and N over the last century represents another transient, which may now be recorded in slope sediments of today.

The most recent shelf export of carbon began during the Holocene transgression, with reappearance of the shelves during the Wisconsin deglaciation between 15,000 and 11,000 BP. At the edge of the Peru shelf at 11°S, for example, the accumulation rate of anchoveta scales in the sediment has increased by an order of magnitude from ~12,000 yr ago to today (DeVries and Pearcy, 1982). Marine kerogen sources of oil, of course, were probably formed on the upper slope during major geological transients of such an export term; accordingly, slope import of phytodetritus is certainly not a new process, and we now consider past depocenters.

5.3 Past Depocenters

During most of the 570 million yr since Cambrian time, the north and south poles have been free of ice. Approximately every 250 × 10⁶ yr, the solar system revolves around the Galaxy, encountering spiral arms of dense cosmic dust that may act to reduce the temperature of the Earth's surface, possibly initiating the present Pleistocene ice age, the Permian, the Ordovician, and the late pre-Cambrian ice ages (Schopf, 1980). At similar intervals, periods of orogenesis have preceded these ice ages (Brooks, 1970). The orogenic episodes could provide volcanic dust, alter atmospheric forcing of paleo–ocean currents, and extract CO_2 from the atmosphere in the process of weathering such igneous inputs, all leading to cooling of the Earth's surface as well.

Migration of continents in and out of the polar regions, with increased reflection of solar radiation at these latitudes, has also been invoked as one of the causes of the most recent Permian and Pleistocene glaciations. Part of Pangaea was located at the South Pole 225 × 10⁶ yr ago (Dietz and Holden, 1970), but subsequent withdrawal of these land masses led to the mild Mesozoic era. After separation of Antarctica from Australia at the end of the Cretaceous 65 × 10⁶ yr ago, the cooling trend of the Tertiary began, with the earliest evidence of Antarctic ice-rafting of till found about 30 × 10⁶ yr ago during the late Oligocene (Hambrey and Harland, 1981). Small mountain glaciers appeared in Alaska 10 × 10⁶ yr ago, while formation of the Antarctic ice sheet took place 5 × 10⁶ yr ago at the beginning of the Pliocene, and occurrence of continental ice sheets in the northern hemisphere started by 3 × 10⁶ yr ago (Imbrie and Imbrie, 1979). Over the last 0.7 × 10⁶ yr of the Pleistocene ice age, seven glacial periods have occurred, modulated by variations in the eccentricity, axial tilt, and precession of the earth's orbit around the sun (Hays *et al.*, 1976).

A major oceanic consequence of these ice ages is the lowering of sea level during glaciation; for example, 15,000 yr ago the sea level (see Fig. 164) was 100–150 m less than the present shoreline (Emery and Uchupi, 1972), such that most of the continental shelves were then dry land. Similarly, during the late Ordovician and Permian ice ages, mass extinction of shallow-water organisms occurred (Berry and Wilde, 1978; Raup and Sepkoski, 1982), suggesting that a major portion of the continental shelves was exposed during these glaciations. During the Permian, the two supercontinents, Gondwana and Laurasia, had not broken apart (Dietz and Holden, 1970): no Tethys Sea stretched between these land masses from North America to Asia, and over this 50-million-yr ice age the sea level dropped (Fig. 159), from more than 250 m above the present shore line to at least 40 m below (Schopf, 1974). Other low sea stands evidently occurred in the late pre-Cambrian ice age (Vail *et al.*, 1978) and during the Ordovician (Fig. 159), when the Sahara Desert was located at the South Pole and was covered with glacial ice.

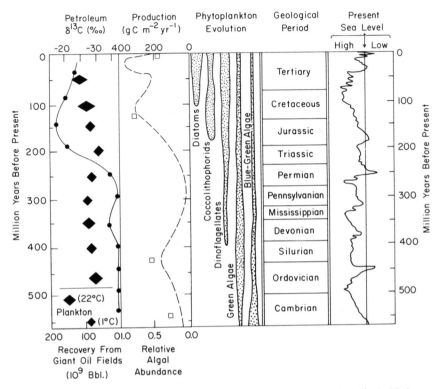

FIG. 159. Phanerozoic shelf production, algal abundance, and ^{13}C content of oil (♦) in relation to sea level, phytoplankton evolution, and oil recovery. [After Degens (1969), Tappan and Loeblich (1970), Moody (1975), Vail et al. (1978), Schopf (1980), and Walsh (1983); © 1983 Pergamon Journals Ltd., with permission.]

With return of the shelves after the Permian ice age, an additional source of external energy (Margalef, 1978) from tidal and wind mixing of shallow seas was made available for the evolution of the negatively buoyant diatoms (Lipps, 1970). Tidal stirring and wind mixing on shallow, flooded continental shelves during interglacial periods are required to prevent the fast-growing, nonmotile diatoms from sinking out of the water column with a w_s of as much as 10 m day^{-1} for individual cells (Smayda, 1970; Eppley et al., 1978), and at 100 m day^{-1} if scavenged within macroaggregates (Alldredge, 1979; Shanks and Trent, 1980). At low stirring rates in experimental enclosures (Eppley et al., 1978; Oviatt, 1981), for example, the more primitive, slower-growing flagellates (Lipps, 1970) with their own energy subsidy (i.e., flagellae) replace diatoms as the dominant phytoplankton. A similar seasonal succession of diatoms and dino-flagellates is observed in the transition from well-mixed to stratified conditions of the coastal water column (Walsh *et al.*, 1974, 1977; Falkowski *et al.*, 1980). In this respect, the weakly motile coccolithophores of intermediate growth rates (Margalef, 1978) evolved in the Jurassic before the diatoms (Fig. 159), while prokaryotic and eukaryotic picoplankton and nanoplankton have probably existed in the sea since pre-Cambrian times (Longhurst, 1985b).

Although the fossil record of diatoms does not begin until the Cretaceous, these organisms were obviously evolving before this time (Hart, 1963; Round and Crawford, 1981), during the 160-million-yr ice-free Mesozoic era of increased shelf area. By the end of the Cretaceous, sea level (Fig. 159) was ~200 m higher than at present (Bond, 1978), the shelf area was threefold larger than now (Schopf, 1980), and all of the continents had separated (Dietz and Holden, 1970). Most of Europe was then under water, and the Tethys Sea was again similar in extent to that of the previous major Ordovician marine invasion of North America, leading to 4000–6000 m of accumulated Cretaceous sediments in the Gulf of Mexico, Utah, Wyoming, and Montana (Reeside, 1957). Deduced from changes in carbon content of these continental slope deposits (Schopf, 1980), the primary production of nearby waters increased from a low of 45 g C m^{-2} yr^{-1} in the pre-cambrian and perhaps in the Permian (Tappan and Loeblich, 1970) (i.e., before the evolution of coccolithophores and diatoms) to a peak of 320 g C m^{-2} yr^{-1} at the end of their blooms in the Cretaceous (Fig. 159). Production then declined during the Holocene to 188 g C m^{-2} yr^{-1} (Schopf, 1980), which is the average shelf production before 1972 (Platt and Subba Rao, 1975).

A switch in marine food chains, from a blue-green-based or dinoflagellate-based system to one derived from first coccolithophorids and subsequently to one from faster-growing diatoms (Fig. 159), should initially lead to higher fixation of carbon in the absence of much grazing stress. Alternating periods of high and low primary production during the Cenozoic, for example, may have led to successive radiations of marine mammals (Lipps and Mitchell, 1976), after evolution of their prey (i.e., herbivorous copepods feeding on diatom chains) (Parsons, 1979). Such a food web (Parsons, 1979), based on large phytoplankton within pulsed production systems and containing an initial large phytodetrital sink, should also have resulted in an increased biological export of carbon to the slope (Walsh *et al.*, 1981) during the Cretaceous.

The enormous calcareous chalk deposits (Tappan, 1968) of the North Sea, such as the white cliffs of Dover, were, in fact, deposited at the outer edge of the Cretaceous continental shelf (Hakansson *et al.*, 1974), while the organic-rich deposits of the Cretaceous South Atlantic were formed on the upper slope (Thiede and Van Andel, 1977), and the ubiquitous Cretaceous biogenic cherts of the deep sea (Weaver and Wise, 1974) presumably represent only 10–15% of the diatomaceous oozes deposited on the continental margin (Heath, 1974). Similarly, after a change in conditions leading to formation of ocean-wide, organic-rich shales in the Cambrian, Ordovician, and Silurian times (Tissot and Welte, 1978), the black shale deposits of the late Permian Cordilleran seaway, an antecedent of the North American Tethys Sea, were restricted to depocenters at the edge of the Permian continental shelf (Claypool et al., 1978).

The major sapropelic marine deposits of 20% carbon content—the same as present organic deposits off Peru and Southwest Africa—are of Cretaceous origin (Ryan and Cita, 1977). Similar smaller deposits are found from Pleistocene interglacial periods (Ryan, 1972), that is, with emergence of the continental shelves, extending from the continental slope to deeper waters of the Mediterranean (McCoy, 1974), and associated with stimulation of coastal production from the Nile River discharge (Rossignal-Strick et al., 1982; Rossignal-Strick, 1983). Assuming a fossil slope depocenter area of 50×10^6 km², a mean 4% organic content of Cretaceous sapropels (Ryan and Cita, 1977), and a sedimentation rate of 0.4 cm yr^{-1} (i.e., the same as that now off Peru, Southwest Africa, and California), 6.2×10^9 tons C yr^{-1} could have been sequestered as shelf export during the high sea levels of the Cretaceous (Vail et al., 1978). Alternatively, analogous to present food chains, only 75% of the algal production might have been consumed by the newly evolved Cretaceous shelf food chain. A primary production rate of 320 g C m^{-2} yr^{-1} and a shelf area of 7.2×10^7 km² (Schopf, 1980), would then allow an export of $\sim 5.8 \times 10^9$ tons C yr^{-1} for burial during Cretaceous anoxic events (Hallam, 1977). This is perhaps fivefold to 10-fold the annual export estimated for the last few decades.

Major oil fields on the continental slope are thought, in fact, to have been formed during the Cretaceous (Dow, 1978; Tissot and Welte, 1978), with an order of magnitude more organic matter stored in the 20% carbon sapropel source deposits of this period than in all known coal and oil reserves (Ryan and Cita, 1977). Formation of oil or gas from these organic deposits then depends, of course, on a number of other factors, such as decomposition, polycondensation, insolubilization to carbon geopolymers (kerogen), increases in temperature, and subsequent mobilization and entrapment in appropriate reservoirs. For example, Tertiary sediments of 5000–10,000 m thickness overlie Cretaceous deposits in some areas such as the Louisiana (Dow, 1978) and Utah (Tissot et al., 1978) regions of the past Tethys Sea.

A combination of this organic source material, diagenesis, inorganic overburden, catagenesis, and available ancient reef reservoirs can then lead to giant Cretaceous oil fields, as recently found in the Gulf of Mexico (Arthur and Schlanger, 1979). Conversely, within regions of small overburden, extensive Cretaceous gas fields ($\sim 96\%$ CH$_4$) are found at shallow depth (~ 300 m) in western Canada and Siberia (Tissot and Welte, 1978). At least 65% of the number and ultimate recovery (Fig. 159) of the giant oil fields [$> 5 \times 10^8$ barrels (bbl)] formed since Cambrian times are, in fact, of Cretaceous age or younger (Moody, 1975). The fossil record of ^{13}C and other isotopes within organic and inorganic sedimentary reservoirs provides an index of the changing amount and possible species content of slope import of phytodetritus over the geological time scale.

5.3.1 Natural Stable Isotopes

Like the fractionation of the ^{14}C and ^{13}C isotopes of carbon during photosynthesis, biological cycling of the elements of nitrogen, oxygen, and sulfur also results in depletion of the heavier isotopes, ^{15}N, ^{18}O, and ^{34}S, within organic matter. The departure of the ratio of these isotopes to the lighter, more abundant forms of ^{14}N, ^{16}O, and ^{32}S within samples and the respective standards of air, Pee Dee Belemnite (PDB), and Canyon Diablo Troilite (CDT), is described by δ ^{15}N, δ ^{18}O, and δ ^{34}S expressions of definition similar to Eqs. (205) and (206). In addition to biological fractionation of these isotopes, different concentrations of these stable forms (i.e., they do not undergo radioactive decay) also result from both physical processes of evaporation and varying temperature effects on solubility of gases and growth of organisms.

Harold Urey's early studies of isotope fractionation (Urey, 1947) led to the initiation of studies of the natural occurrence of both nitrogen (Hoering, 1955) and oxygen (Emiliani, 1954) stable isotopes in the mid-1950s, with subsequent major emphasis on biological transformation of the former and physical consequences of the latter. Kinetic fractionation of nitrogen compounds occurs at each process of this complicated cycle, that is, nitrogen fixation (Delwiche and Steyn, 1970), denitrification (Wellman et al., 1968), nitrification (Miyake and Wada, 1971), or reduction of nitrite to form cell protein (Gaebler et al., 1966). As a result, the δ ^{15}N of nitrate in the sea has a mean of 5–6‰, although a large range can be found (Miyake and Wada, 1967; Cline and Kaplan, 1975), compared to about 0‰ for nitrogen gas (N_2) in seawater and in the atmosphere (Sweeney et al., 1978) and +6.4‰ for particulate nitrogen of oceanic plankton (Saino and Hattori, 1980). With elimination of the lighter ^{14}N during catabolism, however, the ^{15}N content of the particulate nitrogen left behind within higher trophic levels increases, such that an increment of δ ^{15}N of 1–3 per mil occurs at each step of the food web (DeNiro and Epstein, 1981; Macko et al., 1982).

Excretory nitrogen released by zooplankton, for example, has a δ ^{15}N of −3.2‰, compared to that of +3.0‰ for the particular prey ingested by these copepods (Checkley and Entzeroth, 1985). Similarly, recall that in Section 4.2.3, birds eating fish in the Bering Sea were enriched in δ ^{15}N by 4‰, compared to those mainly eating zooplankton. The diets of prehistoric humans (Schoeninger et al., 1983) have also been delineated by this technique, analogous to the similar ^{14}C activity of human protein and terrestrial foods. The sources of nitrogen in ground water (Kreitler, 1979), California coastal waters (Sweeney and Kaplan, 1980), and warm core rings within the Mid-Atlantic Bight

(Altabet *et al.*, 1986) have also been delineated with $^{15}N/^{14}N$ ratios, but geochemical studies have mainly been confined to recent sediments (Peters *et al.*, 1978; Sweeney *et al.*, 1978; Macko *et al.*, 1984; Walsh *et al.*, 1985).

In contrast, $^{18}O/^{16}O$ ratios have been extensively used in geological studies, and only rarely in studies of photosynthesis, or of water-mass exchange (Fairbanks, 1982). During evaporation and glaciation, for example, the lighter ^{16}O is removed from the ocean in the water vapor and precipitated as fresh water or ice, leaving the heavier ^{18}O behind in the sea, to be incorporated within $CaCO_3$ tests by calcareous organisms, such as foraminifera. A significant temperature effect on the growth and ^{18}O fractionation of foraminifera also occurs, with increased ^{18}O concentrations at lower temperatures—the same trend as that of seawater during glaciation. About two-thirds of the oxygen isotope signature in fossil foraminifera may be due to changes in continental ice volume, and one-third is the result of direct temperature variations (Kennett, 1982), with the assumption that other biological fractionation is small.

A polynomial regression (Craig, 1957; Shackleton, 1967) is used to extract the paleotemperature T from the isotopic ratios of oxygen in seawater, $\delta\ ^{18}O_s$, and in calcite of the foraminifera, $\delta\ ^{18}O_c$, by

$$T = 16.9 - 4.38(\delta\ ^{18}O_c - \delta\ ^{18}O_s) + 0.1(\delta\ ^{18}O_c - \delta\ ^{18}O_s)^2 \quad (212)$$

In Fig. 160 of oxygen isotope data of foraminifera since the end of the Cretaceous, from cores in the North and subantarctic Pacific (Savin, 1977; Kennett, 1982), the Modern and Tertiary temperatures were calculated assuming respective values for $\delta\ ^{18}O_s$ of -0.08 and $-1.00‰$. The increase of $\delta\ ^{18}O$ in fossil calcite during the middle of the Miocene and at the start of the Pliocene (Fig. 160), as well as during the last two glaciations (see Fig. 165), is associated with increased ^{18}O concentrations left behind in seawater during lower stands of sea level. In these situations, continental shelf export of carbon would also not occur, because these ecosystems were then dry land.

Such a change in organic carbon fluxes would impact both the $\delta\ ^{13}C$ content of dissolved inorganic carbon left behind in seawater and the $\delta\ ^{34}S$ content of the sedimentary and oceanic reservoirs of sulfur, as monitored by FeS_2 (pyrite) and $CaSO_4 \cdot 2H_2O$ (gypsum) deposits. Similar to the stoichiometric expression [Eq. (113)] for protein synthesis and photosynthesis, a redox reaction can be written (Garrels and Perry, 1974) for the exchange of oxygen between the carbon and sulfur cycles as

$$4FeS_2 + 8CaCO_3 + 7MgCO_3 + 7SiO_4 + 31H_2O$$
$$= 8CaSO_4 \cdot 2H_2O + 2Fe_2O_3 + 15CH_2O + 7MgSiO_3 \quad (213)$$

In a long-term geological context, without production or consumption of gases, Eq. (213) indicates an inverse relationship between the oxygenated and reduced forms of both carbon [i.e., carbonate ($CaCO_3$) and organic matter (CH_2O)] and sulfur (i.e., gypsum and pyrite).

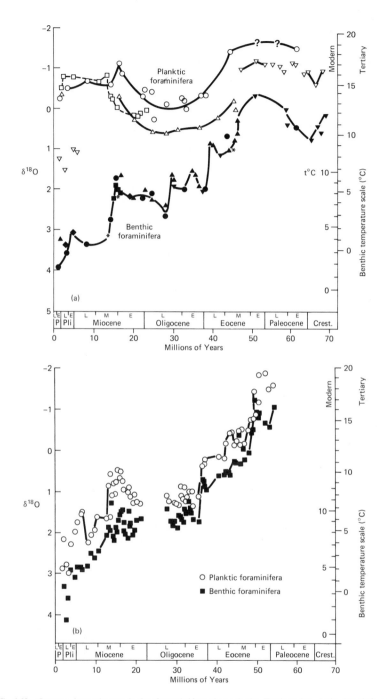

FIG. 160. Oxygen isotopic trends for foraminifera through the Tertiary low latitudes (left panel) and high latitudes (right panel), with estimates of paleotemperature. [After Savin (1977) and Kennett (1982); © Prentice-Hall.]

The δ ^{13}C content of marine carbonates and δ ^{34}S content of marine evaporites of gypsum are usually assumed to be deposited without fractionation of either the ^{13}C/^{12}C or ^{34}S/^{32}S ratios of ambient seawater. The δ ^{13}C content of seawater, and particularly of benthic foraminifera, is instead altered by changes in the decomposition and burial rates of organic matter. If more ^{13}C-deficient organic carbon, originally fractionated by photosynthesis, is buried, for example, less biogenic CO_2 is returned to the water column from oxidative decomposition, or respiration. As a result, the inorganic carbon ^{13}C content of seawater then increases, as monitored by the calcite tests of coccolithophores or foraminifera.

Assuming a δ ^{13}C of $+1.3\%o$ for seawater, that is, carbonates, and $-23.7\%o$ for organic carbon, for example, a simulated increase in carbon burial of 1.2×10^7 tons C yr^{-1} leads to a δ ^{13}C increment of $\sim1.50\%o$ for seawater within 0.5×10^6 yr (Kump and Garrels, 1986). A greater burial rate of 1.2×10^9 tons C yr^{-1} would lead to a seawater increase in δ ^{13}C of $0.75\%o$ within ~2500 yr. Such burial increases would also lead, of course, to less CO_2 and fewer nutrients in seawater, as well as an increase of O_2 in the atmosphere and a concomitant decline of CO_2 in this reservoir—we will return to this subject in Sections 5.3.3 and 6.1.1.

A settling flux of organic carbon to marine sediments is required as an energy input for bacterial reduction there of sulfate to pyrite, schematically represented (Kump, 1986) by

$$2Fe_2O_3 + 8SO_4^{2-} + 15CH_2O + CO_2 = 4FeS_2 + 16HCO_3^- + 7H_2O \qquad (214)$$

in which the heavier ^{34}S is left behind in anoxic pore waters, to be later evaporated, forming gypsum. Pyrite sulfur accordingly has a δ ^{34}S of approximately $-15\%o$, compared to $+20\%o$ for gypsum (Garrels and Lerman, 1984). Inspection of Eqs. (213) and (214) suggests an inverse relationship between the two isotope fractionation processes (Veizer *et al.*, 1980; Lindh, 1983), with more ^{13}C and less ^{34}S left behind in seawater during increased burial of organic carbon. In this situation, less organic substrate would then be available for the sulfate-reducing bacteria involved in Eq. (214), as well as for their oxic counterparts, involved in the reverse reaction of Eq. (113).

Burial rates of CH_2O and FeS_2 are a complex subject (Berner, 1982; Raiswell and Berner, 1985), but over a sufficiently blurred Phanerozoic time scale (Fig. 161), both δ ^{13}C and δ ^{34}S isotope ratios of seawater suggest that the burial rate of organic matter may have increased up to the Permian, that is, a period of shelf withdrawal. It then apparently declined until the Cretaceous, with another increment in response to the advent of the diatoms (Fig. 159); it then decreased again to the Holocene (Fig. 161). Assuming a marine origin of petroleum, the variation of crude oil δ ^{13}C (Degens, 1969) over the same geological time (Fig. 159) can similarly be attributed to possible changes in temperature, CO_2 content, or phytoplankton species within the surface waters. However, one must eventually

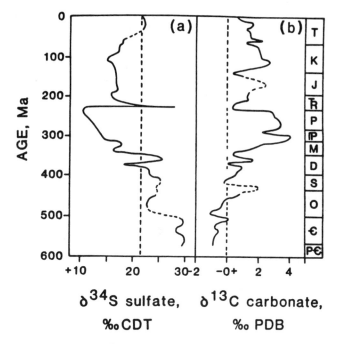

δ^{34}S sulfate,   δ^{13}C carbonate,
‰CDT   ‰ PDB

FIG. 161. Phanerozoic records of isotopic ratios of (a) marine sulfate and (b) inorganic carbon over the last 600 million yr. [After Holser (1984); © Springer-Verlag.]

examine shorter time scales of the last 100–200 million yr, or even the last 10,000–20,000 yrs, to resolve possible consequences of the changes of shelf export.

5.3.2 Oil Reservoirs

Unlike ^{15}N, ^{13}C does not appear to be much further fractionated as organic carbon is passed up the marine food web (Fry and Sherr, 1984), but similar to ^{18}O, a significant temperature effect (Fig. 159) occurs (Sackett *et al.*, 1965; Sackett, 1986). The depletion of plankton ^{13}C from an isotopically heavy δ ^{13}C value of − 16‰ at the equator to a lighter value of approximately − 30‰ near Antarctica (Sackett *et al.*, 1965; Degens *et al.*, 1968; Eadie, 1972) is the opposite effect of temperature on ^{18}O fractionation. It instead may partially reflect temperature effects on lipid synthesis (Smith and Morris, 1980), the solubility of CO_2, and its δ ^{13}C fractionation with HCO_3^- (i.e., − 9.2‰ at 0°C and − 6.8‰ at 30°C) (Deuser and Degens, 1967). CO_2 is the preferential source of carbon during photosynthesis (Deuser *et al.*, 1968), dark fixation of CO_2 also occurs in the Southern Ocean (Horrigan, 1981), and the δ ^{13}C of total inorganic carbon (CO_2,

HCO_3^-, CO_3^{2-}) in surface water is accordingly approximately $+2.5‰$ at 30°C near the equator and $+1.5‰$ at 0°C in the Antarctic (Kroopnick *et al.*, 1977).

Extraction of CO_2 from the atmosphere by coastal diatoms, which are not all grazed and therefore are perhaps sequestered in slope depocenters, is thought to be a major organic carbon sink in the present global CO_2 cycle (Walsh *et al.*, 1981). Similarly, the intense phytoplankton blooms of the Cretaceous (Tappan, 1968; Tappan and Loeblich, 1970) sequestered atmospheric carbon in the form of coccolithophorid and diatom oozes, before grazers evolved to remove some of the surplus production (Tissot and Welte, 1978). The removal of CO_2 from the late Cretaceous atmosphere (see Fig. 163c) by the newly evolved algae is thought to have induced (Worsley, 1974) the inferred 12°C temperature decline (Fig. 160) of the Tertiary ocean (Schopf, 1980), a reversal of the "greenhouse" effect postulated at the end of the nineteenth century (Chamberlin, 1899). A higher CO_2 content of the early atmosphere is, in fact, one of the postulates (Barron *et al.*, 1981) for hypothesizing both a warmer ocean and a milder climate in the Cretaceous $65-140 \times 10^6$ yr ago, and for the unusually light δ ^{13}C values of fossil green algae in the pre-cambrian (Degens, 1969). Increased burial rates of organic matter would then lead to reduced CO_2 concentrations in both seawater and air, until new steady-state values of these carbon reservoirs were reached.

Continued deterioration of Tertiary climate would stimulate polar cooling of seawater (Lipps, 1970), with increased surface solubility of CO_2 at higher latitudes as a paleochemical sink, in addition to the biological extraction of atmospheric CO_2. Such a high-latitude CO_2 storage mechanism is also advocated as a chemical sink in present models of anthropogenic release of fossil-fuel CO_2 (Broecker, 1979; Broecker *et al.*, 1980; Broecker and Peng, 1982). Separation of Greenland and Europe and the formation of the circum-Antarctic Basin $\sim 50-60 \times 10^6$ yr ago then allowed sinking of bottom water at polar latitudes, thereby ventilating the deep regions of the ocean with lower-temperature water of increased nutrient and CO_2 content (Berger, 1970; Berggren and Hollister, 1974), implying a reduction in the δ ^{13}C of marine carbonates (Fig. 161). These temperature and CO_2 effects would also lead to plankton and organic sediments (Hunt and Degens, 1967) with less ^{13}C and not more, as is implied by the 5‰ enrichment of δ ^{13}C from the Triassic to the Tertiary crude oils (Fig. 159).

Laboratory cultures of modern phytoplankton species at 18°C, a pH of 7.5, and excess HCO_3^- conditions (Wong, 1976) exhibit, however, a $5-6‰$ δ ^{13}C difference between the isotopically heavier diatoms and the lighter green algae, coccolithophores, and golden algae. Similarly at 15°C, a pH of 8.3, and with air as the carbon source, the heavier diatoms were enriched by $10-12$ δ ^{13}C, compared to dinoflagellates or blue-greens (P. Falkowski, personal communication). These observations suggest that the apparent ^{13}C changes of fossil organic carbon

(Fig. 159) might partially be a function of algal species succession over geological time.

The cell wall of diatoms consists of a silica frustule, while coccolithophores have calcium carbonate tests, in contrast to the greens, dinoflagellates or chrysophytes. The ash content (Table XII) of these siliceous and calcareous organisms (Parsons *et al.*, 1961; Aaronson *et al.*, 1980) is accordingly ~40% dw (Bacillariophyceae), and 20% dw (Chrysophyceae). As a result of such structural differences, only ~4% of the dry weight of pelagic diatom cells is lipid (Table XII), compared to 15% for chlorophytes, 19% for dinoflagellates, 30% for haptophytes, and 40% for chrysophytes (Aaronson *et al.*, 1980). Lipids are deficient in ^{13}C compared to the rest of algal cell constituents, are left behind after decarboxylation of amino acids, and form one of the precursors of kerogen (Degens, 1969).

The range in percentage of protein (27–39%) and carbohydrate (17–38%) content among these algal groups is less than that of lipid (Table XII), reflecting other structural properties, such as diatom C/N ratios of 5–6 compared to dinoflagellate C/N ratios of 7–8 within natural populations (Walsh *et al.*, 1974). Assuming (1) that lipids, carbohydrates, and proteins constitute all of the organic carbon pool of these phytoplankton (Table XII), and (2) that the respective δ ^{13}C values of these cell components ($-30.0‰$, $-20.5‰$, and $-17.5‰$) for Peruvian diatoms (Hunt and Degens, 1967) apply to the other groups as well, theoretical values can be calculated for the expected ^{13}C content of the total organism within each algal group.

Upon elimination of ^{13}C-enriched carbohydrates and proteins of phytodetritus during diagenesis, the fossil carbon is depleted by about 6‰ of δ ^{13}C, compared to recent marine sediments (Degens, 1969). With these assumptions, a constant diagenetic elimination of ^{13}C over the geological record would lead to crude oil (Table XII) of δ ^{13}C values similar to hypothetical diatom remains in the Tertiary and hypothetical coccolithophore remains in the Triassic (Fig. 159). Furthermore, dinoflagellates contain longer-chain fatty acids than diatoms (Wood, 1974), while green algae such as *Chlorella* have fatty acids lighter in ^{13}C than diatoms (Parker, 1962; Hunt and Degens, 1967), such that inclusion of other components of the organic pool might lead to a greater difference in the proposed ^{13}C enrichment between diatoms and the other isotopically lighter phytoplankton (Table XII).

Assuming a constant species composition of phytodetritus since the Cretaceous, any further isotope changes should reflect differences in the amount of shelf organic matter, removed and sequestered within slope depocenters. Such transitions would be accompanied by concomitant shifts in the ^{13}C content of seawater over this same time period. Calcite is enriched by 1.7‰ of δ ^{13}C at 2°C and by 0.9‰ at 25°C, after inorganic precipitation from seawater (Rubinson and

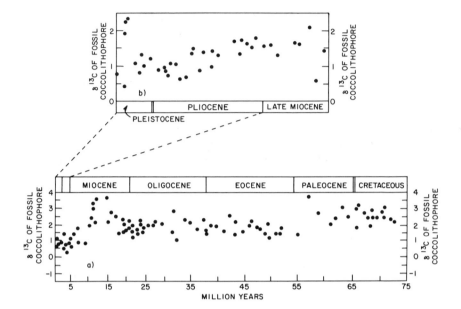

FIG. 162. The δ ¹³C isotope ratio of fossil coccolithophores on the slope from the Cretaceous to the Pleistocene over the last 75 million yr [After Margolis *et al.* (1977), Kennett *et al.* (1979), and Walsh (1983); © 1983 Pergamon Journals Ltd., with permission.]

Clayton, 1969). Biological formation of calcite tests by coccolithophores is thought, however, to incorporate HCO_3^- from seawater with little fractionation. The ¹³C content of fossil coccolithophores as well as of foraminifera may thus reflect the ¹³C content of ~97% of the previous inorganic carbon pool of seawater (Kroopnick *et al.*, 1977), if differential fractionation has not occurred during nanofossil deposition since the Jurassic.

Examination of the ¹³C content of fossil coccolithophores over time from mainly the lower slope (1214–1591 m), south of New Zealand and Tasmania (Margolis *et al.*, 1977) in Fig. 162a, and from midslope (1068 m), north of New Zealand (Kennett *et al.*, 1979) in Fig. 162b, suggests, in fact, a clear depletion of ¹³C. The fossil phytoplankton, and presumably the seawater, had a δ ¹³C of 2–3‰ at the end of the Cretaceous, compared to 0.5–1‰ at the end of the Pleistocene. Recall that the same trend was evident within the longer Phanerozoic time series depicted in Fig. 161. Such an apparent trend of a long-term depletion of ¹³C within fossil coccolithophores is interrupted during high sea levels of the middle Miocene (Fig. 162a), that is, in a major marine transgression as indicated by ¹⁸O/¹⁶O ratios during this period (Fig. 160). If more diatom export and burial of ¹³C-deficient carbon within slope depocenters had occurred, as a consequence of an increase of shelf area in response to a higher Miocene sea level of perhaps 120 m (Vail and Hardenbol, 1979), the ocean would have been enriched for a while in ^{13}C. This is evidently reflected by more positive δ ^{13}C values of the calcareous tests (Kroopnick *et al.*, 1977) of both phytoplankton and benthic foraminifera. Using the isotope records of δ ^{13}C and δ ^{34}S in marine carbonates and evaporites over the last 100 million yr (Lindh, 1983; Shackleton, 1985), and a time-dependent rate of sea-floor spreading (Kominz, 1984), the simulated burial rate of organic carbon (Kump, 1986) peaked during this Miocene event of high sea level (Fig. 163a). It was the inverse of pyrite formation (Fig. 163b), while continued depression of atmospheric CO_2 levels (Fig. 163c) and a peak in O_2 content (Fig. 163d) occurred.

Extensive deposition of phosphorites occurred on the continental margins off Florida and North Carolina at this time in the Miocene, suggesting that nutrients were also stored then (Riggs, 1984). Increased storage of carbon and nutrients may also have taken place during another Pliocene event of high sea level (Riggs, 1984). Within more recent time, the ^{13}C content of benthic foraminifera, at depths of 2500–4000 m, was similarly more depleted in specimens from the late Pleistocene glacial period (Shackleton, 1977) than from the present Holocene interglacial. We now examine these recent data in more detail within the next section.

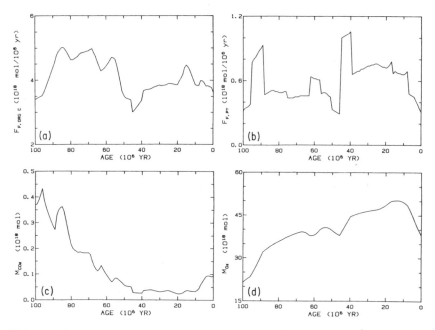

FIG. 163. The simulated global burial rates of (a) organic carbon and (b) pyrite, with respect to atmospheric changes of (c) CO_2 and (d) O_2 over the last 100 million yr. [After Kump (1986).]

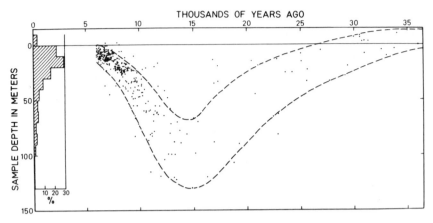

FIG. 164. Sea-level fluctuation over the last 35,000 yr. [After Milliman and Emery (1968); © 1968 AAAS.]

5.3.3 Glacial/Interglacial Oscillation

About 115,000–125,000 yr ago, during the last Eemian (Sangamon) interglacial (Emiliani and Shackleton, 1974), sea level may have been somewhat higher than it is today (Veeh, 1966). Within just 10,000 yr, sea level may then have dropped 60 m (Steinen *et al.*, 1973) and may have remained low for the next 100 millennia, with a minimum of 140 m below present level occurring only ~15,000 yr ago (Milliman and Emery, 1968). A few ¹⁴C dates suggest (Fig. 164) that another interglacial period may have occurred 35,000 yr BP. These older dates could be spurious, however, since the age limit of this technique is about 35,000–50,000 yr—recall that the half-life of ¹⁴C is ~5570 yr. Samples dated by ¹⁴C as $21-31 \times 10^3$ yr old, for example, were found by uranium/thorium dating to actually be $10-40 \times 10^4$ yr old (Gill, 1974). Over the last 5000 yr, rise in sea level (Fig. 164) has been minimal (Bloom, 1977); melting of the remaining glacial ice would result in a sea-level increment of 42–50 m above the present level (Emery and Uchupi, 1984).

Such changes of sea level can be detected in the δ ¹⁸O ratios both of benthic calcareous organisms in the sea and of water stored as glacial ice. With termination of the previous Illinoian (Saale) glacial sequence (termination II of Fig. 165) about 125,000 yr ago, for example, the δ ¹⁸O ratio of a benthic species of foraminifera, *Uvigerina peregrina*, declined from 4.54‰ at 856 cm of a sediment core to 3.17‰ at 824 cm within this core taken at 2573 m on the continental rise off West Africa (Shackleton, 1977)—that is, a warming of seawater and a decline in ice volume had evidently occurred. Similarly, after termination of the last Wisconsin (Wurm) glacial sequence (termination I of Fig. 165) about

11,000 yr ago, the δ ¹⁸O of the same species declined from 4.28‰ at 61 cm of the sediment core to 3.05‰ at 31 cm, marking the onset of the present Holocene interglacial period.

During most of Wisconsin time, a 2162-m ice core, taken through the Antarctic ice sheet at Byrd Station (~80°S), exhibited (Fig. 166) δ ¹⁸O ratios of approximately −40‰, from 75,000 yr ago until 15,000 yr ago; that is, there is little indication of an interglacial period 35,000 yr BP (Epstein *et al.*, 1970). The recent interglacial/glacial transitions of sea level (Fig. 164) and δ ¹⁸O ratios of *U. peregrina* (Fig. 165) were then accompanied, however, by a 5-mil decline in δ ¹⁸O content of this ice sheet over the 4000-yr deglaciation interval (Fig. 166) as well. Over the same glacial/interglacial transitions of the shelf/slope habitat, the δ ¹³C ratio of *U. peregrina,* and presumably of seawater, also increased from

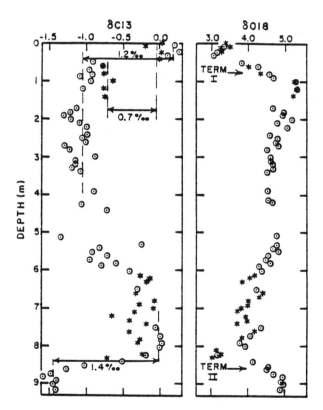

FIG. 165. The δ ¹³C and δ ¹⁸O isotope ratios in foraminifera from the eastern North Atlantic over the last 150,000 yr. The circles are analyses made on *Uvigerina peregrina* and the asterisks are for analyses made on *Panulina wuellerstorfi*. [After Shackleton (1977) and Broecker (1982a, b); © 1982 Pergamon Journals Ltd., with permission.]

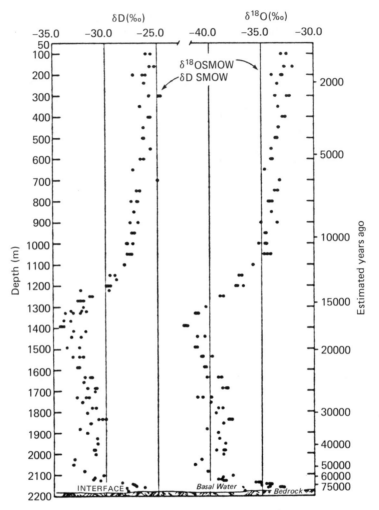

FIG. 166. Oxygen ($^{18}O/^{16}O$) and hydrogen (D/H) isotope ratios, in ice samples from the deep core at Byrd Station, Antarctica. [After Epstein *et al.* (1970); © 1970 AAAS.]

-1.03 to $-0.38‰$ during termination II and from -0.77 to $+0.10‰$ during termination I (Fig. 165): ^{13}C-deficient organic matter had evidently been removed from the sea during each of these deglaciations.

Over the last 2500 yr of the Holocene, the $\delta\ ^{18}O$ and $\delta\ ^{13}C$ ratios of *U. peregrina* (Fig. 167) have exhibited short-term excursions of respectively 0.6‰ and 0.3‰, with perhaps a periodicity of 50–100 yr (Gao *et al.*, 1985). Part of the interglacial/glacial trends in Fig. 165 may thus be noise, associated with biological fractionation, or temperature effects, but at least the same species of benthic foraminifera was used. Furthermore, isotope analyses of six additional cores from deeper depths of 2600–4100 m, ranging from 44°N to 44°S latitude, suggest a mean increment of 0.7‰ of δ ^{13}C within specimens of *U. peregrina* over the 4000-yr transition period from Wisconsin to Holocene time (Broecker, 1982a,b). Assuming a δ ^{13}C ratio of marine organic matter of $-21.0‰$, which is somewhat less than Kump's (1986) model in the last section, a burial of 1.4×10^{12} tons of organic carbon is required to account for such a δ ^{13}C increase of 0.7 mil in the benthic foraminifera (Broecker, 1982a).

Over the first 1000 yr of the Wisconsin deglaciation, a burial rate of 1.4×10^{9} tons C yr^{-1} in this simple model (Broecker, 1982a) is very similar to a result of 1.2×10^{9} tons C yr^{-1} for a 0.75-mil shift of δ ^{13}C from Kump's (1986) more complex calculation—recall, from Section 5.2.3, that estimates of

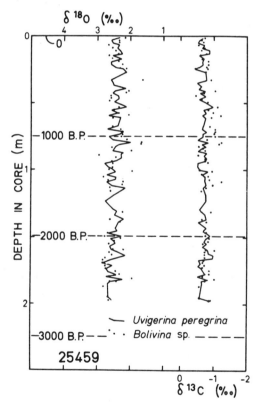

FIG. 167. The δ ^{13}C and δ ^{18}O isotope ratios in the same species of foraminifera, *Uvigerina peregrina*, and *Bolivina* spp. from the San Pedro Basin over the last 2500 yr. [After Gao *et al.* (1985); © 1985 Pergamon Journals Ltd., with permission.]

present carbon burial range from 0.1 to 0.5 × 10^9 tons C yr^{-1}. Assuming that only about 10% of the 150,000-yr record of sediments within the core from 2573 m on the continental rise off West Africa (Fig. 165) was deposited during the ~20,000-yr interval of both the Holocene and Eemian interglacials, a mainly glacial sedimentation rate of ~960 cm per 150,000 yr, or 0.0064 cm yr^{-1}, is obtained. In contrast, during the last 2000 yr of sediment deposition within the core from 912 m on the California slope (Fig. 167), one obtains an interglacial sedimentation rate of 1500 cm per 2000 yr, or 0.7500 cm yr^{-1}—that is, an increase of two orders of magnitude in sedimentation and presumably in burial rate of organic carbon.

Both sediment cores of Figs. 165 and 167 were taken seaward of upwelling ecosystems at mid-latitude, which today have similar productivity (Table III), such that an increased flux of carbon to the continental slope sediments presumably occurred during the Holocene as a result of the apparent increase of sedimentation rate. However, the organic carbon content of glacial sediments from the same 2573-m core on the northwest African rise (Muller and Suess, 1979) is much higher (2–3% dw) in the Wisconsin part of the core than within sediments of ~0.5% dw carbon from either the Holocene or Eemian interglacial periods. Reconciliation of the implied burial increment of organic carbon from the inorganic δ ^{13}C data, of the data on the changing amount of surviving refractory carbon, and of the apparent increment in sedimentation rates leads to at least two scenarios. Either the preservation of organic matter within these interglacial sediments was poor after the initial carbon loading, in addition to the difference in water-column survival rates while sinking to the bottom depths of the two cores (Figs. 165 and 176), or the burial rate declined from the deglaciation period of 15,000–11,000 BP to present Holocene time.

Examination of the aragonite remains of pteropods on the same northwest African rise (Berger, 1977) suggests, in fact, that a burial peak of organic carbon occurred there ~13,500 yr ago, ~1500 yr after the deglaciation began and some 4000 yr after the glacial maximum (Fig. 166). Increased burial of organic carbon in the sea leads to reduction of CO$_2$ dissolved in water, as recorded by the δ ^{13}C signal of *U. peregrina* (Fig. 165). The decline in release of biogenic CO$_2$ within the pore waters of these sediments also leads to increased preservation of rare pteropod tests—that is, an increase of the aragonite compensation depth.

During the last deglaciation, this preservation depth of fossil pteropods increased to 3500 m off northwest Africa, compared to 1500 m during the Wisconsin glaciation and 500 m during the Holocene (Berger, 1977). The preservation spike of aragonite here only lasted ~1000 yr, with rapid dissolution beginning again about 13,000 BP (i.e., presumably remineralization of organic debris began again). Such preservation spikes have also been found in the equatorial Pacific (Berger, 1977) and the Caribbean (Broecker, 1971), with the peak abundance of fossil pteropods associated with the maximum change in the δ ^{18}O ratio during the middle of the deglaciation event (i.e., Figs. 165 and 166).

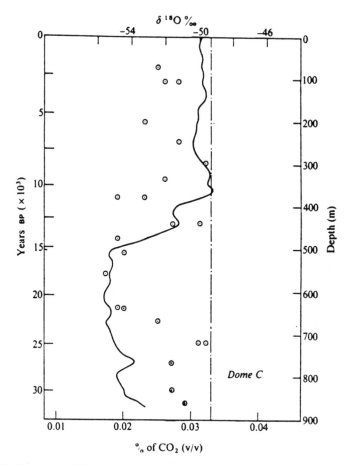

FIG. 168. The percent of CO_2 (v/v) in the air of the bubbles from the dome C ice core against depth (in meters of ice equivalent) and with respect to the isotopic values of δ $^{18}O‰$ (continuous line) and the present-day atmospheric CO_2 content (0.033%), indicated by the dashed line. [After Delmas *et al.* (1980); © 1980 Macmillan Journals Ltd., with permission.]

Increased burial of organic carbon in the sea would also lead to a reduction of CO_2 in the atmosphere, since ~90% of the total inorganic carbon within these two reservoirs is found within the ocean. Recent analyses of prior atmospheric CO_2 concentrations, trapped within frozen air bubbles, from other Greenland and Antarctic ice cores (Delmas *et al.*, 1980; Neftel *et al.*, 1982) provide hints to the possible changes of carbon fluxes within the atmosphere, ocean, and sediment reservoirs. During the last deglaciation at 75°S in the Antarctic, for example, the atmospheric CO_2 content may have declined from 290 parts per million (ppm) about 13,000 BP at 472 m of the ice core to 210 ppm about 11,000 BP at 410 m (Fig. 168).

A smaller decline of CO_2 over this time period may also have been detected in the Byrd core at 80°S and in a Greenland core from 77°N (Neftel *et al.*, 1982), but the samples from depths of 400–1200 m were of poor quality, exhibiting a large scatter in the CO_2 values. A more recent Greenland ice core, taken in 1981 at 65°N (Dansgaard *et al.*, 1982), indicates a similar atmospheric CO_2 decline of ~40 ppm (Stauffer *et al.*, 1985a) within an initial rapid warming period, as shown by the δ ^{18}O ratio of the ice, centered about 11,000–13,000 BP during this deglaciation (Oeschger, 1985). The initial decline of atmospheric CO_2 by 40 ppm was, in fact, predicted by Keir (1983), based on a model of both reduced vertical circulation and increased carbon burial, as a function of rising sea level during the last deglaciation.

A second and perhaps even more significant trend occurs later within these fossil CO_2 data, that of an apparent increase of atmospheric CO_2 in the Holocene after the end of deglaciation; it was also predicted by the model (Keir, 1983). Within another 2000 yr, atmospheric CO_2 content evidently returned to concentrations of ~270 ppm (Fig. 168) and remained at this level until the Industrial Revolution (see Figs. 170 and 171). Presumably organic carbon was no longer being buried at the same rate, aragonitic pteropods were no longer being preserved, and CO_2 had increased in both the ocean and the atmosphere, as a result of increased remineralization of organic matter, and perhaps of a decline in primary production, although not necessarily that of phytoplankton (see Table XIV).

The atmospheric CO_2 concentrations may have risen from ~200 ppm during the coldest part of the Wisconsin glaciation, 15,000–20,000 yr ago, to preindustrial Holocene values of ~300 ppm. Based on the ice cores (Fig. 168) at 75 and 80°S, as well as at 65 and 77°N, this may have also occurred rapidly over 1000–2000 yr, as a result of reduced photosynthesis in the ocean at the beginning of the present interglacial period (Broecker, 1982a,b). Rapid removal of nutrients, in organic form, may have occurred by increased burial of the shelf/slope organic matter at the continental margins, after rising sea level and concomitant shelf inundation during the initial deglaciation. This process of nutrient depletion may then have led to a subsequent reduction in oceanic photosynthesis and in CO_2 extraction from the atmosphere 4000 yr later, by the beginning of Holocene (Broecker, 1982a,b).

As soon as sea level began to rise 15,000 yr BP (Fig. 164), increased primary production of outer shelf waters would presumably have led to increased carbon export, as calculated in the present budgets of the Bering Sea, Mid-Atlantic Bight, and Gulf of Mexico in Chapter 4. Increased burial of diatomaceous phytodetritus may then have taken place, as inferred from δ ^{13}C records (Fig. 165), pteropod preservation spikes, and reduction of atmospheric CO_2 (Fig. 168) during deglaciation. With a C/P ratio of 150 : 1 by weight for slope sediments (Morse and Cook, 1978), such a burial flux of 1.4×10^9 tons C yr^{-1} during the first 1000 yr of the deglaciation would have led to an annual organic phosphorus loss of $\sim 1 \times 10^7$ tons P yr^{-1} from the sea. The dissolved inorganic phosphorus (PO_4) reservoir of the present ocean is $\sim 1 \times 10^{11}$ tons P, such that all phosphorus would be removed from the sea in 10,000 yr (i.e., the duration of the Holocene or Eemian interglacial periods) if this burial rate had continued with no resupply of PO_4 from sedimentary reservoirs.

A comparison of the estimated phosphate content of the deep glacial ocean (Broecker, 1982a), 3.2 μg-at. PO_4 l^{-1}, with today's interglacial mean of 2.2 μg-at. PO_4 l^{-1} suggests that 50% of the present ocean phosphorus reservoir was indeed buried as organic matter within 5000 yr, at a rate of 1×10^7 tons P yr^{-1}. A warming trend of $\sim 2°C$ since the Wisconsin glaciation would have decreased dissolved O_2 solubility in the sea by 5%, thereby increasing significantly both the volume of oxygen-deficient seawater and the rate of marine denitrification. Formation of extensive phosphorite deposits may have occurred after times of such nitrogen depletion from the sea, leading to a net nitrogen loss, concomitant lower primary production, and precipitation of excess phosphorus as carbonate fluoroapatite deposits (Piper and Codispoti, 1975).

A number of large inorganic phosphorite deposits were formed in Cambrian, Ordovician, Devonian, Permian, and Cretaceous periods in relation to major marine transgressions, further adding to the organic removal rate of phosphorus, via burial of organic matter, during interglacial periods (Broecker, 1982a)—recall the Miocene phosphorite deposits as well. The Permian Phosphoria deposit alone contains 7×10^{11} tons P within an area of 3.5×10^5 km^2. This is equivalent to about seven times the total amount of dissolved phosphorus in the sea today contained in only 0.1% of its present surface area, the area of coastal upwelling ecosystems (Ryther, 1969) or salt marshes (Table II).

Unlike the carbon and nitrogen cycles, phosphorus fluxes between the atmosphere and the dissolved PO_4 ocean reservoir (1×10^{11} tons P) are negligible. Other than previous removal of organic phosphorus by anchovy-feeding sea birds (Fig. 121) to form guano deposits on a few islands (Hutchinson, 1950), the transfer of phosphorus from ocean to land and back again has to depend on weathering and uplifting at very long geological time scales; we shall see in the next chapter that pristine rivers delivered 10-fold less phosphorus to the sea 100 yr ago. Without significant renewal to the sea of phosphorus at the longer time scale of 5000 yr, one wonders why the rest of the dissolved phosphate of the ocean was not stripped after 10,000 BP, with concomitant further increase of atmospheric CO_2.

Sea level at the end of the last deglaciation had increased by 100 m from 15,000 to 10,000 BP—that is, within 40 m of today's value (Fig. 164). The U.S. Atlantic shoreline 10,000 yr BP retreated at perhaps 200 m yr^{-1}, compared to about 5 m yr^{-1} 5000 yr ago, and only 1.5 m yr^{-1} over the last ~ 200 yr (Kraft et al., 1979; Belknap and Kraft, 1981). Shelf ecosystems at a present water depth

Table XIV.

Comparison of Carbon and Nitrogen Fluxes in
Modern and Glacial Ecosystems[a,c]

	Modern		Glacial[b]	
	Marine	Marsh	Marine	Marsh
Net primary production	26.37	2.00	20.84	28.40
Nitrogen fixation	1.04	2.69	0.47	38.20
Sediment organic carbon sink	0.71	0.20	0.20	2.84
Denitrification loss	8.50	7.80	0.00	110.80
Methane emission	0.83	3.90	0.39	55.38

[a] Net primary production and sediment organic carbon sink in 10^9 tons yr^{-1}; nitrogen fixation, denitrification, and methane emission in 10^7 tons yr^{-1}. Values based on areas given in Table II.

[b] Assumes that total marsh area increased to 2.84×10^7 km² after disappearance of shelves, coral reefs, seaweed beds, and estuaries, with no export to the continental slopes.

[c] After Walsh (1984); © *BioScience*.

of 40 m could thus still have been as much as 100–500 km from the shelf break supply of nutrients 10,000 yr ago, with consequent lower overall primary production. Recall the annual primary production of 50 g C m^{-2} yr^{-1} of the inner Bering shelf, compared to 165 g C m^{-2} yr^{-1} on the outer shelf, discussed in Section 4.2. At such large distances from the shelf break, most of the inner-shelf phytodetritus would also not have escaped consumption to be buried in slope depocenters. More of the shelf primary production would then have been derived from the rapid return of recycled nutrients during 10,000 BP, compared to the major shelf-break source during 15,000 BP.

In addition to the suspected changes in shelf production, export, nutrient cycling, and the adjacent slope storage of carbon, nitrogen, and phosphorus during the Holocene transgression of present shelves, the transition from glacial to interglacial habitats would have had other impacts on global ecology. With shelf exposure during glaciations, most shelf carbon production, denitrification, and export of organic matter would have stopped. Rivers would have instead discharged their dissolved nutrients and suspended solids to the open ocean rather than to estuaries or shelves, while the coastal marshes might then have covered more than 14 times their present area (Table XIV). Ancient freshwater peats, oysters, and even mastodon teeth, for example, are found on the outer part of the present mid-Atlantic continental shelf (Whitmore *et al.*, 1967). Here productive macrophytes of ancient marshes might have doubled the total glacial primary production and carbon storage (Table XIV), causing the lower atmospheric CO_2 concentrations of the Wisconsin period and possibly inducing even greater changes in the nitrogen cycle.

If a nitrogen outwelling rate of 37 g N m^{-2} yr^{-1}, derived from present South Carolina marshes (Kjerfve and McKellar, 1980), is applied to an increased area of 2.84 × 10^{13} m^2 (Table XIV), as much as 1 × 10^9 tons N yr^{-1} might have been released to the sea from glacial marshes. This is more than 10 times the present estimated anthropogenic inputs. The glacial value of deep-water nitrate might then have been 48 μg-at. NO$_3$ l^{-1}, rather than today's 32 μg-at. NO$_3$ l^{-1} (Walsh, 1984). During interglacial periods, the nitrogen fixed and buried previously by glacial marshes might also then be returned to the ocean, after shore erosion by the slowly rising sea, counteracting the prior nutrient depletion during deglaciation.

As we shall see in the next chapter, the atmospheric methane concentrations (4.8 × 10^9 tons) have increased by 150 parts per billion (ppb) over the last decade (see Fig. 177), or 1.7% annually (Rasmussen and Khalil, 1981), contributing to as much as 23% of the CO_2 greenhouse effect (Lacis et al., 1981). Changes in atmospheric CH_4 input are thought to be mainly biogenic, with only 10% attributed to fossil fuel production and consumption (Sheppard et al., 1982). At present rates per unit area (Cicerone and Shetter, 1981; Harriss and Sebacher, 1981), extensive glacial marshes might have emitted almost 50% of today's total CH_4 fluxes. Such an emission rate would act as a negative feedback mechanism, returning global element cycles to their interglacial state by enhancing the greenhouse effect of the glacial atmosphere.

Admittedly, these scenarios for the past transfer of carbon, nitrogen, and phosphorus, to and from the sea, are still speculative, despite the enormous amount of recent information embodied in the above data sets. But it is clear that since the Industrial Revolution, humans have altered the carbon input to the atmosphere by 10 times the "rapid changes" at the beginning of these deglaciation and interglacial periods, as well as the input of nitrogen and phosphorus to the ocean in both dissolved and particulate forms. If the present burial rate of organic matter on the continental slope is the same as that estimated for ~13,000–15,000 BP, then it might absorb some of these anthropogenic fluxes; if not, global carbon and nitrogen budgets would remain unbalanced. Continued burial of organic matter on the continental slope at these rates, moreover, does require anthropogenic resupply of nutrients to avert another decline in marine primary production and a further increase in atmospheric CO_2 concentrations.

Despite enormous changes in the burial of organic carbon in the Cretaceous, or of phosphorite in the Permian, atmospheric CO_2 and O_2 levels are thought to have changed little over the last several 100 million years (Fig. 163). Such homeostasis is attributed to the effects of the relatively rapid biological processes of marine photosynthesis and respiration (Garrels and Perry, 1974). Enhanced primary production of tropical forests during interglacial periods (Shackleton, 1977), for example, would also mitigate the increasing atmospheric CO_2 levels at the beginning of the Holocene, with a drawdown of CO_2 by the terrestrial biosphere—that is, another negative feedback of Gaian dimensions (Lovelock, 1979).

If these natural processes were to be disrupted or overwhelmed by humanity's direct carbon and nitrogen interactions with the atmosphere, however, future global oscillations of these elements might become greater, perhaps leading to faster extinction of marine and terrestrial species. With a 10-fold increase of world population from 260 million people in A.D. 14 (Botkin and Keller, 1982) to 3.6 billion humans by 1970, humanity now directly impacts the global ecology of Earth by overfishing, wetlands eradication, deforestation, and other anthropogenic accelerations of the carbon, nitrogen, and phosphorus cycles.

6 Alteration

It was the best of times, it was the worst of times, it was the age of wisdom, it was the age of foolishness. . . . For the streets were so full of dense brown smoke that scarcely anything was to be seen. . . . We drove slowly through the dirtiest and darkest streets that ever were seen in the world (I thought), and in such a distracting state of confusion that I wondered how the people kept their senses.

(C. Dickens, 1853, 1859)

Then a strange blight crept over the area and everything began to change. Some evil spell had settled on the community: mysterious maladies swept the flocks of chickens; the cattle and sheep sickened and died. Everywhere was a shadow of death. . . . On the farms the hens brooded, but no chicks hatched. The farmers complained that they were unable to raise any pigs—the litters were small and the young survived only a few days. The apple trees were coming into bloom but no bees droned among the blossoms, so there was no pollination and there would be no fruit. . . . Even the streams were now lifeless. Anglers no longer visited them, for all the fish had died. . . . In the gutters under the eaves and between the shingles of the roofs, a white granular powder still showed a few patches; some weeks before it had fallen like snow upon the roofs and the lawns, the fields and streams. . . . No witchcraft, no enemy action had silenced the rebirth of new life in this stricken world. The people had done it themselves.

(R. Carson, 1962)

The times in which we now live represent an anomaly, both climatologically and ecologically. For 90% of the past 2 million yr, the Earth's climate has been colder than today's Holocene interglacial period. During the 10,000 yr since the last major glaciation, humanity has emerged as a runaway species, so far escaping environmental constraints on population abundance that the "normal" cycling of several elements has been accelerated.

Human abundance is projected to double by the year 2000, with continued consumption of oil and coal. Present release rates of CO_2 to the atmosphere from fossil-fuel burning, cement production, and deforestation are perhaps large enough to double atmospheric carbon dioxide content by the year 2035. The increases of this and other gases—methane (CH_4), nitrous oxide (N_2O), and the freons (CCl_3F, CCl_2F_2)—may result in a concomitant 3°C rise in ocean temperature, almost 10 times that over the last century and similar to the temperature increase of the sea since the last major glaciation (Hansen et al., 1981). Over the last 100 yr, the ocean's mean temperature may have risen 0.5°C (Fig. 169).

Three times less area of the Earth is now covered with ice sheets than during the glacial periods, resulting in about a 3% increase in the ocean's volume since then (Milliman and Emery, 1968). A further melting of ice sheets, similar to the 0.1-cm yr^{-1} sea-level changes over the last century, would result in a sea-level rise of only 0.3–0.6 m by the year 2050 (Gornitz et al., 1982). If the West Antarctic Ice Sheet were to slide off into the Southern Ocean in response to an accelerated anthropogenic temperature rise, however, sea level would quickly rise some 6 m, drowning most coastal cities.

The human perturbation of natural terrestrial nutrient cycles and adjacent aquatic ecosystems began over 2000 yr ago. The rate of nitrogen and carbon deposition in Italian lake sediments increased 10-fold from Roman road construction and deforestation in 171 B.C. (Hutchinson, 1970). This was a small-scale phenomenon in these times, since European forests stretched so far in the drainage basins of the Rhine and Danube rivers that "no one in western Germany would claim to have reached the eastern edge of the forest even after traveling for 60 days, or to have discovered where it ends" (Caesar, 53 B.C.).

After Roman times, around A.D. 1500, similar increases in aquatic sedimentation rates occurred on the upper slope of the Black Sea. A 10-fold change took place between 800 B.C.—A.D. 200 and Medieval times (Degens et al., 1976), presumably in response to conversion of previous Hercynian forests to steppes (Kempe, 1979). Similar deforestation of eastern North America during its colonial period led to a 10-fold increase in suspended solids in the rivers (Meade, 1982) and perhaps in estuarine nitrogen. With completion of colonial deforestation by 1850, subsequent farming, and today's addition of urban wastes, for example, the mean particulate nitrogen in Lake Erie sediments increased sevenfold between 1850 and 1970 (Kemp et al., 1974).

From 1930 to 1970 the nitrogen content of western Lake Erie increased by an order of magnitude to a winter–spring maximum of 30–60 μg-at. N l^{-1} (Burns, 1976), similar to the nitrogen content of the Mississippi River in 1970 and 10 times that of the pristine Amazon, Zaire, or Yukon rivers (Walsh et al.,

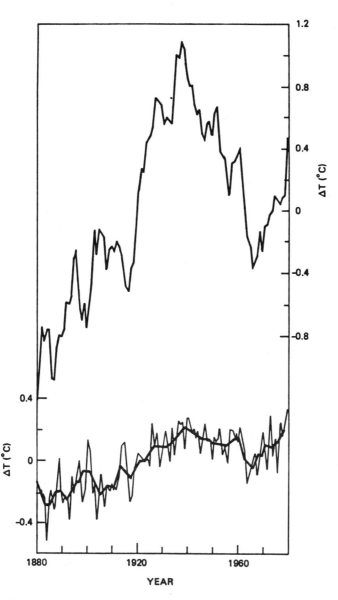

FIG. 169. Measured surface temperature (°C) trends for the period 1880–1980 as a global mean (lower curves, annual and 5-yr running mean) and for stations poleward (upper curve, 5-yr running mean) of latitude 64°N. [After Hansen *et al.* (1983); © 1983 AAAS.]

1981). By this time, phytoplankton numbers had increased 20-fold in the western basin (Leach and Nepszy, 1976), and the mean annual primary production of Lake Erie was about 250 g C m^{-2} yr^{-1}—over two to four times that of the relatively oligotrophic Lakes Huron and Superior (Vollenweider et al., 1974), but similar to that of most continental shelves (Table III). As much as one-third of the lake's annual carbon production in 1970–1971 went unconsumed, accumulating instead on the bottom (Kemp et al., 1976).

Estimates of the sedimentation rate within Lake Erie at that time were 10 times that of the mean Holocene rate (Sly, 1976). In the last decade, the amount of carbon and nitrogen sequestered within organic matter on the bottom of Lake Erie may have doubled again (Fisher and Matisoff, 1982). The magnitude of these carbon and nitrogen perturbations has grown from local phenomena, such as Lake Erie, to impacts on a global scale. To place human activities in proper perspective, we now consider global fluxes of carbon and nitrogen, between their atmospheric, terrestrial and marine reservoirs with respect to burning fossil fuels, changes in land use, coastal eutrophication, and overfishing.

6.1   Atmospheric Forcing

Above an Earth without life, the atmospheric CO_2 content would be ~98% of the gaseous composition, similar to those of Mars and Venus (Table XV). In contrast, an Earth with the present biota has an atmospheric CO_2 content of only ~0.03%, an oxygen content of ~21%, and a nitrogen content of 78%; trace gasses such as methane, nitrous oxide, and freons constitute as little as ~10^{-4}%, 10^{-5}%, and 10^{-8}% of the gaseous composition. As a result of an increased CO_2 content above a silent Earth envisioned by Rachel Carson (1962), the near-surface temperature might be ~290°C instead of 13°C today (Table XV).

Such a hypothetical increase of the Earth's temperature would result from an absorption of the outgoing blackbody radiation at 8–13 μm infrared wavelengths by CO_2 as well as by CH_4, N_2O, CCl_2F_2, and CCl_3F. At 1980 concentrations of

Table XV.

Atmospheric Composition (%) and Surface Temperature (°C) of Venus, Mars, Today's Earth, and an Earth without Biota[a]

	Venus	Mars	Earth	Abiotic Earth
Carbon dioxide	98.00	95.00	0.03	98.00
Nitrogen	1.90	2.70	78.00	1.90
Oxygen	Trace	0.13	21.00	Trace
Argon	0.10	2.00	1.00	0.10
Surface temperature	477	−53	13	290

[a] After Lovelock (1979); © 1979 Oxford University Press.

CO_2 (339 ppm), CH_4 (1.6 ppm), N_2O (0.3 ppm), CCl_2F_2 (0.00028 ppm), and CCl_3F (0.00018 ppm), perhaps half of this radiation trapped in the atmosphere is effected by CO_2 molecules and the rest by these trace gasses. Within year-to-year variations, smoothed by the simplest filter of a 5-yr running mean, the global mean temperature of the ocean appears to have actually increased $\sim 0.5°C$ from 1880 to 1980 (Fig. 169).

Over the same century, between the London "particulars" of dense coal smoke described by Charles Dickens and the hypothetical fallout described by Rachel Carson, the chemical content of the atmosphere has changed considerably. In addition to natural, time-dependent signals of varying biological response to seasonal, decadal, or millenial changes of the physical habitat, human activities of deforestation, fossil-fuel combustion, and chemical synthesis have reached such proportions since the Industrial Revolution that anthropogenic chemical signals can also be detected within the atmosphere—that is, increases of CO_2, CH_4, N_2O, CCl_3F, and CCl_2F_2. Similar to experimental manipulation of agricultural plots or of small ponds, human indirect perturbation experiments on land can be traced through the atmosphere to the sea, allowing further analysis of shelf food webs, since the atmosphere and ocean are coupled systems. Additional direct perturbation experiments on the shelf ecosystems over the last 100 yr (i.e., overfishing at the top of the food web and eutrophication at the bottom) perhaps provide the necessary boundary conditions for solving the role of continental shelves in maintaining an Earth's atmosphere, in contrast to that of Venus or of Mars (Table XV).

6.1.1 Carbon Dioxide

The carbon dioxide content above Earth evidently decreased from a possible Cretaceous atmosphere of ~ 1300 ppm CO_2 85–100 million yr ago (Fig. 163c) to a Wisconsin atmosphere of ~ 200 ppm CO_2 15–20 thousand yr ago (Fig. 168). Since then, the atmospheric CO_2 content increased to 275 ppm about 10,000 yr ago (Fig. 168) and remained at that level until A.D. 1800 (Oeschger, 1985), with another increase in concentration after onset of the Industrial Revolution between 1750 and 1850 (Neftel et al., 1985). From 1850 to 1958 (Fig. 170), atmospheric CO_2 concentrations apparently increased from 275 ppm measured within another ice core, taken at 76°S, to 315 ppm at the beginning of the accurate air measurements, taken at Mauna Loa, Hawaii. Furthermore, over the next 25 yr (Fig. 171), CO_2 concentrations increased above Hawaii by about 1.0 ppm yr^{-1} to 340 ppm in 1982. This annual rate of increase is ~ 20 times the "rapid" atmospheric increment of 0.05 ppm yr^{-1} of CO_2 established between 13,000 and 11,000 BP during the last deglaciation, subsequently corroborated by other deep sea cores (Shackleton et al., 1983), and is about twice that measured from 1850 to 1950 in ice cores.

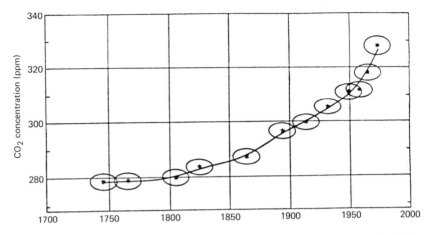

FIG. 170. Atmospheric CO_2 concentrations (ppm) measured in glacier ice formed during the last 200 yr, calibrated against the Mauna Loa record (FIG. 171) for the youngest gas sample. [After Neftel *et al.* (1985); © 1985 Macmillan Journals Ltd., with permission.]

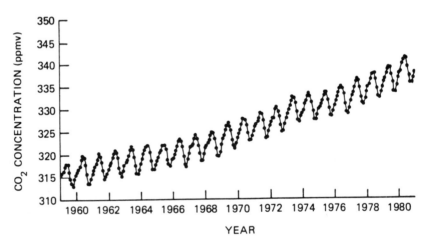

FIG. 171. Trend in atmospheric CO_2 concentration (ppm) at Mauna Loa, Hawaii over the last 25 yr. [After Keeling *et al.* (1982); © 1982 Oxford University Press.]

Ignoring, for a moment, the interannual and seasonal variations of the Mauna Loa data (Fig. 171), the long-term trend of increasing CO_2 in the atmosphere is attributed mainly to the exponential consumption of fossil fuel over the past century (Fig. 172). Following the original suggestions of the possible greenhouse effect of atmospheric gasses by Jean Fourier in 1827 and John Tyndall in 1861, for example, Savante Arrhenius first made the calculation that if the at-

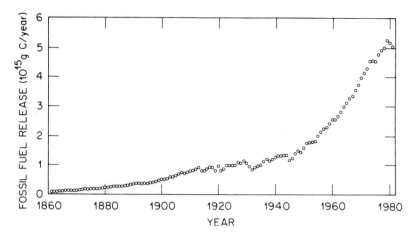

FIG. 172. Fossil-fuel CO_2 emissions ($\times$ 10^9 tons yr^{-1}) since the Industrial Revolution: 1860–1982. [After Keeling (1973), Marland and Rotty (1984).]

mospheric concentration of CO_2 in his time would increase threefold, the Earth's temperature might warm by 9°C (Arrhenius, 1896); like Chamberlin (1899), he thought that CO_2 concentrations were lower in the glacial atmosphere. Burning of coal in his and Charles Dickens's time led to emissions of only <0.5 $\times$ 10^9 tons C yr^{-1} to the atmosphere (Fig. 172), however, compared to 5.2 $\times$ 10^9 tons C yr^{-1} in 1980 from burning oil, natural gas, and coal—a 10-fold increase.

The fate of CO_2 emitted from burning of fossil fuel, from cement production, and from deforestation since the Industrial Revolution is unknown and remains a hotly debated issue. Meteorologists, agronomists, terrestrial ecologists, limnologists, and oceanographers are presently unable to specify either the steady-state annual fluxes of carbon in a global budget, or the nature of transients between the major storage pools (atmosphere, land, and ocean) over the last century. At the present rate of increase in fuel consumption of 4.3% yr^{-1}, a doubling of the CO_2 content of the atmosphere and a concomitant 2–3°C increase in the ocean's temperature could occur by the year 2035. This is somewhat less than Arrhenius's (1896) predictions, based on water vapor as well, but neither calculation includes the additional input of trace gases. From a number of time-series observations of atmospheric CO_2 at Mauna Loa (Fig. 171) and another 21 areas (e.g., Point Barrow, South Pole, and Station P) over the last 25 yr (Keeling *et al.*, 1982; Komhyr *et al.*, 1985), it is generally agreed that of the 5.2 $\times$ 10^9 tons C yr^{-1} now emitted from consumption of fossil fuel and cement production, about half remains in the carbon pool of the atmosphere (Fig. 173).

A rapid equilibrium exists at the air–sea interface between CO_2 in the atmosphere and in the upper ocean, such that CO_2 can easily penetrate the upper mixed layer (~75 m) of the sea. The observed gradients of CO_2 partial pressure

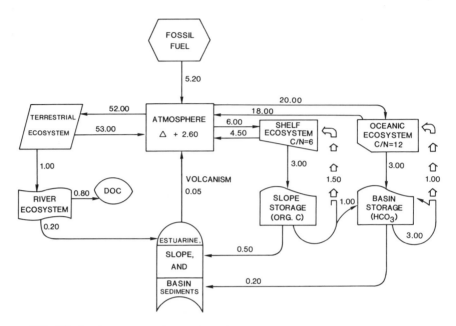

FIG. 173. Steady-state annual fluxes (10^9 tons C yr^{-1}) of a postindustrial 1980 carbon budget.

(Takahashi and Azevedo, 1982) between the atmosphere and the sea are such that a net uptake of CO_2 occurs over most of the open ocean, except for the equatorial region between 10°N and 20°S (Table XVI), as a result of primary production in the surface mixed layer. Because of the photosynthetic sink of surface waters, however, average total inorganic carbon concentrations actually increase with depth to a maximum at ~1500 m (Hoffert et al., 1981). It is thus not clear how the CO_2 of the upper mixed layer could diffuse down across the main thermocline into the deeper open ocean, but it certainly sinks with both biogenic particle formation and during bottom water formation at the polar boundaries of the Antarctic and Norwegian–Labrador–Greenland Seas (Table XVI).

The dissolved flux of CO_2 into the ocean from the atmosphere at the poles is apparently balanced by equatorial outgassing (Table XVI), albeit to and from different depths in the sea, such that a net marine CO_2 uptake of 2.1 × 10^9 tons C yr^{-1} (Broecker et al., 1979) must occur via particle settling. If an annual marine primary production of ~26 × 10^9 tons C yr^{-1} (Table II) is partitioned into 20 × 10^9 tons C yr^{-1} production of the open ocean (~3 × 10^8 km^2) and 6 × 10^9 tons C yr^{-1} production of the continental shelves (~3 × 10^7 km^2), the source term for at least the net uptake of CO_2 in the open ocean emerges from disciplinary rhetoric. Approximately 90% of the open ocean's primary production is consumed in the upper 100 m of the water column, with the fixed particulate carbon recycled back to the atmosphere as CO_2 from respiration (Fig. 173) and the fixed particulate nitrogen returned to seawater as dissolved nitrogen compounds from excretion. Recall that the ^{15}N estimates of the daily nitrogen source of primary production suggest that 90% of this nitrogen demand is met by such recycled ammonium, urea, or amino acids.

Most of the organic detrital particles that sink out of the euphotic zone ($\sim$100 m) of the open ocean, $\sim$2 $\times$ 10^9 tons C yr^{-1} (Fig. 173), are then oxidized within the oxygen minimum layer between 1000 and 2000 m, to be eventually sequestered within the inorganic bicarbonate (HCO_3^-) cycle. A small amount of this biogenic CO_2 is slowly diffused upward toward the euphotic zone of the gyres (Hoffert et al., 1981) at presumably the same rate as the "new" nitrate, supporting this net extraction of CO_2 from the atmosphere. With such penetration of atmospheric CO_2 to the deep sea beneath the main thermocline via a biological "pump", it can then be stored as additional HCO_3^- after dissolution of the calcium carbonate sediments below the main lysocline at $\sim$3700 m (Broecker et al., 1979).

It is estimated (Table XVI) that $\sim$37% of the CO_2 emitted from the burning of fossil fuel, $\sim$2.1 $\times$ 10^9 tons C yr^{-1}, might be sequestered as HCO_3^- in this fashion, after outgassing of CO_2 from the equatorial ocean is subtracted from the polar inputs. Approximately 0.5 $\times$ 10^9 tons C yr^{-1} is a "missing" sink within such a global chemical carbon budget, however, without consideration of any other source terms of CO_2. Note that the continental shelves were also not considered by Takahashi and Azevedo (1982) in Table XVI, since few measurements were made in these regions.

Table XVI.

Net CO_2 Exchange Across the Air–Sea Interface[a]

Region	Area $\times 10^6$ km^2 (%)		ΔpCO_2[b] (μatm)	Flux[c] ($\times 10^9$ tons C yr^{-1})
Temperate gyres	219	(61.0)	-14	-2.3
Antarctic (>50°S)	62	(17.0)	-7	-0.3
Equatorial (10°N–20°S)	40	(11.0)	$+30$	$+0.9$
Norwegian–Labrador–Greenland Seas	7.2	(2.0)	-100	-0.4
Mediterranean Sea	3.0	(0.8)	0	0
Red Sea	0.4	(0.1)	$+25$	$+0.01$
Asiatic seas	8.0	(2.0)	?	?
Arctic seas	14.0	(4.0)	?	?
Other seas	7.4	(2.1)	?	?
Total	361	(100.0)	-8	-2.1

[a] After T. Takahashi, personal communication (1982).
[b] $\Delta pCO_2 = pCO_2$ (seawater) $- pCO_2$ (air).
[c] Flux assumes a gas-exchange coefficient of 0.0615 mol CO_2 m^{-2} μatm^{-1} yr^{-1}.

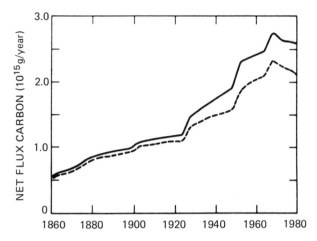

FIG. 174. The annual net release of carbon ($\times$ 10^9 tons C yr^{-1}) from the terrestrial biota and soils, based on different assumptions of tropical biomass. [After Houghton (1986); © Springer-Verlag.]

The land biota used to be considered as only a sink for atmospheric CO_2, but as a result of changing patterns of deforestation, the terrestrial ecosystem has now become a net source of CO_2 within global biological carbon budgets. When forests are cleared to introduce agricultural crops, for example, trees are either rapidly oxidized by burning (Wong, 1978), or slowly oxidized by decay, in conjunction with some of the previous soil organic matter. Tropical deforestation is now the main source of terrestrial carbon to the atmospheric and oceanic reservoirs of CO_2 (Houghton *et al.*, 1985).

The amount of such a terrestrial carbon flux to the atmosphere, or ocean, is still a subject of some controversy (Woodwell *et al.*, 1978; Broecker *et al.*, 1979), similar to previous arguments about the amount of harvestable carbon as fish from the sea (Alversen *et al.*, 1969; Ryther, 1969). Recent analyses (Houghton, 1986) have converged, however, to an estimate of a terrestrial source of $1-2 \times 10^9$ tons C yr^{-1}, based on changes in land use (Fig. 174). Other recent studies of δ ^{13}C content in tree rings (Stuiver *et al.*, 1984; Peng and Freyer, 1986) suggest that the input of CO_2 from terrestrial ecosystems might be only 0.8×10^9 tons C yr^{-1}, compared to earlier ^{13}C/^{12}C estimates of $1.2-1.7 \times 10^9$ tons C yr^{-1} (Stuiver, 1978; Peng *et al.*, 1983).

Part of this carbon input from the land may be in the intermediate forms of both dissolved organic carbon (DOC) and particulate detrital carbon, leached into riverine systems, rather than all in the form of CO_2 gas directly released to the atmosphere. As much as 0.8×10^9 tons C yr^{-1} in the form of DOC may be released annually to the shelves from the world's river systems (Richey *et al.*, 1980). An oxygen budget of the New York Bight (see Table XXII) suggests, however, that this terrestrial DOC source could be respired within coastal waters and eventually released as CO_2 to the atmosphere, rather than just being diluted by offshore waters of lower DOC content.

Furthermore, DOC within the Amazon River has a recent ^{14}C bomb label (Hedges et al., 1986), but the age of subsurface marine DOC is instead ~3400 yr (Williams et al., 1969), implying slow transfer of surviving terrestrial DOC to the deep ocean. Estimates of the sedimentation of particulate matter of terrestrial origin within estuarine ecosystems suggest that an accumulation rate of 0.2×10^9 tons C yr^{-1} (Deuser, 1979) might occur within these nonmarine sediments (Fig. 173). If the land is indeed a direct source of atmospheric CO_2 of only $\sim 1 \times 10^9$ tons C yr^{-1}, a total flux of at least 1.5×10^9 tons C yr^{-1} could be "missing," or overlooked. Such a steady-state global carbon budget contains fossil-fuel and land source terms, as well as atmospheric, DOC, estuarine, and open-ocean sinks, but no shelf biota as another sink term.

Phytoplankton populations of the continental shelf, in contrast to those of the open ocean, do not appear to be nitrogen-, or grazer-, limited for parts of the year. This is evidenced by the spatial and temporal outbreaks of algal blooms within spring coastal waters, discussed in Chapter 3. In further contrast to experimental data for the oligotrophic open ocean, ^{15}N estimates of nitrogen uptake by coastal phytoplankton suggest that 50% of the daily nitrogen demand of photosynthesis is then met by nitrate off Peru and Mexico (MacIsaac and Dugdale, 1972), in the Middle Atlantic Bight (Conway and Whitledge, 1979), in the southeast Bering Sea (Dagg et al., 1982), off northwest Africa (Whitledge, 1978), off California (Eppley et al., 1979), and in the Antarctic (Olson, 1980).

Since nitrate supply is considered an estimate of available "new" production (Dugdale and Goering, 1967; Eppley and Peterson, 1980), since fish yield is not a significant particulate nitrogen loss, and since plankton biomass is not continually increasing on the continental shelf, an upper bound of ~50% of the primary production could be exported from the shelf water column for part of each year. This is opposed to ~10% loss from the open-ocean water column (Fig. 173), and suggests that the maximum shelf export might be $\sim 3 \times 10^9$ tons C yr^{-1}. The C/N content of phytoplankton exported from nitrogen *unlimited* blooms of the coastal zone is ~5–6 : 1, however, in contrast to either higher values of 10–12 : 1 for the detrital particles left behind in the open ocean, or in slope bottom sediments with a C/N content of 8 : 1 (Fig. 126e).

This change in C/N ratios, the sediment-trap fluxes, the carbon accumulation rates of surficial sediments, and the models of particle trajectories all tell us that an unknown amount of this maximum shelf export is consumed, before it reaches the sea bottom of the adjacent continental slope. If, for example, 50% of the possible carbon export from the shelves were left behind as respired CO_2 above the main thermocline of slope regions, another annual particle sink of 1.5×10^9 tons C yr^{-1} beneath the slope thermocline would provide a balanced global carbon budget (Fig. 173). Additional data on sediment metabolism of carbon provide some insight into how we might partition this shelf biotic sink into both buried organic particles and more biogenic CO_2, released by benthic remineralization and entrained laterally within the open-ocean storage of HCO_3^- below the oxygen minimum layer.

Three major processes account for carbon consumption in ocean sediments: sulfate reduction, nitrate reduction, and oxygen metabolism. Accounting for each mole of H_2S produced in the sea, basically 2 mol organic carbon are oxidized to CO_2 during sulfate reduction—recall this process sketched in Eqs. (213) and (214). About 67.5×10^7 tons yr^{-1} of organic carbon may be consumed in the production of 90×10^7 tons S yr^{-1}, now thought to be released by marine sulfur bacteria in slope and other sediments (Ivanov, 1977). Similarly, nitrate reduction involves oxidation by denitrifying bacteria of essentially 2 mol C for each mole of N_2 produced, or 9.4×10^7 tons C yr^{-1} consumed for the 5.5×10^7 tons N_2 yr^{-1} now estimated to be evaded from slope ecosystems (Table II). Within oxygenated waters, however, only 1.3 mol O_2 is required by aerobic bacteria to turn 1 mol plankton carbon into CO_2. This means that 20×10^7 tons C yr^{-1} may now be oxidized from extrapolation of sediment oxygen uptake rates of 16 mg O_2 m^{-2} day^{-1} within most slope regions (Smith and Teal, 1973; Smith, 1978; Murray and Grundmanis, 1980).

With these assumptions, 1×10^9 tons C yr^{-1} might now be consumed in slope depocenters to be laterally sequestered as biogenic CO_2 within part of the open ocean's bicarbonate sink (Fig. 173). The remaining 0.5×10^9 tons C yr^{-1} is now estimated to accumulate today on continental slopes, to be perhaps dispersed downslope by slumping events to the continental rise, for either additional oxidation or for burial. If a possible slope burial of ~0.5×10^9 tons C yr^{-1} is not also contributing to observed increases of atmospheric CO_2 (Fig. 171), more nutrients must be supplied to shelf ecosystems in order to replace those removed with the burial of organic carbon on the slope. Increased seasonal oscillations of atmospheric CO_2 values suggest, in fact, that part of the biosphere's productivity may have been enhanced during the last decade.

A seasonal maximum of atmospheric CO_2 at Mauna Loa in May and a minimum in October (Fig. 171), as well as similar trends at Point Barrow, at Station P in the North Pacific, and in the North Atlantic (Pearman and Hyson, 1980), have mainly been attributed to sequential processes of photosynthesis and respiration within terrestrial plant communities of the northern hemisphere (Lieth, 1963). Seasonal heating and cooling of ocean waters—that is, changes in solubility of CO_2—would instead lead to a release of CO_2 to the atmosphere from the sea during summer and uptake in winter, mitigating these biospheric effects. Oceanic temperature-induced oscillations of atmospheric CO_2 in the northern hemisphere both have 10% of the amplitude of the biosphere-induced oscillations and are out of phase (Pearman et al., 1983). In contrast, temperature and biospheric oscillations of CO_2 in the southern hemisphere are of similar amplitude, with perhaps a dominance of seasonal variance by the oceanic source (Mook et al., 1983; Keeling et al., 1984).

Between 30 and 70°N latitude, the seasonal amplitude of atmospheric CO_2 is $\sim 12-14$ ppm, whereas from 10 to 90°S latitude the seasonal change is <2 ppm (Komhyr et al., 1985). Initial attempts (Hall et al., 1975) to detect long-term changes in such seasonal signals of biospheric activity failed. However, more recent analyses of the CO_2 time series at Mauna Loa, Station P, and Point Barrow (Pearman and Hyson, 1980; Cleveland et al., 1983; Keeling et al., 1984; Komhyr et al., 1985) indicate an increase in seasonal amplitude of $1-2\%$ yr^{-1}, that is, increased plant activity.

Assuming that changes in seasonal use of fossil fuels are small, as much as 70% of such an increase is attributed to biotic effects (Gillette, 1982), perhaps the result of CO_2-enhanced terrestrial photosynthesis (Cleveland et al., 1983). The percentage increase of CO_2 seasonality within the northern hemisphere is twice that of the annual increase of CO_2 levels, however, such that fertilization of the terrestrial biosphere by higher CO_2 concentrations is probably a small factor. Furthermore, most of the change in terrestrial carbon fluxes is taking place within the tropics, where little seasonal signature occurs. Increased terrestrial respiration during warmer winters could lead to the same amplitude increment as from enhanced photosynthesis, but data are not yet available to test this hypothesis.

Large-scale general circulation models (GCMs) have assumed thus far that marine biotic effects on seasonal atmospheric CO_2 fluxes are small (Fung et al., 1983), with no consideration of the continental shelves (recall Table XVI). Comparison of the seasonal CO_2 cycles at weather ships in the midocean (e.g., 66°N, 2°E, or 50°N, 145°W) and at coastal stations between 40 and 70°N suggests, however, that the seasonal amplitude of atmospheric CO_2 is now $3-5$ ppm greater at the edge of the North American continent than above the open ocean (Komhyr et al., 1985). This could just reflect a zonal gradient of atmospheric CO_2 between the gyres and midcontinents as a result of the terrestrial CO_2 fluxes (Fung et al., 1983), or an additional removal of CO_2 by spring blooms within coastal waters.

In the southeastern Bering Sea, the pCO_2 of surface waters in March is 339 ppm, similar to that of the atmosphere, and is reduced to <125 ppm by the end of the spring bloom in June, when a mean CO_2 invasion rate of 0.7 g C m^{-2} day^{-1} occurs (Codispoti et al., 1982, 1986a). As we shall see in Section 6.2.1, the nitrogen content of the Rhine River (see Fig. 180) has doubled over the last decade, with perhaps part of the increased amplitude of the seasonal CO_2 signal within the northern hemisphere attributed to coastal eutrophication and prolongment there of the spring blooms (see Figs. 184–186). Recall from Table III that the impacted shelf ecosystems, eutrophied by humans, are located between 30

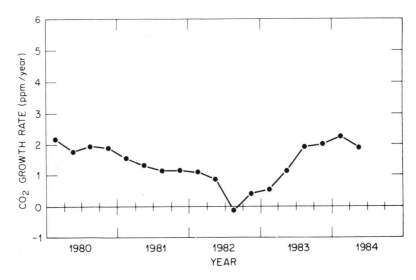

FIG. 175. Year-to-year global atmospheric CO_2 change (ppm yr^{-1}) by season, plotted in the forward year. Data are from the 24-station global flask network of NOAA/GMCC in which the seasonal global means are area weighted; the decade-averaged CO_2 growth rate is ~1.4 ppm yr^{-1}. [After Komhyr *et al.* (1985); © American Geophysical Union.]

and 60°N, exhibiting almost twice the annual primary production of other shelf ecosystems at similar distance from the equator.

The marine biota appear to be important over longer time scales of the atmospheric CO_2 cycle as well. During the last 1982–1983 El Niño , for example, the atmospheric increase of CO_2 stopped by June 1982 (Fig. 175), and then surpassed the mean annual rate of 1.4 ppm yr^{-1} by June 1983 (Komhyr *et al.*, 1985). It is possible that 2–3 × 10^9 tons C was stored within an unknown reservoir during 1982, since fossil fuel-burning remained about the same. Alternatively, the other inputs of CO_2 from equatorial outgassing (Table XVI) and deforestation (Fig. 174) may have declined. The most likely candidate is a relaxation of the rate of equatorial upwelling during the 1982 El Niño , since similar atmospheric CO_2 oscillations have been linked to the other 1965 and 1972 El Niños (Bacastow, 1976; Bacastow *et al.*, 1980). Furthermore, measurements of the $^{13}C/^{12}C$ ratio of atmospheric CO_2 rule out a major land source (i.e., tropical deforestation) for most of the equatorial ocean (Keeling *et al.*, 1984).

Estimates of the CO_2 released from equatorial outgassing range from 0.9 × 10^9 tons C yr^{-1} (Table XVI) to 10.9 × 10^9 tons C yr^{-1} (Bolin and Keeling, 1963). The actual outgassing flux is, of course, a function of the rate of equatorial upwelling, as well as of the surface temperature, and the biologically mediated ΔpCO_2, that is, the difference in partial pressure of CO_2 between the surface ocean and the lower atmosphere. Decreased upwelling of cold water would lower the input of nutrients (Fig. 176) as well as of CO_2, thereby lowering the primary production and increasing the surface temperature (Fig. 176), both of which would act to increase the flux of CO_2 to the atmosphere from the sea. If grazing, rather than nutrient supply, limits the primary production of the equatorial upwelling regions, however, in contrast to continental shelves (Walsh, 1976), then a change in the upwelling rate of CO_2 will be the major determinant here of variation in the outgassing flux.

Assuming a primary production of 1234 g C m^{-2} yr^{-1} and a coastal upwelling area of 1.8 × 10^5 km^2 off Peru (Chavez and Barber, 1985), an annual fixation of 0.2 × 10^9 tons C yr^{-1} is estimated, which is about half of that of the impacted shelf ecosystems between 30 and 60°N latitude (Table III). In contrast, the equatorial upwelling region extends from ~90°W to the International Date

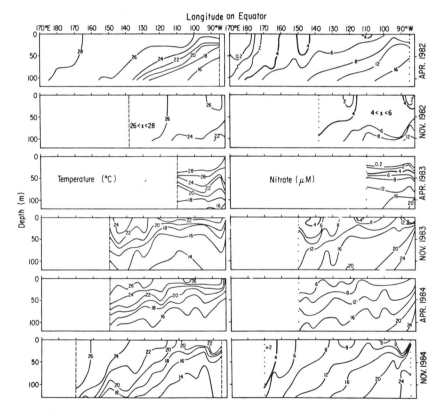

FIG. 176. Nitrate (μg-at. l^{-1}) and temperature (°C) profiles on an equatorial transect from 82°W to 170°E. The high surface nitrates during April 1982, November 1983, April 1984, and November 1984 extend typically from the coast of South America to the dateline. The November 1982 and April 1983 sections are during the last 1982–1983 El Niño and illustrate the interannual variability. [After Barber and Chavez (1986); © 1986 Macmillan Journals Ltd., with permission.]

Line at 180°W (Fig. 176), over perhaps 10° in width (Fig. 19), for a total surface area of 1.3 × 10^7 km^2 (Wyrtki, 1981). Assuming an annual primary production here of 176 g C m^{-2} yr^{-1} (Chavez and Barber, 1985), instead of 73 g C m^{-2} yr^{-1} for this region (Koblentz-Mishke et al., 1970), an areal fixation of 1.9 × 10^9 tons C yr^{-1} is estimated (i.e., 10-fold that of the Peru upwelling system).

The 1982–1983 El Niño, like the 1972 one discussed in Section 4.5.3, propagated eastward, arriving at the Galapagos Islands in August 1982 and off the Peru coast at 5°S by late September 1982. By July 1983, nutrient-rich conditions had returned to the surface waters off both Peru and the Galapagos Islands (Barber and Chavez, 1986). During March 1983, within the nutrient-depleted waters of the equator at 95°W (Fig. 176), the surface primary production of the equatorial upwelling ecosystem had only been reduced by fivefold; in contrast, coastal primary production at 5°S during May 1983 was reduced 20-fold (Barber and Chavez, 1983). In terms of total carbon extraction from the euphotic zone, the equatorial upwelling ecosystem may have been reduced to ~0.40 × 10^9 tons C yr^{-1}, compared to 0.01 × 10^9 tons C yr^{-1} within the coastal system during the 1982–1983 El Niño .

Examination of the atmospheric CO_2 time series (Fig. 175) suggests, moreover, that less than normal CO_2 was transferred to the atmosphere when the Kelvin wave of the 1982–1983 El Niño was crossing the equatorial Pacific. In contrast, more than normal CO_2 outgassed, as much as 3 × 10^9 tons C, when the wave propagated south to Peru, where the upwelling rate of CO_2 is 10-fold that of the offshore divergence (Table IV). The reduction in primary production of the equatorial upwelling ecosystem under grazer control apparently did not increase the flux of CO_2 to the atmosphere, as may have occurred within the coastal upwelling ecosystem under nutrient limitation. Return of equatorial upwelling to its normal rate by November 1983 (Fig. 176) may have also led to increased outgassing of CO_2 in the last half of 1983 (Fig. 175). Using δ ^{18}O isotope time series from annually banded corals at the Galapagos, Fanning, and Canton Islands (Druffel, 1985), one may eventually be able to unravel the biological, chemical, and physical factors leading to the decline of atmospheric carbon dioxide, as well as of methane (Khalil and Rasmussen, 1986), during these events.

6.1.2 Methane and the Freons

The other greenhouse carbon compounds, methane and the freons, have very different origins from carbon dioxide. One is the constituent of a primordial anoxic aquatic ecosystem (i.e., CH_4), while the halogenated carbon compounds were synthesized by humans within the last 60 yr. The latter are capable of lowering the ozone content of the atmosphere, producing a return to the ultraviolet radiation conditions of pre-Cambrian times.

In 1928, scientists at General Motors developed a relatively inert gas for use in thermal transfers: the freons. Since 1933, significant inputs of chlorofluorocarbons have been added to the atmosphere as a by-product of the use of solvents, refrigerants, and spray-can propellants (Molina and Rowland, 1974; Cunnold *et al.*, 1983a,b). Upon reaching the stratosphere, these compounds become reactive, with destruction of ozone (O_3) by release of the chlorine atoms within the freons. Although present in small amounts, the contribution of CCl_2F_2 and CCl_3F to the absorption of infrared radiation is also respectively one-third and one-fifth that of CH_4.

Like bomb tritium and ^{14}C, the freons have been used as tracers to estimate mixing rates within various regions of the ocean (Gammon *et al.*, 1982; Bullister and Weiss, 1983; Weiss *et al.*, 1985). Since 1975, the annual production and release of freons have not changed, with a decline in the atmospheric increase of these gases from 1980 to 1985 (Rasmussen and Khalil, 1986). Of the numerous chlorocarbon compounds in the atmosphere, moreover, only methyl chloride (CH_3Cl), at concentrations of $10^{-7}\%$, is thought to be of natural origin, and we thus consider other greenhouse gases, which are the consequence of accelerated natural cycles. We note, in passing, that each October the ozone concentrations above the South Pole become depleted, with a secular decline over the last decade, which may eventually be related to the anthropogenic release of freons.

Methane is the second most abundant form of carbon in the atmosphere, it is the most effective greenhouse gas, and it is mainly of biological origin, emitted from ruminants (Rasmussen and Khalil, 1981), termites (Zimmerman and Greenberg, 1983), rice paddies (Cicerone and Shetter, 1981), and other anoxic environments (Martens *et al.*, 1986). Biomass burning in the tropics may also be a recent source of methane (Crutzen *et al.*, 1979), while future warming of coastal regions could lead to an outgassing, from methane hydrates within continental-slope sediments, of $1-2$ ppm CH_4 over the next century (Bell, 1982). The present rate of increase of atmospheric CH_4 is $2-3$ ppm per 100 yr, a balance between these inputs and destruction by the hydroxyl radical (OH^-) in the troposphere.

Ice-core samples of methane (Craig and Chou, 1982; Khalil and Rasmussen, 1982; Stauffer *et al.*, 1985b) indicate that preindustrial concentrations in the atmosphere were half those of today (Fig. 177). Since carbon monoxide (CO) is also emitted from the burning of fossil fuels, and competes with CH_4 for oxidation by OH, part of the recent increase of CH_4 may be related to a similar increase of CO (Khalil and Rasmussen, 1984) and concomitant decline of tropospheric OH. Conversely, the observed decline of methane during the 1982–1983 El Niño may have resulted, however, from increased production of OH as a result of greater quantities of water vapor over warmer waters at the sea surface (Khalil and Rasmussen, 1986).

Increments in the populations of farm animals and termites (e.g., there are now about 15 cattle for each human), a 45% increase in area of rice paddies

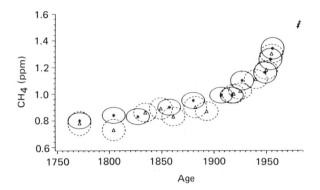

FIG. 177. Measured CH_4 concentrations (ppm) over the last two centuries from the same polar ice core shown in Fig. 170. [After Stauffer *et al.* (1985b); © 1985 AAAS.]

Table XVII.
Sources and Sinks of Tropospheric Methane[a]

Sources ($\times 10^7$ tons yr^{-1})		Sinks ($\times 10^7$ tons yr^{-1})	
Cattle	15.0		
Termites	5.0		
Human	0.4		
Rice paddies	5.5		
Marshes	3.0		
Lakes	0.6		
Tundra	0.2	OH depletion	20.5
Ocean	0.9	NO_3 depletion	2.2
Fossil fuel	3.2	Stratospheric transport	5.5
Total	33.8	Total	28.2

Net tropospheric increment of 5.6×10^7 tons CH_4 yr^{-1}

[a] After Senum and Gaffney (1985); © American Geophysical Union.

(Enhalt and Schmidt, 1978), and a greater carbon loading to anoxic sediments of wetlands and continental shelves could all lead to higher CH_4 concentrations in the atmosphere (Fig. 177). The δ ^{13}C and δ ^{14}C ratios of atmospheric CH_4 suggest (Senum and Gaffney, 1985) that about 90% of its sources are modern (Table XVII), with natural ecosystems perhaps contributing $\sim 5 \times 10^7$ tons CH_4 yr^{-1}. This is about half that estimated in Table II, and 10-fold less than during a glacial scenario (Table XIV). Uncertainties exist in such estimates, however; for example, Enhalt (1979) estimated an annual flux from these ecosystems of $\sim 27 \times 10^7$ tons CH_4 yr^{-1}. as a result of seasonal (Martens *et al.*, 1986) and interannual (Khalil and Rasmussen, 1986) changes in methane fluxes.

When bacterial decomposition of organic compounds exceeds the influx of oxygen to the water column and sediments, local anoxic habitats develop. Consumption of the oxides of nitrogen (NO_3, NO_2, N_2O, NO) and sulfur (SO_4) then proceeds, with liberation of other gases, N_2 (denitrification) and H_2S (sulfate reduction), besides CH_4 (methanogenesis). Nitrous oxide, N_2O, was once thought to also be released to the atmosphere during these anaerobic processes, but it is now considered to be mainly a product of oxidative reactions.

6.1.3 Nitrous Oxide

Unlike carbon, most of the Earth's nitrogen within the hydrosphere and atmosphere (Fig. 178) is found as the free form of N_2 in the latter (3.8×10^{15} tons). Of the small amount dissolved in the ocean (9×10^{11} tons), about 75% is in the more stable fixed form of nitrate. If the biological process of denitrification did not return N_2 to the atmosphere each year from both the ocean and the even smaller terrestrial nitrogen pool (8×10^{10} tons), lightning would lead to fixation of free atmospheric nitrogen.

It would subsequently be stored as NO_3 in a sea of much higher salinity, on an Earth without life (Table XV). From a Gaian viewpoint, a reluctantly reactant N_2 is an excellent diluent of O_2 in the atmosphere; above an atmospheric O_2 content of 25%, lightning would also lead to worldwide conflagrations, consuming even moist tropical rain forests and arctic tundra (Lovelock, 1979).

Any change in the sea's small storage pool of nitrogen must thus reflect changes in either input from rainfall, dry deposition, nitrogen fixation by phytoplankton, and land runoff, or in losses from burial, formation of nitrous oxide, and release of N_2 to the atmosphere during denitrification (Fig. 178). Gaseous N_2O is formed during the oxidation (Elkins et al., 1978) of reduced nitrogen compounds, either during nitrification (Cohen and Gordon, 1978) or by combustion of fossil fuels (Weiss, 1981). During nitrification, the ocean may also lose nitrogen in the form of nitric oxide, NO (Lipschultz et al., 1981).

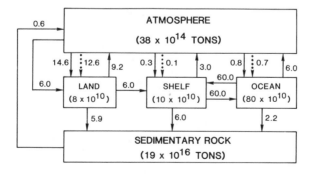

FIG. 178. A postindustrial, 1980 global nitrogen budget, showing contents of the major reservoirs and estimates of fluxes ($\times$ 10^7 tons N yr^{-1}) between them. Discrepancies between the present outputs and inputs of specific reservoirs reflect time lags in the global nitrogen budget, which may indicate a transient state of the present interglacial period.

Present estimates of the ocean's N_2O loss to the atmosphere vary by an order of magnitude (Hahn, 1981), but a reasonable value appears to be 0.5×10^7 tons N yr^{-1} (Liu, 1979). Recent annual increases of 0.4% atmospheric N_2O may have affected both ozone concentrations (Crutzen, 1981) and, together with increases of the other trace gases, methane (1.7%) and the freons (9–10%), the greenhouse response to CO_2 changes between 1970 and 1980 (Lacis et al., 1981). An atmospheric N_2O increase of 6 ppb over the last decade is equivalent to 10% of the greenhouse warming of CO_2; 200 ppm of added CO_2 plus all the trace gases, however, is equivalent to a 300 ppm CO_2 input.

Nitrous oxide is destroyed above the troposphere, within the stratosphere, by photolysis at a rate of $\sim 0.9 \times 10^7$ tons N yr^{-1}, with a lifetime of ~ 180 yr (Johnston et al., 1979). Fossil fuel burning may release 0.2×10^7 tons N yr^{-1} (Weiss and Craig, 1976; Pierotti and Rasmussen, 1976), compared to perhaps 0.2–0.4×10^7 tons N yr^{-1} added from biomass burning and cultivation of soils (Crutzen, 1983). Deforestation may be an additional source of N_2O (Bowden and Bormann, 1986), with an unknown amount derived from undisturbed terrestrial communities. At an atmospheric increase of 0.7 ppb yr^{-1} (3.5×10^6 tons N yr^{-1}), the total N_2O input must be $\sim 1.2 \times 10^7$ tons N yr^{-1}, suggesting that about 50% is supplied by human activities, 40% by the ocean, and 10% from land (Fig. 178).

Within a steady-state nitrogen budget, the loss of nitrogen from the sea by N_2O evasion would be balanced by one or more of the inputs of rainfall, fixation, or runoff. Previous global nitrogen budgets (Emery et al., 1955; Holland, 1973), for example, had suggested that the input of combined nitrogen to the sea exceeded burial losses, with a steady state maintained by the loss of the excess nitrogen through biological denitrification (i.e., evasion of N_2). But more recent analyses (Fogg, 1982; McElroy, 1983) suggest that just the N_2 losses to the atmosphere may exceed the total nitrogen input to the sea, without considering the additional N_2O evasion and organic N burial losses, implying that transient states of the marine nitrogen reservoir might occur every 10,000 yr (McElroy, 1983).

In anoxic slope waters adjacent to productive coastal regions, for example, consumption of shelf carbon export by anaerobic denitrifying bacteria produces a total N_2 loss of 5.5×10^7 tons N yr^{-1} from the eastern tropical North Pacific (Codispoti and Richards, 1976), off the Peru–Chile coasts (Codispoti and Packard, 1980), and in the Arabian Sea (Wajih et al., 1982). Denitrification rates within the low carbon sediments on the Bering (Koike and Hattori, 1979), Chukchi (Haines et al., 1981), and Washington (Christensen et al., 1987) shelves suggest an annual loss of 1.1 tons N km^{-2} yr^{-1}. In contrast, 7.4 tons N km^{-2} yr^{-1} may be lost in nearshore, carbon-rich sediments off the Belgian and Japanese coasts (Billen, 1978; Koike and Hattori, 1979), as well as within estuarine systems such as Danish fjords and Narragansett Bay (Seitzinger et al., 1985; Sorensen et al., 1979).

Salt marshes (Kaplan et al., 1979) have even higher loss rates (40.0 tons N km^{-2} yr^{-2}), while lake denitrification losses are similar to shelf losses (Pheiffer-Madsen, 1979). The annual loss of N$_2$ from the marine regions is thus estimated to be 8.5 × 10^7 tons N yr^{-1} (Tables II and XIV), compared to another 9.1 × 10^7 tons N yr^{-1} from estuaries and wetlands, that is, the land reservoir of Fig. 178. The gaseous loss of 9.0 × 10^7 tons N yr^{-1} from the oceans in the form of N$_2$O and N$_2$ (Fig. 178) is similar, moreover, to the combined burial loss of nitrogen on continental slopes of 6.0 × 10^7 tons N yr^{-1}, discussed in Section 5.2.3, and the additional 2.2 × 10^7 tons N yr^{-1} within the deep sea (Fig. 178).

Recent estimates of N$_2$ extraction from the atmosphere by nitrogen-fixing algae, *Oscillatoria* spp., in the open sea are only 0.5 × 10^7 to 1.0 × 10^7 tons N yr^{-1} (Capone and Carpenter, 1982; Fogg, 1982), compared with 4.2 × 10^7 tons N yr^{-1} fixed by humans for agricultural fertilizer in 1975. By 1980, industrial fixation of nitrogen, as monitored by the United Nations, had increased to 6.0 × 10^7 tons N yr^{-1} (Fig. 178). Summing known N$_2$ fixation rates from all marine regions suggests an input of at most 1.1 × 10^7 tons N yr^{-1}, compared with denitrification losses of 8.5 × 10^7 tons N yr^{-1} (Table II). If nitrogen fixation by bacterial endosymbionts within floating diatom mats (Martinez *et al.*, 1983) is globally significant, however, these estimates (Fig. 178) may represent serious underestimates.

Salt- and freshwater marsh ecosystems, together with estuaries, now fix twice as much total atmospheric N$_2$ (2.75 × 10^7 tons N yr^{-1}) as do marine regions, despite two orders of magnitude less area (Table II). Nuisance blooms of nitrogen fixers, such as *Anabaena* spp., dominate both eutrophic and oligotrophic lakes, where as much as 30% of the nitrogen demanded by annual primary production is met by nitrogen fixation (1.88 × 10^7 tons N yr^{-1}). Terrestrial ecosystems fix an additional 10 × 10^7 tons N yr^{-1} (Capone and Carpenter, 1982).

If all the nitrogen now fixed by humans and by natural lake, marsh, estuarine, marine, and terrestrial ecosystems (21.7 × 10^7 tons N yr^{-1}) were directly transferred to the aquatic ecosystems, denitrification and nitrous oxide losses (18.2 × 10^7 tons N yr^{-1}) would be offset by this input. The total aquatic burial losses of 14.1 × 10^7 tons N yr^{-1} could similarly be offset by a precipitation input of 13.4 × 10^7 tons N yr^{-1} (Fig. 178). Passive transfer of atmospheric nitrogen to the sea from rainfall and dry deposition is now thought to be small (0.8 × 10^7 tons N yr^{-1}, McElroy, 1983), however, although past estimates ranged as high as 8.2 × 10^7 tons N yr^{-1} (Soderlund and Svensson, 1976). Land run off is the most likely vehicle for transfer of the terrestrial nitrogen to the sea, which brings us to the subject of coastal eutrophication.

6.2 Eutrophication

Despite the intensive fishing pressure described in the next section, fish constitute only 10% of the meat intake of the United States (Fig. 179)—recall the

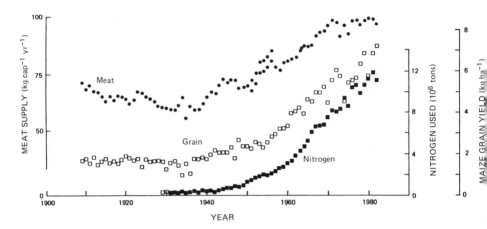

FIG. 179. A time series of the use of nitrogen fertilizer (10^6 tons yr^{-1}), grain yield (kg ha^{-1} yr^{-1}), and meat supply (kg capita^{-1} yr^{-1}) within the United States from 1909 to 1982. [After National Research Council (1986) from Agricultural Statistics.]

similar δ ^{14}C label of human protein in Europe within Section 5.2.1. Between 1909 and 1938, the meat supply per capita actually declined and the United States became a net importer of food in the 1930s. The development of hybrid grains and the use of legume alfalfas increased agricultural productivity from 1938 to 1950, but the major increment of grain yields and meat supply came from the application of fertilizers (Boyer, 1982), with an exponential increase of fertilization of food crops between 1950 and 1980 (Fig. 179).

6.2.1 Nutrient Loading

As a result of similar agrarian efforts in Europe, the leaching rate of soil nitrogen from British arable lands, subject to intense fertilization, has increased fourfold between the 1930s and the 1970s (Forster *et al.*, 1982). There have been severe effects on potable water supplies. The nitrate content of British (Wilkinson and Greene, 1982) and French (Meybeck, 1982) freshwater streams exceeds, at times, both U.S. Public Health Service and World Health Organization standards (715–1615 μg-at. NO$_3$-N l^{-1}).

Between 1850 and 1950, the nitrate content of the Rhine River (Fig. 180) had only increased to ~50 μg-at. NO$_3$ l^{-1}. It doubled by 1960, and again over the last decade to ~250 μg-at. NO$_3$ l^{-1}; the total nitrogen content is now >320 μg-at. N l^{-1} (Van Bennekom and Salomons, 1981). The nitrogen content of the Seine River increased fourfold to 400 μg-at. N l^{-1} from 1965 to 1975 (Meybeck, 1982), that of the Elbe River increased from 270 μg-at. N l^{-1} in 1962 to 470 μg-at. N l^{-1} in 1970 (Lucht and Gillbricht, 1978), while that of the Thames River doubled to >500 μg-at. N l^{-1} between 1968 and 1978 (Wilkinson and

Greene, 1982). Similarly, by this time the Scheldt River contained an amazing 800 μg-at. N l^{-1} (Wollast, 1983).

As a result of these anthropogenic inputs, the mean winter nitrate content of the North Sea shelf in the 1960s was already at least two to four times higher off the Thames estuary, the Wash, and the Rhine estuary than at the edge of the shelf (Johnston, 1973). The winter phosphate content off the Dutch coast also doubled from 1961 to 1978 (Van Bennekom and Salomons, 1981). Indeed, as much as 350 μg-at. NO$_3$ l^{-1}, ten times the concentration of NO$_3$ in deep slope water, is now found both in the Scheldt estuary and 10 km off the Belgian coast (Mommaerts *et al.*, 1979).

The nitrate content of the Mississippi River (Fig. 181) has at least doubled to 150 μg-at. NO$_3$ l^{-1} during the spring flood (Ho and Barrett, 1977) over the past 10 yr. The mean concentrations have increased from 40 μg-at. N l^{-1} in 1905 (Gunter, 1967), 1935 (Riley, 1937), and 1965 to >80 μg-at. N l^{-1} in 1980 (Walsh *et al.*, 1981). Data for other U.S. rivers that drain heavily populated areas, such as the Ohio (Wolman, 1971), Potomac and Susquehanna (Carpenter *et al.*, 1969), Delaware (Kiry, 1974; Sharp *et al.*, 1982), and Hudson (Deck, 1981), suggest that their nitrogen content has also increased over the past 25–50 yr from sewage, fertilizers, and nitrate release to groundwater after deforestation (Likens *et al.*, 1978).

For example, more than 70 μg-at. NO$_3$ l^{-1} is now found in the Hudson, Potomac, Susquehanna, and Delaware rivers, as well as the Po, Vistula (Van Bennekom and Salomons, 1981), Loire, Rhone, Garonne (Kempe, 1982), Huang-He (Meybeck, 1982), and Yangtze rivers (J. Edmond, personal communication). Similarly, U.S. rivers such as the Columbia (Park *et al.*, 1972) that drain less populated areas in the south and west used to contain ten times less inorganic nitrogen (7–10 μg-at. N l^{-1}) than the above rivers before 1970. But

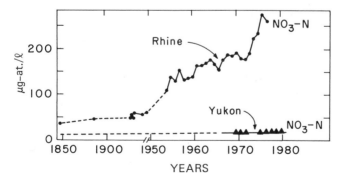

FIG. 180. Concentrations of nitrate (μg-at. l^{-1}) at the mouth of the Yukon River over the last decade (unpublished data) and within the Rhine River over the last century. [After Van Bennekom and Salomons (1981).]

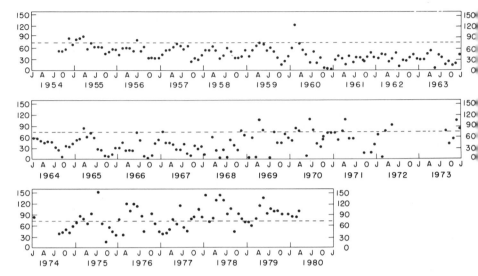

FIG. 181. Monthly nitrate content (μg-at. l^{-1}) of the Mississippi River at 450 km upstream of the mouth, from 1954 to 1980. [After Walsh *et al.* (1981); © 1981 Macmillan Journals Ltd., with permission.]

now the dissolved nitrogen contents of the Altamaha (Walsh *et al.*, 1981), Alabama, Brazos, Sacramento, Columbia (Meybeck, 1982), and Pamlico (Hobbie *et al.*, 1975) Rivers are also ≥30 μg-at. N l^{-1}.

The location and discharge of the world's 30 largest rivers (Table III) suggest that the freshwater flux to the sea from underdeveloped areas (286×10^3 m^3 sec^{-1}) is similar to that from developed areas (234×10^3 m^3 sec^{-1}), having six times the human population and far more anthropogenic nitrogen. If we assume that half the world's total river discharge (4×10^{16} l yr^{-1}) has not yet been affected by deforestation, sewage, and fertilizers in these underdeveloped countries and that the low nitrate content (≤10 μg-at. NO$_3$-N l^{-1}) of the Amazon (Ryther *et al.*, 1967), Congo and Niger (Van Bennekom *et al.*, 1978), Orinoco, Negro, Mackenzie, and Yukon (Fig. 180) Rivers (Meybeck, 1982) is typical, nitrate input to the shelves from such drainage systems is 0.3×10^7 tons NO$_3$-N yr^{-1}.

If we also assume that the other half of the world's river discharge now has at least 60 μg-at. NO$_3$ l^{-1} due to population distribution, food demands, and other consequences of the Industrial Revolution, the annual nitrate input from developed terrestrial ecosystems would be 1.7×10^7 tons NO$_3$-N yr^{-1}. The total present input of 2.0×10^7 tons NO$_3$-N yr^{-1} from major river runoff would be at least three times that estimated more than 25 yr ago (Emery *et al.*, 1955) from data taken more than 50 yr ago (Clarke, 1924). Other estimates of present river nitrogen range as high as 3.5×10^7 tons N yr^{-1} (Delwiche and Likens, 1977). Using an estimate of 0.8×10^7 tons NO_3-N yr^{-1} in river runoff, Garrels *et al.* (1973) suggested that only 10% of the nitrogen fertilizers, previously applied to fields, had been released from the soil to streams by 1970. The nutrient time series over the past 25 yr, however, suggest that NO_3 concentrations in major rivers have at least doubled within the past decade. The present estimate of riverine nitrogen loading would be 33% of the industrial fixation of nitrogen in 1980 (Fig. 178). If these data reflect increased mobilization of nitrogen from soil pools, we would expect future increases of primary production in the coastal zone as has apparently occurred off the Netherlands (Postma, 1978) and within Lake Erie.

Other elements besides nitrogen must be supplied to the coastal zone for increased algal primary production, of course. By 1970, about 1.3×10^7 tons P yr^{-1} were being mined for fertilizer (Garrels *et al.*, 1973) in an N/P weight ratio of 4 : 1, compared to human nitrogen fixation from the atmosphere. As indicated by the dissolved N/P ratio of 12 : 1 within rivers (Kempe, 1982; Meybeck, 1982), however, the leaching rate of the more insoluble phosphorus compounds from arable lands is less than that of nitrogen.

The phosphorus input to the sea from land runoff has nevertheless increased 10-fold since the Industrial Revolution. Pristine rivers such as the Amazon, Zaire, Yukon, Mackenzie, and Orange contain 0.3 μg-at. PO_4 l^{-1}, whereas most developed rivers have at least 3.0 μg-at. PO_4 l^{-1}, with as much as 40 μg-at. P l^{-1} found in the Scheldt and 80 μg-at. P l^{-1} in the Thames (Meybeck, 1982). At an annual river discharge of 4×10^{16} l yr^{-1}, the pre-industrial input of phosphorus was probably 0.04×10^7 tons P yr^{-1}. Assuming that half this discharge contains 3.0 μg-at. PO_4 l^{-1}, the present agrarian flux is estimated to be 0.22×10^7 tons P yr^{-1}.

Rivers are not the only source of anthropogenic nitrogen and phosphorus to coastal waters. Analyzing present nitrogen loadings from coastal sewage outfalls, waste dumping, and river input to the shelves of the New York Bight and the Mediterranean, Japan, Baltic, North, and Irish Seas, Segar (1985) suggests a range of 10–30 kg N yr^{-1} person^{-1} for each of these areas; the total urban nitrogen loading to Lake Erie was estimated to be 15 kg N yr^{-1} person^{-1} (Sly, 1976). Extrapolating an estimate of 20 kg N yr^{-1} for the agricultural, sewage, and other waste demands needed to support each of the 3 billion persons in North America, Asia, and Europe leads to an estimate of 6×10^7 tons anthropogenic N yr^{-1} (Fig. 178).

Taking into account the nitrogen fluxes from domestic wastes, as well as from feed lots, food-processing wastes, and a 20% soil leaching rate from fertilizers, Van Bennekom and Salomons (1981) have made a similar estimate of 6.4×10^7 tons N yr^{-1}. A possible dissolved organic nitrogen (DON) loading of 0.6×10^7 DON yr^{-1} from river discharge (Ittekot *et al.*, 1983) is *not* included within either of these estimates. In any case, such an anthropogenic nitrogen input of 6×10^7 tons N yr^{-1} is 10 times the preindustrial input and occurs mainly in the northern hemisphere, where the seasonal drawdown of atmospheric CO_2 is evidently increasing.

In coastal ecosystems dominated by the discharge of domestic wastes, the N/P ratio of urban effluent is only 2.5 : 1 from sewage treatment and phosphorus detergents. As much as 25–50% of land-derived phosphorus may be from this source (Ryther and Dunstan, 1971). Applying a human excretion rate of only 0.54 kg P yr^{-1} person^{-1} (Vollenweider, 1968) to the 3 billion people in developed countries gives a sewage flux of 0.16×10^7 tons P yr^{-1}, suggesting a total contemporary input of at least 0.4×10^7 tons P yr^{-1}, a 10-fold increase since the Industrial Revolution. What then are the consequences of this increasing nutrient perturbation, which parallels that of nutrient addition on land (Fig. 179), for the coastal food web—that is, are these nutrients passed to higher trophic levels and/or buried within the sediments?

6.2.2 Algal Response

At Helgoland, located between waters of the central North Sea and the estuarine effluent of the Elbe River, a 23-yr time series of meteorological, hydrographic, nutrient, particulate carbon, and phytoplankton species composition data is available from almost daily measurements, taken between 1962 and 1984 (Hagmeier, 1978; Berg and Radach, 1985). Within the Elbe-influenced waters, the winter (November–March) phosphate concentrations increased from 0.88 to 1.36 μg-at. PO_4 l^{-1} over 23 yr, like the Dutch coastal waters over the same time period, compared to an increase from 0.69 to 1.07 μg-at. PO_4 l^{-1} within the saltier shelf waters of the central North Sea. Similarly, the winter rate of increase of nitrate within Elbe-influenced waters was 0.68 μg-at. NO_3 l^{-1} yr^{-1}, compared to 0.61 μg-at. NO_3 l^{-1} yr^{-1} for shelf waters, with a combined increment from 15.99 μg-at. NO_3 l^{-1} in the winter of 1962 to 21.69 μg-at. NO_3 l^{-1} in 1984.

Silicate concentrations within the lower Rhine River are ~90 μg-at. SiO_4 l^{-1} (i.e., N/Si ratio of 3.6 : 1) and may induce phytoplankton species succession within coastal waters (Wollast, 1983), since only diatoms and silicoflagellates require silicon for their frustules. Recall from Section 3.2.1 that silicon appeared to be a controlling nutrient, in addition to nitrogen, within the Peru upwelling model (Fig. 58), which assumed an N/Si ratio of 1.5 : 1 for selection of the limiting nutrient. Between 1966 and 1984, dissolved silicic acid accordingly declined from 12.77 μg-at. SiO_4 l^{-1} to 2.46 μg-at. SiO_4 l^{-1} within Elbe-influenced waters (i.e., a dilution of coastal waters took place), while no trend was detected in silicon concentrations of the adjacent shelf waters, west of Helgoland.

The nondiatomaceous phytoplankton within the Elbe-influenced waters concomitantly increased their biomass from 2.24 μg C l^{-1} to 36.56 μg C l^{-1} within

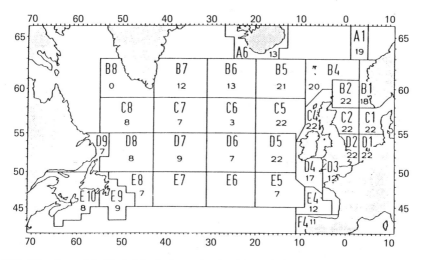

FIG. 182. Chart of the North Atlantic Ocean showing the standard areas for the Continuous Plankton Recorder (CPR) Surveys. In each area the number of years until 1968 is given for which the area has been adequately sampled. [After Colebrook (1972).]

23 yr, compared to an increment of 2.77 μg C l^{-1} to 16.03 μg C l^{-1} over the same period for these organisms within the saltier waters. No significant decline of diatom biomass occurred within the Elbe-influenced waters, however, while their standing stocks declined from 4.92 μg C l^{-1} in 1962 to 2.71 μg C l^{-1} within the shelf waters in 1984. Since 1965, the apparent decline of diatom biomass has been a general phenomenon within the North Sea (Reid, 1975, 1977, 1978).

The Helgoland time series is located within subarea D1 of the Continuous Plankton Recorder (CPR) Survey (Fig. 182), initiated within the North Sea (Hardy, 1939) in 1948 on a routine basis for zooplankton, and in 1958 for phytoplankton, with an extension of these measurements to the North Atlantic in 1961. Designed primarily to sample zooplankton, the CPR is towed at a depth of 10 m behind a merchant vessel at monthly intervals, continuously collecting plankton on a moving silk mesh of 270 μm aperture, to be stored in formalin. Most phytoplankton species pass through this coarse mesh and microflagellates would disintegrate in this preservative (Reid, 1975). However, information on three levels of greenness (Robinson, 1970), on 16 species of chain-forming diatoms, and on seven species of large dinoflagellates (e.g., *Ceratium tripos*) provides an index of seasonal and long-term cycles of the net phytoplankton. The spring blooms on the continental margins, for example, are clearly depicted by these records (Fig. 183), in contrast to the smaller amplitude of the oceanic production cycle—that is, subareas C1 and C8 versus C6, or E9 and D3 versus

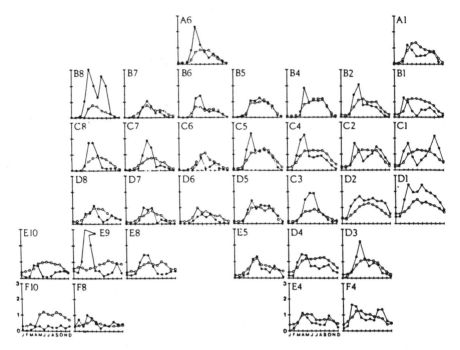

FIG. 183. Mean seasonal cycles of phytoplankton (filled circles) and copepods (open circles), averaged over the years 1948 to 1976, in all CPR areas of the North Atlantic given in Fig. 182. Phytoplankton data are given in arbitrary units of greenness, while copepod data are given as the logarithmic means of their numbers. [After Colebrook (1979); © Springer-Verlag.]

E8 and E5—recall that the seasonal amplitude of atmospheric CO_2 at the weather ships was 3–5 ppm less than at coastal stations.

Note that the mean amount of phytoplankton greenness in subarea D1 (Fig. 184) is greater than in the other six subareas over 50–55°N latitude across the Atlantic Ocean (Fig. 183), presumably as a result of the eutrophication observed in the Helgoland time series. Decomposition of this data set into a 16-yr time series from 1958 to 1973 (Reid, 1978) indicates that while algal color has increased in subarea D1, diatom abundance has declined, and *Ceratium* spp. biomass here has remained the same most years (Fig. 184). The same trend has occurred off the east coast of England in subareas D2 (Fig. 184) and C2 (Fig. 185), and off the west coast of Ireland in subarea D5 (Reid, 1977), where the rapidly sinking spring blooms of phytoplankton have been observed (Billet *et al.*, 1983; Lampitt, 1985).

The increase of nondiatomaceous phytoplankton within subarea D1 is attributed to silicon limitation (Gieskes and Kraay, 1977). After a short spring bloom of diatoms, the phytoplankton here consist almost entirely of small flagellates (Van Bennekom *et al.*, 1975), which would not be enumerated within the CPR

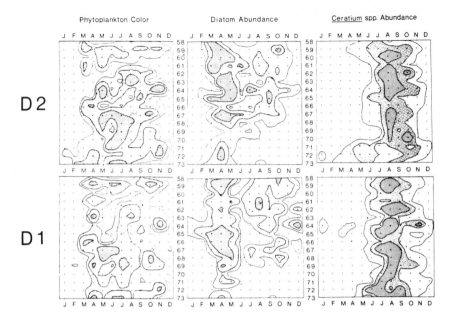

FIG. 184. Phytoplankton abundance from 1958 to 1973 in CPR standard areas D2 and D1 of the southern North Sea. [After Reid (1978); © International Council for Exploration of the Sea, with permission.]

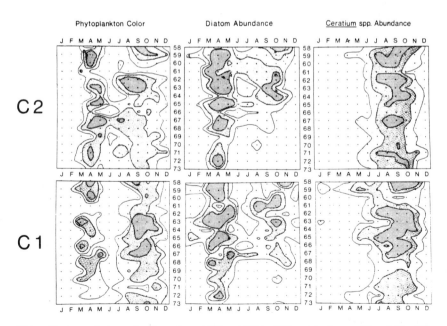

FIG. 185. Phytoplankton abundance from 1958 to 1973 in CPR standard areas C2 and C1 of the central North Sea. [After Reid (1978); © International Council for Exploration of the Sea, with permission.]

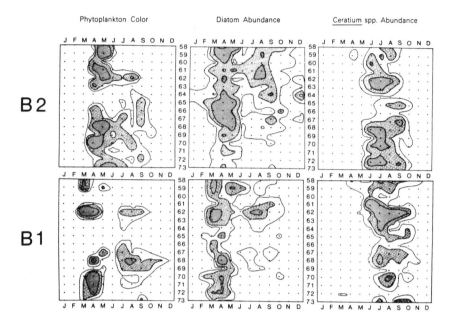

FIG. 186. Phytoplankton abundance from 1958 to 1973 in CPR standard areas B2 and B1 of the northern North Sea. [After Reid (1978); © International Council for Exploration of the Sea, with permission.]

species counts but would contribute to the color index (Reid, 1975). As one proceeds toward the northern North Sea, and away from the eutrophic coastal waters, the interannual change in color within subareas B1 and B2 is less apparent (Fig. 186). As in the southern regions, however, the autumn bloom of diatoms was curtailed in the deeper part of the North Sea.

6.2.3 Herbivore Impact

Over the same time period, the biomass of a number of zooplankton species declined as well; between 1948 and 1973, for example, the biomass of the smaller *Pseudocalanus elongatus* declined by an order of magnitude (Colebrook, 1978). Other species increased (e.g., *Acartia clausi*), while some showed no trend (e.g., *Temora longicornis*) (Glover *et al.*, 1972). A favorite food (Hardy, 1924) of the herring, *Clupea harengus*, was the offshore, larger copepod, *Calanus finmarchicus*, which appeared again within the D2 subarea of the southern North Sea (Fig. 187) in 1950 (Burd and Cushing, 1962), after its disappearance from the English Channel in 1931 (Russell *et al.*, 1971).

The herring growth increment in 1950 was equal to that of the previous 20 yr (Cushing, 1975). The year-class recruitment of herring began at 3 yr of age after

1952, instead of at an age of 4 yr previously (Cushing and Burd, 1957)—that is, the herring population then also became more vulnerable to fishing pressure. The *Calanus/Pseudocalanus* abundance ratios in the southern North Sea have increased 10-fold since 1948, with perhaps initially a change in the pathways of energy flow to higher trophic levels—that is, instead of small copepods to invertebrate predators, large copepods to fish after 1950 (Cushing, 1983).

The total herring stocks of the North Sea have since been overfished, however, with a 10-fold decline in biomass, from 2.5 $\times$ 10^6 tons during 1936–1937 and 1947–1951 to 0.2 $\times$ 10^6 tons by 1975 (Burd, 1978), after the introduction of a purse-seine fishery in 1962—see the decline of the Norwegian stocks in Fig. 189b. Herring yields used to account for more than 50% of the North Sea catch, but these landings were less than 10% by 1974 (Hempel, 1978). Based on the food web leading to this pelagic predator (Fig. 188), one can invoke complex scenarios of predator removal and food chain disruption, in addition to eutrophication, to explain the recent plankton changes of the North Sea (e.g., Figs. 184–187).

For example, although this food web is complex, the diet of the adult herring consists mainly of juvenile sand eels, *Ammodytes marinus* and *A. tobianus,* during the March–April spring bloom (Wyatt, 1976). These juvenile fish, in turn, eat mainly the appendicularians, *Oikopleura dioica,* during the preceding 2 months, when, as post-yolk-sac larvae, they and the plaice larvae constitute most of the mortality of these appendicularian populations (Wyatt, 1971, 1974).

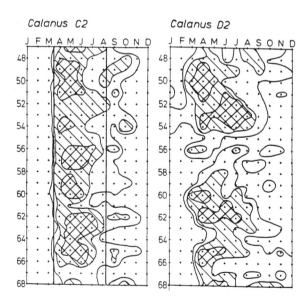

FIG. 187. Monthly means of the abundance of *Calanus finmarchicus* from 1948 to 1968 within the southwestern (D2) and western (C2) North Sea. [After Colebrook (1972).]

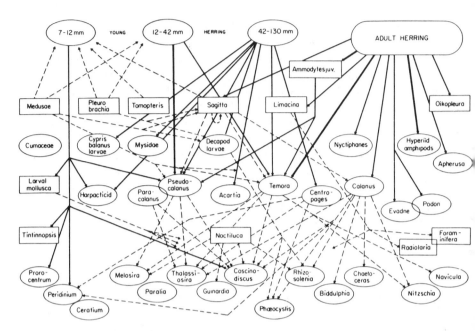

FIG. 188. Feeding relations of the North Sea herring, *Clupea harengus*, during different stages of its life history. [After Hardy (1924) and Wyatt.]

Finally, the appendicularians eat microflagellates by filtering these small algae out of the water column, with *Oikopleura* capable of retaining particles as small as 0.1 μm in size (Jorgensen, 1966), in contrast to particle selection of larger diatoms by the herbivorous copepods.

Following the decline of the herring and mackerel in the North Sea, large catches of sand eels were instead taken in the 1970s, rising from 1×10^5 tons yr^{-1} in 1957 to 5×10^5 tons yr^{-1} in 1974 (Hempel, 1978). A similar shift in abundance of these fish species occurred within the Mid-Atlantic Bight (Sherman *et al.*, 1981) after overfishing there of herring (Table XVIII): a release of natural predatory stress on *Ammodytes* spp. had occurred within both ecosystems. If increased populations of larval sand eels were to consume more appendicularians, the grazing stress on the microflagellates would also be relieved, with a subsequent increase in these plant populations. Since the *Ammodytes* larvae prefer appendicularians to copepod nauplii, or copepodids, their predator stress on some of the copepods would decline as well, allowing these small herbivores to consume more diatoms, with a subsequent decline in these plant populations after 1965!

Finally, when in doubt, blame the weather—part of the plankton shifts in the North Sea have also been attributed to changes in circulation at climatic time scales (Russell *et al.*, 1971; Cushing, 1975). Advection of *Calanus* into the

North Sea after 1950, and a subsequent lack of predation by herring after 1962, could have led to an increased grazing stress by these larger copepods on the diatoms by this time as well, that is, to a decline in their abundance (Figs. 184–186). The populations of *Calanus* in the North Sea, north of 55°N, remained the same within the eastern half in subareas B1 and C1 between 1962 and 1978, but decreased in abundance within the western half (Cushing, 1983), however, suggesting that validation data for this hypothesis are as difficult to obtain as for the *Ammodytes* hypothesis. In an attempt to sort out the different impacts of eutrophication, overfishing, and climate changes, the simpler food web of Peru (see Fig. 190), in contrast to that of the North Sea (Fig. 188), is considered next.

6.3 Overfishing

In a marine sequel to Rachel Carson's "Silent Spring," Farley Mowat (1984) has eloquently described "five centuries of death-dealing" in the decimation of sea birds, cod, salmon, haddock, herring, mackerel, capelin, flatfish, lobsters, oysters, other shellfish, seals, walrus, and whales, since the arrival of civilized humans in North American waters. Note the decline of New York menhaden stocks over just the last century in Fig. 189d. In the earlier pursuit of marine sources of both food and oil during the late fifteenth and early sixteenth centuries, entrepreneurs such as John Cabot and Jacques Cartier were awed at the amount of finfish on the Grand Banks and in the St. Lawrence River.

Their gleeful prospectus of shelf ecosystems respectively "swarming with fish" and "richest in every kind of fish" makes one wonder if European shelf waters were already significantly depleted of fish stocks by then, since a major commodity of the medieval Hanseatic League was cod and herring. In response to a presumed migration, or change in spawning grounds, of herring from the Baltic to the North Sea in the early fifteenth century, for example, coastal cities prospered in the Netherlands at the expense of towns on the Baltic (Nash, 1929).

Their Basque colleagues, having exhausted the European stocks of baleen whales by 1450, harvested peak yields of ~2500 whales yr^{-1} from Canadian coastal waters between 1515 and 1560 (Mowat, 1984). This fishing for the Atlantic right whale, *Eubalaena glacialis,* continued in the western Atlantic until the loss of the Basque whaling fleet off England in 1588; it had been conscripted as part of the Spanish Armada. The destruction of the Basque whaling fleet, however, like that of the Yankee whaling fleet three centuries later by the Confederate privateers *Alabama* and *Shenandoah* during the American Civil War, occurred too late to avert overfishing of these cetaceans at local or global scales.

The last whale of this Atlantic fishery was killed off Long Island in 1918 (Mowat, 1984). This favorite prey of the Basques had become so rare by then that Roy Chapman Andrews began his career at the American Museum of Natu-

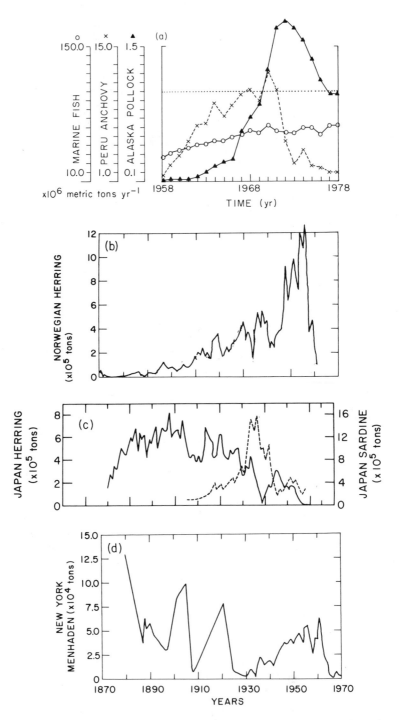

FIG. 189. (a) The world catch ($\times$ 10^6 tons yr^{-1}) over the last 20 yr with respect to that estimated from primary production ($\cdots$) and the individual yields of fisheries for the Peru anchovy and Alaska pollock, and over the last century (b-d) the individual yields of fisheries for the Norwegian herring, Japanese herring, Japanese sardine ($---$), and New York menhaden. [After Walsh (1978), (1983); © 1978 International Council for Exploration of the Sea, © 1983 Pergamon Journals Ltd., with permission.]

ral History by digging the bones of another harpooned right whale out of the Long Island surf in 1907 (Andrews, 1929). Right whales are now the most endangered cetacean species off eastern North America, with perhaps a population of 100–300 whales in the western North Atlantic (Reeves *et al.*, 1978), 10-fold less than just the annual harvest taken 400 yr earlier.

Like previous descriptions by Smith, Cabot, and Cartier of the rich coastal ecosystems of eastern North America, enthusiasts of the nineteenth century (Buchanan, 1886) described the food web off Peru as "no waters in the ocean so teem with life as those on the west coast of South America." Charles Darwin had been similarly impressed after an earlier cruise of the *Beagle* to these waters during July 6–19, 1835, writing in Callao that "the ocean teems with life . . . here in the Pacifick" (Keynes, 1979). We have seen the consequences of careless human harvest of this resource in Section 4.5.3. The purpose of this section is to ascertain the fate of primary production after overfishing occurred within this ecosystem by 1972 (Fig. 1).

6.3.1 Fishing Pressure

The yield of world fisheries has slowly increased over the 20 yr between 1958 and 1978 (Fig. 189a), despite intensive fishing pressure, and many stocks display signs of overexploitation (Gulland, 1983). The anchovy harvest off the coast of Peru, for example, increased from 700,000 tons yr^{-1} in 1958 to a maximum of ~12 million tons yr^{-1} in 1970, but then declined precipitously to ~2 million tons yr^{-1} in 1973 and to <1 million tons yr^{-1} between 1977 and 1983. The collapse of the anchovy fishery, located in the upwelling ecosystem off the coast of Peru, is attributable to overfishing and the diminished survival of anchovy larvae during warm-water intrusions (El Niños) southward along the coast of Peru in both 1972 and 1976 (Walsh *et al.*, 1980).

The very large yield of anchovy (*Engraulis ringens*) in the heyday of the fishery is usually attributed to high primary production in the upwelling system and the high efficiency of the food chain in which the commercial fish feed directly on phytoplankton. In reality, there has been debate (Cushing, 1969; Ryther, 1969) for more than a decade about the relative importance in the maintenance of the anchovy stocks of a short food web (phytoplankton → anchovy), with ~10% ecological efficiency, and of a longer food web (phytoplankton → zooplankton → anchovy), with efficiency of ~1%, but functioning over an area perhaps 10 times as great. The simulation model of the daily nitrogen flux through the Peruvian upwelling system in Section 3.2.1 suggested that the longer and more widely dispersed food web might be of greater importance during the stronger winter upwelling off Peru.

A better understanding of the trophodynamics of this upwelling system is now possible by examining annual carbon budgets before and after overfishing, in the decade between 1966–1969 and 1976–1979, respectively. The 1966 primary production data at 15°S (Table XVIII) are high, because they were collected while the ship was drifting within a rich chlorophyll area at the shelf break. The 1976–1977 data at 15°S are low, because repeated time-series stations were taken within a poor chlorophyll area near the coast (Fig. 59). Other data in Table XVIII were averaged from transects taken across the shelf. High chlorophyll and abundant anchovy eggs are found within 50 km of the coast, suggesting that a combined productivity average of this region is a reasonable estimate of the yearly carbon input to the Peru upwelling food chain. At a rate of at least 3 g C m^{-2} day^{-1} ($\bar{X}$ = 3.7 g C m^{-2} day^{-1} in Table XVIII), the annual primary production is a minimum of ~1000 g C m^{-2} yr^{-1} off Peru (see Table XIX), in contrast to ~200–500 g C m^{-2} yr^{-1} for shelf ecosystems at higher latitudes (Table III).

An upwelling zone of ~50 km offshore width, extending along the Peru–Chile coast of ~2000 km length (4–22°S), and a peak fish catch of 10 million tons yr^{-1}, harvested over the whole area (1 × 10^5 km^2), suggests an anchovy yield of at least 100 tons km^{-2} yr^{-1}. Such a harvest was an order of magnitude larger than that presently derived from the North Sea, the Mid-Atlantic Bight, or the Bering Sea. With a conversion factor of 0.06 g C = 1 gww, a harvest of 100 tons km^{-2} yr^{-1} (100 g m^{-2} yr^{-1}) suggests, however, a yield to humans of only 6 g C m^{-2} yr^{-1} of anchovy off Peru from a primary production of at least ~1000 g C m^{-2} yr^{-1}. Recall from Section 6.1.1 that the present calculation assumes an upwelling area and annual primary production of about half that used by Chavez and Barber (1985)—that is, the food-web efficiency would be even less, using their assumptions.

Table XVIII.

Mean Daily Primary Production over 20 yr within Peru Shelf Waters
from 1966 to 1985

Year	g C m^{-2} day^{-1}	Area
1966	3.3	5–12°S
1966	6.3	15°S
1969	5.2	15°S
1969	1.3	10–15°S
1974	4.5	7–8°S
1975	3.8	6–12°S
1976	3.2	10–15°S
1976	1.7	15°S
1977	1.9	15°S
1977	7.3	9–10°S
1977	4.3	5–12°S
1978	4.3	15°S
1983	3.3	5–12°S
1984	3.8	5–12°S
1985	2.2	5–12°S
1985	2.2	15°S
	$\bar{X}$ = 3.7	

Table XIX.

Changes in Annual Carbon Fluxes of the Peru Upwelling Ecosystem between 1966–1969 and 1976–1979[a]

Parameter	1966–1969 (g C m^{-2} yr^{-1})	1976–1979 (g C m^{-2} yr^{-1})
Primary production	1570[b]	1351[b]
Copepod production	100[c]	100[c]
Euphausiid production	2.5[c]	5[c]
Anchovy yield	60[b,c]	6[b,c]
Sardine yield	0.4[b]	4[b]
Hake yield	0.05[c]	0.7[c]
Bacterioplankton production	215[c]	22[c]
Microbenthic production	2[b,c]	9[b,c]
Meiobenthic production		2.5[b,c]
Macrobenthic production	0.5[b,c]	0.1[c]
Detrital carbon production	320[c]	720[c]
Sinking loss	105[c]	698[c]
Sediment storage/export	82[c]	591[c]

[a] After Walsh (1981b); © 1981 Macmillan Journals Ltd., with permission.
[b] Measured.
[c] Calculated.

As humans were the major consumer of the anchovy during its peak and declining yields, this harvest may be a reasonable estimate of fish secondary production over the whole region. It suggests a food chain efficiency of <1% before the fishery collapsed. The yield from the anchovy fishery in 1976–1979 was ~1 million tons yr^{-1}, or an implied food chain efficiency to humans of only 0.1% after overfishing, if the catch then also approximated secondary production of the anchovy. However, the same maximum estimate of secondary production, taken from a 90% smaller upwelling region, would represent a larger yield to humans per unit area, that is, a greater implied food-chain efficiency.

For example, if most of the peak yield had been derived only from the area of high anchovy egg and larvae abundance (Walsh *et al.*, 1980), north of Callao on the broad shelf between 6 and 10°S, then the annual harvest would have been 60 g C m^{-2} yr^{-1} of fish, before overfishing occurred (Table XIX). If the anchovy in this area also consumed phytoplankton half the year (DeMendiola, 1974) and zooplankton, with a growth efficiency of 20% (Table XIX), during the other half of the year, the fish yield would then have been 10% of their food input from this smaller region between 6 and 10°S. A more detailed carbon budget for the food web off Peru (Fig. 190) may determine the fate of the primary production along 90% of the rest of the Peru coast before overfishing, if most of the annual yield was actually derived from faculative herbivorous fish within an area of only 1 × 10^4 km^2.

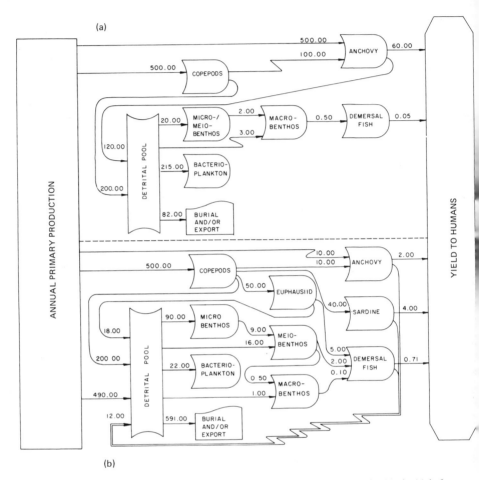

FIG. 190. A budget for the flux of carbon (g C m⁻² yr⁻¹) through the Peru food web: (a) before
(~1966–1969) and (b) after (~1976–1979) overfishing of anchovy. [After Walsh (1981b); © 1981
Macmillan Journals Ltd., with permission.]

6.3.2 Algal Response

Information on oxygen content, particle fluxes to the bottom, seabed respiration,
organic content of the sediments, and nutrient depletion of the water column aids
in the construction of a more complete carbon budget before (Fig. 190a) and
after overfishing (Fig. 190b). Previous analysis of the demersal part of the nearshore food chain before overfishing suggested that the input of carbon to the
bottom consisted mainly of fecal pellets rather than of phytodetritus (Walsh,
1975). In terms of carbon consumption, the decomposition of the fecal pellets

could be attributed to caprophagy and/or bacterial activity in both the water column and the sediments.

Additional calculations had indicated that the oxygen demand of this detrital carbon flux was such that most of the fecal pellets could be remineralized within the upper 60 m of the water column, similar to the North Sea (Krause, 1981) and North Carolina (Hofmann *et al.*, 1981) shelves. After overfishing and the decline of grazing pressure from adult anchovy, however, a large flux of uneaten phytodetritus may have sunk out of the water column, when nutrient depletion occurred at the edge of the shelf. The fatty-acid composition of the shelf sediment at 12°S during 1980 indicated mainly phytoplankton input, for example, with a minor contribution from bacteria (D. J. Smith *et al.*, 1983).

The estimate for bacterioplankton production (Table XIX) from consumption of fecal pellets within the upper part of the water column before overfishing (Fig. 190a) was obtained from a steady-state budget of the vertical decline of oxygen at 15°S in 1966–1969; 4.5 ml O_2 l^{-1} above 20 m and 0.5 ml O_2 l^{-1} below. As a result of oxygen depletion in upwelled source water during March to April 1976, 1977, and 1978, however, the average O_2 concentration at 15°S was ~1 ml O_2 l^{-1} in the upper 20 m and 0.5 ml O_2 l^{-1} below—that is, a vertical gradient of ~0.04 ml O_2 l^{-1} m^{-1} after overfishing, in contrast to 0.4 ml O_2 l^{-1} m^{-1} a decade earlier. The local production of aerobic bacteria was thus estimated to be an order of magnitude less (Table XIX) in the second carbon budget (Fig. 190b), with the remaining detrital carbon then sinking to where more denitrification (Codispoti and Packard, 1980) occurred in deeper water, and more sulfate reduction in the bottom sediments.

The aerobic carbon production of the benthos before overfishing (Fig. 190a) was estimated from both P/B estimates and bottom respiration measurements at 15°S in March to April 1969 (Pamatmat, 1971). The mean macrobenthos biomass across the shelf in 1969 (Rowe, 1971) was 0.5 g C m^{-2}, suggesting a secondary production of 0.5 g C m^{-2} yr^{-1} (Table XIX) and an ingestion demand of 5.0 g C m^{-2} yr^{-1} from a P/B ratio of 1 and a growth efficiency of 10%. Approximately 20% of the measured ~25 g C m^{-2} yr^{-1} bottom respiration off Peru in 1969 (Pamatmat, 1971) might thus be due to macrobenthos. The rest is attributed to micro- and meiobenthos, as in the bottom metabolism of other shelves (Smith *et al.*, 1973; Smith, 1978).

The estimated total input of carbon to the shelf bottom before overfishing (Table XIX) is similar to a flux of 88 g C m^{-2} yr^{-1} at 50 m measured with floating sediment traps on the narrow shelf at 15°S in 1977 (Staresinic *et al.*, 1983). If a higher sinking loss was applied (without other losses) to an early model of the nearshore Peru upwelling system (Walsh and Dugdale, 1971), the predicted amount of downstream chlorophyll in the offshore region was less than that observed. The nearshore area at 15°S contained the same amount of low chlorophyll (<2 μg l^{-1}) during 1966, 1969, 1976, 1977, and 1978 as that found inshore at 10°S during 1960–1969 and 1976–1977, suggesting that an estimate of a sinking loss of only 10% of nearshore primary production is reasonable, as higher phytoplankton biomass is found downstream, that is, offshore.

The production of the benthic food web after overfishing (Fig. 190b) was first estimated from sulfate reduction estimates of 90 g C m^{-2} yr^{-1} for anaerobic metabolic demand at 15°S in surface sediments. The sulfur bacterium, *Thioploca*, was found on the bottom in large numbers along the whole Peru coast during 1976, and the near-bottom oxygen concentration at 15°S on the 100-m isobath was <0.1 ml O_2 l^{-1} in 1976–1977, compared with ~0.2–1.0 ml O_2 l^{-1} in 1969. The content of the secondary nitrite maximum increased from 3–6 μg-at. NO_2 l^{-1} in 1966–1969 (Fiadeiro and Strickland, 1968) to 14 μg-at. NO_2 l^{-1} in 1977 (Codispoti and Packard, 1980), and finally to as much as 23 μg-at. NO_2 l^{-1} in 1985 (Codispoti *et al.*, 1986b).

Similarly, no sulfate reduction was detected in the water column in 1969 (Goering and Pamatmat, 1971) at 15°S, whereas H_2S was found there in March to April 1976 (Dugdale *et al.*, 1977) and at 8–9°S during November 1977. Based on an average 1977 meiobenthic biomass of 0.6 g C m^{-2} across the shelf and a P/B ratio of 4 : 1 reflecting a higher turnover rate of these small organisms, the secondary production of the meiobenthos might have been 2.5 g C m^{-2} yr^{-1} (Table XIX), that is, 28% of the sulfate reduction estimate of microbial production. Quantitative macrobenthic data are not available for 1976–1978, but their biomass and production (Table XIX) were then presumably smaller because more extensive anoxic conditions occurred on the Peru shelf.

Since the primary electron acceptor in the oxidation of organic matter within the sea is oxygen, then nitrate in the process of denitrification, and finally sulfate in the production of H_2S and FeS_2, the above chemical sequence suggests that a larger-than-normal organic loading to the bottom of the shelf and upper slope had occurred by 1976–1977 (Fig. 190b). The total detrital flux from the water column after overfishing in 1976–1979 is, in fact, estimated (Table XIX) to be 70% of the assumed 1000 g C m^{-2} yr^{-1} primary production. After the 1976 El Niño , such detrital loss (Fig. 190b) is almost an order of magnitude larger than that estimated for 1966–1969 (Fig. 190a), but similar to an estimate of 64% detrital loss (Shushkina *et al.*, 1978) in 1974 at 7–8°S after the 1972 El Niño . Within the geological record, a 10-fold increase in sedimentation rate is associated with a doubling of the organic carbon content of the sediments (Muller and Suess, 1979). If the above carbon budgets are correct, a similar process may have occurred off Peru in the decade from 1966–1969 to 1976–1979.

The distribution of organic carbon in the surface sediments (Fig. 125) suggests that before overfishing in 1968 and 1969 the sediment carbon off Peru increased both seaward at the slope and poleward. Without consideration of changes in food chain dynamics, a primary production rate of 1000 g C m^{-2} yr^{-1} and a previous sedimentation rate of 0.1 cm yr^{-1} can be calculated (Muller and Suess, 1979) as a carbon accumulation rate within the sediments of 114 g C m^{-2} yr^{-1} off Peru, which is similar to that from the carbon budget before overfishing (Table XIX).

The location of the 2% carbon isopleth in the sediments of the wide shelf north of Callao in 1968–1969 (Fig. 125) also suggests that less carbon used to be lost from the water column at 10°S than at 15°S. These spatial trends continued after overfishing in 1976 and 1977 (Rodinova *et al.*, 1973; C. Delgado, L. Flores, E. Suess, P. Muller, and G. Rowe, unpublished data), but the carbon content of the surface sediments may have doubled recently, compared to subsurface concentrations (Fig. 158). The same primary production and a sixfold increase in sedimentation rate (Table XIX) lead to an empirical carbon accumulation rate (Muller and Suess, 1979) of 684 g C m^{-2} yr^{-1} in the sediments, compared with that of 591 g C m^{-2} yr^{-1} estimated from the overfishing budget (Fig. 190b).

At 15°S, 4.5% carbon was the highest surface sediment value observed on the shelf in March to April 1967, but 10% carbon was found at the same stations with the same methods in 1976. Furthermore, within 22 cores taken across the 15°S Peru shelf and slope in 1977–1978, the mean subsurface C/N value at 50 cm depth of these cores was 7.76, compared with 7.53 in the surface sediments, suggesting little diagenetic change. However, the mean percent carbon was 2.89 at 50 cm depth and 4.82 at the surface (see Fig. 158 for an example), implying a recent increase in carbon flux to the sediments that could have occurred within the past decade. Inadequate spatial sampling implies that areas of high organic content may have been missed in these 1968–1978 surveys, but the horizontal and vertical trends of the sediment carbon data are at least consistent with the carbon budgets of the water column. The first carbon budget (Fig. 190a) for the food web before overfishing may be more applicable to northern Peru waters during 1966–1969. The second budget (Fig. 190b) may instead reflect the food web both before overfishing off southern Peru and afterward off northern Peru during 1976–1979.

6.3.3 Herbivore Impact

The predicted change in dominance of pelagic and demersal fish species after overfishing (Fig. 190b) reflects the decline in these carbon budgets of anchovy ingestion demands on both phytoplankton and zooplankton (DeMendiola, 1974). The increase of zooplankton biomass after the 1972 and 1976 El Niños was mainly in the standing stocks of omnivorous euphausiids (Shushkina *et al.*, 1978). In contrast, there was little apparent change in their biomass after the 1957 El Niño (Brinton, 1962), when the anchovy fishery was just beginning. The sardine, *Sardinops sagax,* and hake, *Merluccius gayi,* both eat zooplankton off Peru, suggesting that the stocks of these fish should also increase after the anchovy decline. Similar changes occurred in demersal species abundance within the North Sea after the pelagic fish stocks were overfished (Burd, 1978). Following the 1976 El Niño , sardine indeed constituted ~67% of the pelagic yield of 1.5 million tons yr^{-1} in 1977. During 1978 and 1979, the catch of mackerel, sardine, pipefish, and horse-mackerel was still ~50% of the total pelagic yield of 1–2 million tons yr^{-1}, and after this period, the sardine dominated the catch (Fig. 1). Furthermore, the potential demersal yield of "virgin" hake stocks was estimated to be 0.75 million tons yr^{-1} after the 1972 El Niño . This is similar to the change predicted in the second carbon budget (Fig. 190b) of a yield of 0.71 g C m^{-2} yr^{-1}, or 1.2 million tons yr^{-1} of demersal fish along the whole coast (1 × 10^5 km^2). Predictably, the hake stocks are now commercially exploited off Peru, with 0.50 × 10^6 tons yr^{-1} of hake landed in 1978, only 0.19 × 10^6 tons yr^{-1} in 1979 (Gulland, 1983), and as little as ~0.3 × 10^3 tons $month^{-1}$ landed at the port of Paita (5°S) in 1982, before commencement of the last El Niño (Barber and Chavez, 1983).

Despite initial restrictions in fishing, however, the same amount of plankton was not consumed by euphausiids, sardine, mackerel, and hake in 1976 as by anchovy in 1966. Instead, the oxygen content of the euphotic zone and seabed has since declined, the nitrite and sulfide content of the water column has increased, and the organic content of the bottom sediments may also have increased after overfishing. If 10^3 tons km^{-2} yr^{-1} of anchovy had been produced on the wide shelf north of Callao (1 × 10^4 km^2) and 10^2 tons km^{-2} yr^{-1} along the rest of the coast (9 × 10^4 km^2), the combined secondary production of 19 million tons yr^{-1} before overfishing is the same as that originally estimated in 1969 (Cushing, 1969; Ryther, 1969). The carbon budget for the Peru food chain after overfishing (Fig. 190b) is thus a more reasonable description of the food chain along most of the Peru coast before and after overfishing, with the exception of the 6–10°S area.

If this second budget of Table XIX did apply to 15°S during 1966–1969 as well (i.e., before overfishing), then about half of the daily production of phytoplankton was always lost as sinking detritus to the upper slope, after nutrient depletion within offshore Peruvian waters, rather than cropped inshore by zooplankton and anchovy. Recall that, from Sections 4.2–4.4, the annual carbon budgets of some of the shelf food webs in the Mid-Atlantic Bight, the Gulf of Mexico, and the Bering Sea similarly suggested that ~50% of their primary production may be exported to the continental slope as phytodetrital carbon from these shelves, thus constituting together a major organic sink of the global CO_2 budget of ~0.5 × 10^9 tons C yr^{-1} (Fig. 173). An annual detrital carbon loss of ~500 g C m^{-2} yr^{-1} at 15°S, and a C/N ratio of 5 : 1 implies that the particulate nitrogen export in the form of detritus is 100 g N m^{-2} yr^{-1}, in contrast to ~1 g N m^{-2} yr^{-1} in the form of fish yield (Fig. 190b).

To maintain a nitrogen mass balance, in an annual steady state, of the food web on this Peru shelf, at least ~100 g N m^{-2} yr^{-1} of dissolved nitrogen must be returned to the shelf to replace the detrital and fish nitrogen loss. Furthermore, an annual primary production of ~1000 g C m^{-2} yr^{-1} at 15°S requires 200 g N m^{-2} yr^{-1}, 40% of which could be supplied by uptake of ammonium from nitrogen recycling (Whitledge, 1978) and 60% by physical input of nitrate from upwelling, because river run off is negligible. An estimate of the nitrate supply can be made with a simple budget, $\partial(NO_3)/\partial t = w\ \partial(NO_3)/\partial z$, using an average upwelling rate w of 10 m day^{-1} and an observed vertical nitrate gradient of 5 mg-at. NO$_3$ m^{-3} over the 50-m depth interval of the nearshore upwelling region off Peru; this daily flux sums to 5 g NO$_3$ m^{-3} yr^{-1}. Over a 20- to 30-m euphotic zone, such an input of 100–150 g NO$_3$ m^{-2} yr^{-1} would be sufficient to balance the "new" nitrogen demands of photosynthesis, fish removal, and detrital loss from the shelf at 15°S, similar to the mid-Atlantic nutrient budget in Section 4.3.5.

Although the anchovy fishery of the Peru upwelling ecosystem was one of the world's most "studied," "advised," and "managed" coastal resources, societal conflicts led to its possible demise. Along the whole coast, the fish yield may now be only 10 tons km^{-2} yr^{-1}, similar to that from the Mid-Atlantic Bight, North Sea, or Bering Sea, despite a three- to sevenfold higher primary production off Peru. It is possible that the Peru upwelling ecosystem will either return to the same level of secondary production as that before 1972, or maintain the present decline in food chain yield, with a greater export in the form of detrital carbon rather than of fish carbon. The only clupeid fishery (Fig. 189c) that has rebounded thus far is perhaps that for the Japanese sardine, *Sardinops melanosticta* (Gulland, 1983), while the only underexploited stocks are the Argentine clupeids, as a result of extensive cattle production rather than prescient management techniques.

The individual estimates in the Peru carbon budgets (Fig. 190) may be questionable, but it seems that most of both the larval survival and the production of adult anchovy used to occur within a 10% area of the Peru shelf. This is about the same percent area of the Bering Sea from which most stocks of the Alaska pollock are harvested (Fig. 101). The hypothesis that greater changes in shelf trophodynamics have occurred within the 6–10°S area than in the 15°S region seems to reflect both the impact of overfishing by humans and the alteration of larval survival by nature off Peru—recall the southerly migrations of anchovy discussed in Section 4.5.3. The decline of anchovy grazing pressure has apparently led to increases in the plankton biomass available to stocks of sardine and hake, to increased carbon loading and sulfate reduction at the "downstream" seabed, and to a decline in oxygen and nitrate content of the water column. A similar disruption of the summer trophodynamics of the Mid-Atlantic Bight, as well as ensuing anoxia, occurred during 1976; these events are the subject of the next section.

6.4 Climate

As a result of either eutrophication or relaxation of grazing pressure, an increase of either microflagellate or diatom populations on the outer continental shelf would lead to additional export to the adjacent slope. Early diagenesis there would lead to releases of biogenic CO_2, to be stored in the sea's inorganic bicarbonate cycle (Broecker *et al.*, 1979), and of nutrients, to be diffused from pore water of the sediments into the overlying water column (Froelich *et al.*, 1982) Away from local anoxic regions, between the Gulf of Mexico and Georges Bank, for example, this process results in the oxygen minimum layer located near the slope bottom at >300 m depths, where low oxygen contents of 2.5–3.0 ml O_2 l^{-1} are usually found (Richards and Redfield, 1954; Emery and Uchupi, 1972). These values are similar to those for the oxygen minimum layer in the open Atlantic Ocean at greater depths of $\sim$1500 m.

Greater oxygen depletion within enclosed fjords and other coastal regions is a common feature of areas of restricted flow (Richards, 1965), such as in the Baltic Sea, or within the Cariaco Trench off Venezuela. Eutrophication and prevention of oxygen renewal by vertical stratification of the water column led to a decline in O_2 content of the Baltic Sea, for example, from 30% O_2 saturation within bottom waters in 1900 to anoxic conditions by 1970 (Fonselius, 1969). Similar anoxic events have been observed off New South Wales (Rainer and Fitzhardinge, 1981), in the northern Adriatic Sea (Fedra *et al.*, 1976; Degobbis *et al.*, 1979), in Tokyo Bay (Tsuji *et al.*, 1973), in Chesapeake Bay (Seliger *et al.*, 1985), in Mobile Bay (May, 1973), in Tampa Bay (Santos and Simon, 1980), and in the Bay of Vilaine (Rossignal-Strick, 1985).

Anoxic regions of open coastal zones are usually restricted to upwelling systems (Brongersma-Sanders, 1957), as on the Peruvian shelf, where intense carbon production, sinking of organic matter, and subsequent oxygen demands may override the greater flushing rate of these areas (Walsh, 1981b). However, low-oxygen water of varying spatial extent has also been observed over past years within the open, less-productive coastal area of the New York Bight (Fig. 191). A $60 million loss (United States Court of Appeals, 1979) of shellfish resulted from the most recent anoxic event along the New Jersey coast during the summer of 1976 (Sindermann and Swanson, 1979).

The development of this 1976 anoxia has been attributed to increased anthropogenic carbon loading from urban areas adjacent to the New York Bight. It may also reflect an unusual climatological regime that restricted renewal of oxygen to the bottom waters, in conjunction with an unusual abundance and subsequent respiratory demand of the dinoflagellate *Ceratium tripos* beneath the pycnocline. In an attempt to distinguish between human-induced and natural generic causes of oxygen depletion within the New York Bight, historical data extending back to 1910 were analyzed (Falkowski *et al.*, 1980).

As a result of this climatological analysis, a causal chain of seasonal events

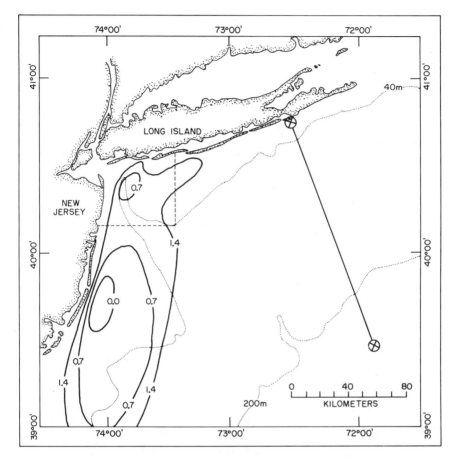

FIG. 191. The mean distribution of anoxia during July 1976 in relation to the New York Bight Apex
(– – –), and the location of the offshore Shinnecock transect as shown between circled X's. [After
Falkowski *et al.* (1980) with permission under DOE Contract No. DE-AC02-76 CH00016.]

that led to the observed 1976 anoxia has been identified: namely, a warm winter
with large runoff; a low frequency of spring storm events (i.e., a deep spring
thermocline); persistent southerly summer winds with few reversals; a large au-
tochthonous carbon load (e.g., *Ceratium tripos*); and low grazing pressure by
zooplankton. A simulation model (Stoddard, 1983; Stoddard and Walsh, 1986)
then suggested that anoxia could have occurred off the New Jersey coast in the
summer of 1976 without any allochthonous carbon loading from eutrophication
by New York City. Occasional anoxia in this open shelf system could result from
natural physical forcing and biological response at climatic time scales, similar
to changes of *C. tripos* populations within the North Sea (Reid, 1975, 1977).

Table XX.

Oxygen Content of Bottom Waters in the New York Bight[a]

Period	Cold-pool cross-sectional area (km²)	O₂ solubility (ml/l)		Cold-pool O₂ stock (× 10³ m³/m)	Percent of May 1975
		Cold pool	Midshelf surface		
May 1975	2.83	7.09	6.58	20.06	100%
June 1975	1.90	7.00	5.63	13.30	66%
May 1976	0.98	6.79	6.10	6.65	33%
June 1976	0.82	6.84	5.75	5.61	28%

[a] After Falkowski et al. (1980) with permission under DOE Contract No. DE-AC02-764CH00016.

6.4.1 Physical Transition

As the New York Bight ecosystem progresses into the summer stratified mode (Fig. 8), the subpycnoclinal waters hold an initial spring stock of heat, salt, and nutrients, as well as of oxygen. The spring stock of oxygen in the bottom layer of "cold pool" water (Ketchum and Corwin, 1964) becomes an important variable in any consideration of summer anoxia. This stock is a function of the bottom water mass volume and its saturation capacity. A simple comparison is made between the May–June 1975 and May–June 1976 cold-pool stocks (Table XX) to demonstrate the oxygen disadvantage of the 1976 summer aphotic zone, even before carbon loading and plankton respiration are considered.

The spring stock of O_2 in the bottom water is relatively unaffected by those vertical mixing processes that depend on the vertical gradient, since in the case of oxygen, the saturation content during spring is less in the surface layer, thus acting to reduce the vertical gradient. For example, the oxygen capacity was 10% greater in the bottom waters than in the surface waters in May 1976 and 20% greater in June 1976 (Table XX). Vertical diffusion processes do not provide a source of oxygen to the bottom layer until this saturation difference is consumed *in situ* below the pycnocline. The difference in the spring stock between the two years is primarily due to the difference in the bottom water volumes of the cold pool.

The March surface-temperature anomaly (Fig. 192) of the New York Bight reflects the vertically homogeneous heat content of the winter shelf water column (Walsh et al., 1978) and is an index of both previous local winter mixing and the amount of heat extracted by winter's end. A negative March temperature anomaly indicates a colder-than-normal winter and a larger-than-normal volume of cold shelf water going into the vernal heating cycle. Delays in spring runoff and/or a decrease in occurrence of spring storms act to increase the spring vertical heat transfer on the shelf. In general, cold winters precede shallow summer thermoclines (10–20 m) south of Long Island, whereas deeper summer thermoclines (15–30 m) are observed in the same area after the warm winters (see, e.g., Ketchum and Corwin, 1964).

In March 1976, anomalously warm sea-surface temperatures (Fig. 192) were associated with anomalously warm air temperatures and above-normal insolation (Diaz, 1979). The 1976 February and March mean air temperatures above New York were the second warmest in the last 100 yr. Anomalous atmospheric patterns for February–June 1976 consisted of winds of much more westerly and southerly origin, with very few major cyclones during this period (Diaz, 1979). An unusually large freshwater runoff was coupled with the warm winter of 1976 and the high frequency of southwesterly winds (Armstrong, 1979).

River discharges of the Hudson Valley were early and high during the winter–spring of 1976. The spring maximum occurred in February, 2 months before the normal April maximum at the Battery, and constituted a record for the last 30 yr. The rainfall in May 1976 was higher than average, 33 mm above normal in Central Park (Swanson et al., 1978), and the 1976 discharge from the Hudson River at Troy, New York, was then much higher than the mean 30-yr freshwater input (Fig. 193). The 1976 May discharge of fresh water was exceeded only four times in the last 30 yr.

Ketchum et al. (1951) and Ketchum and Keen (1955) have reported on freshwater accumulation in the New York Bight. They found only 6–10 days of accumulation of fresh water in the Apex at any one time, whereas in May 1976 there was nearly 2 months of accumulation. This led to a large difference, between 1975 and 1976, in the salinity contribution to spring stratification of the

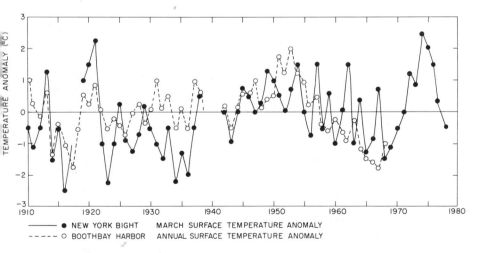

———— ● NEW YORK BIGHT   MARCH SURFACE TEMPERATURE ANOMALY
– – – – ○ BOOTHBAY HARBOR   ANNUAL SURFACE TEMPERATURE ANOMALY

FIG. 192. The annual surface temperature anomaly off Boothbay Harbor, Maine (open circles, dashed line), and the March anomaly within the New York Bight (filled circles, solid line) from 1910 to 1978. [After Taylor et al. (1957), Colton (1972), Diaz (1979), and Falkowski et al. (1980); with permission under DOE Contract No. DE-AC02-764 CH00016.]

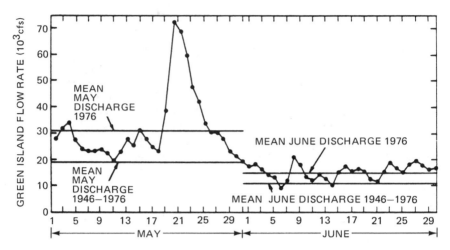

FIG. 193. The runoff of the Hudson River for May–June 1976, in comparison with the mean discharge over the last 30 yr. [After Malone (1978), and Falkowski *et al.* (1980); with permission under DOE Contract No. DE-AC02-764 CH00016.]

water column. The amount of fresh water in the bight apex was ~12 times greater in May and ~5 times greater in June of 1976 than in the same months in 1975.

During the stratified season, the New York Bight waters are typically characterized by a sharp pycnocline reaching σ_t gradients of 0.2 m^{-1} (Walsh *et al.*, 1978). This vertical density structure is usually controlled by a temperature gradient, which can change by 17°C over 15 m (Bowman and Wunderlick, 1976). The sea-surface temperatures begin to rise in April and peak in early August (Fig. 8). This seasonal temperature structure has been summarized by Bigelow (1933), Beardsley *et al.* (1976), Bowman and Wunderlick (1976), and Hopkins and Garfield (1979).

The June 1976 thermocline was thicker than normal with a constant slope. The thermocline (Fig. 194) had a fairly low decrease of ~0.3°C m^{-1} within the 10- to 30-m depth band off Long Island at the 40-m isobath in June 1976, compared to a 0.7°C m^{-1} decrease here over the 10- to 20-m band in June 1975. A critical factor leading to this broad thermocline was the small number of cyclones which occurred (Diaz, 1979) during spring 1976. Consequently, the high runoff and small amount of wind mixing allowed the shelf waters to stratify approximately a month earlier than normal.

The warm winter and high runoff in February thus resulted in March shelf bottom waters that were warmer and fresher than normal. The summer cold pool (Bigelow, 1933; Ketchum and Corwin, 1964) in 1976 was deeper in the water column, it was warmer, and it extended across most of the shelf. This contrasts to the thicker, less extended, and colder distribution of the cold pool in 1975

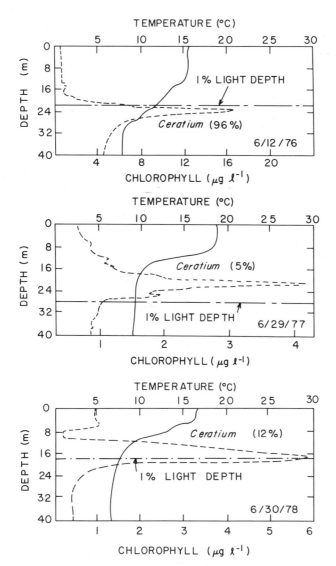

FIG. 194. The nearshore phytoplankton distribution in relation to the depths of the seasonal thermocline and euphotic zone during June 1976, 1977, and 1978 off Long Island. [After Falkowski *et al.* (1980); with permission under DOE Contract No. DE-AC02-764 CH00016.]

(Mayer *et al.,* 1979). The interannual temperature of the cold pool varies from 5 to 10°C during August–September at a bottom depth of 60 m, south of Long Island. When the cold pool in this area is warm, low bottom O_2 values are found off New Jersey and vice versa (Fig. 195).

With establishment of a deep pycnocline, the vertical exchange between sur-

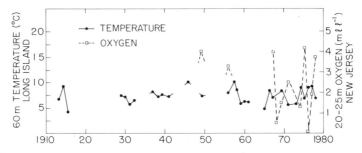

FIG. 195. The bottom temperature at 60 m south of Long Island and the near-bottom oxygen content at 20–25 m east of New Jersey during August-September 1910–1978. [After Bigelow (1915, 1927, 1933), Clarke (1940), Ketchum et al. (1951), Ketchum and Corwin (1964), Cresswell (1967), Colton et al. (1968), Ogren and Chess (1969), Whitcomb (1970), Alexander and Alexander (1976), Bowman and Wunderlick (1976), and Falkowski et al. (1980); with permission under DOE Contract No. DE-AC02-764 CH00016.]

Table XXI.

Contingency Table for Anoxia within the New York Bight[a]

	Warm winter (March)	High Hudson River flow (Feb.–Apr.)	Onshore transport (April)	Deep thermocline (June)	Large *Ceratium* population (June)	Steady SW wind (summer)	Extent of anoxia
1966	No	Yes	No	No	No	No	No
1967	Mild	No	No	?	?	Yes	No
1968	No	No	No	?	Yes	No	Mild
1969	No	Yes	Yes	?	?	No	No
1970	Mild	No	No	?	?	No	No
1971	Mild	No	Yes	?	?	Yes	Mild
1972	Mild	No	No	No	No	Yes	No
1973	Mild	Yes	No	?	No	No	No
1974	Yes	No	Yes	Yes	Yes	No	Mild
1975	Yes	No	No	No	No	Yes	No
1976	Yes	Yes	Yes	Yes	Yes	Yes	Large

[a] After Falkowski et al. (1980); with permission under DOE Contract No. DE-AC02-764CH00016.

face and bottom regions of the water column becomes more "diffuse," that is, slower. The most significant difference between 1975 and 1976, in terms of mixing, was that the energy available for mixing in 1976 was stored as potential energy in a broad pycnocline, creating a third layer. The volume of this layer was generated at the expense of the bottom water rather than of the surface layer. A two-layered flow regime (Beardsley and Butman, 1974) then became three-layered, with inclusion of onshore flow within the thermocline (16–30 m) that responds to offshore wind forcing (Hopkins, 1974). As the spring water column stratifies in this mode, a persistent onshore convergence develops in the pycnocline with upwelling circulation dynamics, at both midshelf (Han *et al.*, 1980) and near the coast (Walsh *et al.*, 1978), which erodes the bottom water volume and advects plankton onshore (Table XXI).

6.4.2 Algal Response

Within several cross-shelf transects during April–May 1976, the subsurface chlorophyll maximum increased from 4–5 μg chl l^{-1} offshore to 15–30 μg chl l^{-1} inshore (Fig. 194), within an onshore salinity tongue of 32–32.5‰ water (the 25.0–25.5 σ_t band). The vertical locations of these chlorophyll maxima were carefully determined at each station with a continuous chlorophyll pump and an *in situ* fluorometer equipped with a depth sensor. These data strongly suggest an onshore subsurface accumulation of the phytoplankton populations in the New York Bight during the spring of 1976, similar to transport of phytoplankton by estuarine circulation within Chesapeake Bay (Tyler and Seliger, 1978).

In contrast, during April 1978, the surface isopleths of the 32–32.5‰ salinity band were 60 km closer to the coast than in 1976, reflecting less runoff during the colder winter of 1978. The April 1978 inshore tongue of 32–32.5‰ salinity contained subsurface maxima of only 4–10 μg chl l^{-1} and two orders of magnitude fewer *C. tripos* populations than in 1976 (compare Figs. 196b and c). The 1976 anomaly in spring phytoplankton composition and abundance is attributed to an earlier onset of species succession and a larger volume of *Ceratium*-occupied water exposed to subsurface onshore accumulation.

A comparison of *C. tripos* distributions in "normal" shallow pycnocline years with those during deep or broad pycnocline years (Fig. 194) provides a description of the differences in timing of seasonal succession within the phytoplankton niche as regulated by changes in the circulation field. Recall the seasonal regularity of *Ceratium* spp. abundance within different regions of the North Sea (Figs. 184–186). The "normal" seasonal, onshore progression of *C. tripos* across the New York shelf can be similarly seen in a composite of March to August sections (Fig. 196), taken south of Long Island (Fig. 191) in 1975, 1977, and 1978, that is, during years of no anoxia.

At the end of a "normal" winter, $\sim$1–5 $\times$ 10^2 cells l^{-1} of *C. tripos* are usually found offshore within the upper 20 m of the water column (Fig. 196a); similar levels of *C. tripos* were found in this area in March 1975, 1977, 1978, and 1979. During midwinter of the "abnormal" year 1976, surface populations of *C. tripos* were again only predominant in the outer part of the New York Bight (Fornshell *et al.*, 1977). By early March, however, 1–5 $\times$ 10^3 cells l^{-1} were found inshore within the New York Bight Apex and off Fire Island (Malone, 1978). Zooplankton nets were then clogged by *Ceratium* from Cape May, New Jersey, to Montauk Point, New York.

With onset of spring stratification and depletion of the surface nutrients, maxima of *C. tripos* are found lower in the water column, but usually at the

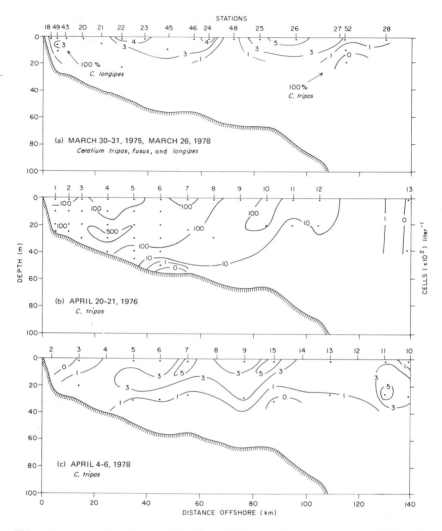

FIG. 196. Cross-shelf distributions of *C. tripos* (× 10² cells l⁻¹) in "normal" years of 1975, 1977, and 1978 and during the *Ceratium* bloom of 1976 off Long Island. [After Falkowski *et al.* (1980); with permission under DOE Contract No. DE-AC02-764 CH00016.]

same amount of their winter abundance. There is an onshore gradient of *C. tripos*, however, suggesting an onshore accumulation of the populations at mid-depth by April–May (Figs. 196c and d). Although more cells were found in the anoxic year of April 1976 (Fig. 196b) than in 1978 (Fig. 196c), the cross-shelf pattern of *C. tripos* distribution is similar in both years. The inshore May populations of *C. tripos* are always larger than the April stocks, suggesting seasonal increase of these dinoflagellates in this area.

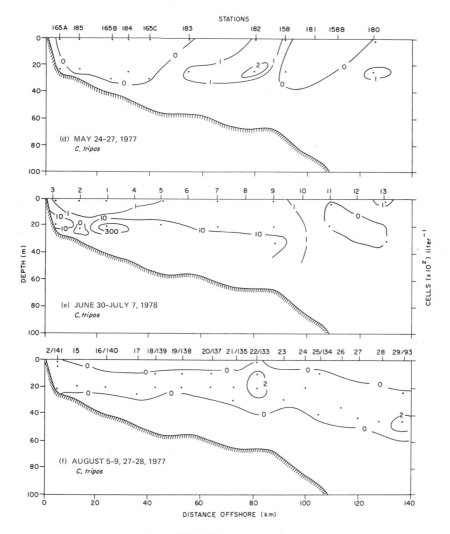

FIG. 196. (*continued*).

By June 1976, the nearshore phytoplankton of the subsurface chlorophyll maximum (Fig. 194) was 99% *C. tripos*, with an abundance of up to 5×10^5 cells l^{-1}. Similarly, *C. tripos* constituted 65% of the phytoplankton population during June 1964 (Grumman *et al.*, 1975), but only 12% in June 1978 (Fig. 196e). In both August 1976 (Thomas *et al.*, 1979) and August 1977, however, very few *C. tripos* were found across the shelf (Fig. 196f). *Ceratium* populations were also less than 1% of the phytoplankton in September 1971 (Nuzzi, 1973).

The disappearance of these dinoflagellates from the New York Bight by late summer is part of the normal species succession pattern (Lillick, 1937, 1940); the phenomenon of most importance is thus the length of their residence time during early summer.

The normal wind pattern within the New York Bight during early summer is an air flow from the southwest, with frequent reversals from the northeast. Over the last 15 yr, however, there were several cases (June 1967, July 1971, July 1972, July 1975, and June 1976—see Table XXI) in which more than 12 days of steady wind stress occurred from the southwest direction (Tingle *et al.*, 1979). After a month of such anomalously strong and steady June winds from the southwest in 1976, the Hudson River plume was shifted toward the east as indicated by salinity patterns. A reversal to the northeast of the normal southwest surface (3 m) currents occurred by June 10, 1976 (Swanson *et al.*, 1978).

This wind event was of sufficient duration to allow a longshore flow reversal within the thermocline, in addition to the onshore–offshore Ekman response of the upper layer. The current velocities, found in the thermocline, are usually about $0.5-1.0$ cm sec^{-1} of residual drift to the southwest off Long Island and New Jersey (Bumpus, 1973). During June 1976, the flow at $16-30$ m off New Jersey was instead $1.0-2.0$ cm sec^{-1} toward New York and Long Island, while the bottom flow was still sluggish to the southwest (Mayer *et al.*, 1979). From bottom current-meter observations over the New Jersey shelf, Mayer *et al.* (1979) estimated the spring–summer travel time of bottom water through the New York Bight to be on the order of 100 days in 1975 and 500 days in 1976.

The biological impact of such a flow reversal was to retain and apparently concentrate *C. tripos* populations within the northwest sector of the New York Bight (Malone, 1978). An additional spatial pattern was a change in the depth of the chlorophyll maximum of *C. tripos* in relation to the pycnocline. The populations outside the New York Bight apex were still found within the pycnocline and above the 1% light level. However, *C. tripos* populations at the same depth, but beneath the Hudson River plume, were located below both the 1% light level and the pycnocline within the apex, that is, the area where the 1976 anoxia event was first observed (Fig. 197). The chlorophyll content of the *C. tripos* maximum within the pycnocline was also greater beneath the Hudson River plume than outside the apex and may thus reflect onshore accumulation within this area in response to the June 1976 events of wind and current reversal.

A change in seasonal circulation modes thus sets the stage for both the introduction and the retention of *C. tripos* populations in the New York Bight. As a result of heavy runoff in February and weak March winds, earlier onset of stratification was associated with a deep pycnocline and fewer nutrients in April 1976 (Walsh *et al.*, 1978). These conditions would favor those species of phytoplankton adapted to low light and nutrient regimes. The early onset of spring abundance and the unusual persistence of *Ceratium tripos* throughout the summers of 1968 (Gold, 1970), 1974 (Weaver, 1979), and 1976 suggest that initiation of the

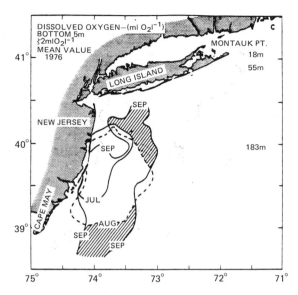

FIG. 197. Temporal progression of anoxia within the New York Bight: July–September 1976. [After Stoddard (1983).]

normal seasonal phytoplankton species succession was simply advanced in time during the spring of these years. Occurrence of extensive anoxia only in 1976 (Table XXI), however, suggests that physical events later in the season are of equal importance in leading to anoxic conditions.

6.4.3   Herbivore Impact

Before the water column stratifies, there is sometimes an abrupt shift in spring dominance of net phytoplankton from fast-growing diatoms to these slower-growing *Ceratium tripos,* as in March 1966 (Mandelli *et al.,* 1970). This suggests that some mechanism, other than just differential growth of phytoplankton, is a primary influence on seasonal phytoplankton species succession. Malone (1978) came to the same conclusion in consideration of the spring decline in diatoms and the relative increase of *Ceratium* on the continental shelf during March 1976. The usual spring transition of phytoplankton species from diatom-dominated netplankton to the chlorophyte-dominated nanoplankton community of the bight apex is not associated with changes in the rates of biomass-specific photosynthesis (Malone, 1977a). The disappearance of the winter–spring diatom bloom is attributed to sinking on both the continental shelf (Walsh *et al.,* 1978) and within the bight apex (Malone and Chervin, 1979). Species that do not sink out (that is, motile forms) will accumulate in the region of maximum stability in the pycnocline.

In a stratified water column, phytoplankton species composition (i.e., abundance of *C. tripos*) in the subsurface chlorophyll maximum is greatly influenced by the light regime (Steemann-Nielsen, 1934; Nordli, 1957; Mandelli *et al.*, 1970) relative to the position of the pycnocline. The June 1976 chlorophyll maximum at the 40-m isobath was between the 0.3 and 1% light depths, the ratio of nanoplankton to netplankton chlorophyll biomass was 0.01–0.10, and ceratian abundance was as high as 5 × 10⁵ cells l⁻¹ at depth (Fig. 194). In contrast, the June 1977 chlorophyll maximum was found within the 1–5% light depths, the ratio of nanoplankton to netplankton biomass was 10, and the ceratian populations were only a maximum of 1 × 10⁴ cells l⁻¹.

Similarly, during June 1978, the chlorophyll maximum was above the 1% light level, beneath a shallow thermocline at the same 40-m isobath as the previous observations. Only 3 × 10⁴ cells l⁻¹ of *C. tripos* were then found at 20 m, and this species constituted 12% of the total phytoplankton. The nanoplankton (<20 μm) made up 27% of the algae in 1978, with *Chaetoceros* spp. comprising more than 50% of the netplankton (Falkowski *et al.*, 1980). Moreover, only 4–6 μg chl l⁻¹ were found in June 1977 and 1978, in contrast to 15 μg chl l⁻¹ in June 1976. These data suggest that large increases in the summer populations of *Ceratium tripos* within the New York Bight occur in years of deep thermoclines.

At optimal light intensities, the maximum growth rates of *Ceratium* (0.3 doublings day⁻¹) in the field (Apstein, 1911) and in the laboratory (Nordli, 1957) are much lower than those of nanoplankton (1–2 doublings day⁻¹). At suboptimal light intensities, field populations of *Ceratium* grow as slowly as 0.03 doublings day⁻¹ (Elbrachter, 1973). Estimates of carbon- and nitrogen-specific growth rates, as well as those obtained by direct cell counting, all suggest that the doubling time of *C. tripos* within the subsurface chlorophyll maximum of the New York Bight is 20–30 days, 10-fold longer than that of the nanoplankton. A bloom of *C. tripos,* not nanoplankton, was observed in 1976, however. The selective removal mechanism of the summer nanoplankton community is considered to be grazing, rather than sinking of the spring diatom bloom.

Coastal copepods, some of the major herbivores of this ecosystem, do not appear to eat *C. tripos* but will ingest the smaller phytoplankton. The water column contained a high percentage of nanoplankton (97%) in August 1977, and shipboard gut fluorescence measurements indicated that the summer copepod community was ingesting these small algal cells of nearshore waters (Walsh *et al.*, 1978). In contrast, grazing experiments of the previous year (1976) suggested that *Centropages typicus*, a codominant of the summer zooplankton community in the inner New York Bight, did not eat the large *Ceratium tripos* (M. Dagg, personal communication). Laboratory studies also show that both *Centropages* and *Pseudocalanus* spp. will not ingest ceratians (Elbrachter,

1973). Furthermore, in additional grazing experiments, *Temora longicornis,* the other summer codominant copepod of the New York Bight, will not consume *Ceratium* during blooms within Bedford Basin (Conover, 1978) and Bras d'Or Lake (Hargrave and Geen, 1970), Nova Scotia. What then was the fate of the 1976 *C. tripos* bloom in the New York Bight?

6.4.4 Oxygen Budget

Photosynthetic oxygen production and atmospheric reaeration are the two sources of oxygen above the pycnocline. More specifically, gas exchange at the air–sea interface (Z_0) acts to maintain saturation values of oxygen in the mechanically mixed surface layer (Z_m), which in the New York Bight is ∼15 m. This means effectively that the upper boundary condition for a vertical distribution of O_2 is that of full saturation at $Z = Z_m$ rather than Z_0. With a surface exchange coefficient of 2×10^{-5} cm sec^{-1} in a stagnant-film model (Garside *et al.,* 1979), the usual respiration demands of the surface mixed layer (Table XXII) would have to be increased more than 10-fold before exceeding the oxygen resupply rate of reaeration.

Photosynthetic production, moreover, may contribute oxygen as far down as $Z = Z_e$ (euphotic-zone depth), depending on the distribution of phytoplankton.

Table XXII.

Summer Oxygen Budget (ml O_2 m^{-2} day^{-1}) for the New York Bight Apex during 1976[a]

Above the pycnocline			
Computed respiration		Observed respiration	
Phytoplankton	460	Water column	
Metazoan	550		
Sludge	1000		
Dissolved organic carbon	2000		
	4010		4050
Gross photosynthetic input	4640		
Time scale for anoxia ≈ infinity			
Below the pycnocline			
Computed respiration		Observed respiration	
Fecal pellets	300	Benthic	360
		Water column	1350
		C. tripos	5400
			7110
Vertical diffusive input		Time scale for anoxia	
With K_z of 1.0 cm^2 sec^{-1}	7000	∼600 days	
With K_z of 0.1 cm^2 sec^{-1}	700	∼30 days	

[a] After Falkowski *et al.* (1980); with permission under DOE Contract No. DE-AC02-764 CH00016.

Mean gross photosynthesis, calculated from carbon fixation rates within the apex (Malone, 1976; Thomas et al., 1979), is estimated to be 4640 ml O_2 m^{-2} day^{-1} (Table XXII). Phytoplankton respiration is assumed to be 10% of the daily gross primary production (Falkowski and Owens, 1978), or approximately 460 ml O_2 m^{-2} day^{-1} within the euphotic zone.

The respiration rate of the zooplankton herbivore community (500 ml O_2 m^{-2} day^{-1}) is estimated from shipboard measurements (0.07 μl O_2 $animal^{-1}$ hr^{-1} at 10–20°C) of adult *Centropages typicus* during June 1978 (S. Howe, personal communication) and the mean abundance of nearshore copepods (3 × 10^5 animals m^{-2}) during June 1975–1976 (Judkins et al., 1980). Similarly, the oxygen demand of the carnivorous zooplankton (48 ml O_2 m^{-2} day^{-1}) is estimated from the oxygen uptake rate (0.5 μl O_2 $animal^{-1}$ hr^{-1} at 26°C) of medium-size (100 μg dw) *Sagitta hispida* (Reeve et al., 1970) and the nearshore abundance (4 × 10^3 animals m^{-2}) of *Sagitta* spp. at the same time (Judkins et al., 1980). Finally, when the respiration rate at 8°C (0.5 μl O_2 $animal^{-1}$ hr^{-1}) of recently hatched (100 μg dw) flounder larvae (Laurence, 1977) is applied to the mean biomass of yellow-tail flounder larvae within a patch (130 animals m^{-2}) in June 1972 (Smith et al., 1978), an upper bound for oxygen demand of larval fish is 1.6 ml O_2 m^{-2} day^{-1}. With these assumptions, the cumulative metazoan contribution to oxygen demands of the water column of this budget is ~550 ml O_2 m^{-2} day^{-1} (Table XXII).

The annual input of sewage sludge and dredge spoils to the apex of the New York Bight is equivalent to an oxygen demand added to the euphotic zone of ~1000 ml O_2 m^{-2} day^{-1} (Segar and Berberian, 1976). This input does not fluctuate annually by more than 15% (Mueller et al., 1976). Most of this part of the urban oxygen demand is probably consumed within the water column (Segar and Berberian, 1976), because during the years of both high (1975, 1977) and low (1974, 1976) bottom oxygen concentrations in the bight apex the benthic respiration demands remained the same (Table XXII), ~360 ml O_2 m^{-2} day^{-1} (Thomas et al., 1976, 1979).

The additional urban carbon loading of this ecosystem is in the form of dissolved organic carbon (DOC), which constitutes 90% of the daily carbon flux from the Hudson River (Segar and Berberian, 1976). Additional DOC is produced by extracellular release from phytoplankton at a maximum rate of ~20% of the net daily primary production, 1.4 g C m^{-2} day^{-1}, or at most 0.02 mg DOC l^{-1} day^{-1} (Thomas et al., 1979). Values of DOC range from 8–10 mg DOC l^{-1} at the mouth of the Hudson River (Mueller et al., 1976; Thomas et al., 1979) and 15 mg DOC l^{-1} at the sewage sludge dump (G. Harvey, personal communication) to ~1 mg DOC l^{-1} at the edge of the shelf (J. Sharp, personal communication). The oxygen demand of DOC is estimated to be ~2000 ml O_2 m^{-2} day^{-1} (Segar and Berberian, 1976) within the bight apex, where about 50% of the DOC standing crop appears to be consumed (Table XXII).

Independent estimates of the separate respiratory demands of the budget

(Table XXII) agree fairly well with the observed total utilization rates in the absence of *Ceratium* from the New York Bight. The apex surface layer in this budget contains a cumulative oxygen demand of sewage and sludge wastes, dissolved organic carbon, nonceratian phytoplankton respiration, and metazoan respiration (1000 + 2000 + 460 + 550, or 4010 ml O_2 m^{-2} day^{-1}). This total computed oxygen demand is close to the observations of the surface water-column respiration (~4050 ml O_2 m^{-2} day^{-1}) without *Ceratium*. Such an oxygen demand is just balanced by a gross photosynthetic production (net + DOC production + respiration) of ~4640 ml O_2 m^{-2} day^{-1} within the euphotic zone.

Below the pycnocline, the two major oxygen sinks are those of benthic respiration and detrital oxidation. In the absence of *C. tripos,* the subpycnocline respiration rate (Thomas *et al.*, 1979) of the water column is estimated to be 1350 ml O_2 m^{-2} day^{-1}. This reflects the oxygen demands of both particulate material, sinking out of the lower water column, and the DOC, suspended within this layer. The benthic oxygen demand of 360 ml O_2 m^{-2} day^{-1} (Thomas *et al.*, 1976, 1979) is comprised of both particulate matter that already has sunk to the bottom and that of the benthic infauna. The sum of these respiration demands is 1710 ml O_2 m^{-2} day^{-1} (Table XXII).

The detrital component of this budget consists of fecal pellets, phytodetritus, and urban particulate wastes. If we use an assimilation efficiency of 60% for zooplankton, with the assumption that 50% of the phytoplankton standing stock is removed daily from the summer euphotic zone (Walsh *et al.*, 1978; Thomas *et al.*, 1979), the ensuing fecal pellets of the herbivores would have an oxygen demand of ~600 ml O_2 m^{-2} day^{-1}. Chervin (1978) estimated, however, that the zooplankton ingest as much as 90% of the total particulate carbon, including fecal pellets, within the bight apex each day, suggesting a fecal pellet demand of only ~300 ml O_2 m^{-2} day^{-1} on the bottom. This compares to the measured bottom respiration rate of ~360 ml O_2 m^{-2} day^{-1}. The herbivores may thus ameliorate bottom oxygen consumption in the New York Bight, rather than contribute to anoxia, as is the case in Tokyo Bay (Seki *et al.*, 1974).

When *Ceratium* is not abundant below the surface layer, 75% of the oxygen demand of the water column is located above the pycnocline (Thomas *et al.*, 1979). In August 1976, the ratio of net ^{14}C production/respiration was ~1.5 : 1 in waters above the pycnocline. This suggests that after respiratory demands are subtracted from the gross primary production, more oxygen is evolved from phytoplankton photosynthesis than is consumed by heterotrophic processes. During such times the dissolved oxygen in the water column is sustained near saturation levels above the pycnocline, while oxygen is slowly depleted near the bottom. However, if deep populations of *Ceratium tripos* are not consumed, continue to respire, and are concentrated by currents in a local area of the New York Bight, then these dinoflagellates could create an oxygen deficit beneath the pycnocline.

With a measured mean carbon content of 25 ng C cell^{-1} for *C. tripos,* a Q_{10}

of 2.3, and a power law of weight-dependent respiration (Banse, 1976), Falkowski and Howe (1976) computed a respiration rate of $1.4 \times 10^{-4} \mu l \, O_2$ cell^{-1} hr^{-1} at 10°C. In June 1977, the respiration of *C. tripos* was directly measured in the New York Bight, using both an oxygen polarographic electrode (Falkowski and Owens, 1978) and an electron transport activity assay (Packard, 1971); the mean of the measured respiration rates was also about $1.4 \times 10^{-4} \mu l$ O_2 cell^{-1} hr^{-1} at 10°C. Using this specific respiration rate and *C. tripos* abundance, Falkowski and Howe (1976) estimated that the integrated respiration beneath the pycnocline of *Ceratium* was 5400 ml O_2 m^{-2} day^{-1} during the summer of 1976 (Table XXII). This value is ~15 times that of the mean benthic respiration rate of 360 ml O_2 m^{-2} day^{-1}.

The effects of a deep thermocline and prolonged abundance of *C. tripos* populations would thus appear to be a shift in the normal water column oxygen demand from above to below the pycnocline. Just mineralization, rather than continued respiration, of the ceratian biomass (3255 mg at C m^{-2}) within 20 m of the bottom during June 1976, for example, would have consumed about 70% of the initial mean oxygen levels (Falkowski and Howe, 1976). The benthic respiration was 13–14% of the total respiration (water column and benthos) in August 1976, compared to 6–7% in August 1975, implying that, indeed, there had been an increased carbon input to the benthos during 1976 (Thomas *et al.*, 1979). Some dead *C. tripos* cells were observed on the bottom in July, but not in August 1976 (Mahoney, 1979). These organisms are subject to both parasitism (Arndt, 1967) and ingestion by ciliates and isopods (Elbrachter, 1973). It is possible that their bloom was terminated within the apex or along the New Jersey coast, with dead cells falling to the bottom, rather than being flushed seaward.

To prevent anoxia, the respiration demands in the bottom layer must be met by resupply of oxygen from either vertical diffusion across the pycnocline, or by horizontal advection. Assuming no horizontal advection in a one-dimensional model, the previous respiration demands, and a K_z of 0.1 cm^2 sec^{-1}, the layer beneath the pycnocline would become anoxic within ~30 days (Table XXII). In contrast, if an eddy diffusion coefficient of 1.0 cm^2 sec^{-1} is used across the pycnocline of the same model, anoxia would not occur for 600 days, even in the presence of *C. tripos*. These simple calculations suggest that (1) *C. tripos* respiration could have a major influence on bottom-water O_2, (2) small changes in eddy diffusivity could make large differences in bottom-water O_2 fluxes, (3) normal benthic respiration does not result in anoxia, and (4) anoxia could result in 3–5 weeks, if *C. tripos* is abundant below the pycnocline and the eddy diffusivity is low. Previous estimates of the time scale for anoxia were on the order of 2–3 weeks (Segar and Berberian, 1976).

Using an apparent K_z of 0.1 cm^2 sec^{-1} for summer conditions (P. Biscaye, personal communication), the rate of oxygen supply is sufficient to meet the benthic oxygen demand and part of the oxygen demand of the subpycnocline water column in normal years (Table XXII). In abnormal years, if a *C. tripos* demand of 5400 ml O_2 m^{-2} day^{-1} is added to the background respiration of the aphotic waters of the bight apex, the total oxygen demand of the bottom layer could be as high as 7110 ml O_2 m^{-2} day^{-1}. Under these circumstances, anoxia is likely to occur within about 30 days, provided that the *C. tripos* populations and their respiration demands are not flushed from the New York Bight. Longshore advective processes could both transport *C. tripos* out of the bight apex and laterally renew the bottom oxygen stocks (Han *et al.*, 1980). The apparent spread of anoxia south along the New Jersey coast from the apex during July and the continuation of anoxia in the New York Bight until September 1976 (Fig. 197) suggest, however, that the near bottom flow was weak—that is, the low-oxygen water was not renewed until the fall overturn of this system.

6.4.5 Mesoscale Coupling

To further evaluate the physical and biological processes affecting the time-dependent, spatial changes of low-oxygen bottom water in the New York Bight (Fig. 197), a two-layered ecosystem model was then developed to analyze the causes of the anoxic episode in 1976 (Stoddard, 1983). Using a modeling framework similar to other phytoplankton–nutrient–oxygen models (see, e.g., Walsh, 1975; DiToro and Connolly, 1980) and to the examples described thus far in this monograph, the simulations involved numerical integration of the usual nonlinear, partial differential equations. The model was focused on the time period of stratified conditions of the water column during May–October, over a two-layered grid of the New York Bight, from Montauk Point, Long Island, to Cape May, New Jersey (Fig. 198). After replication of the 1976 anoxic event (see Figs. 201–203), the predicted responses of the model's variables were next evaluated for a change in the urban waste inputs (see Fig. 205). Those estimated for 1974 (Mueller *et al.*, 1976) constituted a base run, while a 10-fold increment of carbon and nitrogen inputs ($10 \times$) was used for analysis of the impact of increased waste loading on eutrophication and oxygen depletion within the New York Bight.

The state variables of this more complex model included dissolved nitrogen (ammonium and nitrate), dissolved silicon, dissolved organic carbon, and dissolved oxygen, as well as particulate organic carbon, phytoplankton (nanoplankton and *C. tripos*), and zooplankton (copepods and chaetognaths) above and beneath the pycnocline, over the continental shelf of the New York Bight. Sources of ammonium nitrogen included microbial remineralization of organic nitrogen, macrobenthic regeneration of ammonium, waste inputs, and zooplankton excretion. Loss terms for this form of nitrogen included nitrification and nutrient uptake by the two groups of phytoplankton. The sinks of dissolved and particulate organic carbon (POC) included consumption by bacterial decomposition and settling of POC. The sources of dissolved organic carbon included

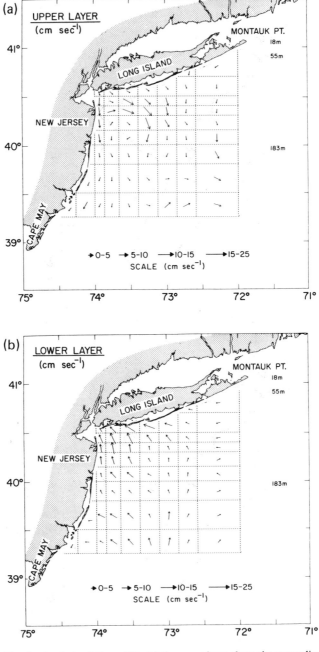

FIG. 198. The simulated circulation within (a) the upper layer above the pycnocline and (b) the lower layer of a diagnostic circulation model (Han *et al.* (1980), under wind forcing from the west of 1.22 dyn cm⁻² during May 18–23, 1976. [After Stoddard and Walsh, 1986).]

photosynthetic excretion and loading from waste discharges. In the model, simulated POC was the sum of phytoplankton carbon and detrital sewage/sludge carbon.

Using the rate parameters of Table XXII and of the preceding text, the simulated interactions of the dissolved oxygen pool reflected oxidation of particulate and dissolved organic carbon, as well as nitrification, respiration of phytoplankton and zooplankton, seabed oxygen demand, atmospheric reaeration, and photosynthetic oxygen production (see Fig. 200). Photosynthesis was also the source term for production of phytoplankton biomass as part of the POC, while the sink terms included respiration, grazing by zooplankton, settling, and excretion of DOC. The temperature-dependent growth rates of the two phytoplankton groups were regulated by both light and nitrogen limitation, with the extinction coefficient calculated as a function of total POC, similar to equations presented in Sections 3.1.1–3.1.4.

The time-dependent physical forcing functions of the model included water temperature, incident solar radiation and photoperiod, depth of the pycnocline, and seasonal stratification of the water column. The model also considered parameterization of the advection and eddy dispersion processes over both the horizontal and vertical dimensions. A velocity field of u, v (Fig. 198), and w (Fig. 199) was obtained from a quasi-steady-state diagnostic model of circulation

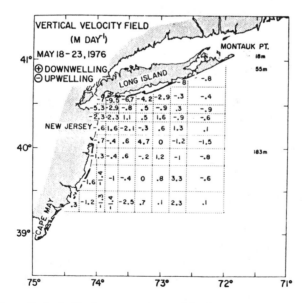

FIG. 199. The vertical velocities between two layers of the diagnostic circulation model shown in Fig. 198. [After Stoddard and Walsh, 1986).]

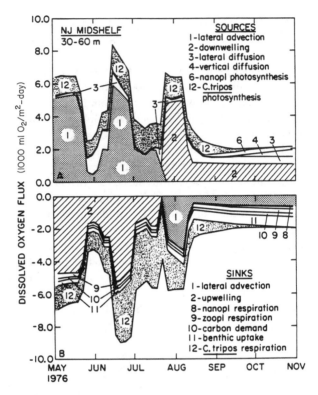

FIG. 200. The time-dependent sources and sinks of dissolved oxygen below the pycnocline on the New Jersey shelf from May to October 1976. [After Stoddard (1983).]

within the New York Bight (Han *et al.*, 1980), with discharge from the Hudson River included as a boundary condition, similar to the baroclinic model of Section 2.7. The wind forcing and density field of the model were updated nine times between May 18 and August 13, 1976. The details of the boundary conditions, of the time-dependent K_z and spatially varying K_x, K_y, of the numerical methods, and of the 10 state equations can be found in Stoddard (1983). Figure 200 illustrates the sources and sinks of dissolved oxygen for this model.

A qualitative causal chain of events was identified (Table XXI), which may lead to aperiodic anoxic events of varying spatial extent within the New York Bight. For example, all of these events occurred in 1976, but most did not occur in 1966—a spatial snapshot of simulated anoxic conditions on August 15, 1976, is presented in Fig. 201. The time-dependent results of this quantitative model from the Bight Apex (Fig. 202) and from nearshore waters (5–30 m) off New Jersey (Fig. 203) suggest, moreover, that anoxia would have occurred off the New Jersey coast in July 1976 (Fig. 197) without the oxygen demands of anthro-

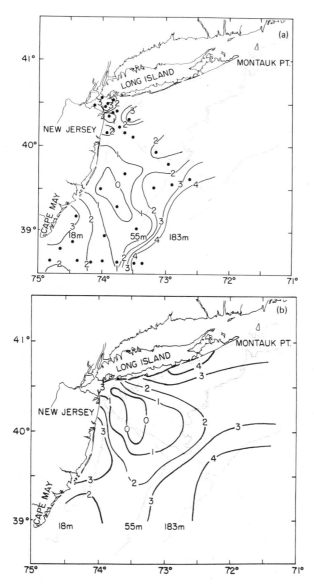

FIG. 201. A comparison of (a) observed bottom oxygen data collected during August 24 to September 9, 1976, with the (b) spatial field of the lower layer of the model on August 15, 1976. [After Stoddard (1983).]

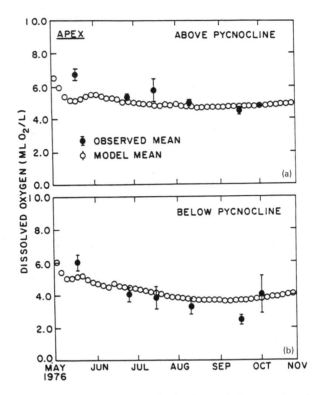

FIG. 202. A comparison of the observed and simulated mean O_2 concentrations above and below the pycnocline within the apex of the New York Bight from May to October 1976. [After Stoddard (1983).]

pogenic carbon loading from New York City. In these unusual years, when physical conditions of the shelf habitat allow early onset of phytoplankton species succession, the slow seasonal buildup of *C. tripos* populations represents a large summer carbon pool that is not removed by herbivores. Under these conditions of the coastal food web, if a steady summer wind regime also allows *C. tripos* populations to accumulate beneath the Apex pycnocline, just the oxygen demand of these organisms exceeds the rate of resupply within the aphotic zone.

Even in the absence of *Ceratium tripos,* however, the water-column respiration during summer is an order of magnitude greater inside the apex of the New York Bight than off Long Island. Most of this increased oxygen demand results from the urban carbon wastes, sewage sludge, and DOC, with an associated nitrogen loading (Fig. 204) of the bight apex. Stimulated by the increased availability of nitrogen (Ryther and Dunstan, 1971; Garside *et al.,* 1976; Garside and Malone, 1978), growth of phytoplankton in the summer could simultaneously

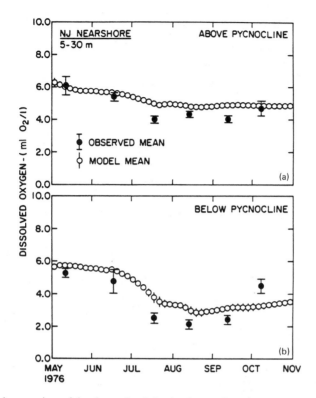

FIG. 203. A comparison of the observed and simulated mean O_2 concentrations above and below the pycnocline off the coast of New Jersey from May to October 1976. [After Stoddard (1983).]

both reduce the nitrogen load and offset part of the oxygen demand of the large carbon input to the apex.

Because eutrophication occurs as a result of the present sewage-treatment processes, part of the urban nitrogen loading in the apex could be effectively transformed into phytoplankton. After releasing O_2 to the water column, the nanoplankton are normally grazed during stratified conditions of the water column (Malone and Chervin, 1979). While $\sim 1.0\%$ dw organic carbon is found in some of the apex sediments (Fig. 123), little of this benthic oxygen demand is usually due to remineralization of the motile summer phytoplankton community (Officer and Ryther, 1977). Most of their carbon is instead stored within the longer-lived components of the metazoan part of the food web—that is, herbivores that eventually respire the algal carbon that they have eaten, above the pycnocline.

Such a scenario assumes, however, that light limitation by other particulates

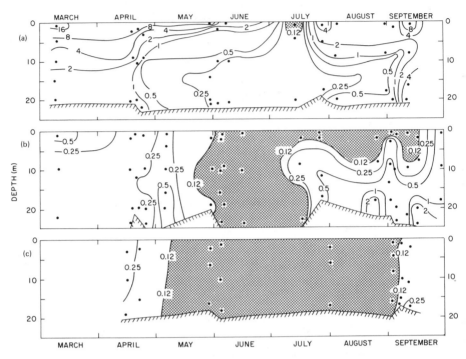

FIG. 204. The seasonal vertical structure of nitrate (μg-at. l^{-1}) at the 20-m isobath off the (a) Hudson, (b) Delaware, and (c) Chesapeake estuaries during March–September. [After Walsh *et al.* (1987c); © 1987 Pergamon Journals Ltd., with permission.]

in the water column does not negate the benefit of anthropogenic nitrogen loadings. We thus increased 10-fold (1) the flux of both nitrogen and sewage/sludge carbon across the Sandy Hook–Rockaway boundary (Fig. 205); (2) the existing loading of the apex dump sites; and (3) the existing loading from coastal sewage outfalls (Stoddard and Walsh, 1986). The phytoplankton species composition was then assumed to be all microflagellates (i.e., no *Ceratium* spp.), This numerical experiment is in contrast to experiments (g) and (h) of Table X, in which only the nitrogen loading of the Hudson River effluent was increased 10-fold in Section 5.1.3.

The water quality impact of such a 10-fold increase in present sewage carbon loading from the estuary and the apex dump sites is significant and is contrary to the above scenario. Maximum sewage carbon concentrations of 3.1–5.2 mg C l^{-1} would then account for over 90% of the total POC pool, above the pycnocline within the inner apex. Away from the inner apex, mixing and advection reduce sewage carbon concentrations of the surface layer to 0.1 mg C l^{-1}, but these are still 50% of the total POC within a radius of about 60 km. On a depth-integrated basis, such an increased carbon loading results in a mean sewage carbon content of 49 g C m^{-2} over the entire apex water column, with a range of 50–100 g C m^{-2} in the inner apex to 10 g C m^{-2} in the outer apex.

Light attenuation in the apex results, of course, from absorption and scattering by both detritus and living phytoplankton, with the phytoplankton accounting for only a minor component of total light extinction. Without an increment of sewage carbon, the model yields euphotic-zone depths (1% light level) of 7–10 m in the inner apex and 25–30 m beyond. These agree with Secchi disk obser-

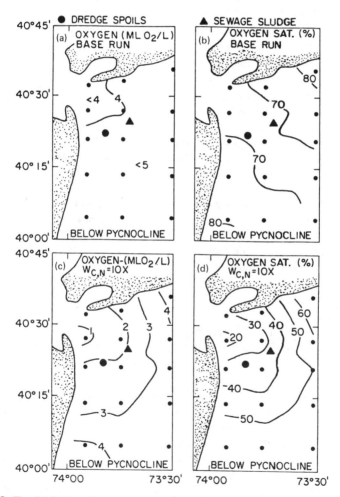

FIG. 205. The distribution of oxygen concentration and percent saturation beneath the pycnocline of the apex of the New York Bight under (a, b) 1974 loadings of carbon and nitrogen, compared to (c, d) a 10-fold increase of these inputs. [After Stoddard and Walsh (1986).]

vations in August 1974, when the 1% light depth ranged from 7–8 m near the Hudson River plume to 25–40 m within the outer apex. As a result of increased turbidity of the water column, however, the 10-fold increment of sewage carbon input to the apex results in substantial reduction of light penetration. Near the apex dump sites, for example, the 1% light depth is reduced to 2–3 m for the 10× case. Mixing and dilution of the 10-fold increment of sewage carbon also results in euphotic zones of 15–20 m depths over the outer apex.

The simulated 10-fold increase of carbon and nitrogen loading thus leads to negligible changes of the chlorophyll stocks over the apex of the New York Bight (Stoddard and Walsh, 1986). Within the Hudson River plume, the model's phytoplankton biomass is actually reduced by 10–20%, as a result of the 60–70% reduction of primary production, after increased light limitation in these model segments. Beyond the influence of the estuary and the apex dump sites, the chlorophyll biomass increased slightly (5–10%), relative to the base run. Such a population increment at the outer boundary of the apex resulted from enhanced primary production, as a consequence of the higher nitrogen levels and a reduction in turbidity.

Without much of an increase of oxygen production from photosynthesis, but with a 10-fold increase in oxygen demand from oxidation of sewage carbon, the impact of the simulated increase of urban waste is found to be most significant near the Hudson River estuary and the dump sites. Dissolved oxygen concentrations here are reduced to 3.5 ml O_2 l^{-1} (60–65% saturation) within the surface layer of the model (Fig. 205). Over the inner apex, subpycnocline anoxic conditions (Fig. 205) of <1 ml O_2 l^{-1} and 17% saturation result from the marked increase of the water column oxygen demands from 168–487 ml O_2 m^{-2} day^{-1} of the base run to 638–1554 ml O_2 m^{-2} day^{-1} of the 10× case. This is a minimal estimate, since seabed oxygen consumption was assumed constant for both the base run and the 10× case—that is, sewage carbon that sank to the bottom (e.g., Fig. 131) was neglected.

The simulated 10-fold increase of waste inputs thus resulted in both subpycnocline anoxia (Fig. 205) and a reduction of primary production, within a radius of 30 km of the apex ocean dump sites and the Hudson River estuary. The simulated POC content of this numerical experiment was then similar to observed estimates, within the apex and the New Jersey midshelf (80–125 g C m^{-2}), during the *Ceratium* bloom in June 1976. This last simulation result suggests that any detrital carbon loading to the apex water column, of the order of 50–100 g C m^{-2}, would be sufficient to initiate anoxic events. This estimate might be useful as a possible criterion for regulatory decision-making, regarding site designations for ocean waste disposal in the New York Bight.

Models, like data sets, can always be improved. In future evaluation of the assimilative capacity of these coastal waters for urban waste loading, the present model structure could be refined further to account for the farfield fate of sewage-related nitrogen inputs and phytoplankton production, (e.g., Figs. 147c and 148c of Section 5.1.3). Terms for algal mortality and for accumulation of nonliving algal biomass in the detrital carbon pool could also be included. In addition, seabed oxygen consumption and benthic remineralization of ammonium could be explicitly computed as a function of both the detrital carbon flux to the sediments and the remineralization of organic matter at the sediment–water interface (DiToro and Connolly, 1980). The results of the present model (Stoddard and Walsh, 1986) appear sufficiently robust, however, to avoid fine-tuning of its assumptions, until a larger, unaliased data set becomes available, such as from *in situ* or remote biological sensors.

Prediction of such events has both immediate and future societal value—that is, removal of shellfish before loss to anoxia, and the best mode of sewage treatment (Ketchum *et al.*, 1981). As one moves from prediction of meteorological to biological processes of the coastal food web, however, increasing sources of error arise in such prognostications (Tingle *et al.*, 1979). Nevertheless, sufficient information on the appropriate time and space scales of continental shelf processes is beginning to emerge, which suggests that delineation of cause and effect, within a perturbation response of the coastal zone, is a feasible goal over the next few decades.

After eutrophication in the North Sea, overfishing off Peru, and climatic changes in circulation within the Mid-Atlantic Bight, for example, there have been observed increases in the biomass of, respectively, microflagellates, diatoms, and dinoflagellates. Simulation models, involving oxygen, nitrogen, and carbon budgets, suggest that such increases of algal biomass have led to anoxia, denitrification, and increased production of hydrogen sulfide, as well as perhaps of methane and carbon dioxide. Presumably, the future global impacts of gas exchange between marine and atmospheric reservoirs are derived from a combination of such mesoscale responses, documented from these three shelf ecosystems.

7 Summary

She noticed a curious appearance in the air: it puzzled her very much at first, but after watching it a minute or two, she made it out to be a grin, and said to herself "It's the Cheshire Cat: now I shall have somebody to talk to!" . . . and she went on. "Would you tell me please, which way I ought to go from here?" "That depends a good deal on where you want to get to," said the Cat. . . . "—so long as I get somewhere," Alice added as an explanation. "Oh, you're sure to do that" said the Cat, "if you only walk long enough" . . . and this time it vanished quite slowly, beginning with the end of the tail, and ending with a grin, which remained some time after the rest of it had gone.

(Carroll, 1865)

Reconstruction of a Cheshire Cat from just its grin, like the description of a marine ecosystem deduced from a few shipboard, satellite, or *in situ* measurements, is an appropriate objective of simulation analysis, discussed at great length in this monograph. One wonders if there are now more models than data sets! Simulation models, like isolated current-meter data, nutrient measurements, or plankton tows, can only provide an accurate reconstruction, however, if the object of the study is already known. If one "walks long enough" in an iterative process of model confrontation with hopefully unaliased measurements (Walsh, 1972), a reasonable, if not unique, description of an unknown Cheshire Cat—that is, natural system—emerges from this process. A decade earlier, David Cushing and I wrote in the preface of another book (Cushing and Walsh, 1976) that "marine ecology is a diverse subject composed of many facts, many concepts, and few testable theories." After writing the present monograph, I feel that the words attributed to the Cheshire Cat by Lewis Carroll (1865) may now be more appropriate. Perhaps we are now "walking long enough" in our global perturbation experiments. One hopes, however, that we don't run over the edge of the continental shelf, like herded mastodons, during this obligatory peregrination!

The open sea, as discussed in Chapter 1, is an extensive, unproductive region of the ocean, in contrast to the other 20% of ocean surface area comprising the continental shelves (Figs. 206 and 207) and slopes. A recent study of Warm Core Ring 82B, shed from the north wall of the Gulf Stream into New England slope waters during February 1982, provides insight, or theories, into the mechanisms of enhanced primary production at the continental margins. The 82B Ring is analogous to a giant rotating controlled pollution experiment (CEPEX) bag, in which a fertilization experiment was performed by nature on the ambient Sargasso Sea food web (Fig. 23).

Outbreaks of cold, dry air from the land during winter lead to increased buoyancy extraction from the original Sargasso Sea water and convective overturn of such a ring (Joyce and Stalcup, 1985), as the ring moves closer to the continental margin. With a cooling rate of $\sim0.04°C$ day^{-1} for Ring 82B during March–April 1982 (Evans et al., 1985), a surface mixed layer depth of ~400 m was observed (Schmitt and Olson, 1985). This is in contrast to March mixed layer depths of ~250 m (Fig. 208) in the adjacent Sargasso Sea (Levitus, 1982).

Introduction of nutrient-rich water within the center of Ring 82B (Fox and Kester, 1986) by convective overturn of the water column leads, of course, to higher algal biomass. As much as 150 mg chl m^{-2} was found over the Ring thermostad (i.e., the isothermal layer) (Seitz, 1967), compared to only 40 mg chl m^{-2} in the Sargasso Sea (Hitchcock et al., 1985). Although the surface algal concentration of Ring 82B was initially less than that of the surrounding slope water, vertical mixing was sufficiently intense to allow survival of diatom populations at 300 m, with a depth-integrated biomass as great as that on the mid-Atlantic continental slope (Nelson et al., 1985).

As much as 6 μg-at. NO$_3$ l^{-1} was initially measured in the slope water euphotic zone (McCarthy and Nevins, 1986b). A carbon fixation of 1.5 g C m^{-2} day^{-1} for this spring bloom (Hitchcock et al., 1985) is almost equal to that of shelf waters in the Mid-Atlantic Bight (Brown et al., 1985). Similar enhancement of algal biomass and production within other rings has been observed in both the East Australian Current (Tranter et al., 1980, 1983) and off New Zealand (Bradford et al., 1982).

The lack of such deep convective mixing at Station P in the North Pacific may maintain the close seasonal coupling between herbivores and their algal food (Evans and Parslow, 1985). The mixed layer depth (Fig. 208) in March is only ~100 m at Station P (Dodimead et al., 1963). In winter, these organisms are presumably not dispersed over a deeper water column to allow spatial uncoupling of predator and prey, that is, without subsequent time lags in their population dynamics during spring (Fig. 21).

In contrast, the microzooplankton (64–333 μm) in April 1982 consumed <1% of the daily primary production of Ring 82B (Roman, 1984), where 90% of the algal biomass was initially <10 μm (G. Hitchcock, personal communication). The biomass of the microzooplankton did not increase significantly by

FIG. 206. The distribution of continental shelves in the northern hemisphere.

June 1982 within Ring 82B (Roman *et al.*, 1985). Reproduction of smaller copepods by then, however, led to a consumption of 10–40% of the daily algal production, in which 30–40% of the chlorophyll biomass was then netplankton (20–153 μm), analogous to shelf food webs.

The macrozooplankton (>333 μm) biomass of Ring 82B was initially less than that of the surrounding slope waters in March 1982, but increased threefold by June 1982, mainly as a function of *in situ* growth rather than of immigration (Wiebe *et al.*, 1985). A concomitant shift in trophic structure occurred, with marked increase of large carnivores (e.g., chaetognaths, ctenophores, and euphausiids), from March to June 1982 (Davis and Wiebe, 1985)—recall the similar seasonal cycle of these organisms within shelf ecosystems in Section

FIG. 207. The distribution of continental shelves in the southern hemisphere.

1.3.5. Similarly, importance of the saprovore component of the Ring 82B food web changed with time, from a bacterial secondary production of 1–2% of the daily algal carbon fixation in April 1982 to about 20% in June 1982 (Ducklow, 1986), analogous to seasonal estimates for the New York shelf (Ducklow *et al.*, 1982).

A second nutrient enrichment mechanism of warm core rings evidently takes place after these anticyclonic eddies lose some of the original kinetic energy of the western boundary current, by frictional decay within the surrounding slope water. About 2 months after ring formation, over which a horizontal translation of ~125 km can occur (Evans *et al.*, 1985), relaxation of the depressed thermocline may lead to an upwelling rate of ~1 m day^{-1} at ring center (Franks *et*

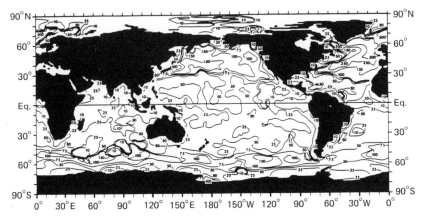

FIG. 208. The March surface mixed-layer depths of the ocean, based on a σ_t criterion of 0.125. [After Levitus (1982).]

al., 1986). This vertical velocity is similar to that in either the Bering Sea or within the cyclonic spin-off eddies of western boundary currents, formed when the current is at the edge of the continental shelf—recall Sections 1.5.3 and 4.2.

With seasonal stratification of the ring water column, surface waters are stripped of nutrients (Garside, 1985). The resupply rate of nitrate is then dominated by this apparent upwelling process, prolonging dominance of diatoms in the center of the ring (Nelson *et al.*, 1985), and supplying perhaps 40–100% of the measured uptake of nitrate by phytoplankton (Nelson *et al.*, 1987). This vertical flux results in an order of magnitude larger nitrate gradient across the nutricline of the ring than in the original Sargasso Sea water (McCarthy and Nevins, 1986a,b), such that the ring primary production in June 1982 still ranged from 0.3 to 1.0 g C m^{-2} day^{-1}, similar to that of the previous April (Hitchcock *et al.*, 1985).

The extent and importance of these western boundary sites of intensified primary production within slope waters and the amount of possible transport of particulate matter to depth is unknown. Within the open ocean, interannual changes in winter ventilation of surface waters may be the cause of year-to-year variation of primary production within the Sargasso Sea (Menzel and Ryther, 1961b; Jenkins and Goldman, 1985) and the North Pacific Ocean (McGowan and Hayward, 1978). Similar variability of physical forcing is to be expected within the more productive waters at the continental margins (Figs. 206 and 207), with episodic transfer of particulate matter to depth.

Based on modeled and measured chemical composition of particles caught in the deep sea, for example, as much as 50% of the annual import may be composed of unaltered plant cells, generated at times and places of high production

(Watson and Whitfield, 1985). Furthermore, when episodes of enhanced vertical mixing of nutrients into the euphotic zone were simulated for the Panama Basin, primary production varied only threefold, while the vertical particle flux changed by an order of magnitude (Bishop and Marra, 1984). Another simulation model of wind-pulsed nitrate fluxes to the surface mixed layer of the oligotrophic Mediterranean Sea (Klein and Coste, 1984) suggested that "new" nitrogen fluxes were equal to those of the total carbon production—that is, no "recycled" nitrogen was required.

Conventional wisdom of a one-layered euphotic zone of the sea is now shifting to a two-layered system (Knauer et al., 1983; Karl et al., 1984; Coale and Bruland, 1987). In such a paradigm, high vertical fluxes of particulate matter take place just above the nutricline, while an upper layer of ~10% "new" production contains small particles of low sinking rates. Resolution of arguments about oxygen and ^{14}C estimates of primary production in the oligotrophic central gyres (Schulenberger and Reid, 1981; Platt, 1984; Jenkins and Goldman, 1985; Platt and Harrison, 1985) may hinge on measurements of the episodic nature of vertical nitrate fluxes within this bottom part of the euphotic zone.

Equating "new production" with particle export (Eppley and Peterson, 1980) implies that an upward revision of the rate of nitrate supply to the euphotic zone will lead to an increased flux of biogenic particles, sinking out of a two-layered euphotic zone. If the particle size remains small, as in the open ocean, however, significant export will only occur at the ocean's margins, where larger diatoms, dinoflagellates, and macroaggregates usually occur. The transition of the Warm Core Ring 82B food web, from an open ocean to a coastal community, suggests that marine ecosystems must be structured by vertical input rates of >1 m day^{-1} (10^{-3} cm sec^{-1}—recall Table IV) to support diatomaceous seed populations. As precursors of infusorial earth, these diatoms are capable of opportunistically converting episodic pulses of dissolved nitrate into particulate nitrogen for export to sediments.

These same ecosystems at the continental margins are the most likely candidates for converting anthropogenic nutrient loading into sedimentary organic matter as well. An upwelling velocity of ~0.1 m day^{-1} within the equatorial Pacific divergence (Table IV), for example, is evidently sufficient to introduce enough flux of CO_2 to serve as a global source of outgassing from surface waters (Table XVI). Such a vertical motion is not great enough to support diatom populations, however, to strip the concomitant flux of nitrate. In contrast, upwelling rates of >1 m day^{-1} are found in slope waters of western boundary currents, within eastern boundary currents, and on other continental shelves (Fig. 199), where populations of large diatoms occur. Sinking fluxes of phytodetritus from these productive ecosystems have evidently impacted the global ecology of Earth in the past, and may do so in the future.

The principles of fluid mechanics and physical chemistry advanced in the seventeenth and eighteenth centuries, of mathematics and natural history in the nineteenth century, and of the earth sciences in the twentieth century provide refutable hypotheses for analysis of the massive data sets now being collected from continental shelves. Moored instruments, satellite sensors, and modeling capabilities, developed within the last decade, are mainly responsible for this exciting turn of events. Of equal importance, however, is the realization that the results of humanity's inadvertent tracer experiments will become more detectable in the next decade. They may possibly provide unwelcome validation data for theories of the consequences of enhanced rates of biogeochemical cycling of elements on Earth. As the shallow extension of 70% of our watery planet (Figs. 206 and 207), the continental shelves warrant our continued attention as a source of energy, food, and recreation, and perhaps for early detection of global change.

References

Aagaard, K., A. T. Roach, and J. D. Schumacher (1985). On the wind-driven variability of the flow through Bering Strait. *J. Geophys. Res.* **90**, 7213–7221.

Aaronson, S., T. Berner, and Z. Dubinsky (1980). Microalgae as a source of chemicals and natural products. *In* "Algal Biomass" (G. Sheief, C. J. Soeder, and M. Balaban, eds.), pp. 711–726. Elsevier, Amsterdam.

Abbott, M. R., and P. M. Zion (1985). Satellite observations of phytoplankton variability during an upwelling event. *Cont. Shelf Res.* **6**, 661–680.

Aiken, J. (1981). A chlorophyll sensor for automatic, remote operation in the marine environment. *Mar. Ecol. Prog. Ser.* **4**, 235–239.

Alberts, J. (1970). Inorganic controls of dissolved phosphorous in the Gulf of Mexico. Ph.D. Thesis, Florida State Univ., Tallahassee.

Aleem, A. A., and N. Dowidar (1967). Phytoplankton production in relation to nutrients along the Egyptian Mediterranean coast. *Proc. Int. Congr. Trop. Oceanogr., Univ. Miami, Coral Gables* pp. 305–327.

Alexander, J. E., and E. Alexander (1976). "Chemical Properties," Bight Atlas Monogr. No. 2, pp. 1–47. MESA, New York. Alexander, J. E., J. H. Steele, and E. F. Corcoran (1962). The seasonal cycle of chlorophyll in Florida Straits. *Proc. Annu. Gulf Caribb. Fish. Inst.* **14**, 63–67.

Alldredge, A. L. (1979). The chemical composition of macroscopic aggregates in two neritic seas. *Limnol. Oceanogr.* **24**, 855–866.

Altabet, M. A., A. R. Robinson, and L. J. Walstad (1986). A model for the vertical flux of nitrogen in the upper ocean: simulating the alteration of isotopic ratios. *J. Mar. Res.* **44**, 203–225.

Alton, M. S. (1972). Characteristics of the demersal fauna inhabiting the outer continental shelf and slope off the northern Oregon Coast. *In* "The Columbia River Estuary and Adjacent Waters" (A. T. Pruter and D. L. Alversen, eds.) pp. 583–636. Univ. of Washington Press, Seattle.

Alversen, D. L., A. R. Longhurst, and J. A. Gulland (1969). How much food from the sea? *Science* **168**, 503–505.

Andersen, K. P. and E. Ursin (1977). A multispecies extension to the Beverton and Holt theory of fishing, with accounts of phosphorus circulation and primary production. *Medd. Dan. Fisk. Havunders.***7**, 319–435.

Anderson, G. C. and R. E. Munson (1972). Primary productivity studies using merchant vessels in the North Pacific Ocean. *In* "Biological Oceanography of the Northern North Pacific Ocean" (A. Y. Takenouti, ed.) pp. 245–252. Idemitsu Shoten, Tokyo.

Anderson, G. C., T. R. Parsons, and K. Stephens (1969). Nitrate distribution in the subarctic northeast Pacific Ocean. *Deep-Sea Res.* **16**, 329–334.

Anderson, J. S. (1977). Primary productivity associated with sea ice at Godhaven, Disko, West Greenland. *Ophelia* **16**, 205–220.

Anderson, L. W., and B. M. Sweeney (1978). Role of inorganic ions in controlling sedimentation rate of a marine centric diatom *Ditylum brightwelli*. *J. Phycol.* **14**, 204–214.

Anderson, R. F., R. F. Bopp, K. O. Buessler, and P. E. Biscaye (1987). Mixing of particles and organic constituents in sediments from the continental shelf and slope off Cape Cod: SEEP-I results. *Cont. Shelf Res.* (in press).

Andrews, J. C., and P. Gentien (1982). Upwelling as a source of nutrients for the Great Barrier Reef ecosystem: a solution to Darwin's question? *Mar. Ecol. Prog. Ser.* **8**, 257–269.

Andrews, R. C. (1929). "Ends of the Earth." Putnam, New York.

Ankar, S. (1980). Growth and production of *Macoma baltica* in a northern Baltic soft bottom. *Ophelia* **1**, 31–41.

Apstein, C. (1911). Biologische Studie über *Ceratium tripos* var. *subsalsa*. *Ostf. Wiss. Meeres.* **12**, 137–162.

Arenholz, D. W. (1981). Recruitment and exploitation of Gulf menhaden, *Brevoortia patronus*. *Fish. Bull.* **79**, 325–336.

Armstrong, D. W. (1974). Some dynamics of carbon, nitrogen, and phosphorus in the marine shelf environment of the Mississippi Fan. M.S. Thesis, Texas A&M Univ., College Station.

Armstrong, R. S. (1979). Bottom oxygen and stratification in 1976 and previous years. *In* "Oxygen Depletion and Associated Benthic Mortalities in the New York Bight, 1976" (C. J. Sindermann and R. L. Swanson, eds.) *NOAA Prof. Pap.* No. 11, 137–148.

Arndt, E. A. (1967). Untersuchungen an populationen von *Ceratium tripos* f. *subsalsum* ostenf im gebeit der südküste der Mecklen. burger Bucht. *Wiss. Z. Univ. Rostock, Math.-Naturwiss. Reihe* **16**(9/10), 1199–1206.

Arrhenius, S. 1896. On the influence of carbonic acid in the air upon the temperature of the ground. *London Edinburgh Dublin Philos. Mag. J. Sci.* **41**, 237–276.

Arthur, M. A., and S. O. Schlanger (1979). Cretaceous oceanic anoxic events as causal factors in development of reef-reservoired giant oil fields. *Am. Assoc. Pet. Geol. Bull.* **63**, 870–885.

Arthur, R. S. (1965). On the calculation of vertical motion in eastern boundary currents from determinations of horizontal motion. *J. Geophys. Res.* **70**, 2799–2803.

Atkinson, L. P., G. A. Paffenhofer, and W. M. Dunstan (1978). The chemical and biological effect of a Gulf Stream intrusion off St. Augustine, Florida. *Bull. Mar. Sci.* **28**, 667–679.

Bacastow, R. (1976). Modulation of atmospheric carbon dioxide by the Southern Oscillation. *Nature (London)* **261**, 116–118.

Bacastow, R., J. A. Adams, C. D. Keeling, J. D. Moss, T. P. Whorf, and C. S. Wong (1980). Response of atmospheric carbon dioxide to the weak 1975 El Niño. *Science* **210**, 66–68.

Bacon, M. P., D. W. Spencer, and P. G. Brewer (1976). $^{210}Pb/^{226}Ra$ and $^{210}Po/^{210}Pb$ disequilibria in seawater and suspended particulate matter. *Earth Planet. Sci. Lett.* **32**, 277–296.

Bailey, K. M., and J. Yen (1983). Predation by a carnivorous marine copepod, *Euchaeta elongata*, on eggs and larvae of the Pacific hake. *J. Plankton Res.* **5**, 71–82.

Bainbridge, V. (1960). The plankton of the inshore waters off Freetown, Sierra Leone. *G. B. Colon. Off. Fish. Publ.* No. 13, 1–48.

Bakkala, R. G. (1981). Population characteristics and ecology of yellowfin sole. *In* "The Eastern Bering Sea Shelf: Oceanography and Resources," (D. W. Hood and J. A. Calder, eds.) Vol. 1, pp. 553–574. Univ. of Washington Press, Seattle.

Bakkala, R. G., and L. L. Low (1983). Condition of ground fish resources of the eastern Bering Sea and Aleutian Islands region in 1982. *NOAA Tech. Mem., Natl. Mar. Fish. Serv.* **F/NAWC-42**, 59–63.

Bakun, A. (1973). Coastal upwelling indices, west coast of North America, 1946–71. *NOAA Tech. Rep.* **NMFS SSRF-671**, 1–103.

Bakun, A. (1978). Guinea Current upwelling. *Nature (London)* **271**, 147–150.

Ballester, A., and R. Margalef (1967). Produccion primaria. *Mem. Soc. Cienc. Nat. La Salle* **25**, 209–221.

Balsam, W. L. (1982). Carbonate dissolution and sedimentation on the mid-Atlantic continental margin. *Science* **217**, 929–931.

Balsam, W. L., and L. E. Heusser (1976). Direct correlation of sea surface paleotemperatures, deep circulation, and terrestrial paleoclimates: foraminiferal and palynological evidence from two cores off Chesapeake Bay. *Mar. Geol.* **21**, 121–147.

Baly, E. C. (1935). The kinetics of photosynthesis. *Proc. R. Soc. London, Ser. B* **117**, 218–239.

Bannister, R. C. (1978). Changes in plaice stocks and plaice fisheries in the North Sea. *Rapp. P-V. Reun. Cons. Int. Explor. Mer* **172**, 86–101.

Bannister, T. T. (1974a). Production equations in terms of chlorophyll concentration, quantum yield, and upper limit to production. *Limnol. Oceanogr.* **19**, 1–12.

Bannister, T. T. (1974b). A general theory of steady state phytoplankton growth in a nutrient saturated mixed layer. *Limnol. Oceanogr.* **19**, 13–30.

Banse, K. (1976). Rates of growth, respiration, and photosynthesis of unicellular algae as related to cell size—a review. *J. Phycol.* **12**, 135–140.

Barber, R. T., and F. P. Chavez (1983). Biological consequences of El Niño. *Science* **222**, 1203–1210.

Barber, R. T., and F. P. Chavez (1986). Ocean variability in relation to living resources during the 1982–83 El Niño. *Nature (London)* **319**, 279–285.

Barber, R. T., and J. H. Ryther (1969). Organic chelators: factors affecting primary production in the Cromwell Current upwelling. *J. Exp. Mar. Biol. Ecol.* **3**, 191–199.

Barber, R. T., R. C. Dugdale, J. J. MacIsaac, and R. L. Smith (1971). Variations in phytoplankton growth associated with the source and conditioning of upwelling water. *Invest. Pesq.* **35**, 171–193.

Barron, E. J., S. L. Thompson, and S. H. Schneider (1981). An ice-free Cretaceous? Results from climate model simulations. *Science* **212**, 501–508.

Beardsley, R. C., and B. Butman (1974). Circulation on the New England continental shelf: response to strong winter storms. *Geophys. Res. Lett.* **1**, 181–184.

Beardsley, R. C., and C. N. Flagg (1976). The water structure, mean currents, and shelf-water/slope water front on the New England continental shelf. *Mem. R. Soc. Liege* **6**, 209–225.

Beardsley, R. C., W. C. Boicourt, and D. V. Hansen (1976). Physical oceanography of the Middle Atlantic Bight. *ASLO Spec. Symp.* No. 2, 20–34.

Beardsley, R. C., C. A. Mills, J. A. Vermersch, W. S. Brown, N. Pettigrew, J. Irish, S. Ramp, R. Schlitz, and B. Butman (1983). Nantucket Shoals Flux Experiment (NSFE 79). Part 2: Moored Array Data Report. *WHOI Tech. Rep.* **WHOI-83-13.**

Beardsley, R. C., D. C. Chapman, K. H. Brink, S. R. Ramp, and R. Schlitz (1985). The Nantucket Shoals Flux Experiment (NSFE 79). Part I: A basic description of the current and temperature variability. *J. Phys. Oceanogr.* **15**, 713–748.

Becacos-Kontos, T. (1968). The annual cycle of primary production in the Saronicos Gulf (Aegean Sea) for the period November 1963–October 1964. *Limnol. Oceanogr.* **13**, 485–489.

Beebe, W. (1926). "The Arcturus Adventure." Putnam, New York.

Beebe, W. (1934). "Half Mile Down." Harcourt, New York.

Beers, J. R., D. M. Steven, and J. B. Lewis (1968). Primary productivity in the Caribbean Sea off Jamaica and the tropical North Atlantic off Barbados. *Bull. Mar. Sci.* **18**, 86–104.

Beers, J. R., and G. L. Stewart (1971). Microzooplankters in the plankton communities of the upper waters of the eastern tropical Pacific. *Deep-Sea Res.* **18**, 861–865.

Beers, J. R., M. R. Stevenson, R. W. Eppley, and E. R. Brooks (1971). Plankton populations and upwelling off the coast of Peru, June 1969. *Fish. Bull.* **69**, 859–876.

Belknap, D. F., and J. C. Kraft (1981). Preservation potential of transgressive coastal lithosomes on the U.S. Atlantic shelf. *Mar. Geol.* **42**, 429–442.

Bell, P. R. (1982). Methane hydrate and the carbon dioxide question. *In* "Carbon Dioxide Review: 1982" (W. C. Clark, ed.), pp. 401–406. Oxford Univ. Press, London and New York.

Benninger, L. K., R. C. Aller, K. J. Cochran, and K. K. Turekian (1979). Effects of biological sediment mixing on the ²¹⁰Pb chronology and trace metal distribution in a Long Island Sound sediment core. *Earth Planet. Sci. Lett.* **43**, 241–259.

Benninger, L. K., and S. Krishnaswami (1981). Sedimentary processes in the inner New York Bight: evidence from excess ²¹⁰Pb and ²³⁹,²⁴⁰Pu. *Earth Planet. Sci. Lett.* **53**, 158–174.

Benoit, G. J., K. K. Turekian, and L. K. Benninger (1979). Radiocarbon dating of a core from Long Island Sound. *Estuarine Coast. Mar. Sci.* **9**, 171–180.

Berg, J., and G. Radach (1985). Trends in nutrient and phytoplankton concentrations at Helgoland Reede (German Bight) since 1962. ICES C.M./L:2/Sess. R, pp. 1–16.

Berger, R., and W. F. Libby (1966). UCLA radiocarbon dates V. *Radiocarbon* **8**, 467–497.

Berger, R., R. E. Taylor, and W. F. Libby (1966). Radiocarbon content of marine shells from the California and Mexican west coast. *Science* **153**, 864–866.

Berger, R., and W. F. Libby (1967). UCLA radiocarbon dates VI. *Radiocarbon* **9**, 477–504.

Berger, W. H. (1970). Biogenous deep-sea sediments: fractionation by deep-sea circulation. *Geol. Soc. Am. Bull.* **81**, 1385–1401.

Berger, W. H. (1977). Deep-sea carbonate and the deglaciation preservation spike in pteropods and foraminifera. *Nature (London)* **269**, 301–304.

Berggren, W. A., and C. D. Hollister (1974). Paleogeography, paleobiogeography and the history of circulation in the Atlantic Ocean. *In* "Studies in Paleo-oceanography" (W. W. Hay, ed.), *Soc. Econ. Paleonto. Mineral. Spec. Publ.* No. 20, 126–186.

Berman, T., D. W. Townsend, S. Z. El-Sayed, C. C. Trees, and Y. Azov (1984). Optical transparency, chlorophyll, and primary productivity in the eastern Mediterranean near the Israeli coast. *Oceanol. Acta* **7**, 367–372.

Berman, T., P. D. Walline, A. Schneller, J. Rothenberg, and D. W. Townsend (1985). Secchi disk depth record: a claim for the eastern Mediterranean. *Limnol. Oceanogr.* **30**, 447–448.

Bernard, F. R. (1974). Annual biodeposition and gross energy budget of mature Pacific oysters, *Crassostrea gigas. J. Fish. Res. Board Can.* **31**, 185–190.

Berner, R. A. (1982). Burial of organic carbon and pyrite sulfur in the modern ocean: its geological and environmental significance. *Am. J. Sci.* **282**, 451–473.

Berrit, G. R. (1964). Observations oceanographiques cotieres a Pointe-Noire de 1953 a 1963. *Cah. ORSTOM, Ser. Oceanogr.* **2**, 31–56.

Berry, W. N., and P. Wilde (1978). Progressive ventilation of the oceans—An explanation for the distribution of the lower Paleozoic black shales. *Am. J. Sci.* **278**, 257–275.

Bessenov, N. M., and M. V. Fedosov (1965). Primary productivity in the shelf waters off the West African coast. *Oceanology* **5**, 88–93.

Betzer, P. R., W. J. Showers, E. A. Laws, C. D. Winn, G. R. DiTullio, and P. M. Kroopnick (1984). Primary productivity and particle fluxes on a transect of the equator at 153°W in the Pacific Ocean. *Deep-Sea Res.* **31**, 1–11.

Bien, G. S., N. W. Rakestraw, and H. E. Suess (1960). Radiocarbon concentration in Pacific Ocean water. *Tellus* **12**, 436–443.

Bien, G. S., N. W. Rakestraw, and H. E. Suess (1965). Radiocarbon in the Pacific and Indian oceans and its relation to deep water movement. *Limnol. Oceanogr.* **10**, R25-R37.

Bien, G. S., and L. J. Pandolfi (1972). La Jolla natural radiocarbon measurements VI. *Radiocarbon* **14**, 363–379.

Bigelow, H. B. (1915). Exploration of the coast waters between Nova Scotia and Chesapeake Bay, July and August 1913, by the U.S. Fisheries schooner Grampus. Oceanography and Plankton. *Bull. Mus. Comp. Zool.* **59**, 152–359.

Bigelow, H. B. (1926). Plankton of the offshore waters of the Gulf of Maine. *Bull. Bur. Fish.* **40**, Part II, 1–509.

Bigelow, H. B. (1927). Physical oceanography of the Gulf of Maine. *Bull. Bur. Fish.* **40**, 511–1027.

Bigelow, H. B. (1933). Studies on waters of the continental shelf, Cape Cod to Chesapeake Bay. 1. The cycle of temperature. *Pap. Phys. Oceanogr. Meteorol.* **2**, 1–135.

Bigelow, H. B., and W. W. Welsh (1925). Fishes of the Gulf of Maine. *Bull. U.S. Bur. Fish.* **40**, 1–567.

Bigelow, H. B., and M. Sears (1939). Studies of the waters of the continental shelf, Cape Cod to Chesapeake Bay. III. A volumetric study of the zooplankton. *Mem. Comp. Zool.* **54**, 183–378.

Billen, G. (1978). A budget of nitrogen recycling in North Sea sediments off the Belgian coast. *Estuarine Coast. Mar. Sci.* **7**, 127–146.

Billet, D. S., K. S. Lampitt, A. L. Rice, and R. F. Mantoura (1983). Seasonal sedimentation of phytoplankton to the deep-sea benthos. *Nature (London)* **302**, 520–522.

Biscaye, P. E., R. F. Anderson, and B. L. Deck (1987). Fluxes of particles and constituents to the eastern U.S. continental slope and rise: SEEP I. *Cont. Shelf Res.* (in press).

Bishop, J. K., J. J. Edmond, D. R. Ketten, M. P. Bacon, and W. B. Silker (1977). The chemistry, biology, and vertical flux of particulate matter in planktonic environments. *Limnol. Oceanogr.* **21**, 14–23.

Bishop, J. K., and J. Marra (1984). Variations in primary production and particle flux through the base of the euphotic zone at the site of the Sediment Trap Intercomparison Experiment (Panama Basin). *J. Mar. Res.* **42**, 189–206.

Bishop, S. S., J. A. Yoder, and G. A. Paffenhofer (1980). Phytoplankton and nutrient variability along a cross-shelf transect off Savannah, Georgia, U.S.A. *Estuarine Coast. Mar. Sci.* **11**, 359–368.

Bjerknes, J. (1964). Atlantic air-sea interaction. *Adv. Geophys.* **10**, 1–82.

Bjerknes, J. (1966). Survey of El Niño 1957–1958 in its relation to tropical Pacific meteorology. *Inter-Am. Trop. Tuna Comm. Bull.* **12**, 1–62.

Blackburn, M., R. M. Laurs, R. W. Owen, and B. Zeitzschel (1970). Seasonal and areal changes in standing stocks of phytoplankton, zooplankton, and micronekton in the eastern tropical Pacific. *Mar. Biol. (Berlin)* **7**, 14–31.

Blackburn, T. H., and K. Henriksen (1983). Nitrogen cycling in different types of sediments from Danish Waters. *Limnol. Oceanogr.* **28**, 477–493.

Blackman, F. F. (1905). Optima and limiting factors. *Ann. Bot.* **19**, 281–295.

Blackman, R. B., and J. W. Tukey (1958). "The Measurement of Power Spectra." Dover, New York.

Blaxter, J. H., and M. E. Staines (1971). Food searching potential in marine fish larvae. *Proc. Eur. Mar. Biol. Symp., 4th* pp. 467–485.

Bloesch, J., and N. M. Burns (1980). A critical review of sedimentation trap techniques. *Schweiz. Z. Hydrol.* **42**, 15–55.

Blomquist, S., and L. Hakanson (1981). A review of sediment traps in aquatic environments. *Arch. Hydrobiol.* **91**, 101–132.

Bloom, A. L. (1977). Atlas of sea level curves. *Int. Geol. Corr. Proj.* **61**, 1–123.

Bodungen, B. V., K. Brockel, V. S. Smetacek, and B. Zeitzschel (1981). Growth and sedimentation of the phytoplankton spring bloom in the Bornholm Sea (Baltic Sea). *Kiel. Meeresforsch., Sonderh.* **5**, 460–490.

Bodungen, B. V., V. S. Smetacek, M. M. Tilzer, and B. Zeitzschel (1986). Primary production and sedimentation during spring in the Antarctic Peninsula region. *Deep-Sea Res.* **33**, 177–194.

Bogorov, V. G. (1941). The biological seasons of plankton in the different seas. *Dokl. Akad. Nauk SSSR* **31**, 403–406.

Bogorov, V. G., M. D. Vinogradov, N. M. Voronina, I. P. Kanaeva, and I. A. Suetova (1968). Distribution of zooplankton biomass within the surficial layer of the world ocean. *Dokl. Akad. Nauk SSSR* **182**, 1205–1207.

Bolin, B., and C. D. Keeling (1963). Large-scale atmospheric mixing as deduced from the seasonal and meridional variation of carbon dioxide. *J. Geophys. Res.* **68**, 3899–3920.

Bond, G. (1978). Speculation on real sea-level changes and vertical motions of continents at selected times in the Cretaceous and Tertiary Periods. *Geology* **6**, 247–250.

Booth, B. C., J. Lewin, and R. E. Norris (1982). Nanoplankton species predominant in the subarctic Pacific in May and June 1978. *Deep-Sea Res.* **29**, 185–200.

Booth, J. S. (1979). Recent history of mass-wasting on the upper continental slope, northern Gulf of Mexico, as interpreted from the consolidation states of the sediment. *In* "Geology of Continental Slopes" (L. J. Doyle and O. H. Pilkey, eds.), SEPM Spec. Publ. 27, pp. 153–164. Soc. Econ. Paleontol. Mineral., Tulsa, Oklahoma.

Bothner, M. H., E. C. Spiker, P. P. Johnson, R. R. Rendings, and P. J. Aruscavage (1981). Geochemical evidence for modern sediment accumulation on the continental shelf off southern New England. *J. Sediment. Petrol.* **51**, 281–292.

Botkin, D. B., and E. A. Keller (1982). "Environmental Studies." Merrill, Columbus, Ohio.

Bowden, W. B., and F. H. Bormann (1986). Transport and loss of nitrous oxide in soil water after forest clear-cutting. *Science* **233**, 867–869.

Bowen, V. T., and H. D. Livingston (1981). Radionuclide distribution in sediment cores retrieved from marine radioactive waste dumpsites. *In* "Impacts of Radionuclide Releases into the Marine Environment," IAEA, Vienna.

Bowman, M. J., and L. D. Wunderlick (1976). Hydrographic properties. *MESA N.Y. Bight Atlas Monogr.* No. 1, 1–78.

Boyer, J. S. (1982). Plant productivity and environment. *Science* **218**, 443–448.

Bradford, M. R., R. A. Heath, F. H. Chang, and C. H. Hay (1982). The effect of warm-core eddies on oceanic productivity off northeastern New Zealand. *Deep-Sea Res.* **29**, 1501–1516.

Brandhorst, W., J. G. Simpson, M. Careno, and O. Rojas (1968). Anchoveta resources in northern Chile in relation to environmental conditions from January to February 1965. *Arch. Fisch.* **19**, 167–235.

Brinton, E. (1962). The distribution of Pacific euphausiids. *Bull. Scripps Inst. Oceanogr.* **8**, 226–240.

Bristow, M. F., B. Bundy, R. Furtek, and J. Baker (1979). "Airborne Laser Fluorosensing of Surface Water Chlorophyll a," Rep. EPA-600/4-79-048. Environ. Prot. Agency, Las Vegas, Nevada.

Broecker, W. S. (1971). Calcite accumulation rates and glacial to interglacial changes in ocean mixing. *In* "The Late Cenozoic Glacial Ages" (K. K. Turekian, ed.), pp. 239–265. Yale Univ. Press, New Haven, Connecticut.

Broecker, W. S. (1979). A revised estimate for the radiocarbon age of North Atlantic Deep Water. *J. Geophys. Res.* **84**, 3218–3226.

Broecker, W. S. (1982a). Glacial to interglacial changes in ocean chemistry. *Prog. Oceanogr.* **11**, 151–197.

Broecker, W. S. (1982b). Ocean chemistry during glacial time. *Geochim. Cosmochim. Acta* **46**, 1689–1705.

Broecker, W. S., R. Gerard, M. Ewing, and B. C. Heezen (1960). Natural radiocarbon in the Atlantic Ocean. *J. Geophys. Res.* **71**, 5827–5936.

Broecker, W. S., and E. A. Olson (1961). Lamont radiocarbon measurements VIII. *Radiocarbon* **3**, 176–204.

Broecker, W. S., T. Takahashi, H. J. Simpson, and T. H. Peng (1979). Fate of fossil fuel carbon dioxide and the global carbon budget. *Science* **206**, 409–418.

Broecker, W. S., T.-H. Peng, and R. Engh (1980). Modeling the carbon system. *Radiocarbon* **22**, 565–598.

Broecker, W. S., and T.-H. Peng (1982). "Tracers in the Sea," pp. 368–372. Eldigio Press, New York.

Brongersma-Sanders, M. (1957). Mass mortality in the sea. *In* "Treatise on Marine Ecology and Paleoecology" (J. W. Hedgepeth, ed.), Vol. 1, pp. 941–1010. *Mem. Geol. Soc. Am.* No. 67.

Brooks, C. E. (1970). "Climate Through the Ages." Dover, New York.

Brouardel, J., and E. Rink (1963). Mesure de la production organique en Mediterranee dans les parages de Monaco a l'aide du ^{14}C. *Ann. Inst. Oceanogr. (Paris)* **40**, 109–164.

Brown, O. B., R. H. Evans, J. W. Brown, H. R. Gordon, R. C. Smith, and K. S. Baker (1985). Phytoplankton blooming off the U.S. East Coast: a satellite description. *Science* **229**, 163–167.

Brown, P. C., and J. G. Field (1986). Factors limiting phytoplankton production in a nearshore upwelling area. *J. Plankton Res.* **8**, 55–68.

Bruns, E. (1958). "Ozeanologie," Vol. 1. VEB Dtsch. Verlag Wiss., Berlin.

Brush, G. S., E. A. Martin, R. S. DeFries, and C. A. Rice (1982). Comparisons of ^{210}Pb and pollen methods for determining rates of estuarine sediment accumulation. *Quat. Res. (N.Y.)* **18**, 196–217.

Bsharah, L. (1957). Plankton of the Florida Current. V. Environmental conditions, standing crop, seasonal and diurnal changes at a station 40 miles east of Miami. *Bull. Mar. Sci. Gulf Caribb.* **7**, 201–251.

Buchanan, J. 1886. On similarities in the physical geography of the great oceans. *Proc. R. Geogr. Soc.* **8**, 753–770.

Bue, C. (1970). Stream flow from the United States into the Atlantic Ocean during 1931–60. *U.S. Geol. Surv. Pap.* No. 1899, 1–36.

Bulfinch, D. L., M. T. Ledbetter, B. B. Elwood, and W. L. Balsam (1982). The high-velocity core of the Western Boundary Undercurrent at the base of the U.S. continental rise. *Science* **215**, 970–973.

Bullister, J. L., and R. F. Weiss (1983). Anthropogenic chlorofluoromethanes in the Greenland and Norwegian Seas. *Science* **221**, 265–268.

Bumpus, D. F. (1973). A description of the circulation on the continental shelf of the east coast of the United States. *Prog. Oceanogr.* **6**, 111–157.

Burchall, J. (1968). An evaluation of primary productivity studies in the continental shelf regions of the Agulhas current near Durban. *Invest. Rep. Oceanogr. Res. Inst. S. Afr.* **21**, 1–44.

Burd, A. C. (1978). Long-term changes in North Sea herring stocks. *Rapp. P.-V. Reun. Cons. Int. Explor. Mer* **172**, 137–153.

Burd, A. C., and D. H. Cushing (1962). Growth and recruitment in the herring of the southern North Sea. II. Recruitment to the North Sea herring stocks. *Fish. Invest. London, Ser. 2* **23**, 1–71.

Burnett, W. C., and O. A. Schaeffer (1980). Effect of ocean dumping on ^{13}C/^{12}C ratios in marine sediments from the New York Bight. *Estuarine Coast. Mar. Sci.* **11**, 605–611.

Burns, N. M. (1976). Temperature, oxygen, and nutrient distribution patterns in Lake Erie, 1970. *J. Fish. Res. Board Can.* **33**, 485–511.

Byers, R. J. (1963). The metabolism of twelve aquatic laboratory microecosystems. *Ecol. Monogr.* **33**, 281–306.

Caesar, G. J. (53 BC) "De bello Gallico," VI, p. 25. (Transl. by A. Wiseman and P. Wiseman, p. 226. Godine, Boston, Massachusetts, 1980.)

Caperon, J. (1967). Population growth in micro-organisms limited by food supply. *Ecology* **48**, 715–722.

Caperon, J. (1968). Population growth response of *Isochrysis galbana* to nitrate variation at limiting concentrations. *Ecology* **49**, 866–872.

Caperon, J., and J. Meyer (1972). Nitrogen-limited growth of marine phytoplankton II. Uptake kinetics and their role in nutrient-limited growth of phytoplankton. *Deep-Sea Res.* **19**, 619–632.

Capone, D. G., and E. J. Carpenter (1982). Nitrogen fixation in the marine environment. *Science* **217**, 1140–1142.

Carpenter, E. J., and R. L. Guillard (1971). Interspecific differences in nitrate half-saturation constants for three species of marine phytoplankton. *Ecology* **52**, 183–185.

Carpenter, J. H., D. W. Pritchard, and R. C. Whaley (1969). Observations of eutrophication and nutrient cycles in some coastal plain estuaries. *In* "Eutrophication: Causes, Consequences, and Correctives," pp. 210–224. Natl. Acad. Sci., Washington, D.C.

Carpenter, R. C., J. T. Bennett, and M. L. Peterson (1981). ^{210}Pb activities in and fluxes to sediments of the Washington continental slope and shelf. *Geochim. Cosmochim. Acta* **45**, 1155–1172.

Carpenter, R. C., M. L. Peterson, and J. T. Bennett (1982). ^{210}Pb-derived sediment accumulation and mixing rates for the Washington continental slope. *Mar. Geol.* **48**, 135–164.

Carroll, L. 1865. "Alice's Adventures in Wonderland." Macmillan, London.

Carson, R. (1962). "Silent Spring," pp. 2–3. Houghton, Boston, Massachusetts.

Chamberlin, T. C. 1899. An attempt to frame a working hypothesis of the cause of glacial periods on an atmospheric basis. *J. Geol.* **7**, 575, 667, 751.

Chang, S. (1974). An evaluation of the eastern Bering Sea fishery for Alaska pollock (*Theragra chalcogramma*, Pallas): population dynamics. Ph.D. Thesis, Univ. of Washington, Seattle.

Chapman, D. C., J. A. Barth, R. C. Beardsley, and R. G. Fairbanks (1986). On the continuity of mean flow between the Scotian shelf and the Middle Atlantic Bight. *J. Phys. Oceanogr.* **16**, 758–772.

Charney, J. G. (1955). The generation of oceanic currents by wind. *J. Mar. Res.* **14**, 477–498.

Chavez, F. P., and R. T. Barber (1985). Plankton production during El Niño. *In* "International Conference on the TOGA Scientific Programme," pp. 23–32. W.M.O., Geneva.

Checkley, D. M. (1980). The egg production of a marine planktonic copepod in relation to its food supply: laboratory studies. *Limnol. Oceanogr.* **25**, 430–446.

Checkley, D. M., and L. C. Entzeroth (1985). Elemental and isotopic fractionation of carbon and nitrogen by marine planktonic copepods and implications to the marine nitrogen cycle. *J. Plankton Res.* **7**, 553–568.

Cheney, W., and D. Kincaid (1980). "Numerical Mathematics and Computing." Brooks/Cole, Monterey, California.

Chervin, M. B. (1978). Assimilation of particulate organic carbon by estuarine and coastal copepods. *Mar. Biol. (Berlin)* **49**, 265–275.

Christensen, J. P., G. T. Rowe, and C. H. Clifford (1983). The possible importance of primary amino nitrogen in nitrogen regeneration by coastal marine sediments in Buzzards Bay, Massachusetts. *Int. Rev. Ges. Hydrobiol.* **68**, 501–512.

Christensen, J. P., W. M. Smethie, and A. H. Devol (1987). Benthic nutrient regeneration and denitrification on the Washington continental shelf. *Deep-Sea Res.* **34**, 1027–1048.

Churchill, J. H., P. E. Biscaye, and F. Aikman (1987). The character and motion of suspended sediment over the shelf-edge and upper slope off Cape Cod. *Cont. Shelf Res.* (in press).

Cicerone, R. J., and J. D. Shetter (1981). Sources of atmospheric methane: measurements in rice paddies and a discussion. *J. Geophys. Res.* **86**, 7203–7209.

Clarke, F. W. (1924). The data of geochemistry. *U.S. Geol. Bull.* No. 770, 1–841.

Clarke, G. L. (1940). Comparative richness of zooplankton in coastal and offshore areas of the Atlantic. *Biol. Bull. (Woods Hole, Mass.)* **78**, 226–255.

Clarke, G. L., E. L. Pierce, and D. G. Bumpus (1943). The distribution and reproduction of *Sagitta elegans* on Georges Bank in relation to hydrographic conditions. *Biol. Bull. (Woods Hole, Mass.)* **85**, 201–226.

Clarke, G. L., and E. Denton (1962). Light and animal life. *In* "The Seas" (M. N. Hill, ed.), Vol. 1, pp. 456–468. Wiley (Interscience), New York.

Clarke, G. L., G. C. Ewing, and C. J. Lorenzen (1970). Spectra of backscattered light from the sea obtained from aircraft as a measure of chlorophyll concentration. *Science* **167**, 1119–1121.

Clarke, M. E. (1978). Some aspects of the feeding biology of larval walleye pollock, *Theragra chalcogramma* (Pallas) in the southeast Bering Sea. M.S. Thesis, Univ. of Alaska, Fairbanks.

Claypool, G. E., A. H. Love, and E. K. Mangham (1978). Organic geochemistry, incipient metamorphism, and oil generation in black shale members of the Phosphoria Formation, western interior United States. *Am. Assoc. Pet. Geol. Bull.* **62**, 98–120.

Cleveland, W. S., A. F. Freeny, and T. E. Graedel (1983). The seasonal component of atmospheric CO_2: information from new approaches to the decomposition of seasonal time series. *J. Geophys. Res.* **88**, 10934–10946.

Cline, J. D., and I. R. Kaplan (1975). Isotopic fractionations of dissolved nitrate during denitrification in the tropical North Pacific Ocean. *Mar. Chem.* **3**, 271–299.

Coachman, L. K., K. Aagaard, and R. B. Tripp (1975). "Bering Strait: The Regional Oceanography." Univ. of Washington Press, Seattle.

Coachman, L. K., and R. L. Charnell (1979). On lateral water mass interaction—a case study, Bristol Bay, Alaska. *J. Phys. Oceanogr.* **9**, 278–297.

Coachman, L. K., and K. Aagaard (1981). Reevaluation of water transports in the vicinity of Bering Strait. *In* "The Eastern Bering Sea Shelf: Oceanography and Resources" (D. W. Hood and J. A. Calder, eds.), Vol. 1, pp. 95–110. Univ. of Washington Press, Seattle.

Coachman, L. K., and J. J. Walsh (1981). A diffusion model of cross-shelf exchange of nutrients in the Bering Sea. *Deep-Sea Res.* **28**, 819–837.

Coale, K. H., and K. W. Bruland (1986). Oceanic two-layer euphotic zone as elucidated by 234Th/238U disequilibria. *Limnol. Oceanogr.* (Submitted).

Cochran, J. K., and R. C. Aller (1979). Particle reworking in sediments from the New York Bight Apex: evidence from ^{234}Th/^{238}U disequilibrium. *Estuarine Coast. Mar. Sci.* **9**, 739–747.

Codispoti, L. A., and F. A. Richards (1971). Oxygen supersaturation in the Chukchi and East Siberian Seas. *Deep-Sea Res.* **18**, 341–351.

Codispoti, L. A., and F. A. Richards (1976). An analysis of the horizontal regime of denitrification in the eastern tropical North Pacific. *Limnol. Oceanogr.* **21**, 379–388.

Codispoti, L. A., and T. T. Packard (1980). Denitrification rates in the eastern tropical South Pacific. *J. Mar. Res.* **38**, 453–477.

Codispoti, L. A., G. E. Friederich, R. L. Iverson, and D. W. Hood (1982). Temporal changes in the inorganic carbon system of the southeastern Bering Sea during spring 1980. *Nature (London)* **296**, 242–245.

Codispoti, L. A., G. E. Friederich, and D. W. Hood. (1986a). Variability in the inorganic carbon system over the southeastern Bering Sea shelf during spring 1980 and spring-summer 1981. *Cont. Shelf Res.* 5, 133–160.

Codispoti, L. A., G. E. Friederich, T. T. Packard, H. E. Glover, P. J. Kelly, R. W. Spinrad, R. T. Barber, J. W. Elkins, B. B. Ward, F. Lipschultz and N. Lostaunace (1986b). High nitrite levels off northern Peru: a signal of instability in the marine denitrification rate. *Science* **233**, 1200–1202.

Cohen, Y., and L. I. Gordon (1978). Nitrous oxide in the oxygen minimum of the eastern tropical North Pacific. *Deep-Sea Res.* **25**, 509–524.

Colebrook, J. M. (1972). Variability in the distribution and abundance of plankton. *Int. Conf. N.W. Atl. Fish. Spec. Publ.* No. 8, 167–186.

Colebrook, J. M. (1978). Changes in the zooplankton of the North Sea, 1948 to 1973. *Rapp. P.-V. Reun. Cons. Int. Explor. Mer* **172**, 390–396.

Colebrook, J. M. (1979). Continuous plankton records: seasonal cycles of phytoplankton and copepods in the North Atlantic Ocean and the North Sea. *Mar. Biol. (Berlin)* **51**, 23–32.

Colton, J. B. (1972). Temperature trends and the distribution of groundfish in continental shelf waters, Nova Scotia to Long Island. *Fish. Bull.* **70**, 637–657.

Colton, J. B., R. R. Marak, S. E. Nickerson, and R. R. Stoddard (1968). Physical, chemical, and biological observations on the continental shelf, Nova Scotia to Long Island, 1964–1966. *U.S. Fish Wildl. Data Rep.* No. 23.

Colton, J. B., W. G. Smith, W. A. Kendall, P. L. Berrien, and M. P. Fahay (1979). Principal spawning areas and times of marine fishes, Cape Sable to Cape Hatteras. *Fish. Bull.* **76**, 911–915.

Conover, R. J. (1978). Feeding interactions in the pelagic zone. *Rapp. P.-V. Reun. Cons. Int. Explor. Mer* 173, 66–76.

Conway, H. L., and T. E. Whitledge (1979). Distribution, fluxes, and biological utilization of inorganic nitrogen during a spring bloom in the New York Bight. *J. Mar. Res.* **37**, 657–668.

Cooley, J. W., and J. W. Tukey (1965). An algorithm for the machine calculation of complex Fourier series. *Math. Comput.* **19**, 297–301.

Cooney, R. T., and K. O. Coyle (1982). Trophic implications of cross-shelf copepod distribution in the southeastern Bering Sea. *Mar. Biol. (Berlin)* **70**, 187–196.

Cooper, G. R., and C. D. McGillem (1967). "Methods of Signal and System Analysis." Holt, New York.

Cooper, L. H. (1938). Phosphate in the English Channel 1933–8, with a comparison with earlier years, 1916 and 1923–32. *J. Mar. Biol. Assoc. U.K.* **23**, 181–195.

Cooper, L. H. (1956). Hypotheses considering fluctuations in Arctic climate with biological productivity of the English Channel. *Pap. Mar. Biol. Oceanogr., Suppl. Deep-Sea Res.* **3**, 212–223.

Corcoran, E. F., and J. E. Alexander (1963). Nutrient, chlorophyll, and primary production studies in the Florida Current. *Bull. Mar. Sci. Gulf Caribb.* **13**, 527–541.

Coriolis, G. G. (1835). Memoire sur les equations du mouvement relatif des systemes de corps. *J. Ec. Polytech. (Paris)* **15**,142–154.

Corkett, C. J., and I. A. McLaren (1978). The biology of *Pseudocalanus*. *Adv. Mar. Biol.* **15**, 1–231.

Corlett, C. (1958). Measurements of primary production in the western Barents Sea. *Rapp. P.-V. Reun. Cons. Int. Explor. Mer* **144**, 76–78.

Craig, H. (1954). Carbon 13 in plants and the relationships between carbon 13 and carbon 14 variations in nature. *J. Geol.* **62**, 115–149.

Craig, H. (1957). Isotopic standards for carbon and oxygen, and correction factors for mass-spectrometric analysis of carbon dioxide. *Geochim. Cosmochim. Acta* **12**, 133–149.

Craig, H., S. Krishnaswami, and B. L. Somayajulu (1973). $^{210}Pb/^{226}Ra$: radioactive disequilibrium in the deep-sea. *Earth Planet. Sci. Lett.* **17**, 293–305.

Craig, H., and C. C. Chou (1982). Methane: the record in polar ice cores. *Geophys. Res. Lett.* **9**, 1221–1224.

Creager, J. S., and J. McManus (1966). Geology of the southeastern Chukchi Sea. *In* "Environment of the Cape Thompson Region, Alaska" (N. J. Wilimovsky and J. M. Wolfe, eds.), pp. 755–786. USAEC, Washington, D.C.

Cresswell, G. M. (1967). Quasi-synoptic monthly hydrography of the transition regions between coastal and slope water south of Cape Cod, Mass. *WHOI Tech. Rep.* WHOI-67-35, 1–45.

Cromwell, T. (1953). Circulation in a meridional plane in the central equatorial Pacific. *J. Mar. Res.* **12**, 196–213.

Crutzen, P. J. (1981). Atmospheric chemical processes of the oxides of nitrogen, including nitrous oxide. *In* "Denitrification, Nitrification, and Atmospheric Nitrous Oxide" (C. C. Delwiche, ed.), pp. 17–44. Wiley, New York.

Crutzen, P. J. (1983). Atmospheric interactions—homogeneous gas reactions of C, N, and S containing compounds. *In* "The Major Biogeochemical Cycles and Their Interactions" (B. Bolin and R. B. Cook, eds.), pp. 67–112. J. Wiley, New York.

Crutzen, P. J., L. E. Heidt, J. P. Krasnel, W. H. Pollock, and W. Seiler (1979). Biomass burning as a source of atmospheric gases CO_2, H_2, N_2O, NO, CH_3Cl, and COS. *Nature (London)* **282**, 253–256.

Csanady, G. T. (1976). Mean circulation in shallow seas. *J. Geophys. Res.* **81**, 5389–5399.

Csanady, G. T. (1981a). Circulation in the coastal ocean, part 1. *EOS* **62**, 9–11.

Csanady, G. T. (1981b). Shelf circulation cells. *Philos. Trans. R. Soc. London Ser. A* **302,** 515–530.

Csanady, G. T. (1982). "Circulation in the Coastal Ocean." Reidel, Dordrecht, Netherlands.

Csanady, G. T. (1984). Circulation induced by river inflow in well mixed water over a sloping continental shelf. *J. Phys. Oceanogr.* **14,** 1703–1711.

Csanady, G. T. (1986). Mass transfer to and from small particles in the sea. *Limnol. Oceanogr.* **31,** 237–248.

Cunnold, D. M., R. G. Prinn, R. A. Rasmussen, P. G. Simmons, F. N. Alyea, C. A. Cardelino, A. J. Crawford, P. J. Fraser, and R. D. Rosen. (1983a). The atmospheric lifetime experiment. 3. Lifetime methodology and application to three years of $CFCl_3$ data. *J. Geophys. Res.* **88,** 8379–8400.

Cunnold, D. M., R. G. Prinn, R. A. Rasmussen, P. G. Simmons, F. N. Alyea, C. A. Cardelino and A. J. Crawford. (1983b). The atmospheric lifetime experiment. 4. Results for CF_2Cl_2 based on three years of data. *J. Geophys. Res.* **88,** 8401–8414.

Currie, R. I. (1958). Some observations on organic production in the northeast Atlantic. *Rapp. P.-V. Reun. Cons. Int. Explor. Mer* **144,** 96–102.

Cushing, D. H. (1959). The seasonal variation in oceanic production as a problem in population dynamics. *J. Cons. Int. Explor. Mer* **24,** 455–464.

Cushing, D. H. (1968). "Fisheries Biology: A Study in Population Dynamics." Univ. of Wisconsin Press, Madison.

Cushing, D. H. (1969). Upwelling and fish production. *FAO Fish. Tech. Pap.* No. 84, 1–40.

Cushing, D. H. (1971a.) Upwelling and the production of fish. *Adv. Mar. Biol.* **9,** 255–334.

Cushing, D. H. (1971b.) A comparison of production in temperate seas and upwelling areas. *Trans. R. Soc. S. Afr.* **40,** 17–33.

Cushing, D. H. (1975). "Marine Ecology and Fisheries." Cambridge Univ. Press, London and New York.

Cushing, D. H. (1983). Sources of variability in the North Sea ecosystem. *In* "North Sea Dynamics" (J. Sundermann and W. Lenz, eds.), pp. 498–516. Springer-Verlag, Berlin and New York.

Cushing, D. H., and A. C. Burd (1957). On the herring of the southern North Sea. *Fish. Invest. London, Ser. 2* **20,** 1–31.

Cushing, D. H., and J. J. Walsh (1976). "The Ecology of the Seas." Blackwell, Oxford.

Dagg, M. J. (1977). Some effects of patchy food environments on copepods. *Limnol. Oceanogr.* **22,** 99–107.

Dagg, M. J. (1978). Estimated, *in situ,* rates of egg production for the copepod *Centropages typicus* (Kroyer) in the New York Bight. *J. Exp. Mar. Biol. Ecol.* **34,** 183–196.

Dagg, M. J., and D. W. Grill (1980). Natural feeding rates of *Centropages typicus* females in the New York Bight. *Limnol. Oceanogr.* **25,** 597–609.

Dagg, M. J., and J. T. Turner (1982). The impact of copepod grazing on the phytoplankton of Georges Bank and the New York Bight. *Can. J. Fish. Aquat. Sci.* **39,** 979–990.

Dagg, M. J., J. Vidal, T. E. Whitledge, R. L. Iverson, and J. J. Goering (1982). The feeding, respiration, and excretion of zooplankton in the Bering Sea during a spring bloom. *Deep-Sea Res.* **29,** 45–64.

Dagg, M. J., M. E. Clarke, T. Nishiyama, and S. L. Smith (1984). The production and standing stock of food items for larvae of the Alaska pollock, *Theragra chalcogramma,* in the southeastern Bering Sea. *Mar. Ecol. Prog. Ser.* **19,** 7–16.

Dandonneau, Y. (1963). Etude du phytoplankton sur le plateau continentale le Cote d'Ivoire. III. Facteurs dynamics et variations spatio-temporelles. *Cah. ORSTOM Ser. Oceanogr.* **11,** 431–454.

Dansgaard, W., H. B. Clausen, N. Gunderstrup, C. U. Hammer, S. F. Johnsen, P. M. Kristindottir and N. Reeh (1982). A new Greenland deep ice core. *Science* **218,** 1273–1277.

Davis, C. S., and P. H. Wiebe (1985). Macrozooplankton biomass in a warm-core Gulf Stream ring: Time-series changes in size, structure, taxonomic composition and vertical distribution. *J. Geophys. Res.* **90**, 8871–8884.

Davis, J. M., and R. Payne (1984). Supply of organic matter to the sediment in the northern North Sea during a spring phytoplankton bloom. *Mar. Biol. (Berlin)* **78**, 315–324.

Deck, B. L. (1981). Nutrient-element distributions in the Hudson estuary. Ph.D. Thesis, Columbia Univ., New York.

Degens, E. T. (1969). Biogeochemistry of stable carbon isotopes. *In* "Organic Geochemistry" (G. Eglinton and M. I. Murphy, eds.), pp. 304–329. Springer-Verlag, Berlin and New York.

Degens, E. T., M. Behrendt, B. Gotthardt, and E. Reppmann (1968). Metabolic fractionation of carbon isotopes in marine plankton. II. Data on samples collected off the coasts of Peru and Equador. *Deep-Sea Res.* **15**, 1–19.

Degens, E. T., A. Paluska, and E. Eriksson (1976). Rates of soil erosion. *Ecol. Bull.* **22**, 185–191.

Degens, E. T., and V. Ittekot (1984). A new look at clay-organic interactions. *Mitt. Geol.-Palaeonto. Inst. Univ. Hamburg* **56**, 229–248.

Degobbis, D., N. Smodlaka, L. Pojed, L. Skrivanic, and R. Precali (1979). Increased eutrophication of the Northern Adriatic Sea. *Mar. Pollut. Bull.* **10**, 298–301.

de la Vega, G. (1605). "The Florida of the Inca" (J. G. Varner and J. J. Varner, transl. pp. 554–555. Univ. of Texas Press, Austin, 1951).

Delmas, R. J., J. M. Ascencio, and M. Legrand (1980). Polar ice evidence that atmospheric CO_2 20,000 yr BP was 50% of present. *Nature (London)* **284**, 155–157.

Delwiche, C. C., and P. L. Steyn (1970). Nitrogen isotope fractionation in soils and microbial reactions. *Environ. Sci. Technol.* **4**, 929–935.

Delwiche, C. C., and G. E. Likens (1977). Biological response to fossil fuel combustion products. *In* "Global Chemical Cycles and Their Alterations by Man" (W. Stumm, ed.), pp. 89–98. Dahlem Konf., Berlin.

DeMaster, D. J. (1979). The marine budgets of silica and ^{32}silicon. Ph.D. Thesis, Yale Univ., New Haven, Connecticut.

DeMendiola, B. R. (1974). Food of the larval anchoveta, *Engraulis ringens*. *In* "The Early Life History of Fish" (J. H. Blaxter, ed.), pp. 277–286. Springer-Verlag, Berlin and New York.

DeNiro, M. J., and S. Epstein (1981). Influence of diet on the distribution of nitrogen isotopes in animals. *Geochim. Cosmochim. Acta* **45**, 341–351.

Denman, K. L., and T. Platt (1976). The variance spectrum of phytoplankton in a turbulent ocean. *J. Mar. Res.* **34**, 593–601.

Denman, K. L., A. Okubo, and T. Platt (1977). The chlorophyll fluctuation spectrum in the sea. *Limnol. Oceanogr.* **22**, 1033–1038.

Dessier, A., and J. R. Donguy (1985). Planktonic copepods and environmental properties of the eastern equatorial Pacific: seasonal and spatial variation. *Deep-Sea Res.* **32**, 1117–1134.

Deuser, W. G. (1979). Marine biota, nearshore sediments, and the global carbon balance. *Org. Geochem.* **1**, 243–247.

Deuser, W. G. (1986). Seasonal and interannual variations in deep-water particle fluxes in the Sargasso Sea and their relation to surface hydrography. *Deep-Sea Res.* **33**, 225–246.

Deuser, W. G., and E. T. Degens (1967). Carbon isotope fractionation in the system CO_2 (gas)–CO_2 (aqueous)–HCO_3^- (aqueous). *Nature (London)* **215**, 1033–1035.

Deuser, W. G., E. T. Degens, and R. R. Guillard (1968). Carbon isotope relationships between plankton and sea water. *Geochim. Cosmochim. Acta* **32**, 657–660.

Deuser, W. G., and W. H. Ross (1980). Seasonal change in the flux of organic carbon to the deep Sargasso Sea. *Nature (London)* **283**, 364–365.

Devold, F. (1963). The life history of the Atlanto-Scandian herring. *Rapp. P.-V. Reun. Cons. Int. Explor. Mer* **154**, 98–108.

DeVooys, C. G. (1979). Primary production in aquatic environments. *In* "The Global Carbon Cycle" (B. Bolin, E. T. Degens, S. Kempe, and P. Ketner, eds.), pp. 259–292. Wiley, New York.

DeVries, T. J., and W. G. Pearcy (1982). Fish debris in sediments of the upwelling zone of central Peru: A late Quaternary record. *Deep-Sea Res.* **28**, 87–109.

Diaz, H. F. (1979). Atmospheric conditions and comparison with past records. *In* "Oxygen Depletion and Associated Benthic Mortalities in the New York Bight, 1976" (C. J. Sindermann and R. L. Swanson, eds.) *NOAA Prof. Pap.* No. 11, 51–78.

Dickens, C. 1853. "Bleak House." Reprinted by Colonial Press, Clinton.

Dickens, C. 1859. "A Tale of Two Cities." Reprinted by Colonial Press, Clinton, Connecticut.

Dickson, R. R. (1971). A recurrent and persistent pressure anomaly pattern as the principal cause of intermediate scale hydrographic variation in the European Shelf Seas. *Dtsch. Hydrogr. Z.* **24**, 97–119.

Dietz, R. S., and J. C. Holden (1970). Reconstruction of Pangaea: Break-up and dispersion of continents, Permian to present. *J. Geophys. Res.* **75**, 4939–4956.

Digby, P. S. (1953). Plankton production in Scoresby Sound, East Greenland. *J. Anim. Ecol.* **22**, 289–322.

Digby, P. S. (1954). The biology of marine planktonic copepods of Scoresby Sound, East Greenland. *J. Anim. Ecol.* **23**, 298–338.

Dinesen, I. (1934). "The Deluge at Norderney," p. 4. Smith & Haas, New York.

DiToro, D. M., and J. P. Connolly (1980). "Mathematical Models of Water Quality in Large Lakes. Part 2, Lake Erie," EPA-600/3–80–065. Environ. Prot. Agency, Duluth, Minnesota.

Dodimead, A. J., F. Favorite, and T. Hirano (1963). Review of the oceanography of the subarctic region. *Int. North Pac. Fish. Comm. Bull.* **13**, 1–195.

Dow, W. G. (1978). Petroleum source beds on continental slopes and rises. *Am. Assoc. Pet. Geol. Bull.* **62**, 1584–1606.

Dowidar, N. M. (1984). Phytoplankton biomass and productivity of the southeastern Mediterranean. *Deep-Sea Res.* **31**, 983–1000.

Doyle, L. J., O. H. Pilkey, and C. C. Woo (1979). Sedimentation on the eastern United States continental slope. *In* "Geology of Continental Slopes" (L. J. Doyle and O. H. Pilkey, eds.), SEPM Spec. Publ. No. 27, 119–129.

Droop, M. R. (1968). Vitamin B_{12} and marine ecology IV. The kinetics of uptake, growth, and inhibition in *Monochrysis Luther. J. Mar. Biol. Assoc. U.K.* **48**, 689–733.

Droop, M. R. (1973). Some thoughts on nutrient limitation of algae. *J. Phycol.* **9**, 264–272.

Druffel, E. R. (1985). Detection of El Niño and decade time scale variations of sea surface temperature from banded coral records: implications for the carbon dioxide cycle. *In* "The Carbon Cycle and Atmospheric CO_2: Natural Variation, Archean to Present" (E. T. Sundquist and W. S. Broecker, eds.), AGU Monogr. 32, pp. 111–122. Am. Geophys. Union, Washington, D.C.

Ducklow, H. W. (1986). Mesoscale distributions, temporal changes and production of bacterial biomass in warm-core Gulf Stream Ring 82-B. *Deep-Sea Res.* **33**, 1789–1812.

Ducklow, H. W., D. L. Kirchman, and G. T. Rowe (1982). Production and vertical flux of attached bacteria in the Hudson River plume of the New York Bight as studied with floating sediment traps. *Appl. Environ. Microbiol.* **25**, 769–776.

Ducklow, H. W., S. M. Hill, and W. D. Gardner (1985). Bacterial growth and the decomposition of particulate organic carbon collected in sediment traps. *Cont. Shelf Res.* **4**, 445–464.

Ducklow, H. W., D. A. Purdie, P. J. Williams, and J. M. Davies (1986). Bacterioplankton: a sink for carbon in a coastal marine community. *Science* **232**, 865–867.

Dugdale, R. C. (1967). Nutrient limitation in the sea: dynamics, identification, and significance. *Limnol. Oceanogr.* **12**, 685–695.

Dugdale, R. C., and J. J. Goering (1967). Uptake of new and regenerated forms of nitrogen in primary productivity. *Limnol. Oceanogr.* **12**, 196–206.

458   References

Dugdale, R. C., and J. J. Goering (1970). Nutrient limitation and the path of nitrogen in Peru Current production. *Anton Bruun Rep.* **5,** 5.3–5.8.

Dugdale, R. C., J. J. Goering, R. T. Barber, R. L. Smith, and T. T. Packard (1977). Denitrification and hydrogen sulfide in the Peru upwelling region during 1976. *Deep-Sea Res.* **24,** 601–608.

Duing, W. O., C. N. K. Mooers, and T. N. Lee (1977). Low-frequency variability in the Florida Current and relations to atmospheric forcing from 1972 to 1974. *J. Mar. Res.* **35,** 129–161.

Duntley, S. Q. (1963). Light in the sea. *J. Opt. Soc. Am.* **53,** 214–233.

Eadie, B. J. (1972). Distribution and fractionation of stable carbon isotopes in the Antarctic ecosystem. Ph.D. Thesis, Texas A&M Univ., College Station.

Edwards, R. L., and A. S. Merrill (1977). A reconstruction of the continental shelf area of eastern North America for the times 9500 B.P. and 12500 B.P. *Arch. East. North Am.* **5,** 1–44.

Edwards, R. L., and R. E. Bowman (1979). Food consumed by continental shelf fishes. *In* "Predator-prey systems in fish communities and their role in fisheries management," pp. 387–406. Sport Fish. Inst., Washington, D.C.

EG&G (1979). "Analysis Report," Appendix F, 10th Quarterly Progress Report. N. Engl. O.C.S. Phys. Oceanogr. Program, Waltham, Massachusetts.

Eggimann, D. W., F. T. Manheim, and P. R. Betzer (1980). Dissolution and analysis of amorphous silica in marine sediments. *J. Sediment Pet.* **50,** 215–225.

Eisenstat, S. C., M. H. Schultz, and A. H. Sherman (1976). Considerations in the design of software for sparse Gaussian elimination. *In* "Sparse Matrix Computations" (J. R. Bunch and D. J. Rose, eds.), pp. 1–453. Academic Press, New York.

Ekman, V. W. (1905). On the influence of the earth's rotation on ocean-currents. *Ark. Mat. Astron. Fys.* **2,** 1–52.

Elbrachter, M. (1973). Population dynamics of *Ceratium* in coastal waters of the Kiel Bay. *Oikos* **15,** 43–48.

Elkins, J. W., S. C. Wofsy, M. B. McElroy, C. E. Kolb, and W. A. Kaplan (1978). Aquatic sources and sinks for nitrous oxide. *Nature (London)* **275,** 602–606.

El-Sayed, S. Z. (1972). Primary productivity and standing crop of phytoplankton. *In* "Chemistry, Primary Productivity, and Benthic Algae of the Gulf of Mexico" (V. C. Bushnell, ed.), *Ser. Atlas Mar. Environ., Am. Geogr. Soc.* **22,** 8–13.

El-Sayed, S. Z., and H. R. Jitts (1973). Phytoplankton production in the southeastern Indian Ocean. *In* "The Biology of the Indian Ocean" (B. Zeitzschel, ed.), pp. 131–142. Springer-Verlag, Berlin and New York.

El-Sayed, S. Z., and S. Taguchi (1981). Primary production and standing crop of phytoplankton along the ice-edge in the Weddell Sea. *Deep-Sea Res.* **28,** 1017–1032.

El-Sayed, S. Z., D. C. Biggs, and O. Holm-Hansen (1983). Phytoplankton standing crop, primary productivity, and near-surface nitrogenous nutrient fields in the Ross Sea, Antarctica. *Deep-Sea Res.* **30,** 871–886.

Emery, K. O. (1960). "The Sea off Southern California, A Modern Habitat of Petroleum," pp. 247–253. Wiley, New York.

Emery, K. O. (1968). Relict sediments on continental shelves of the world. *Am. Assoc. Pet. Geol. Bull.* **52,** 445–464.

Emery, K. O. (1969). The continental shelves. *Sci. Am.* **221,** 106–122.

Emery, K. O., W. L. Orr, and S. C. Rittenberg (1955). Nutrient budgets in the ocean. *In* "Essays in Honor of Captain Allan Hancock," pp. 147–157. Univ. of Southern California Press, Los Angeles.

Emery, K. O., and E. E. Bray (1962). Radiocarbon dating of California basin sediments. *Am. Assoc. Pet. Geol. Bull.* **46,** 1839–1856.

Emery, K. O., and E. Uchupi (1972). Western North Atlantic Ocean: Topography, rocks, structure, water, life, and sediments. *Mem. Am. Assoc. Pet. Geol. Mem.* **17,** 1–532.

Emery, K. O., and E. Uchupi (1984). "The Geology of the Atlantic Ocean." Springer-Verlag, Berlin and New York.

Emiliani, C. (1954). Depth habitats of some species of pelagic Foraminifera as indicated by oxygen isotope ratios. *Am. J. Sci.* **252,** 149–158.

Emiliani, C., and N. J. Shackleton (1974). The Brunhes Epoch: isotopic paleotemperatures and geochronology. *Science* **183,** 511–514.

Enhalt, D. H. (1979). Der atmospharische krieslauf von methan. *Naturwissenschaften* **66,** 307–311.

Enhalt, D. H., and U. Schmidt (1978). Sources and sinks of atmospheric methane. *Pure Appl. Geophys.* **116,** 452–464.

Eppley, R. W. (1972). Temperature and phytoplankton growth in the sea. *Fish. Bull.* **70,** 1063–1085.

Eppley, R. W., and J. H. Strickland (1968). Kinetics of marine phytoplankton growth. *In* "Advances in Microbiology of the Sea" (M. R. Droop and E. J. Wood, eds.), pp. 23–62. Academic Press, New York.

Eppley, R. W., and W. H. Thomas (1969). Comparison of half-saturation constants for growth and nitrate uptake of marine phytoplankton. *J. Phycol.* **5,** 375–379.

Eppley, R. W., J. N. Rogers, and J. J. McCarthy (1969). Half saturation constants for uptake of nitrate and ammonium by marine phytoplankton. *Limnol. Oceanogr.* **14,** 912–920.

Eppley, R. W., A. F. Carlucci, O. Holm-Hansen, D. Kiefer, J. J. McCarthy, E. L. Venrick and P. M. Williams (1971). Phytoplankton growth and composition in shipboard cultures supplied with nitrate, ammonium, or urea as the nitrogen source. *Limnol. Oceanogr.* **16,** 741–751.

Eppley, R. W., E. H. Renger, E. L. Venrick, and M. M. Mullin (1973). A study of plankton dynamics and nutrient cycling in the central gyre of the North Pacific Ocean. *Limnol. Oceanogr.* **18,** 534–551.

Eppley, R. W., P. Koeller, and G. T. Wallace (1978). Stirring influences the phytoplankton species composition within enclosed columns of coastal sea water. *J. Exp. Mar. Biol. Ecol.* **32,** 219–239.

Eppley, R. W., E. H. Renger, and W. G. Harrison (1979). Nitrate and phytoplankton production in southern California coastal waters. *Limnol. Oceanogr.* **24,** 483–494.

Eppley, R. W., and B. J. Peterson (1980). Particulate organic matter flux and planktonic new production in the deep ocean. *Nature (London)* **282,** 677–680.

Epstein, S., R. P. Sharp, and A. J. Gow (1970). Antarctic Ice Sheet: Stable isotopic analyses of Byrd Station cores and interhemispheric climatic implications. *Science* **168,** 1570–1572.

Erlenkeuser, H., and H. Willkomm (1973). University of Kiel radiocarbon measurements VII. *Radiocarbon* **15,** 113–126.

Erlenkeuser, H., E. Suess, and H. Willkomm (1974). Industrialization affects heavy metal and carbon isotope concentration in recent Baltic Sea sediments. *Geochim. Cosmochim. Acta* **38,** 823–842.

Esaias, W. E. (1980). Remote sensing of oceanic phytoplankton: present capabilities and future goals. *In* "Primary Productivity in the Sea" (P. G. Falkowski, ed.), pp. 321–337. Plenum, New York.

Estrada, M. (1974). Photosynthetic pigments and productivity in the upwelling region of northwest Africa. *Tethys* **6**(1/2): 247–260.

Evans, G. T., and J. S. Parslow (1985). A model of annual plankton cycles. *Biol. Oceanogr.* **3,** 327–347.

Evans, R. H., K. S. Baker, O. B. Brown, and R. C. Smith (1985). Chronology of Warm-Core Ring 82-B. *J. Geophys. Res.* **90,** 8803–8811.

Everett, D. E. (1971). Hydrologic and quality characteristics of the lower Mississippi River. *Tech. Rep. La. Dep. Public Works* No. 5.

Fairbanks, R. G. (1982). The origin of continental shelf and slope water in the New York Bight and Gulf of Maine: evidence from $H_2^{18}O/H_2^{16}O$ ratio measurements. *J. Geophys. Res.* **87,** 5796–5808.

Falkowski, P. G. (1981). Light-shade adaptation and assimilation numbers. *J. Plankton Res.* **3**, 203–216.

Falkowski, P. G. (1983). Light–shade adaptation and vertical mixing of marine phytoplankton: a comparative field study. *J. Mar. Res.* **41**, 215–237.

Falkowski, P. G., and S. O. Howe (1976). Preliminary report on the possible effects of the *Ceratium tripos* bloom in the New York Bight, March–July 1976. *In* "IDOE Workshop on Anoxia on the Middle Atlantic Shelf during the Summer of 1976" (J. H. Sharp, ed.) pp. 57–62. Natl. Sci. Found., Washington, D.C.

Falkowski, P. G., and T. G. Owens (1978). Effects of light intensity on photosynthesis and dark respiration in six species of marine phytoplankton. *Mar. Biol. (Berlin)* **45**, 289–295.

Falkowski, P. G., T. S. Hopkins, and J. J. Walsh (1980). An analysis of factors affecting oxygen depletion in the New York Bight. *J. Mar. Res.* **38**, 479–506.

Falkowski, P. G., and C. D. Wirick (1981). A simulation model of the effects of vertical mixing on primary productivity. *Mar. Biol. (Berlin)* **65**, 69–75.

Falkowski, P. G., J. Vidal, T. S. Hopkins, G. T. Rowe, T. E. Whitledge, and W. G. Harrison (1983). Summer nutrient dynamics of the Middle Atlantic Bight: primary production and utilization of phytoplankton carbon.*J. Plankton Res.* **5**, 515–537.

Falkowski, P. G., C. N. Flagg, G. T. Rowe, S. L. Smith, T. E. Whitledge, and C. D. Wirick (1987).The fate of the spring phytoplankton bloom: export or oxidation? *Cont. Shelf Res.* (in press).

Fanning, K. A., K. L. Carder, and P. R. Betzer (1982). Sediment resuspension by coastal waters: a potential mechanism for nutrient recycling on the ocean's margins. *Deep-Sea Res.* **29**, 953–965.

Fasham, M. J. (1985). Flow analysis of materials in the marine euphotic zone. *In* "Ecosystem Theory for Biological Oceanography" (R. E. Ulanowicz and T. Platt, eds.), *Can. Bull. Fish. Aquat. Sci.* **213**, pp. 139–162.

Fasham, M. J., and T. Platt (1983). Photosynthetic response of phytoplankton to light: a physiological model. *Proc. R. Soc. London, Ser. B* **219**, 355–370.

Fasham, M. J., P. M. Holligan, and P. R. Pugh (1983). The spatial and temporal development of the spring phytoplankton bloom in the Celtic Sea, April 1979. *Prog. Oceanogr.* **12**, 87–145.

Fedra, K. E., M. Olscher, C. Scherubel, M. Stachowitsch, and R. S. Wurzian (1976). On the ecology of a North Adriatic benthic community: distribution, standing crop, and composition of macrobenthos. *Mar. Biol. (Berlin)* **38**, 129–145.

Fee, E. J. (1969). A numerical model for the estimation of photosynthetic production, integrated over time and depth, in natural waters. *Limnol. Oceanogr.* **14**, 906–911.

Feldman, G. C. (1986). Variability of the productive habitat in the eastern equatorial Pacific. *Proc. Symp. Vertical Motion Equatorial Upper Ocean Its Eff. Upon Living Resour. Atmos.* (in press).

Feldman, G. C., D. Clark, and D. A. Halpern (1984). Satellite color observations of the phytoplankton distribution in the eastern Equatorial Pacific during the 1982–83 El Niño. *Science* **226**, 1069–1071.

Fellows, D. A., D. M. Karl, and G. A. Knauer (1981). Large particle fluxes and the vertical transport of living carbon in the upper 1500 m of the northeast Pacific Ocean. *Deep-Sea Res.* **28**, 921–936.

Fenchel, T. (1969). The ecology of the marine microbenthos. IV. Structure and function of the benthic ecosystems, its chemical and physical factors and the microfauna communities with special reference to the ciliated protozoa. *Ophelia* **6**, 1–182.

Fiadeiro, M., and J. D. Strickland (1968). Nitrate reduction and the occurrence of a deep nitrite maximum in the ocean off the west coast of South America. *J. Mar. Res.* **26**, 187–201.

Finenko, Z. Z., and V. E. Zaika (1970). Particulate organic matter and its role in the productivity of the sea. *In* "Marine Food Chains" (J. H. Steele, ed.), pp. 32–45. Univ. of California Press, Berkeley.

Fischer, H. B. (1980). Mixing processes on the Atlantic continental shelf, Cape Cod to Cape Hatteras. *Limnol. Oceanogr.* **25**, 114–125.

Fisher, J. B., and G. Matisoff (1982). Downcore variation in sediment organic nitrogen. *Nature (London)* **296**, 345–347.

Fitzgerald, G. P. (1968). Detection of limiting or surplus nitrogen in algae and aquatic weeds, *J. Phycol.* **4**, 121–126.

Fleming, R. H. (1939). The control of diatom populations by grazing. *J. Cons. Cons. Int. Explor. Mer* **14**, 210–227.

Flint, R. W., and N. N. Rabelais (1981). "Environmental studies of a marine ecosystem: South Texas Outer Continental Shelf." Univ. of Texas Press, Austin.

Florek, R. J., and G. T. Rowe (1983). Oxygen consumption and dissolved inorganic nutrient production in marine coastal and shelf sediments of the Middle Atlantic Bight. *Int. Rev. Ges. Hydrobiol.* **68**, 73–112.

Fogg, G. E. (1959). Nitrogen nutrition and metabolic patterns in algae. *Symp. Soc. Exp. Biol.* **13**, 106–125.

Fogg, G. E. (1982). Nitrogen cycling in sea waters. *Philos. Trans. R. Soc. London, Ser. B* **296**, 511–570.

Folger, D. W. (1972). Texture and organic carbon content of bottom sediments in some estuaries of the United States. *Geol. Soc. Am. Mem.* No. 133, 391–408.

Fonselius, S. (1969). Hydrography of the Baltic deep basin III. *Fish. Board Swed. Ser. Hydrogr. Rep.* No. 23, 1–97.

Fornshell, J. A., D. D. Frydenland, and P. Christensen (1977). Zoogeography of the genus *Ceratium* on the east coast of North America from Cape Canaveral to Cape May. *Int. Congr. Protozool., 5th, New York Abstr.*

Forsberg, F. J. (1963). Phytoplankton production in the southeastern Pacific. *Nature (London)* **200**, 87–88.

Forster, S. S., A. C. Cripps, and A. Smith-Carington (1982). Nitrate leaching to ground water. *Philos. Trans. R. Soc. London, Ser. B* **296**, 477–489.

Fourier, J. B. (1822). "Analytical Theory of Heat." (Transl. by A. Freeman. Cambridge Univ. Press, Cambridge, 1878.)

Fournier, R. O., J. Marra, R. Bohrer, and M. Van Det (1977). Plankton dynamics and nutrient enrichment of the Scotian shelf. *J. Fish. Res. Board Can.* **34**, 1004–1018.

Fox, M. F., and D. R. Kester (1986). Nutrient distributions in warm core ring 82B. *Deep-Sea Res.* **33**, 1761–1772.

Francis, R. C., and K. M. Bailey (1983). Factors affecting recruitment of selected gadoids in the northeast Pacific and east Bering Sea. *In* "From Year to Year" (W. S. Wooster, ed.), pp. 35–60. Univ. of Washington Press, Seattle.

Frank, K. T., and W. C. Leggett (1983). Survival value of an opportunistic life-stage transition in capelin *(Mallotus villosus). Can. J. Fish. Aquat. Sci.* **40**, 1442–1448.

Franks, P. J., J. S. Wroblewski, and G. R. Flierl (1986). Prediction of phytoplankton growth in response to the frictional decay of a warm core ring. *J. Geophys. Res.* (in press).

Froelich, P. N., M. Bender, N. A. Luedtke, G. R. Heath, and T. DeVries (1982). The marine phosphorus cycle. *Am. J. Sci.* **282**, 475–511.

Frost, B. W., M. R. Landry, and R. P. Hassett (1983). Feeding behavior of large calanoid copepods *Neocalanus cristatus* and *N. plumchrus* from the subarctic Pacific Ocean. *Deep-Sea Res.* **30**, 1–14.

Fry, B. (1981). Natural stable carbon isotope tag traces Texas shrimp migration. *Fish. Bull.* **79**, 337–346.

Fry, B., and E. B. Sherr (1984). Del ^{13}C measurements as indicators of carbon flow in marine and freshwater ecosystems. *Contrib. Mar. Sci.* **27**, 13–48.

Fucik, K. W. (1974). The effect of petroleum operations on the phytoplankton ecology of the Louisiana coastal waters. M.S. Thesis, Texas A&M Univ., College Station.

Fuhrmann, J., and F. Azam (1980). Bacterioplankton secondary production estimates for coastal waters of British Columbia, Antarctica, and California. *Appl. Environ. Microbiol.* **39,** 1085–1095.

Fujita, Y. (1970). Photosynthesis and plant pigments. *Bull. Plank. Soc. Jpn.* **17,** 20–31.

Fulton, J. (1973). Some aspects of the life history of *Calanus plumchrus* in the Strait of Georgia. *J. Fish. Res. Board Can.* **30,** 811–815.

Fulton, J. (1978). Seasonal and annual variations of net zooplankton at Ocean Station P, 1965–76. *Rep. Fish. Mar. Serv. (Can.)* No. 49.

Fung, J., K. Prentice, E. Mathews, J. Lerner, and G. Russell (1983). Three-dimensional tracer model study of atmospheric CO_2: response to seasonal exchanges with the terrestrial biosphere. *J. Geophys. Res.* **88,** 1281–1294.

Gaebler, O. H., T. G. Vitti, and R. Vukmirovich (1966). Isotope effects in metabolism of ^{14}N and ^{15}N from unlabelled dietary protein. *Can. J. Biochem.* **44,** 1249–1257.

Gale, J. (1972). Availability of carbon dioxide for photosynthesis at high altitudes: theoretical considerations. *Ecology* **53,** 494–497.

Gammon, R. H., J. Cline, and D. Wisegarver (1982). Chlorofluoromethanes in the northeast Pacific ocean: measured vertical distribution and applications as transient tracers of upper ocean mixing. *J. Geophys. Res.* **87,** 9441–9454.

Gao, L., K. O. Emery, and L. D. Keigwin (1985). Late Quaternary stable isotope paleoceanography off southern California. *Deep-Sea Res.* **32,** 1469–1484.

Garrels, R. M., F. T. MacKenzie, and C. Hunt (1973). "Chemical Cycles and the Global Environment." William Kaufmann, Los Altos, California.

Garrels, R. M., and E. A. Perry (1974). Cycling of carbon, sulfur, and oxygen through geologic time. *In* "The Sea, Vol. 5" (E. D. Goldberg, ed.), pp. 303–336. Wiley (Interscience), New York.

Garrels, R. M., and A. Lerman (1984). Coupling of sedimentary sulfur and carbon cycles—an improved model. *Am. J. Sci.* **284,** 989–1007.

Garrett, C. (1979). Topographic Rossby waves off east Australia: identification and role in shelf circulation. *J. Phys. Oceanogr.* **9,** 244–253.

Garrod, D. J., and A. D. Clayden (1972). Current biological problems in the conservation of deep-sea fishery resources. *Symp. Zool. Soc. London* No. 29, 161–184.

Garside, C. (1985). The vertical distribution of nitrate in oceanic surface seawater. *Deep-Sea Res.* **32,** 723–732.

Garside, C., T. C. Malone, O. A. Roels, and B. F. Shorfstein (1976). An evaluation of sewage-derived nutrients and their influence on the Hudson Estuary and the New York Bight. *Estuarine Coast. Mar. Sci.* **4,** 281–289.

Garside, C., and T. C. Malone (1978). Monthly oxygen and carbon budgets of the New York Bight Apex. *Estuarine Coast. Mar. Sci.* **6,** 93–104.

Garside, C., G. Han, J. P. Thomas, J. J. Walsh, and T. E. Whitledge (1979). Assimilative capacity for nitrogen. "Assimilative Capacity of U.S. Coastal Waters for Pollutants," (E. D. Goldberg, ed.), *NOAA Tech. Rep. NMFS* pp. 164–168.

Gershanovitch, D. E. (1962). New data on recent sediments of the Bering Sea. *In* "Issledovanye po Progrome Mezhdunarodenovo Geofizicheskovo Goda" (L. G. Vinogradova and M. V. Fedosova, eds.), pp. 128–164. Pischchem Promizdat, Moscow.

Gieskes, W. W., and G. W. Kraay (1975). The phytoplankton spring bloom in Dutch coastal waters of the North Sea. *Neth. J. Sea Res.* **9,** 166–196.

Gieskes, W. W., and G. W. Kraay (1977). Continuous plankton records: changes in the plankton of the North Sea and its eutrophied Southern Bight from 1949 to 1975. *Neth. J. Sea Res.* **11,** 334–364.

Gill, A. E., and A. J. Clarke (1974). Wind-induced upwelling, coastal currents and sea level changes. *Deep-Sea Res.* **21**, 325–345.

Gill, E. D. (1974). Carbon-14 and uranium/thorium check on suggested interstadial high sea level around 30,000 BP. *Search* **5**, 211.

Gillette, D. A. (1982). Decomposition of annual patterns of atmospheric carbon dioxide concentrations: a preliminary interpretation of one year of data at 30 globally distributed locations. *Atmos. Environ.* **16**, 2536–2542.

Glibert, P. M. (1982). Regional studies of daily, seasonal and size fraction variability in ammonium remineralization. *Mar. Biol. (Berlin)* **70**, 209–222.

Glover, R. S., G. A. Robinson, and J. M. Colebrook (1972). Plankton in the North Atlantic: an example of the problems of analyzing variability in the environment. *In* "Marine Pollution and Sea Life" (M. Ruivo, ed.), pp. 439–445. FAO, Rome.

Glynn, P. W. (1973). Ecology of a Caribbean coral reef. The *Porites* reef - flat biotype. II. Plankton community with evidence for depletion. *Mar. Biol. (Berlin)* **22**, 1–21.

Goering, J. J., and M. M. Pamatmat (1971). Denitrification in the sediments of the sea off Peru. *Invest. Pesq.* **35**, 233–242.

Goering, J. J., D. M. Nelson, and J. A. Carter (1973). Silicic acid uptake by natural populations of marine phytoplankton. *Deep-Sea Res.* **20**, 777–789.

Gold, K. (1970). Plankton as indicators of pollution in New York waters. *In* "Water Pollution in the Greater New York Area" (A. A. Johnson, ed.), pp. 93–100. Gordon & Breach, New York.

Goldberg, E. D. (1976). "The Health of the Oceans." UNESCO, Paris.

Goldman, J. C., and E. J. Carpenter (1974). A kinetic approach to the effect of temperature on algal growth. *Limnol. Oceanogr.* **19**, 756–766.

Gordon, A. L., A. F. Amos, and R. D. Gerard (1976). New York Bight water stratification–October 1974. *ASLO Spec. Symp.* No. 2, 45–57.

Gordon, H. R., and D. K. Clark (1980). Remote sensing optical properties of a stratified ocean: an improved interpretation. *Appl. Opt.* **19**, 3428–3430.

Gordon, H. R., D. K. Clark, J. L. Mueller, and W. A. Hovis (1980). Phytoplankton pigments derived from the Nimbus-7 CZCS: initial comparisons with surface measurements. *Science* **210**, 63–66.

Gordon, H. R., D. K. Clark, J. W. Brown, O. B. Brown, and R. H. Evans (1982). Satellite measurement of the phytoplankton pigment concentration in the surface waters of a warm core Gulf Stream ring. *J. Mar. Res.* **40**, 491–502.

Gordon, H. R., and A. G. Morel (1983). "Remote Assessment of Ocean Color for Interpretation of Satellite Visible Imagery." Springer-Verlag, Berlin and New York.

Gordon, H. R., D. K. Clark, J. W. Brown, O. B. Brown, R. H. Evans, and W. W. Broenkow (1983). Phytoplankton pigment concentrations in the Middle Atlantic Bight: comparison of ship determinations and CZCS estimates. *Appl. Opt.* **22**, 20–36.

Gornitz, V., S. Lebedeff, and J. Hansen (1982). Global sea level trend in the past century. *Science* **215**, 611–614.

Govoni, J. J., D. E. Hoss, and A. J. Chester (1983). Comparative feeding of three species of larval fishes in the northern Gulf of Mexico: *Brevoortia patronus, Leiostomus xanthurus,* and *Micropogonias undulatus. Mar. Ecol. Prog. Ser.* **13**, 189–199.

Graf, G., W. Bengtsson, U. Diesner, R. Schultz, and H. Theede (1982). Benthic response to sedimentation of a spring phytoplankton bloom: process and budget. *Mar. Biol. (Berlin)* **67**, 201–208.

Graf, G., R. Schultz, R. Peinert, and L. A. Meyer-Reil (1983). Benthic response to sedimentation events during autumn to spring at a shallow-water station in the western Kiel Bight. *Mar. Biol. (Berlin)* **77**, 235–246.

Green, C. W. (1944). Summer upwelling on the northeast coast of Florida. *Science* **100**, 546–547.

Grice, G. D., and A. D. Hart (1962). The abundance, seasonal occurrence, and distribution of the epizooplankton between New York and Bermuda. *Ecol. Monogr.* **32,** 287–309.

Griffin, W. L., R. D. Lacewell, and J. P. Nichols (1976). Optimum effort and rent distribution in the Gulf of Mexico shrimp fishery. *Am. J. Agric. Econ.* **58,** 644–652.

Grumman Ecosystems Corporation and Lawler, Matusky, and Skelly (1975). "Environmental Conditions in the New York Bight, 1973–1974." Final Report to New York State Atomic and Space Development Authority, Albany.

Gulland, J. A. (1971). Ecological aspects of fishery research. *Adv. Ecol. Res.* **7,** 115–176.

Gulland, J. A. (1983). World resources of fisheries and their management. *In* "Marine Ecology" (O. Kinne, ed.), Vol. 5, pp. 839–1061. Wiley, New York.

Gunter, G. (1967). Some relationships of estuaries to the fisheries of the Gulf of Mexico. *In* "Estuaries" (G. H. Lauff, ed.), pp. 621–638. Am. Assoc. Adv. Sci., Washington, D.C.

Hackney, G. T., and E. B. Haines (1980). Stable carbon isotope composition of fauna and organic matter collected in a Mississippi estuary. *Estuarine Coast. Mar. Sci.* **10,** 703–708.

Haflinger, K. (1981). A survey of benthic infaunal communities of the southeastern Bering Sea. *In* "The Eastern Bering Sea Shelf: Oceanography and Resources" (D. W. Hood and J. A. Calder, eds.), Vol. 2, pp. 1091–1104. Univ. of Washington Press, Seattle.

Hagmeier, E. (1978). Variations in phytoplankton near Helgoland. *Rapp. P.-V. Reun. Cons. Int. Explor. Mer* **172,** 361–363.

Hahn, J. (1981). Nitrous oxide in the oceans. *In* "Denitrification, Nitrification, and Atmospheric Nitrous Oxide" (C. C. Delwiche, ed.) pp. 191–240. Wiley, New York.

Haines, E. B., and W. M. Dunstan (1975). The distribution and relation of particulate organic material and primary productivity in the Georgia Bight, 1973–74. *Estuarine Coast. Sci.* **3,** 431–441.

Haines, J. R., R. M. Atlas, R. P. Griffiths and R. Y. Morita (1981). Denitrification and nitrogen fixation in Alaska continental shelf sediments. *Appl. Environ. Microbiol.* **41,** 412–421.

Hairston, N. G., F. E. Smith, and L. B. Slobodkin (1960). Community structure, population control, and competition. *Am. Nat.* **94,** 421–425.

Hakansson, E., R. Bromley, and K. Perch-Nielsen (1974). Maestrictian chalk of Northwest Europe-A pelagic shelf sediment. *In* "Pelagic Sediments: On Land and Under the Sea" (K. J. Hsu and H. C. Jenkyns, eds.), *Spec. Publ. Int. Assoc. Sediment* **1,** 211–233.

Halim, Y. (1960). Observations on the Nile bloom of phytoplankton in the Mediterranean. *J. Cons.* **26,** 57–67.

Hall, C. A., C. A. Ekdahl, and D. E. Wartenberg (1975). A fifteen-year record of biotic metabolism in the northern hemisphere. *Nature (London)* **255,** 136–138.

Hallagraeff, G. M. (1981). Seasonal study of phytoplankton pigments and species at a coastal station off Sydney: importance of diatoms and the nannoplankton. *Mar. Biol. (Berlin)* **6,** 107–118.

Hallam, A. (1977). Anoxic events in the Cretaceous ocean. *Nature (London)* **268,** 15–16.

Hamai, I., K. Kyushin, and T. Kionshita (1971). Effects of temperature on the body form and mortality in the development and early larval stages of the Alaska pollock (*Theragra chalcogramma* Pallas). *Bull. Hokkaido Univ. Fac. Fish.* **22,** 11–29.

Hambrey, M. J., and W. B. Harland (1981). The evolution of climate. *In* "The Evolving Earth" (L. R. M. Cocks, ed.), pp. 137–154. Cambridge Univ. Press, London and New York.

Hameedi, M. J. (1978). Aspects of water column primary productivity in the Chukchi Sea during summer. *Mar. Biol. (Berlin)* **48,** 37–46.

Hamilton, P., and M. Rattray (1978). A numerical model of the depth dependent, wind driven circulation on a continental shelf. *J. Phys. Oceanogr.* **8,** 437–457.

Hamming, R. W. (1973). "Numerical Methods for Scientists and Engineers." McGraw-Hill, New York.

Hamming, R. W. (1977). "Digital Filters." Prentice-Hall, Englewood Cliffs, New Jersey.

Hamon, B. V., and J. D. Kerr (1968). Time and space scales of variation in the East Australian Current from merchant ship data. *Aust. J. Mar. Freshwater Res.* **19,** 101–106.

Hamre, J. (1978). The effect of recent changes in the North Sea mackerel fishery on stock and yield. *Rapp. P.-V. Reun. Cons. Int. Explor. Mer* **172,** 197–210.

Han, G. C., D. V. Hansen, and J. A. Galt (1980). Steady state diagnostic model of the New York Bight. *J. Phys. Oceanogr.* **10,** 1998–2020.

Hansen, J., D. Johnson, A. Lacis, S. Lebedeff, P. Lee, D. Rind, and G. Russell (1981). Climate impact of increasing atmospheric carbon dioxide. *Science* **213,** 957–966.

Hansen, J., D. Johnson, A. Lacis, S. Lebedeff, P. Lee, D. Rind, and G. Russell (1983). Climatic effects of atmospheric carbon dioxide. *Science* **220,** 873–875.

Hantel, M. (1972). Wind stress curl—the forcing function for oceanic motions. *In* "Studies in Physical Oceanography" (A. L. Gordon, ed.), pp. 121–136. Gordon & Breach, New York.

Happ, G., J. G. Gosselink, and J. W. Day (1977). The seasonal distribution of organic carbon in a Louisiana estuary. *Estuarine Coast. Mar. Sci.* **5,** 695–705.

Harden Jones, F. R. (1968). "Fish Migration." Arnold, London.

Hardy, A. C. (1924). The herring in relation to its animate environment. Part I. The food and feeding habits of the herring with special reference to the east coast of England. *Fish. Invest. Ser. 2* **7,** 1–53.

Hardy, A. C. (1939). Ecological investigation with the continuous plankton recorder: object, plan and methods. *Hull Bull. Mar. Ecol.* **1,** 1–57.

Hargrave, B. T., and G. H. Geen (1970). Effects of copepod grazing on two natural phytoplankton populations. *J. Fish. Res. Board Can.* **27,** 1395–1403.

Hargrave, B. T., and N. M. Burns (1979). Assessment of sediment trap collection efficiency. *Limnol. Oceanogr.* **24,** 1124–1136.

Harkness, D. D., and A. Walton (1969). Carbon-14 in the biosphere and humans. *Nature (London)* **223,** 1216–1218.

Harkness, D. D., and A. Walton (1972). Glasgow University Radiocarbon measurements IV. *Radiocarbon* **14,** 111–113.

Harlow, F. H. (1963). The particle-in-cell method for numerical solution of problems in fluid dynamics. *Proc. Symp. Appl. Math.* **15,** 269.

Harris, G. P. (1980). The measurement of photosynthesis in natural populations of phytoplankton. *In* "The Physiological Ecology of Phytoplankton" (I. Morris, ed.), pp. 129–187. Univ. of California Press, Berkeley.

Harrison, W. G. (1980). Nutrient regeneration and primary production in the sea. *In* "Primary Productivity in the Sea" (P. G. Falkowski, ed.), pp. 433–460. Academic Press, New York.

Harrison, W. G., D. Douglas, P. G. Falkowski, G. T. Rowe, and J. Vidal (1983). Summer nutrient dynamics of the Middle Atlantic Bight: nitrogen uptake and regeneration. *J. Plankton Res.* **5,** 539–556.

Harriss, R. C., and D. I. Sebacher (1981). Methane flux in forested freshwater swamps of the southeastern United States. *Geophys. Res. Lett.* **8,** 1002–1004.

Hart, T. J. (1963). Speciation in marine phytoplankton. *In* "Speciation in the Sea" (J. P. Harding and N. Tebble, eds.), pp. 145–155. Publ. 3, Syst. Assoc., London.

Hartwig, E. O. (1976). Nutrient cycling between the water column and a marine sediment. I. Organic carbon. *Mar. Biol. (Berlin)* **34,** 285–295.

Harvey, H. W., L. N. Cooper, M. V. Lebour, and F. S. Russell (1935). Plankton production and its control. *J. Mar. Biol. Assoc. U.K.* **20,** 407–441.

Hastings, J. R. (1976). A single-layer hydrodynamical-numerical model of the eastern Bering Sea shelf. *Mar. Sci. Comm.* **2,** 335–356.

Hathaway, J. C. (1971). WHOI Data File, Continental Margin Program, Atlantic Coast of the United States. *WHOI Tech. Rep.* WHOI-77-15.

Hathaway, J. C., C. W. Poag, P. C. Valentine, R. W. Miller, D. M. Schultz, F. T. Manheim, F. A. Kohout, M. H. Bothner, and D. A. Sangrey (1979). U.S. Geological Survey core drilling on the Atlantic Shelf. *Science* **206**, 515–527.

Hawley, N. (1982). Settling velocity distribution of natural aggregates. *J. Geophys. Res.* **87**, 9489–9498.

Hay, W. W., and J. R. Southam (1977). Modulation of marine sedimentation by the continental shelves. *In* "The Fate of Fossil Fuel CO_2 in the Oceans" (N. R. Anderson and A. Malahoff, eds.), pp. 569–604. Plenum, New York.

Haykin, S. (1983). "Communication Systems." Wiley, New York.

Hays, J. D., J. Imbrie, and N. J. Shackleton (1976). Variations in the earth's orbit: Pacemaker of the ice ages. *Science* **194**, 1121–1132.

Hazelworth, J. B., and G. A. Berberian (1979). MESA New York Bight Project expanded water column characterization cruise (XWCC 21). *NOAA Data Rep.* **OMPA-7**.

Heald, E. J. (1969). "Atlas of the Principal Fishery Resources on the Continental Shelf from the West Coast of Florida to Texas," Publ. R8854, Inst. Mar. Sci., Univ. of Miami, Coral Gables.

Heath, G. R. (1974). Dissolved silica and deep-sea sediments. *In* "Studies in Paleo-oceanography" (W. W. Hay, ed.), *Spec. Publ. Soc. Econ. Paleontol. Mineral.* No. 20, 77–93.

Heath, G. R., T. C. Moore, and J. P. Dauphin (1977). Organic carbon in deep-sea sediments. *In* "The Fate of Fossil Fuel CO_2 in the Oceans" (N. R. Anderson and A. Malahoff, eds.), Plenum, New York.

Hedges, J. I., and P. L. Parker (1976). Land-derived organic matter in surface sediments from the Gulf of Mexico. *Geochim. Cosmochim. Acta* **40**, 1019–1029.

Hedges, J. I., J. R. Ertel, P. D. Quay, P. M. Grootes, J. E. Richey, A. H. Devol, G. W. Farwell, F. W. Schmidt, and E. Salati (1986). Organic carbon-14 in the Amazon River system. *Science* **231**, 1129–1131.

Heimdal, B. R. (1983). Phytoplankton and nutrients in the waters northwest of Spitsbergen in the autumn of 1979. *J. Plankton Res.* **5**, 901–918.

Heinrich, A. K. (1962a). The life histories of plankton animals and seasonal cycles of plankton communities in the oceans. *J. Cons. Cons. Int. Explor. Mer* **27**, 15–24.

Heinrich, A. K. (1962b). On the production of copepods in the Bering Sea. *Int. Rev. Ges. Hydrobiol.* **47**, 465–469.

Helland-Hansen, B. (1934). The Sognefjord section. Oceanographic observations in the northernmost part of the North Sea and the southern part of the Norwegian Sea. *In* "James Johnstone Memorial Volume," pp. 257–274. Liverpool Univ. Press, Liverpool.

Hellerman, S. (1967). An updated estimate of the wind stress on the world ocean. *Mon. Weather Rev.* **95**, 607–626.

Hemingway, E. (1935). "Green Hills of Africa," pp. 149–150. Scribners, New York.

Hempel, G. (1978). North Sea fisheries and fish stocks—a review of recent change. *Rapp. P.-V. Cons. Int. Explor. Mer* **173**, 145–167.

Hennemeuth, R. C. (1976). Fisheries and renewable resources of the Northwest Atlantic Shelf. *In* "Effects of Energy-Related Activities on the Atlantic Continental Shelf" (B. Manowitz, ed.), BNL 50484, pp. 146–166. Brookhaven Natl. Lab., Upton, New York.

Henrichs, S. M., and J. W. Farrington (1984). Peru upwelling region sediments near 15°S. I. Remineralization and accumulation of organic matter. *Limnol. Oceanogr.* **29**, 1–19.

Henrichs, S. M., J. W. Farrington, and C. Lee (1984). Peru upwelling region sediments near 15°S. 2. Dissolved free and total hydrolyzable amino acids. *Limnol. Oceanogr.* **29**, 20–34. Henriksen, K., J. I. Hansen and T. H. Blackburn (1981). Rates of nitrification, distribution of nitrifying bacteria, and nitrogen fluxes in different types of sediment from Danish Waters. *Mar. Biol. (Berlin)* **61**, 299–304.

Hentschel, E. (1933–1936). Allgemeine Biologie des Sudatlantischen Ozeans. *Wiss. Ergeb. Dtsch. Atl. Exped. "Meteor"* **11**.

Herbland, A., and A. LeBouteiller (1981). The size distribution of phytoplankton and particulate organic matter in the Equatorial Atlantic Ocean: importance of ultraseston and consequences. *J. Plankton Res.* **3,** 659–673.

Herman, A. W. (1983). Vertical distribution patterns of copepods, chlorophyll, and production in northeastern Baffin Bay. *Limnol. Oceanogr.* **28,** 708–719.

Herman, A. W., and K. L. Denman (1979). Intrusions and vertical mixing at the shelf/slope water front south of Nova Scotia. *J. Fish. Res. Board Can.* **36,** 1445–1453.

Hewitt, R. P., G. H. Theilacker, and N. C. Loo (1985). Causes of mortality in young jack mackerel. *Mar. Ecol. Prog. Ser.* **26,** 1–10.

Hicks, D. C., and J. R. Miller (1980). Meteorological forcing and bottom water movement off the northern New Jersey Coast. *Estuarine Coast. Mar. Sci.* **11,** 563–571.

Hildebrand, F. B. (1974). "Introduction to Numerical Analysis." McGraw-Hill, New York.

Hinze, J. O. (1959). "Turbulence." McGraw-Hill, New York.

Hirota, J. (1974). Quantitative natural history of *Pleurobrachia bachei* in La Jolla Bight. *Fish. Bull.* **72,** 295–352.

Hirt, C. W. (1968). Heuristic stability theory for finite-difference equations. *J. Comput. Phys.* **2,** 339–355.

Hitchcock. G. L., and T. J. Smayda (1977). The importance of light in the initiation of the 1972–1973 winter–spring diatom bloom in Narragansett Bay. *Limnol. Oceanogr.* **22,** 126–131.

Hitchcock, G. L., C. Langdon, and T. J. Smayda (1985). Seasonal variations in phytoplankton biomass and productivity of a warm-core Gulf Stream ring. *Deep-Sea Res.* **32,** 1287–1300.

Hjort, J. (1914). Fluctuations in the great fisheries of Northern Europe viewed in the light of biological research. *Rapp. P.-V. Reun. Cons. Int. Explor. Mer* **20,** 1–228.

Ho, C. L., and B. B. Barrett (1977). Distribution of nutrients in Louisiana's coastal waters influenced by the Mississippi River. *Estuarine Coast. Mar. Sci.* **5,** 173–195.

Hobbie, J. E., B. J. Copeland and W. G. Harrison (1975). Nutrient cycling within the Pamlico Estuary. *In* "Estuarine Research," (L. E. Cronin, ed.), Vol. 1, pp. 287–305. Academic Press, New York.

Hobbie, J. E., T. J. Novitsky, P. A. Rublee, R. L. Ferguson, and A. V. Palumbo (1986). Microbiology of Georges Bank. *In* "Georges Bank and Its Surrounding Waters" (R. H. Backus, ed.), MIT Press, Cambridge, Massachusetts. In press.

Hobson, L. A. (1971). Relationships between particulate organic carbon and micro-organisms in upwelling areas off southwest Africa. *Invest. Pesq.* **35,** 195–208.

Hobson, L. A., and R. J. Pariser (1970). The effect of inorganic nitrogen on macromolecular synthesis by *Thalassiosira fluviatilis* H. and *Cyclotella nana* H. grown in batch culture. *J. Exp. Mar. Biol. Ecol.* **6,** 69–76.

Hoering, T. C. (1955). Variations of nitrogen-15 abundance in naturally occurring substances. *Science* **122,** 1233–1234.

Hoffert, M. I., A. J. Callegari, and C. T. Hsieh (1981). A box-diffusion carbon cycle model, with upwelling, polar bottom water formation, and a marine biosphere. *In* "Carbon Cycle Modelling" (B. Bolin, ed.), pp. 287–305. Wiley, New York.

Hoffman, J. J. (1974). A comparison of organic matter in river water and sea-water. M.S. Thesis, Texas A&M Univ., College Station.

Hofmann, E. E., J. M. Klinck, and G. A. Paffenhofer (1981). Concentrations and vertical fluxes of zooplankton fecal pellets on a continental shelf. *Mar. Biol. (Berlin)* **61,** 327–335.

Hoge, F. E., and R. N. Swift (1981). Airborne spectroscopic detection of laser-induced water raman backscatter and fluorescence from chlorophyll a, and other naturally occurring pigments. *Appl. Opt.* **20,** 3197–3205.

Holland, H. D. (1973). Ocean water, nutrients, and atmospheric oxygen. *In* "Proceedings of the Symposium on Hydrogeochemistry and Biogeochemistry" (E. Ingerson, ed.), pp. 68–81. Clarke Boardman, New York.

Holland, H. D. (1978). "The Chemistry of the Atmosphere and Oceans." Wiley (Interscience), New York.

Holligan, P. M., M. Viollier, D. S. Harbour, P. Camus, and M. Champagne-Phillipe (1983). Satellite and ship studies of Coccolithophore production along a continental shelf edge. *Nature (London)* **304**, 339–342.

Holling, C. S. (1973). Resilience and stability of ecological systems. *Annu. Rev. Ecol. Syst.* **4**, 1–23.

Holm-Hansen, O., K. Nishida, V. Moses, and M. Calvin (1959). Effects of mineral salts on short-term incorporation of carbon dioxide in *Chlorella. J. Exp. Bot.* **10**, 109–124.

Holm-Hansen, O., C. J. Lorenzen, R. W. Holmes, and J. D. Strickland (1965). Fluorometric determination of chlorophyll. *J. Cons. Cons. Int. Explor. Mer* **30**, 3–15.

Holm-Hansen, O., J. D. H. Strickland, and P. M. Williams (1966). A detailed analysis of biologically important substances in a profile off southern California. *Limnol. Oceanogr.* **11**, 548–561.

Holmes, R. W. (1970). The Secchi disk in turbid coastal waters. *Limnol. Oceanogr.* **15**, 688–694.

Holser, W. T. (1984). Gradual and abrupt shifts in ocean chemistry during the Phanerozoic time. *In* "Patterns of Change in Earth Evolution" (H. D. Holland and A. F. Trendall, eds.), pp. 123–143. Springer-Verlag, Berlin and New York.

Honjo, S. (1980). Material fluxes and modes of sedimentation in the mesopelagic and bathypelagic zones. *J. Mar. Res.* **38**, 53–97.

Honjo, S. (1982). Seasonality and interaction of biogenic and lithogenic particulate flux at the Panama Basin. *Science* **218**, 883–884.

Honjo, S. (1984). Study of ocean fluxes in time and space by bottom-tethered sediment trap arrays: a recommendation. *In* "Global Ocean Flux Study," pp. 306–324. Natl. Acad. Press, Washington, D.C.

Honjo, S., and M. R. Roman (1978). Marine copepod fecal pellets: production, preservation, and sedimentation. *J. Mar. Res.* **36**, 45–57.

Honjo, S., S. J. Manganini, and J. J. Cole (1982). Sedimentation of biogenic matter in the deep ocean. *Deep-Sea Res.* **26**, 609–625.

Hopkins, T. S. (1974). On time dependent wind induced motions. *Rapp. P.-V. Reun. Cons. Int. Explor. Mer* **167**, 21–36.

Hopkins, T. S., and N. Garfield (1979). Gulf of Maine Intermediate Water. *J. Mar. Res.* **37**, 103–137.

Hopkins, T. S., and D. A. Dieterle (1983). An externally forced barotropic circulation model for the New York Bight. *Cont. Shelf Res.* **2**, 49–73.

Hopkins, T. S., and D. A. Dieterle (1986). Particle dispersal in the New York Bight Apex. *Rapp. P.-V. Reun. Cons. Int. Explor. Mer* **186**, 128–149.

Hopkins, T. S., and L. A. Slatest (1986). The vertical eddy viscosity in the presence of baroclinic flow in coastal waters. *J. Geophys. Res.* **91**, 14269–14280.

Hopkins, T. S., and D. A. Dieterle (1987). Analysis of the baroclinic circulation in the New York Bight with a 3-d diagnostic model. *Cont. Shelf Res.* **7**, 237–265.

Horne, E. P. (1978). Physical aspects of the Nova Scotian shelf-break fronts. *In* "Oceanic Fronts in Coastal Processes" (M. J. Bowman and W. E. Esaias, eds.), pp. 59–68. Springer-Verlag, Berlin and New York.

Horner, R. (1984). Phytoplankton abundance, chlorophyll a, and primary productivity in the western Beaufort Sea. *In* "The Alaskan Beaufort Sea" (P. W. Barnes, D. W. Schell, and E. Reimnitz, eds.), pp. 295–310. Academic Press, New York.

Horrigan, S. G. (1981). Primary production under the Ross Ice Shelf, Antarctica. *Limnol. Oceanogr.* **26**, 378–382.

Horwood, J. W. (1978). Observations on spatial heterogeneity of surface chlorophyll in one and two dimensions. *J. Mar. Biol. Assoc. U.K.* **58**, 487–502.

Hoss, D. E., and W. F. Hettler (1979). Gulf of Mexico fisheries: current state of knowledge and suggested continental-related research. Unpublished manuscript, NMFS SE Fish. Cent., Beaufort, North Carolina.

Houde, E. D. (1978). Critical food concentrations for larvae of three species of subtropical marine fishes. *Bull. Mar. Sci.* **28,** 395–411.

Houghton, R. A. (1986). Estimating changes in the carbon content of terrestrial ecosystems from historical data. *In* "The Changing Carbon Cycle: A Global Analysis" (J. R. Trabalka and D. E. Reichle, eds.), pp. 175–193. Springer-Verlag, Berlin and New York.

Houghton, R. A., R. D. Boone, J. M. Melillo, C. A. Palm, G. M. Woodwell, N. Meyers, B. Moore and D. L. Skole (1985). Net flux of carbon dioxide from tropical forests in 1980. *Nature (London)* **316,** 617–620.

Houghton, R. W., and M. A. Mensah (1978). Physical aspects and biological consequences of Ghanian coastal upwelling. *In* "Upwelling Ecosystems" (R. Boje and M. Tomczak, eds.), pp. 167–180. Springer-Verlag, Berlin and New York.

Houghton, R. W., P. C. Smith, and R. O. Fournier (1978). A simple model for cross-shelf mixing on the Scotian shelf. *J. Fish. Res. Board Can.* **35,** 414–421.

Hovis, W. A., and K. C. Leung (1977). Remote sensing of ocean color. *Opt. Engineer.* **16,** 158–164.

Hovis, W. A., D. K. Clark, F. Anderson, R. W. Austin, W. H. Wilson, E. T. Baker, B. Ball, H. R. Gordon, J. L. Mueller, S. Z. El-Sayed, B. Sturm, R. C. Wrigley, and C. S. Yentsch (1980). Nimbus-7 Coastal Zone Color Scanner: systems description and initial imagery. *Science* **210,** 60–63.

Howarth, R. W., and J. J. Cole (1985). Molybdenum availability, nitrogen limitation, and phytoplankton growth in natural waters. *Science* **229,** 653–655.

Howe, S. O. (1979). Biological consequences of environmental changes related to coastal upwelling: a simulation study. Ph.D. Thesis, Univ. of Washington, Seattle.

Hsueh, Y. (1980). On the theory of deep flow in the Hudson Shelf Valley. *J. Geophys. Res.* **85,** 4913–4918.

Hsueh, Y., and J. J. O'Brien (1972). Steady coastal upwelling induced by an along-shore current. *J. Phys. Oceanogr.* **1,** 180–186.

Hsueh, Y., and C. Y. Peng (1978). A diagnostic model of continental shelf circulation. *J. Geophys. Res.* **83,** 3033–3041.

Hubbs, C. L. (1948). Changes in the fish fauna of western North America correlated with changes in ocean temperature. *J. Mar. Res.* **7,** 459–482.

Hubbs, C. L., and G. I. Roden (1964). Oceanography and marine life along the Pacific coast of Middle America. *In* "Natural Environment and Early Cultures," Handbook of Middle American Indians, pp. 143–186. Univ. of Texas Press, Austin.

Hubbs, C. L., and G. S. Bien (1967). La Jolla natural radiocarbon measurements V. *Radiocarbon* **9,** 261–294.

Hulburt, E. M., and N. Corwin (1969). Influence of the Amazon River outflow on the ecology of the western tropical Atlantic III. The planktonic flora between the Amazon River and the Windward Islands. *J. Mar. Res.* **27,** 55–72.

Humphrey, G. F. (1963). Seasonal variation in phytoplankton pigments in waters off Sydney. *Aust. J. Mar. Freshwater Res.* **14,** 24–36.

Hunt, J. M. (1966). The significance of carbon isotope variations in marine sediments. *In* "Advances in Organic Geochemistry" (S. D. Hobson and G. C. Speers, eds.), pp. 27–35. Pergamon, New York.

Hunt, J. M., and E. T. Degens (1967). Carbon isotope fractionation in living and fossil organic matter. *Stud. Biophy.* **4,** 179–190.

Hunter, J. R. (1972). Swimming and feeding behavior of larval anchovy *Engraulis mordax*. *Fish. Bull.* **70,** 821–838.

Husby, D. M., and C. S. Nelson (1982). Turbulence and vertical stability in the California Current. *CalCOFI Rep.* **23**, 113–129.

Hutchings, L. (1981). The formation of plankton patches in the southern Benguela Current. *In* "Coastal Upwelling" (F. A. Richards, ed.), pp. 496–506. Am. Geophys. Union, Washington, D.C.

Hutchinson, G. E. (1950). The biogeochemistry of vertebrate excretion. *Bull. Am. Mus. Nat. Hist.* **96**, 1–554.

Hutchinson, G. E. (1970). Ianula: an account of the history and development of the Lago de Monterosi, Latium, Italy. *Trans. Am. Philos. Soc.* **60**, 1–178.

Ikeda, T., and S. Motoda (1978). Zooplankton production in the Bering Sea calculated from 1956–1970 *Oshro Maru* data. *Mar. Sci. Comm.* **4**, 329–346.

Ikushima, I. (1967). Ecological studies on the productivity of aquatic plant communities, III. Effect of depth on daily photosynthesis in submerged macrophytes. *Bot. Mag.* **80**, 57–67.

Imamura, A. (1934). Past tsunamis of the Sanriku coast. *Jpn. J. Astron. Geophys.* **11**, 79–93.

Imbrie, J., and K. P. Imbrie (1979). "Ice Ages, Solving the Mystery." Enslow, Hillside, New Jersey.

Incze, L. S. (1983). Larval life history of Tanner crabs, *Chionectes bairdi* and *C. opilio*, in the southeastern Bering Sea and relationships to regional oceanography. Ph.D. Thesis, Univ. of Washington, Seattle.

Incze, L. S., M. E. Clarke, M. J. Dagg, J. J. Goering, A. W. Kendall, T. Nishiyama, A. J. Paul, S. L. Smith, J. Vidal, and P. D. Walline (1986). Pollock eggs, pollock larvae and the planktonic environment. *Proc. Pollock Ecosyst. Rev. Workshop, 1985* Natl. Mar. Fish. Ser., Seattle, Washington. In press.

Isaacs, J. D. (1973). Potential trophic biomasses and trace-substance concentrations in unstructured marine food webs. *Mar. Biol. (Berlin)* **22**, 97–104.

Iseki, K., F. Whitney, and C. S. Wong (1980). Biochemical changes of sedimented matter in sediment traps in shallow coastal waters. *Bull. Plankton Soc. Jpn.* **27**, 27–56.

Iselin, C. O'D. (1939). Some physical factors which may influence the productivity of New England's coastal waters. *J. Mar. Res.* **2**, 75–85.

Ishevskii, G. K. (1964). "The systematic basis of predicting oceanological conditions and the reproduction of the fisheries." Moscow.

Ishida, T. (1967). Age and growth of Alaska pollock in the eastern Bering Sea. *Bull. Hokkaido Reg. Fish. Res. Lab.* **32**, 1–7.

Ittekot, V., O. Martins, and R. Seifert (1983). Nitrogenous organic matter transported by major world rivers. *In* "Transport of Carbon and Minerals in Major World Rivers." (E. T. Degens, S. Kempe, and H. Soliman, eds.), Part 2. *Mitt. Geol.-Palaeontol. Inst. Univ. Hamburg* **55**, 119–127.

Ivanov, M. V. (1977). Influence of microorganisms and microenvironment on the global sulfur cycle. *In* "Environmental Biogeochemistry and Geomicrobiology. I. The Aquatic Environment" (W. E. Krumbein, ed.), pp. 47–61. Ann Arbor Sci. Publ., Ann Arbor, Michigan.

Iverson, R. L., L. K. Coachman, R. T. Cooney, T. S. English, J. J. Goering, G. L. Hunt, M. C. Macauley, C. P. McRoy, W. S. Reeburg, and T. E. Whitledge (1979). Ecological significance of fronts in the Southeastern Bering Sea. *In* "Coastal Ecological Processes" (R. J. Livingston, ed.), pp. 437–468. Plenum, New York.

Izdar, E., T. Konuk, S. Honjo, V. Asper, S. Manganini, E. T. Degens, V. Ittekot,, and S. Kempe (1984). First data on sediment trap experiment in Black Sea deep water. *Naturwissenschaften* **71**, 478–479.

Jamart, B. M., D. F. Winter, K. Banse, G. C. Anderson, and R. K. Lam (1977). A theoretical study of phytoplankton growth and nutrient distribution in the Pacific Ocean off the northwestern U.S. coast. *Deep-Sea Res.* **24**, 753–773.

James, I. D. (1977). A model of the annual cycle of temperature in a frontal region of the Celtic Sea. *Estuarine Coast. Mar. Sci.* **5**, 339–353.

Jenkins, G. M., and D. G. Watts (1968). "Spectral Analysis and Its Applications." Holden-Day, San Francisco, California.

Jenkins, W. J., and J. C. Goldman (1985). Seasonal oxygen cycling and primary production in the Sargasso Sea. *J. Mar. Res.* **43,** 465–491.

Jerlov, N. G. (1976). "Marine Optics." Elsevier, Amsterdam.

Johnson, P. W., and J. M. Sieburth (1979). Chroococcoid cyanobacteria in the sea: a ubiquitous and diverse phototrophic biomass. *Limnol. Oceanogr.* **24,** 298–308.

Johnston, H. S., O. Serang, and J. Podolske (1979). Instantaneous global nitrous oxide photochemical rates. *J. Geophys. Res.* **84,** 5077–5082.

Johnston, R. (1973). Nutrients and metals in the North Sea. *In* "North Sea Science" (E. D. Goldberg, ed.), pp. 293–307. MIT Press, Cambridge, Massachusetts.

Joiris, C., G. Billen, C. Lancelot, M. H. Daro, J. P. Mommaerts, A. Bertels, M. Bossicart, J. Nijs, and J. H. Hecq (1982). A budget of carbon cycling in the Belgian coastal zone: relative roles of zooplankton, bacterioplankton, and benthos in the utilization of primary production. *Neth. J. Sea Res.* **16,** 260–275.

Jones, R. (1973). Density dependent regulation of the numbers of cod and haddock. *Rapp. P.-V. Reun. Cons. Int. Explor. Mer* **164,** 156–173.

Jordan, R. (1971). Distribution of anchoveta *(Engraulis ringens J.)* in relation to the environment. *Invest. Pesq.* **35,** 113–126.

Jorgensen, C. B. (1966). "Biology of Suspension Feeding." Pergamon, Oxford.

Joyce, T., and M. Stalcup (1985). Wintertime convection in a Gulf Stream warm-core ring. *J. Phys. Oceanogr.* **15,** 1032–1042.

Judkins, D. C., C. D. Wirick, and W. E. Esaias (1980). Composition, abundance, and distribution of zooplankton in the New York Bight, September 1974–September 1975. *Fish. Bull.* **77,** 669–683.

Kabanova, J. G. (1964). Primary production and nutrient salt content in the Indian Ocean waters in October–April 1960/61. *Tr. Inst. Okeanol. Akad. Nauk USSR* **64,** 85–93.

Kabanova, J. G. (1968). Primary production of the northern part of the Indian Ocean. *Oceanology* **8,** 214–225.

Kalle, K. (1951). Meereskundlich-chemische Untersuchungen mit Hilfe des Pulfrich-Photometer von Zeiss. VII. Die mikrobe-stimmung des chlorophylls und der Eigenfluoreszens des meerwassers. *Dtsch. Hydrogr. Z.* **4,** 92–96.

Kamba, M. (1977). Feeding habits and vertical distribution of walleye pollock, *Theragra chalcogramma* (Pallas), in early life history stage in Uchiura Bay, Hokkaido. *In* "Fisheries Biological Production in the Subarctic Pacific Region," Spec. Vol., pp. 123–273. Rep. Inst. North Pac. Fish., Hokkaido Univ., Hakkaido, Japan.

Kamykowski, D., and S. J. Zentara (1986). Predicting plant nutrient concentrations from temperature and sigma-t in the upper kilometer of the world oceans. *Deep-Sea Res.* **33,** 89–105.

Kaplan, W., I. Valiela, and J. M. Teal (1979). Denitrification in a salt marsh ecosystem. *Limnol. Oceanogr.* **24,** 726–734.

Karl, D. M., G. A. Knauer, J. H. Martin, and B. B. Ward (1984). Chemolithotrophic bacterial production in association with rapidly sinking particles: implications to oceanic carbon cycles and mesopelagic food webs. *Nature (London)* **309,** 54–56.

Keeling, C. D. (1961). Concentration and isotopic abundances of carbon dioxide in rural and marine air. *Geochim. Cosmochim. Acta* **12,** 133–149.

Keeling, C. D. (1973). Industrial production of carbon dioxide from fossil fuels and limestone. *Tellus* **28,** 174–198.

Keeling, C. D., R. B. Bacastow, and T. P. Whorf (1982). Measurements of the concentration of carbon dioxide at Mauna Loa observatory, Hawaii. *In* "Carbon Dioxide Review: 1982" (W. C. Clark, ed.), pp. 377–385. Oxford Univ. Press, London and New York.

Keeling, C. D., A. F. Carter, and W. G. Mook (1984). Seasonal, latitudinal and secular variations in the abundance and isotopic ratios of atmospheric CO_2. 2. Results from oceanographic cruises in the tropical Pacific Ocean. *J. Geophys. Res.* **89**, 4615–4628.

Keir, R. S. (1983). Reduction of thermohaline circulation during deglaciation: the effect on atmospheric radiocarbon and CO_2. *Earth Planet. Sci. Lett.* **64**, 445–456.

Keller, G. H., D. N. Lambert, and R. H. Bennett (1979). Geotechnical properties of continental slope deposits—Cape Hatteras to Hydrographer Canyon. *In* "Geology of Continental Slopes" (L. J. Doyle and O. H. Pilkey, eds.), *Spec. Publ. Soc. Econ. Paleontol. Mineral.* No. 27, 131–151.

Kelley, J. C. (1976). Sampling the sea. *In* "Ecology of the Seas" (D. H. Cushing and J. J. Walsh, eds.), pp. 361–387. Blackwell, Oxford.

Kelly, P. M., D. A. Campbell, P. P. Micklin, and J. R. Tarrant (1983). Large-scale water transfer in the USSR. *Geojournal* **7**(3), 201–214.

Kemp, A. L., T. W. Anderson, R. L. Thomas, and A. Mudrochova (1974). Sedimentation rates and recent sedimentary history of Lakes Ontario, Erie, and Huron. *J. Sediment Pet.* **44**, 207–218.

Kemp, A. L., R. L. Thomas, C. I. Dell, and J. M. Jaquet (1976). Cultural impact on the geochemistry of sediments in Lake Erie. *J. Fish. Res. Board Can.* **33**, 440–462.

Kempe, S. (1979). Carbon in the rock cycle. *In* "The Global Carbon Cycle" (B. Bolin, E. T. Degens, S. Kempe, and P. Ketner, eds.), pp. 343–378. Wiley, New York.

Kempe, S. (1982). Long-term records of the CO_2 pressure fluctuations in fresh water. *Mitt. Geol.-Palaeontol. Inst. Univ. Hamburg* **52**, 91–332.

Kendall, A. W., and L. A. Walford (1979). Sources and distribution of bluefish, *Pomatomus saltatrix*, larvae and juveniles off the east coast of the United States. *Fish. Bull.* **77**, 213–227.

Kennedy, V. S., and J. A. Mihursky (1972). Effects of temperature on the respiratory metabolism of three Chesapeake Bay bivalves. *Chesapeake Sci.* **13**, 1–22.

Kennett, J. P. (1982). "Marine Geology." Prentice-Hall, Englewood Cliffs, New Jersy.

Kennett, J. P., N. J. Shackleton, S. V. Margolis, D. E. Goodney, W. C. Dudley, and P. M. Kroopnick (1979). Late Cenozoic oxygen and carbon isotope history and volcanic ash stratigraphy: DSDP Site 284, South Pacific. *Am. J. Sci.* **279**, 52–69.

Ketchum, B. H. (1939). The absorption of phosphate and nitrate by illuminated cultures of *Nitzschia closterium*. *Am. J. Bot.* **26**, 399–407.

Ketchum, B. H., A. C. Redfield, and J. C. Ayers (1951). The oceanography of the New York Bight. *Pap. Phys. Oceanogr. Meteorol.* **12**, 3–46.

Ketchum, B. H., and D. J. Keen (1955). The accumulation of river water over the continental shelf between Cape Cod and Chesapeake Bay. *Deep-Sea Res.* **3**, 346–357.

Ketchum, B. H., and N. Corwin (1964). The persistence of "winter" water on the continental shelf south of Long Island, N.Y. *Limnol. Oceanogr.* **9**, 467–475.

Ketchum, B. H., D. R. Kester, and P. K. Park (1981). "Ocean Dumping of Industrial Wastes." Plenum, New York.

Keynes, R. D. (1979). "The Beagle Record." Cambridge Univ. Press, London and New York.

Khalil, M. A., and R. A. Rasmussen (1982). Secular trends of atmospheric methane (CH_4). *Chemosphere* **11**, 877–883.

Khalil, M. A., and R. A. Rasmussen (1984). Carbon monoxide in the Earth's atmosphere: increasing trend. *Science* **224**, 54–56.

Khalil, M. A., and R. A. Rasmussen (1986). Interannual variability of atmospheric methane: possible effects of the El Niño-Southern Oscillation. *Science* **232**, 56–58.

Kierstead, H., and L. Slobodkin (1953). The size of water masses containing plankton blooms. *J. Mar. Res.* **12**, 141–147.

Kihara, K. (1971). Studies on the formation of demersal fishing grounds 2. Analytical studies on the effect of the wind on the spreading of water masses in the eastern Bering Sea. *Bull. Soc. Fr.-Jpn. Oceanogr.* **9**, 12–22.

Killingley, J. S., and W. H. Berger (1979). Stable isotopes in a mollusk shell: Detection of upwelling events. *Science* **205**, 186–188.

Kimball, H. H. (1928). Amount of solar radiation that reaches the surface of the earth on the land and on the sea, and the methods by which it is measured, *Mon. Weather Rev.* **56**, 393–399.

Kinder, T. H., L. K. Coachman, and J. A. Galt (1975). The Bering slope current system. *J. Phys. Oceanogr.* **5**, 231–244.

Kinder, T. H., and J. D. Schumacher (1981). Circulation over the continental shelf of the southeastern Bering Sea. *In* "The Eastern Bering Sea Shelf: Oceanography and Resources" (D. W. Hood and J. A. Calder, eds.), Vol. 1, pp. 53–75. Univ. of Washington Press, Seattle.

King, F. D., and A. H. Devol (1979). Estimates of vertical eddy diffusion through the thermocline from phytoplankton uptake rates in the mixed layer of the eastern Tropical Pacific. *Limnol. Oceanogr.* **24**, 645–651.

Kiry, P. R. (1974). An historical look at the water quality of the Delaware River estuary to 1973. *Contrib. Dep. Limnol., Acad. Nat. Sci. Phila.* **4**, 1–76.

Kjerfve, B., and H. N. McKellar (1980). Time series measurements of estuarine material fluxes. *In* "Estuarine Perspectives" (V. S. Kennedy, ed.), pp. 341–357. Academic Press, New York.

Klein, P. (1986). A simulation study of the interactions between physical and biological processes on Georges Bank. *In* "Georges Bank and Its Surrounding Waters" (R. H. Backus, ed.), MIT Press, Cambridge, Massachusetts. In press.

Klein, P., and B. Coste (1984). Effects of wind-stress variability on nutrient transport into the mixed layer. *Deep-Sea Res.* **31**, 21–37.

Knauer, G. A., J. A. Martin, and K. W. Bruland (1979). Fluxes of particulate carbon, nitrogen, and phosphorus in the upper water column of the northeast Pacific. *Deep-Sea Res.* **26**, 97–108.

Knauer, G. A., D. Hebel, and F. Cipriano (1982). Marine snow, major site of primary production in coastal waters. *Nature (London)* **300**, 630–631.

Knauer, G. A., J. H. Martin, and D. M. Karl (1983). Sediment trap derived estimates of new production in oligotrophic north Pacific Ocean waters: additional evidence for a two layer euphotic zone and high rates of primary production. *Eos* **64**, 1054.

Knox, F., and M. B. McElroy (1984). Changes in atmospheric CO_2: influence of the marine biota at high latitude. *J. Geophys. Res.* **89**, 4629–4637.

Koblentz-Mishke, O. J., V. V. Valkovsinky, and J. C. Kabanova (1970). Plankton primary production of the world oceans. *In* "Scientific Exploration of the South Pacific" (W. S. Wooster, ed.), pp. 189–193. Natl. Acad. Sci., Washington, D.C.

Koide, M., A. Soutar, and E. D. Goldberg (1972). Marine geochronology with ^{210}Pb. *Earth Planet. Sci. Lett.* **57**, 263–277.

Koide, M., K. W. Bruland, and E. D. Goldberg (1973). Th-228/Th-232 and ^{210}Pb geochronologies in marine and lake sediments. *Geochim. Cosmochim. Acta* **37**, 1171–1187.

Koide, M., J. J. Griffin, and E. D. Goldberg (1975). Records of plutonium fallout in marine and terrestrial samples. *J. Geophys. Res.* **80**, 4153–4162.

Koide, M., E. D. Goldberg, and V. F. Hodge (1980). ^{241}Pu and ^{241}Am in sediments from coastal basins off California and Mexico. *Earth Planet. Sci. Lett.* **48**, 250–256.

Koide, M., and E. D. Goldberg (1982). Transuranic nuclides in two coastal marine sediments off Peru. *Earth Planet. Sci. Lett.* **57**, 263–277.

Koike, I., and A. Hattori (1979). Estimates of denitrification in sediments of the Bering Sea shelf. *Deep-Sea Res.* **26**, 409–415.

Kolmogorov, A. N. (1941). The local structure of turbulence in an incompressible viscous fluid for very large Reynolds number. *Dokl. Akad. Nauk SSSR* **30**, 299–303.

Komhyr, W. D., R. H. Gammon, T. B. Harris, L. S. Waterman, T. J. Conway, W. R. Taylor and K. W. Thoning (1985). Global atmospheric CO_2 distribution and variation from 1968–82 NOAA/GMCC flask sample data. *J. Geophys. Res.* **90**, 5567–5596.

Kominz, M. A. (1984). Ocean ridge volumes and sea level change—an error analysis. *In* "Interregional Unconformities and Hydrocarbon Accumulation" (J. S. Schlee, ed.), *Mem. Am. Assoc. Pet. Geol.* **36**, 108–127.

Kotenev, B. N. (1972). Bottom relief and sediments and some features of the geological structure of the continental slope in the eastern Bering Sea. *In* "Soviet Fisheries Investigations in the Northeastern Pacific" (P. A. Moiseev, ed.), pp. 35–62. Isr. Program Sci. Transl., Keter Press, Jerusalem.

Kraft, J. C., E. A. Allen, D. F. Belknap, C. J. John, and E. M. Mauermeyer (1979). Processes and morphological evolution of an estuarine and coastal barrier system. *In* "Barrier Islands from the Gulf of St. Lawrence to the Gulf of Mexico" (S. P. Leatherman, ed.), pp. 149–183. Academic Press, New York,

Krause, M. (1981). Vertical distribution of faecal pellets during FLEX '76. *Helgol. Wiss. Meeresunters* **34**, 313–327.

Kreitler, C. W. (1979). Nitrogen-isotope ratio studies of soils and groundwater nitrate from alluvial fan aquifers in Texas. *J. Hydrol.* **42**, 147–170.

Krey, J. (1973). Primary production in the Indian Ocean. *In* "The Biology of the Indian Ocean" (B. Zeitzschel, ed.), pp. 115–126. Springer-Verlag, Berlin and New York.

Krishnaswami, S., D. Lal, and B. S. Amin (1973). Geochronological studies in Santa Barbara Basin: ^{55}Fe as a unique tracer for particle settling. *Limnol. Oceanogr.* **18**, 763–770.

Kroopnick, P., S. V. Margolis, and C. S. Wong (1977). δ^{13}C variation in marine carbonate sediment as indicator of the CO_2 balance between the atmosphere and ocean. *In* "The Fate of Fossil Fuel CO_2 in the Ocean" (N. R. Anderson and A. Malahoff, eds.), pp. 295–322. Plenum, New York.

Kulm, L. D., and K. F. Scheidegger (1979). Quaternary sedimentation on the tectonically active Oregon continental slope. *In* "Geology of Continental Slopes" (L. J. Doyle and O. H. Pilkey, eds.), *Spec. Publ. Soc. Econ. Paleontol. Mineral.* No. 27, 247–263.

Kump, L. R. (1986). The global sedimentary redox cycle. Ph.D. Thesis, Univ. of South Florida, St. Petersburg.

Kump, L. R., and R. M. Garrels (1986). Modeling atmospheric O_2 in the global sedimentary redox cycle. *Am. J. Sci.* (in press).

Kundu, P. K. (1980). A numerical investigation of mixed-layer dynamics. *J. Phys. Oceanogr.* **10**, 220–236.

Labeyrie, L. D., H. D. Livingston, and V. T. Bowen (1976). Comparison of the distributions in marine sediments of the fall-out derived nuclides ^{55}Fe and ^{239,240}Pu. *In* "Transuranium Nuclides in the Marine Environment," pp. 121–137. IAEA, Vienna.

Lacis, A., J. Hansen, P. Lee, T. Mitchell, and S. Lebedeff (1981). Greenhouse effect of trace gases, 1970–1980. *Geophys. Res. Lett.* **8**, 1035–1038.

Lamb, H. 1879. "Treatise on the Mathematical Theory of the Motion of Fluids." Cambridge Univ. Press, London and New York.

Lampitt, R. S. (1985). Evidence for the seasonal deposition of detritus to the deep-sea floor and its subsequent resuspension. *Deep-Sea Res.* **32**, 885–898.

Lange, R. (1973). ADPIC, a three-dimensional computer code for the study of pollutant dispersal and deposition under complex conditions. *Lawrence Livermore Lab.* [*Rep.*] **UCRL-51462**, 1–60.

Lasker, R. (1975). Field criteria for survival of anchovy larvae: the relation between inshore chlorophyll maximum layers and successful first feeding. *Fish. Bull.* **73**, 453–462.

Lasker, R. (1985). What limits clupeoid production? *Can. J. Fish. Aquat. Sci.* **42**, 31–38.

Lasker, R., H. M. Feder, G. H. Theilacker, and R. C. May (1970). Feeding, growth, and survival of *Engraulis mordax* larvae reared in the laboratory. *Mar. Biol. (Berlin)* **5**, 345–353.

Laurence, G. C. (1974). Growth and survival of haddock (*Melanogrammus aeglefinus*) larvae in relation to planktonic prey concentration. *J. Fish. Res. Board Can.* **31**, 1415–1419.

Laurence, G. C. (1977). A bioenergetic model for the analysis of feeding and survival potential of winter flounder, *Pseudopleuronectes americanus*, larvae during the period from hatching to metamorphosis. *Fish. Bull.* **75**, 529–546.

Laurence, G. C. (1978). Comparative growth, respiration, and delayed feeding activities of larval cod *(Gadus morhua)* and haddock *(Melanogrammus aeglefinus)* as influenced by temperature during laboratory studies. *Mar. Biol. (Berlin)* **50**, 1–7.

Lawacz, W. (1969). The characteristics of sinking material and the formation of bottom deposits in a eutrophic lake. *Mitt. Int. Ver. Theor. Angew. Limnol.* **17**, 319–331.

Leach, J. H., and S. J. Nepszy (1976). The fish community in Lake Erie. *J. Fish. Res. Board Can.* **33**, 622–638.

LeBlond, P. H., and L. A. Mysak (1978). Waves in the Ocean. *Elsevier Oceanogr. Ser. (Amsterdam)* **20**, 1–602.

Le Borgne, R. (1975). Equivalence entre les measures de biovolumes, poids secs, carbone, azote, et phosphore du mesozooplankton de l'Atlantique tropical. *Cah. ORSTOM Ser. Oceanogr.* **13**, 179–196.

LeBrasseur, R. J. (1965). Seasonal and annual variation of net zooplankton at Ocean Station P, 1956–64. *Fish. Res. Board Can. Rep. Ser.* No. 202.

LeBrasseur, R. J., and O. D. Kennedy (1972). Microzooplankton in coastal and oceanic areas of the Pacific subarctic water mass: a preliminary report. *In* "Biological Oceanography of the Northern North Pacific" (A. Y. Takenouti, ed.), pp. 355–365. Idemitsu Shoten, Tokyo.

Lee, C., and C. Cronin (1982). The vertical flux of particulate organic nitrogen in the sea: decomposition of amino acids in the Peru upwelling area. *J. Mar. Res.* **48**, 227–251.

Lee, T. N. (1975). Florida Current spin-off eddies. *Deep-Sea Res.* **22**, 753–766.

Lee, T. N., and D. A. Mayer (1977). Low-frequency current variability and spin-off eddies along the shelf off Southeast Florida. *J. Mar. Res.* **35**, 193–220.

Lee, T. N., L. P. Atkinson, and R. Legeckis (1981). Observations of a Gulf Stream frontal eddy on the Georgia continental shelf, April, 1977. *Deep-Sea Res.* **28**, 347–378.

Lee, T. N., E. Daddio, and G. C. Han (1982). Steady-state diagnostic model of summer mean circulation on the Georgia shelf. *J. Phys. Oceanogr.* **12**, 820–838.

Leetma, A. (1977). Effects of the winter of 1976–1977 on the northwestern Sargasso Sea. *Science* **198**, 188–189.

Leggett, W. C., K. T. Frank, and J. E. Carscadden (1984). Meteorological and hydrographic regulation of year-class strength in capelin *(Mallotus villosus). Can. J. Fish. Aquat. Sci.* **41**, 1193–1201.

Lerman, A. (1981). Controls on river water composition and the mass balance of river systems. *In* "River Inputs to Ocean Systems," pp. 1–4. UNESCO, Geneva.

Leslie, P. H. (1945). On the use of matrices in certain population mathematics. *Biometrika* **33**, 183–212.

Levitus, S. (1982). Climatological atlas of the world ocean. *NOAA Prof. Pap.* No. 13, 1–173.

Lewis, C. S. (1950). "The Lion, the Witch, and the Wardrobe." pp. 1–186. Macmillan, New York.

Lewis, C. S. (1960). "Of Other Worlds" (W. Hooper, ed.), pg. 42. Harcourt, New York.

Li, W. K. (1980). Temperature adaptation in phytoplankton: cellular and photosynthetic characteristics. *In* "Primary Productivity in the Sea" (P. G. Falkowski, ed.), pp. 259–279. Plenum, New York.

Li, W. K. (1985). Photosynthetic response to temperature of marine phytoplankton along a latitudinal gradient (16°N to 74°N). *Deep-Sea Res.* **32**, 1381–1391.

Li, W. K., D. V. Subba Rao, W. G. Harrison, J. C. Smith, J. J. Cullen, B. Irwin, and T. Platt (1983). Autotrophic picoplankton in the tropical oceans. *Science* **219**, 292–295.

Li, Y. H., H. W. Feely, and P. H. Santschi (1979). ^{228}Th-^{228}Ra radioactive disequilibria in the New York Bight and its implications for coastal pollution. *Earth Planet. Sci. Lett.* **42**, 13–26.

Libby, W. F., E. C. Anderson, and J. R. Arnold (1949). Age determination by radiocarbon content: world-wide array of natural radiocarbon. *Science* **109**, 227–228.

Liebig, J. 1840. "Chemistry in Its Application to Agriculture and Physiology." Taylor & Walton.

Lieth, H. (1963). The role of vegetation in the carbon dioxide content of the atmosphere. *J. Geophys. Res.* **68**, 3887–3898.

Likens, G. E., F. H. Bormann, R. S. Pierce, and W. A. Reiners (1978). Recovery of a deforested ecosystem. *Science* **199**, 492–496.

Lillick, L. C. (1937). Seasonal studies of the phytoplankton off Woods Hole, Massachusetts. *Biol. Bull. (Woods Hole, Mass.)* **73**, 488–503.

Lillick, L. C. (1940). Phytoplankton and planktonic protozoa of the offshore waters of the Gulf of Maine. Part II. Qualitative composition of the planktonic flora. *Trans. Am. Philos. Soc.* **31**, 194–237.

Lindall, W. N., and C. H. Saloman (1977). Alteration and destruction of estuaries affecting fishing resources of the Gulf of Mexico. *Mar. Fish. Rev.* **39**, 1–7.

Lindeman, R. L. (1942). The trophic-dynamic aspect of ecology. *Ecology* **23**, 399–418.

Lindh, T. B. (1983). Temporal variations in ^{13}C, ^{34}S and global sedimentation during the Phanerozoic. Ms. Thesis, Univ. of Miami, Coral Gables.

Linick, T. W. (1977). La Jolla natural radiocarbon measurements. VII. *Radiocarbon* **19**, 19–48.

Linick, T. W. (1979). La Jolla radiocarbon measurements VIII. *Radiocarbon* **21**, 186–202.

Linick, T. W. (1980). Bomb-produced carbon-14 in the surface water of the Pacific Ocean. *Radiocarbon* **22**, 599–606.

Lipps, J. H. (1970). Plankton evolution. *Evolution* **24**, 1–22.

Lipps, J. H., and E. Mitchell (1976). Trophic model for the adaptive radiations and extinctions of marine mammals. *Paleobiology* **2**, 147–155.

Lipschultz, F. O., C. Zafiriou, S. C. Wofsky, M. B. McElroy, F. W. Valois, and S. W. Watson (1981). Production of NO and N_2O by soil nitrifying bacteria. *Nature (London)* **294**, 641–643.

Lisitzin, A. P. (1966). "Recent Sedimentary Processes in the Bering Sea." Nauka, Moscow.

Liu, K. (1979). Geochemistry of inorganic nitrogen compounds in two marine environments: the Santa Barbara basin and the ocean off Peru. Ph.D. Thesis, Univ. of California at Los Angeles.

Livingston, H. D., and V. T. Bowen (1979). Pu and ^{137}Cs in coastal sediments. *Earth Planet. Sci. Lett.* **43**, 29–45.

Livingston, R. J., R. L. Iverson, R. H. Estabrook, V. E. Keys, and J. Taylor (1974). Major features of the Apalachacola Bay system: physiography, biota, and resource management. *Fla. Sci.* **37**, 245–271.

Lloyd, I. J. (1970). Primary production off the coast of Northwest Africa. *J. Cons. Cons. Int. Explor. Mer* **33**, 312–323.

Loder, J. W., D. G. Wright, C. Garrett, and B. A. Juszko (1982). Horizontal exchange on central Georges Bank. *Can. J. Fish. Aquat. Sci.* **39**, 1130–1137.

Lohmann, H. (1908). Untersuchungen zur festellung des Vollständigen gehaltes des Meeres an plankton. *Wiss. Meer. Abt. Kiel.* **10**, 129–370.

Long, P. E., and D. W. Pepper (1981). An examination of some simple numerical schemes for calculating scalar advection. *J. Appl. Meteorol.* **20**, 146–156.

Longhurst, A. R. (1967a). Vertical distribution of zooplankton in relation to the eastern Pacific oxygen minimum. *Deep-Sea Res.* **14**, 51–64.

Longhurst, A. R. (1967b). The pelagic phase of *Pleuroncodes planipes* Stimpson (Crustacea, Galatheidae) in the California Current. *Calif. Coop. Oceanic Fish. Invest.* Rep. 11, 142–154.

Longhurst, A. R. (1971). The clupeoid resources of tropical seas. *Oceanogr. Mar. Biol. Annu. Rev.* **9**, 349–385.

Longhurst, A. R. (1983). Benthic-pelagic coupling and export of organic carbon from a tropical Atlantic continental shelf—Sierra Leone. *Estuarine Coast. Shelf Sci* **17**, 261–285.

Longhurst, A. R. (1985a). Relationship between diversity and the vertical structure of the upper ocean. *Deep-Sea Res.* **32**, 1535–1570.

Longhurst, A. R. (1985b). The structure and evolution of plankton communities. *Prog. Oceanogr.* **15**, 1–35.

Longuet-Higgins, M. S. (1965). Some dynamical aspects of ocean currents. *Q. J. R. Meteorol. Soc.* **91**, 425–451.

Lorenzen, C. J. (1966). A method for the continuous measurement of *in vivo* chlorophyll concentration. *Deep-Sea Res.* **13**, 223–227.

Lorenzen, C. J. (1971). Continuity in the distribution of surface chlorophyll. *J. Cons. Cons. Int. Explor. Mer* **34**, 18–23.

Lorenzen, C. J., N. A. Welschmeyer, and A. E. Copping. (1983a). Particulate organic carbon flux in the subarctic Pacific. *Deep-Sea Res.* **30**, 639–643.

Lorenzen, C. J., N. A. Welschmeyer, A. E. Copping, and M. Vernet. (1983b). Sinking rates of organic particles. *Limnol. Oceanogr.* **28**, 766–769.

Lotka, A. J. (1925). "Elements of Physical Biology." Williams & Wilkins Baltimore, Maryland.

Lough, R. G. (1984). Larval fish trophodynamic studies on Georges Bank: Sampling strategy and initial results. *Flod. Rapp.* **1**, 395–434.

Love, C. M. (1974). "Eastropac Atlas," Vol. 8. U.S. Govt. Print. Off., Washington D.C.

Lovelock, J. E. (1979). "Gaia." Oxford Univ. Press, London and New York.

Lowdon, J. A. (1969). Isotopic fractionation in corn. *Radiocarbon* **11**, 391–393.

Lucht, F., and M. Gillbricht (1978). Long-term observations of nutrient contents near Helgoland in relation to nutrient input of the River Elbe. *Rapp. P.-V. Reun. Cons. Int. Explor. Mer* **172**, 358–360.

Lutter, S. (1984). Quantitative Untersuchungen zur sedimentation der Fruhjahrsblute im Balsfjord, Nordnorwegen. M.S. Thesis, Kiel Univ., Kiel.

Machta, L. (1973). The role of the ocean and the biosphere in the CO_2 cycle. *In* "Changing Chemistry of the Oceans" (D. Dryssen and D. Jagner, eds.), pp. 121–145. Wiley (Interscience), New York.

MacIlvane, J. C. (1973). Sedimentary processes on the continental slope off New England. Ph.D. Thesis, MIT, Cambridge, Massachusetts.

MacIsaac, J. J., and R. C. Dugdale (1969). The kinetics of nitrate and ammonium uptake by natural populations of marine phytoplankton. *Deep-Sea Res.* **16**, 45–57.

MacIsaac, J. J., and R. C. Dugdale (1972). Interaction of light and inorganic nitrogen in controlling nitrogen uptake in the sea. *Deep-Sea Res.* **19**, 209–232.

Macko, S. A. (1981). Stable nitrogen isotope ratios as tracers of organic geochemical processes. Ph.D. Thesis, Univ. of Texas, Austin.

Macko, S. A., W. Y. Lee, and P. L. Parker (1982). Nitrogen and carbon isotope fractionation by two species of marine amphipods: laboratory and field studies. *J. Exp. Mar. Biol. Ecol.* **63**, 145–149.

Macko, S. A., L. Entzeroth, and P. L. Parker (1984). Regional differences in nitrogen and carbon isotopes on the continental shelf of the Gulf of Mexico. *Naturwissenschaften* **71**, 374.

Maeda, T. (1977). Relationship between annual fluctuation of oceanographic conditions and abundance of year classes of the yellow-fin sole in the eastern Bering Sea. *In* "Fisheries Biological Production in the Sub-Arctic Pacific Region," Spec. Vol., pp. 259–268. Rep. Inst. North Pac. Fish., Hokkaido Univ., Hokkaido, Japan.

Maeda, T., T. Fujii, and K. Masuta (1967). Studies on the trawl fishing grounds of the eastern Bering Sea. *Nippon Suisan Gakkaishi* **33**, 713–720.

Mahoney, J. (1979). Environmental and physiological factors in growth and seasonal maxima of the dinoflagellate, *Ceratium tripos. Bull. N.J. Acad. Sci.* **24**, 28–38.

Malone, T. C. (1971a). The relative importance of nanoplankton and net plankton as primary producers in the California Current system. *Fish. Bull.* **69**, 799–820.

Malone, T. C. (1971b). The relative importance of nanoplankton and net plankton as primary producers in tropical and neritic phytoplankton communities. *Limnol. Oceanogr.* **16**, 633–639.

Malone, T. C. (1976). Phytoplankton productivity in the apex of the New York Bight: environmental regulations of productivity/chlorophyll a. *ASLO Symp.* No. 2, 260–272.

Malone, T. C. (1977a). Plankton systematics and distribution. *MESA N.Y. Bight Atlas Monogr.* **13**, 1–45.

Malone, T. C. (1977b). Light-saturated photosynthesis by phytoplankton size fractions in the New York Bight, U.S.A. *Mar. Biol. (Berlin)* **42**, 281–292.

Malone, T. C. (1978). The 1976 *Ceratium tripos* bloom in the New York Bight: causes and consequences. *NOAA Tech. Rep. NMFS Circ.* No. 410, 1–14.

Malone, T. C. (1982). Phytoplankton photosynthesis and carbon-specific growth: light-saturated rates in a nutrient rich environment. *Limnol. Oceanogr.* **27**, 226–235.

Malone, T. C., and M. B. Chervin (1979). The production and fate of phytoplankton size fractions in the plume of the Hudson River, New York Bight. *Limnol. Oceanogr.* **24**, 683–696.

Malone, T. C., and P. J. Neale (1981). Parameters of light-dependent photosynthesis for phytoplankton size fractions in temperate estuarine and coastal environments. *Mar. Biol. (Berlin)* **61**, 289–297.

Malone, T. C., T. S. Hopkins, P. G. Falkowski, and T. E. Whitledge (1983). Production and transport of phytoplankton biomass over the continental shelf of the New York Bight. *Cont. Shelf Res.* **1**, 305–337.

Malthus, T. R. 1798. "An Essay on the Principle of Population, as it Affects the Future Improvement of Society, with Remarks on the Speculations of Mr. Godwin, M. Condorcet, and other Writers." J. Johnson, London.

Mandelli, E. F., P. R. Burkholder, T. E. Doheny, and R. Brody (1970). Studies of primary productivity in coastal waters of southern Long Island, New York. *Mar. Biol. (Berlin)* **7**, 153–160.

Mann, K. H. (1976). Production on the bottom of the sea. *In* "Ecology of the Seas" (D. H. Cushing and J. J. Walsh, eds.), pp. 225–250. Blackwell, Oxford.

Margalef, R. (1967). Composicion y ditribucion del fitoplancton. *Mem. Soc. Cienc. Nat. La Salle* **25**, 141–205.

Margalef, R. (1974). Distribution de seston dans la region d'affleurement du Nort-ouest de l'Afrique en mars de 1973. *Tethys* **6**, 77–88.

Margalef, R. (1978). Life forms of phytoplankton as a survival alternative in an unstable environment. *Oceanol. Acta* **1**, 493–509.

Margalef, R., and A. Ballester (1967). Fitoplancton y produccion primaria de la costa catalana, de Junio de 1965 a Junio de 1966. *Invest. Pesq.* **31**, 165–182.

Margolis, S. V., P. M. Kroopnick, and D. E. Goodney (1977). Cenozoic and late Mesozoic paleo-oceanographic and paleoglacial history recorded in circum-Atlantic deep-sea sediments. *Mar. Geol.* **25**, 131–147.

Marland, G., and R. M. Rotty (1984). Carbon dioxide emissions from fossil fuels: a procedure for estimation and results for 1950–82. *Tellus* **36**, 232–261.

Marr, J. C. (1956). The "critical period" in the early life history of marine fishes. *J. Cons. Cons. Int. Explor. Mer* **21**, 160–170.

Marshall, S. M. (1933). The production of microplankton in the Great Barrier Reef region. *Sci. Rep. Great Barrier Reef Exped.* **2**, 111–157.

Martens, C. S., N. E. Blair, C. D. Green, and D. J. Des Marais (1986). Seasonal variations in the stable carbon isotopic signature of biogenic methane in a coastal sediment. *Science* **233**, 1300–1303.

Martinez, L. A., M. W. Silver, J. M. King, and A. L. Alldredge (1983). Nitrogen fixation by floating diatom mats: a source of new nitrogen to oligotrophic ocean waters. *Science* **221**, 152–154.

Matheke, G. E., and R. Horner (1974). Primary productivity of the benthic micro-algae in the Chukchi Sea near Barrow, Alaska. *J. Fish. Res. Board Can.* **31**, 1779–1786.

Mathews, T. D., A. D. Fredericks, and W. M. Sackett (1973). The geochemistry of radiocarbon in the Gulf of Mexico. *In* "Radioactive Contamination of the Marine Environment," pp. 725–734. IAEA, Vienna.

Matsuyama, M. (1973). Organic substances in sediment and settling matter during spring in a meromictic Lake Shigetsu. *J. Oceanogr. Soc. Jpn.* **29**, 53–60.

Maul, G. A. (1977). The annual cycle of the Gulf Loop Current. Part I: Observations during a one-year time series. *J. Mar. Res.* **35**, 29–47.

May, E. B. (1973). Extensive oxygen depletion in Mobile Bay, Alabama. *Limnol. Oceanogr.* **18**, 353–366.

May, R. C. (1974). Larval mortality in marine fishes and the critical period concept. *In* "The Early Life History of Fish" (J. H. Blaxter, ed.), pp. 3–19. Springer-Verlag, Berlin and New York.

May, R. M., and G. F. Oster (1976). Bifurcation and dynamic complexity in simple ecological models. *Am. Nat.* **110**, 573–599.

Mayer, D. A., D. V. Hansen, and D. Ortman (1979). Long term current and temperature observations on the Middle Atlantic shelf. *J. Geophys. Res.* **84**, 1776–1792.

Mayer, D. A., G. C. Han, and D. V. Hansen (1982). The structure of circulation: MESA physical oceanographic studies in New York Bight. *J. Geophys. Res.* **87**, 9579–9588.

McAllister, C. D. (1969). Aspects of estimating zooplankton production from phytoplankton production. *J. Fish. Res. Board Can.* **26**, 199–200.

McAllister, C. D., T. R. Parsons, and J. D. H. Strickland (1960). Primary productivity and fertility at station "P" in the northeast Pacific Ocean. *J. Cons. Cons. Int. Explor. Mer* **25**, 240–259.

McCarthy, J. J. (1972). The uptake of urea by natural populations of marine phytoplankton. *Limnol. Oceanogr.* **17**, 738–748.

McCarthy, J. J., W. R. Taylor, and J. L. Taft (1977). Nitrogenous nutrition of the plankton in the Chesapeake Bay. *Limnol. Oceanogr.* **22**, 996–1011.

McCarthy, J. J., and J. C. Goldman (1979). Nitrogenous nutrition of marine phytoplankton in nutrient-depleted waters. *Science* **203**, 670–672.

McCarthy, J. J., and J. L. Nevins (1986a). Sources of nitrogen for primary production in Warm-Core Rings 79-E and 81-D. *Limnol. Oceanogr.* **31**, 690–700.

McCarthy, J. J., and J. L. Nevins (1986b). Utilization of nitrogen and phosphorus by primary producers in Warm-Core Ring 82-B following deep convective mixing. *Deep-Sea Res.* **33**, 1773–1788.

McCave, I. N. (1975). Vertical flux of particles in the ocean. *Deep-Sea Res.* **22**, 491–502.

McConnaughey, T., and C. P. McRoy (1979). ^{13}C label identifies eelgrass *(Zostera marina)* carbon in an Alaskan estuarine food web. *Mar. Biol. (Berlin)* **53**, 263–269.

McCoy, F. W. (1974). Late Quaternary sedimentation in the eastern Mediterranean Sea. Ph.D. Thesis, Harvard Univ., Cambridge, Massachusetts.

McElroy, M. B. (1983). Marine biological controls on atmospheric CO_2 and climate. *Nature (London)* **302**, 328–329.

McGowan, J. A. (1974). The nature of oceanic ecosystems. *In* "The Biology of the Oceanic Pacific" (C. B. Miller, ed.), pp. 9–28. Oregon State Univ., Corvallis.

McGowan, J. A., and T. L. Hayward (1978). Mixing and oceanic productivity. *Deep-Sea Res.* **25**, 771–793.

McHugh, J. L. (1976). Estuarine fisheries: are they doomed? *In* "Estuarine Processes" (M. Wiley, ed.), Vol. 1, pp. 15–27. Academic Press, New York.

McLain, D. R., and F. Favorite (1976). Anomalously cold winters and the southeastern Bering Sea, 1971–75. *Mar. Sci. Comm.* **2**, 299–334.

McLaren, I. A. (1978). Generation lengths of some temperate marine copepods: estimation, prediction, and implication. *J. Fish. Res. Board Can.* **35**, 1330–1342.

Meade, R. H. (1982). Sources, sinks, and storage of river sediment in the Atlantic drainage of the United States. *J. Geol.* **90**, 235–252.

Mellor, G. L., and T. Yamada (1974). A hierarchy of turbulence closure models for planetary boundary layers. *J. Atmos. Sci.* **31**, 1791–1806.

Menard, H. W., and S. M. Smith (1966). Hypsometry of ocean basin provinces. *J. Geophys. Res.* **71**, 4305–4325.

Menzel, D. W. (1960). Utilization of food by a Bermuda reef fish, *Ephinephelus guttatus. J. Cons. Cons. Int. Explor. Mer* **25**, 216–222.

Menzel, D. W., and J. H. Ryther (1960). The annual cycle of primary production in the Sargasso Sea off Bermuda. *Deep-Sea Res.* **6**, 351–367.

Menzel, D. W., and J. H. Ryther. (1961a). Zooplankton in the Sargasso Sea off Bermuda and its relation to organic production. *J. Cons. Cons. Int. Explor. Mer* **26**, 250–258.

Menzel, D. W., and J. H. Ryther. (1961b). Annual variations in primary production of the Sargasso Sea off Bermuda. *Deep-Sea Res.* **7**, 282–288.

Menzel, D. W., and J. P. Spaeth (1962). Occurrence of ammonium in Sargasso Sea waters and in rain water at Bermuda. *Limnol. Oceanogr.* **7**, 159–162.

Meybeck, M. (1982). Carbon, nitrogen, and phosphorus transport by world rivers. *Am. J. Sci.* **282**, 401–450.

Miller, C. B., B. W. Frost, H. P. Batchelder, M. J. Clemons, and R. E. Conway (1984). Life histories of large, grazing copepods in a subarctic ocean gyre: *Neocalanus plumchrus, Neocalanus cristatus,* and *Eucalanus bungii* in the northeast Pacific. *Prog. Oceanogr.* **13**, 201–243.

Miller, J. M., J. P. Reed, and L. J. Pietrafesa (1984). Patterns, mechanisms and approaches to the study of migrations of estuarine-dependent fish larvae and juveniles. *In* "Mechanisms of Migration in Fishes" (J. D. McCleave, G. P. Arnold, J. J. Dodson, and W. H. Neill, eds.), pp. 209–225. Plenum, New York.

Miller, K. G., and W. B. Curry (1982). Eocene to Oligocene benthic foraminiferal isotopic record in the Bay of Biscaye. *Nature (London)* **296**, 347–350.

Miller, S. M., and H. B. Moore (1953). Significance of nannoplankton. *Nature (London)* **171**, 1121.

Milliman, J. D., and K. O. Emery (1968). Sea levels during the past 35,000 years. *Science* **162**, 1121–1123.

Mills, E. L., and R. O. Fournier (1979). Fish production and the marine ecosystems of the Scotian shelf, Eastern Canada. *Mar. Biol. (Berlin)* **54**, 101–108.

Minoda, T. (1972). Characteristics of the vertical distribution of copepods in the Bering Sea and south of the Aleutian Chain, May–June 1962. *In* "Biological Oceanography of the Northern North Pacific Ocean" (A. Y. Takenouti, ed.), pp. 323–331. Idemitsu Shoten, Tokyo.

Miyake, Y., and E. Wada (1967). The abundance ratio of $^{15}N/^{14}N$ in marine environments. *Rec. Oceanogr. Works Jpn.* **9**, 37–53.

Miyake, Y., and E. Wada (1971). The isotope effect on the nitrogen in biochemical, oxidation-reduction reactions. *Rec. Oceanogr. Works Jpn.* **11**, 1–6.

Molenkamp, C. R. (1968). Accuracy of finite-difference methods applied to the advection equation. *J. Appl. Meteorol.* **7**, 160–167.

Molina, M. J., and F. S. Rowland (1974). Stratospheric sink for chlorofluoromethanes: chlorine atom catalyzed destruction of ozone. *Nature (London)* **249**, 810–812.

Mommaerts, J. P., W. Baeyens, and G. Decadt (1979). Synthesis of research on nutrients in the Southern Bight of the North Sea. *In* "Actions de Recherche Concertées," pp. 215–234. Brussels.

Monin, A. S. (1970). Weather and climate oscillation. *In* "Scientific Exploration of the South Pacific" (W. S. Wooster, ed.), pp. 5–15. Natl. Acad. Sci., Washington, D.C.

Monod, J. (1949). The growth of bacterial cultures. *Annu. Rev. Microbiol.* **3**, 371–394.

Moody, J. D. (1975). Distribution and geological characteristics of giant oil fields. *In* "Petroleum and Global Tectonics" (A. G. Fischer and S. Judson, eds.), pp. 307–320. Princeton Univ. Press, Princeton, New Jersey.

Mook, W. G., M. Koopmans, A. F. Carter, and C. D. Keeling (1983). Seasonal, latitudinal, and secular variations in the abundance and isotopic ratios of atmospheric carbon dioxide: 1. Results from Land Stations. *J. Geophys. Res.* **88**, 10915–10933.

Moore, W. S., K. W. Bruland, and J. Michel (1981). Fluxes of uranium and thorium series isotopes in the Santa Barbara Basin. *Earth Planet. Sci. Lett.* **53**, 391–399.

Mopper, K., and E. T. Degens (1972). Aspects of the biogeochemistry of carbohydrates and proteins in aquatic environments. *WHOI Tech. Rep.* **WHOI-72-68**, 1–118.

Morel, A., and R. C. Smith (1974). Relation between total quanta and total energy for aquatic photosynthesis. *Limnol. Oceanogr.* **19**, 591–600.

Morel, A., and L. Prieur (1977). Analysis of variations in ocean color. *Limnol. Oceanogr.* **22**, 708–722.

Morgan, C. W., and J. M. Bishop (1977). An example of Gulf Stream eddy-induced water exchange in the Mid-Atlantic Bight. *J. Phys. Oceanogr.* **7**, 91–99.

Morrison, J. M., and W. D. Nowlin (1977). Repeated nutrient, oxygen, and density sections through the Loop Current. *J. Mar. Res.* **35**, 105–128.

Morse, J. W., and N. Cook (1978). The distribution and form of phosphorous in North Atlantic ocean deep sea and continental slope sediments. *Limnol. Oceanogr.* **23**, 825–830.

Mosby, H. (1938). Svalbard waters. *Geophys. Norv.* **12**, 1–85.

Motoda, S., and T. Minoda (1974). Plankton of the Bering Sea. *In* "Oceanography of the Bering Sea, With Emphasis on Renewable Resources" (D. W. Hood and E. D. Kelley, eds.), *Occas. Publ. Univ. of Alaska Inst. Mar. Sci.* No. 2, 207–243.

Mowat, F. (1984). "Sea of Slaughter." Atlantic Monthly Press, Boston, Massachusetts.

Mueller, J. A., A. A. Anderson, and J. S. Jeris (1976). Contaminants entering the New York Bight: Sources, mass loads, significances. *ASLO Symp.* No. 2, 162–170.

Muller, P. J., and E. Suess (1979). Productivity, sedimentation rate, and sedimentary organic matter in the ocean. I. Organic carbon preservation. *Deep-Sea Res.* **26**, 1347–1362.

Mullin, M. M., and E. R. Brooks (1976). Some consequences of distributional heterogeneity of phytoplankton and zooplankton. *Limnol. Oceanogr.* **21**, 784–796.

Munk, W. H. (1950). On the wind-driven ocean circulation. *J. Meteorol.* **7**, 79–93.

Munk, W. H. (1966). Abyssal recipes. *Deep-Sea Res.* **13**, 707–730.

Murphy, G. I. (1966). Population biology of the Pacific sardine *(Sardinops caerulea). Proc. Calif. Acad. Sci.* **34**, 1–84.

Murphy, G. I. (1973). Clupeoid fishes under exploitation with special reference to the Peruvian anchovy. *Hawaii Inst. Mar. Biol. Tech. Rep.* No. 30.

Murphy, L. S., and E. M. Haugen (1985). The distribution and abundance of phototrophic ultraplankton in the North Atlantic. *Limnol. Oceanogr.* **30**, 47–58.

Murray, J., and A. F. Renard. 1891. "Report on Deep-Sea Deposits Based on Specimens Collected during the Voyage of H.M.S. Challenger in the Years 1872 to 1876." HM Station. Off., London.

Murray, J. W., and V. Grundmanis (1980). Oxygen consumption in pelagic marine sediments. *Science* **209**, 1527–1530.

Nagle, C. M. (1978). "Climatology of Brookhaven National Laboratory, 1974 through 1977," BNL-50857 UC-11, Environ. Control Technol. Earth Sci., T10–4500. Brookhaven Natl. Lab., Upton, New York.

Namias, J. (1964). Seasonal persistence and recurrence of European blocking during 1958–1960. *Tellus* **16**, 394–407.

Namias, J. (1973). Response of the equatorial countercurrent to the subtropical atmosphere. *Science* **181**, 1244–1245.

Nash, E. G. (1929). "The Hansa: Its History and Romance." Dodd, Mead, New York.

National Research Council (1986). "Global Change in the Geosphere–Biosphere." Natl. Acad. Press, Washington, D.C.

Neftel, A., H. Oeschger, J. Schwander, B. Stauffer, and R. Zumbrunn (1982). Ice core sample measurements give atmospheric CO_2 content during the past 40,000 yr. *Nature (London)* **295**, 220–223.

Neftel, A., E. Moor, H. Oeschger, and B. Stauffer (1985). The increase of atmospheric CO_2 in the last two centuries: evidence from polar ice cores. *Nature (London)* **315**, 45–47.

Nellen, W. (1966). Horizontal and vertical distribution of plankton production in the Gulf of Guinea and adjacent areas from February to May, 1964. *Proc. Symp. Oceanogr. Fish. Res. Trop. Atl., Abidjan* pp. 255–264.

Nelson, D. M., H. W. Ducklow, G. L. Hitchcock, M. A. Brzezinski, T. J. Cowles, C. Garside, R. W. Gould, T. M. Joyce, C. Langdon, J. J. McCarthy, and C. S. Yentsch (1985). Distribution and composition of biogenic particulate matter in a Gulf Stream warm-core ring. *Deep-Sea Res.* **32**, 1347–1369.

Nelson, D. M., J. J. McCarthy, T. M. Joyce, and H. W. Ducklow (1986). Decay of a warm-core ocean eddy results in accelerated nutrient transport and primary production. *Nature (London)* (submitted).

Nelson, W. R., M. C. Ingham, and W. E. Schaaf (1977). Larval transport and year-class strength of Atlantic menhaden, *Brevoortia tyrannus*. *Fish. Bull.* **75**, 23–41.

Neumann, G., and W. J. Pierson (1966). "Principles of Physical Oceanography." Prentice-Hall, Englewood Cliffs, New Jersy.

Nicholson, W. R. (1978). Movements and population structure of Atlantic menhaden indicated by tag returns. *Estuaries* **1**, 141–150.

Nihoul, J. J. (1980). "Marine Turbulence." Elsevier, Amsterdam.

Nihoul, J. J. (1982). "Hydrodynamic Models of Shallow Continental Seas." RIGA, Neupre.

Nihoul, J. J. (1984). A three-dimensional general marine circulation model in a remote sensing perspective. *Ann. Geophys.* **2**, 433–442.

Niiler, P. P. (1975). Deepening of the wind-mixed layer. *J. Mar. Res.* **33**, 405–422.

Niiler, P. P., and E. B. Kraus (1975). One-dimensional models of the upper ocean. *In* "Modelling and Prediction of the Upper Layers of the Ocean" (E. B. Kraus, ed.), pp. 143–172. Pergamon, Oxford.

Nittrouer, C. A., R. W. Sternberg, R. C. Carpenter, and J. T. Bennett (1979). The use of the Pb-210 geochronology as a sedimentological tool: application to the Washington continental shelf. *Mar. Geol.* **42**, 201–232.

Nixon, S. W., C. A. Oviatt, and S. S. Hale (1976). Nitrogen regeneration and metabolism of coastal marine bottom communities. *In* "The Role of Terrestrial and Aquatic Organisms in Decomposition Processes" (J. M. Anderson and A. MacFadyed, eds.), pp. 269–283. Blackwell, Oxford.

Nordli, E. (1957). Experimental studies on the ecology of *Ceratia*. *Oikos* **8**, 200–265.

Nozaki, Y., J. K. Cochran, and K. K. Turekian (1977). Radiocarbon and ^{210}Pb distribution in submersible-taken deep-sea cores from Project Famous. *Earth Planet. Sci. Lett.* **34**, 167–173.

Nuzzi, R. (1973). "The Distribution of Phytoplankton in the New York Bight, September and November 1971. The Oceanography of the New York Bight: Physical, Chemical, Biological." Vol. 1, NYOSL Tech. Rep. 17, 76–108, Montaulk, New York.

Nydal, R., K. Löuseth, and S. Gulliksen (1979). A survey of radiocarbon variation in nature since the Test Ban Treaty. *In* "Radiocarbon Dating" (R. Berger and H. E. Suess, eds.), pp. 313–323. Univ. of California Press, Berkeley.

Nyquist, H. (1928). Certain topics in telegraph transmission theory. *AIEE Trans.* **47**, 617–644.

O'Brien, J. J., and J. S. Wroblewski (1972). An ecological model of the lower marine trophic levels on the continental shelf off West Florida. *Fla. State Univ. Tech. Rep.* **NONR-N00014-67-A-0235002**.

O'Connell, C. P. (1980). Percent of starving northern anchovy larvae *(Engraulis mordax)* in the sea as estimated by histological methods. *Fish. Bull.* **78**, 475–489.

Odum, H. T. (1970). An emerging view of the ecological system at El Verde. *In* "A Tropical Rain Forest" (H. T. Odum and R. F. Pigeon, eds.), pp. I-191A-AI-281. US AEC, Oak Ridge, Tennessee.

Oeschger, H. (1985). The contribution of ice core studies to the understanding of environmental processes. *In* "Greenland Ice Core: Geophysics, Geochemistry, and the Environment" (C. C.

Langway, H. Oeschger and W. Dansgaard, eds.), Monogr. 33, pp. 9–17. Am. Geophys. Union, Washington, D.C.

Oeschger, H., U. Siegenthaler, U. Schotterer, and A. Gugleman (1975). A box-diffusion model to study the carbon dioxide exchange in nature. *Tellus* **27**, 168–192.

Officer, C. B. (1982). Mixing, sedimentation rates, and age dating for sediment cores. *Mar. Geol.* **46**, 261–278.

Officer, C. B., and J. H. Ryther (1977). Secondary sewage treatment versus ocean outfalls: an assessment. *Science* **197**, 1056–1060.

Ogren, L., and J. Chess (1969). A marine kill on New Jersey wrecks. *Underwater Nat.* **6**, 4–12.

O'Leary, M. H. (1981). Carbon isotope fractionation in plants. *Phytochemistry* **20**, 553–567.

Olson, R. J. (1980). Nitrate and ammonium uptake in Antarctic waters. *Limnol. Oceanogr.* **25**, 1064–1074.

O'Reilly, J. E., and D. A. Busch (1984). Phytoplankton primary production on the northwestern shelf. *Rapp. P.-V. Reun. Cons. Int. Explor. Mer* **183**, 255–268.

O'Reilly, J. E., C. Evans-Zeitlin, and D. Busch (1986). Primary production on Georges Bank. *In* "Georges Bank and Its Surrounding Waters" (R. H. Backus, ed.). MIT Press, Cambridge, Massachusetts. In press.

Orlanski, I. (1976). A simple boundary condition for unbounded hyperbolic flows. *J. Comput. Phys.* **21**, 251–259.

Orr, A. P. (1933). Physical and chemical conditions in the sea in the neighborhood of the Great Barrier Reef. *Sci. Rep. Great Barrier Reef Exped.* **2**, 37–86.

Ortner, P. B., R. L. Ferguson, S. R. Piotrowicz, L. Chesal, G. Berberian, and A. V. Palumbo (1984). Biological consequences of hydrographic and atmospheric advection within the Gulf Loop intrusion. *Deep-Sea Res.* **31**, 1101–1120.

Osterberg, C. L. (1975). Radiological impacts of releases from nuclear facilities into aquatic environments—U.S.A. views. *In* "Impacts of Nuclear Releases into the Aquatic Environment," pp. 25–35. IAEA, Vienna.

Ostlund, H. G., and M. Stuiver (1980). GEOSECS Pacific radiocarbon. *Radiocarbon* **22**, 25–53.

Ostwald, W. (1903). Zur theorie der schweberorgange sowie der spezifischen gewichts bestimmungen schwebender organismen. *Arch. Gesamte Physiol.* **3**, 1–94.

Oviatt, C. A. (1981). Effects of different mixing schedules on phytoplankton, zooplankton, and nutrients in marine microcosms. *Mar. Ecol. Prog. Ser.* **4**, 57–67.

Owen, R. W., and B. Zeitzschel (1970). Phytoplankton production: seasonal change in the oceanic eastern tropical Pacific. *Mar. Biol. (Berlin)* **7**, 32–36.

Owens, T. G., P. G. Falkowski, and T. E. Whitledge (1980). Diel periodicity in cellular chlorophyll content in marine diatoms. *Mar. Biol. (Berlin)* **59**, 71–77.

Ozmidov, R. V. (1965). The scales of oceanic turbulence. *Oceanology* **6**, 325–328.

Paasche, E. (1973). Silicon and the ecology of marine plankton diatoms. II. Silicate-uptake kinetics in five diatom species. *Mar. Biol. (Berlin)* **19**, 262–269.

Pace, M. L., J. E. Glasser, and L. R. Pomeroy (1984). A simulation analysis of continental shelf food webs. *Mar. Biol. (Berlin)* **82**, 47–63.

Packard, T. T. (1971). The measurement of respiratory electron transport activity in marine phytoplankton. *J. Mar. Res.* **29**, 235–244.

Packard, T. T., A. H. Devol, and F. D. King (1975). The effect of temperature on the respiratory electron transport system in marine plankton. *Deep-Sea Res.* **22**, 237–249.

Paffenhofer, G. A. (1985). The abundance and distribution of zooplankton on the southeastern shelf of the United States. *In* "Oceanography of the Southeastern U.S. Continental Shelf" (L. P. Atkinson, D. W. Menzel, and K. A. Bush, eds.), *Coast. Estuar. Sci.* **2**, 104–117.

Paine, T. R. (1969). A note on trophic complexity and community stability. *Am. Nat.* **103**, 91–93.

Paluszkiewicz, T., L. P. Atkinson, E. S. Posmentier, and C. R. McClain (1983). Observations of a

Loop Current frontal eddy intrusion onto the west Florida shelf. *J. Geophys. Res.* **88,** 9639–9651.

Pamatmat, M. M. (1971). Oxygen consumption by the seabed IV. Shipboard and laboratory experiments. *Limnol. Oceanogr.* **16,** 536–550.

Panofsky, H. A., and G. W. Grier (1965). "Some Applications of Statistics to Meteorology." Pennsylvania State Univ. Press, University Park.

Park, K. H., C. L. Osterberg, and W. D. Forster (1972). Chemical budget of the Columbia River. *In* "The Columbia River Estuary and Adjacent Ocean Waters" (A. T. Pruter and D. L. Alverson, eds.), pp. 123–134. Univ. of Washington Press, Seattle.

Parker, P. L. (1962). The isotopic composition of fatty acids. *Annu. Rep. Carnegie Inst.* **61,** 187–190.

Parker, P. L., E. W. Behrens, J. A. Calder, and D. Shultz (1972). Stable carbon isotope ratio variations in the organic carbon from Gulf of Mexico sediments. *Contrib. Mar. Sci. Univ. Texas* **16,** 139–147.

Parker, R. A. (1974). Empirical functions relating metabolic processes in aquatic systems to environmental variables. *J. Fish. Res. Board Can.* **31,** 1550–1552.

Parker, R. A. (1977). Radiocarbon dating of marine sediments. Ph.D. Thesis, Texas A&M Univ., College Station.

Parsons, T. R. (1976). The structure of life in the sea. *In* "Ecology of the Seas" (D. H. Cushing and J. J. Walsh, eds.), pp. 81–97. Blackwell, Oxford.

Parsons, T. R. (1979). Some ecological, experimental, and evolutionary aspects of the upwelling ecosystem. *J. S. Afr. Sci.* **75,** 536–540.

Parsons, T. R., K. Stephens and J. D. Strickland (1961). On the chemical composition of eleven species of marine phytoplankton. *J. Fish. Res. Board Can.* **18,** 1001–1016.

Parsons, T. R., and R. J. LeBrasseur (1968). A discussion of some critical indices of primary and secondary production for large scale ocean surveys. *Calif. Coop. Fish. Invest. Rep.* No. 12, 54–63.

Parsons, T. R., and M. Takahashi (1973). Environmental control of phytoplankton cell size. *Limnol. Oceanogr.* **18,** 511–515.

Parsons, T. R., M. Takahashi, and B. T. Hargrave (1977). "Biological Oceanographic Processes." Pergamon, Oxford.

Pastuszak, M., W. R. Wright, and D. A. Patango (1982). One year of nutrient distribution in the Georges Bank region in relation to hydrography, 1975–76. *J. Mar. Res.* **40,** 525–542.

Paulson, C. A., and J. J. Simpson (1977). Irradiance measurements in the upper ocean. *J. Phys. Oceanogr.* **7,** 952–956.

Pearcy, W. G. (1972). Distribution and ecology of oceanic animals off Oregon. *In* "The Columbia River Estuary and Adjacent Waters" (A. T. Pruter and D. L. Alversen, eds.), pp. 351–377. Univ. of Washington Press, Seattle.

Pearman, G. I., and P. Hyson (1980). Activities of the global biosphere as reflected in atmospheric CO_2 records. *J. Geophys. Res.* **85,** 4457–4467.

Pearman, G. I., P. Hyson, and P. J. Fraser (1983). The global distribution of atmospheric carbon dioxide: I. Aspects of observations and modeling. *J. Geophys. Res.* **88,** 3581–3590.

Pearson, K. (1906). A mathematical theory of random migration. *Drapers Co. Res. Mem., Biometric Ser.* **3,** 1–45.

Peinert, R., A. Saure, P. Stegman, C. Stienen, H. Haardt, and V. Smetacek (1982). Dynamics of primary production and sedimentation in a coastal ecosystem. *Neth. J. Sea Res.* **16,** 276–289.

Peng, T. H., W. S. Broecker, H. D. Freyer, and S. Trumbore (1983). A deconvolution of the tree ring based del $\delta^{13}C$ record. *J. Geophys. Res.* **88,** 3609–3620.

Peng, T. H., and W. S. Broecker (1984). Ocean life cycles and the atmospheric CO_2 content. *J. Geophys. Res.* **89,** 8170–8180.

Peng, T. H., and H. D. Freyer (1986). Revised estimates of atmospheric CO_2 variations based on the tree ring $\delta^{13}C$ record. *In* "The Changing Carbon Cycle: A Global Analysis" (J. R. Trabalka and D. E. Reichle, eds.). Springer-Verlag, Berlin and New York. In press.

Pereyra, W. T., J. E. Reeves, and R. G. Bakkala (1976). "Demersal Fish and Shellfish Resources of the Eastern Bering Sea in the Baseline Year 1975." Natl. Mar. Fish. Ser., N.W. Fish. Cent., Seattle.

Perry, M. J., M. C. Talbot, and R. S. Alberts (1981). Photoadaptation in marine phytoplankton: response of the photosynthetic unit. *Mar. Biol. (Berlin)* **62,** 91–101.

Peterman, R. M. (1980). Testing for density-dependent marine survival in Pacific salmonids. *In* "Salmonid Ecosystems of the North Pacific" (M. J. McNeil and D. C. Himsworth, eds.), pp. 1–23. Oregon State Univ. Press, Corvallis.

Peters, K. E., R. E. Sweeney, and I. R. Kaplan (1978). Correlation of carbon and nitrogen stable isotope ratios in sedimentary organic matter. *Limnol. Oceanogr.* **23,** 598–604.

Petersen, W. T., and C. B. Miller (1975). Year-to-year variations in the planktology of the Oregon upwelling zone. *Fish. Bull.* **73,** 642–653.

Petersen, W. T., and C. B. Miller (1976). "Zooplankton Along the Continental Shelf Off Newport, Oregon 1969–1972: Distribution, Abundance, Seasonal Cycle, and Year-to-Year Variations." ORESU-T-76-002. OSU Sea Grant Program Publ., Corvallis, Oregon.

Petersen, W. T., and C. B. Miller (1977). Seasonal cycle of zooplankton abundance and species composition along the central Oregon coast. *Fish. Bull.* **75,** 717–724.

Petersen, W. T., C. B. Miller, and A. Hutchinson (1979). Zonation and maintenance of copepod populations in the Oregon upwelling zone. *Deep-Sea Res.* **26,** 467–494.

Pheiffer-Madsen, P. (1979). Seasonal variation of denitrification rate in sediment determined by use of ^{15}N. *Water Res.* **13,** 461–465.

Phillips, O. M. (1966). "The Dynamics of the Upper Ocean." Cambridge Univ. Press, London and New York.

Piacsek, S. A., and G. P. Williams (1970). Conservation properties of convection difference schemes. *J. Comput. Phys.* **6,** 392–405.

Pierotti, D., and R. A. Rasmussen (1976). Combustion as a source of nitrous oxide. *Geophys. Res. Lett.* **3,** 265–267.

Pingree, R. D., P. M. Holligan, G. T. Mardell, and R. N. Head (1976). The influence of physical stability on spring, summer, and autumn phytoplankton blooms in the Celtic Sea. *J. Mar. Biol. Assoc. U.K.* **56,** 845–873.

Pingree, R. D., P. M. Holligan, and G. T. Mardell (1978). The effects of vertical stability on phytoplankton distribution in the summer on the northwest European shelf. *Deep-Sea Res.* **25,** 1011–1028.

Piper, D. Z., and L. A. Codispoti (1975). Marine phosphorite deposits and the nitrogen cycle. *Science* **188,** 15–18.

Platt, T. (1972). Local phytoplankton abundance and turbulence. *Deep-Sea Res.* **19,** 183–187.

Platt, T. (1984). Estimation of the molar flux ratio of oxygen:carbon in the pelagic zone of the oligotrophic ocean. *Eos* **64,** 1062.

Platt, T., and K. L. Denman (1975). Spectral analysis in ecology. *Annu. Rev. Ecol. Syst.* **6,** 189–210.

Platt, T., and D. V. Subba Rao (1975). Primary production of marine microphytes. *In* "Photosynthesis and Productivity in Different Environments" (J. P. Cooper, ed.), pp. 249–80. Cambridge Univ. Press, London and New York.

Platt, T., and A. D. Jassby (1976). The relationship between photosynthesis and light for natural assemblages of coastal marine phytoplankton. *J. Phycol.* **12,** 421–430.

Platt, T., K. L. Denman, and A. D. Jassby (1977). Modeling the productivity of phytoplankton. *In* "The Sea" (E. D. Goldberg, I. N. McCave, J. J. O'Brien and J. H. Steele, eds.), Vol. 6, pp. 807–856. Wiley (Interscience), New York.

Platt, T., C. L. Gallegos, and W. G. Harrison (1980). Photoinhibition of photosynthesis in natural assemblages in marine phytoplankton. *J. Mar. Res.* **38,** 687–701.

Platt, T., and A. W. Herman (1983). Remote sensing of phytoplankton in the sea: surface layer chlorophyll as an estimate of water-column chlorophyll and primary production. *Int. J. Remote Sensing* **4**, 343–351.

Platt, T., and W. G. Harrison (1985). Biogenic fluxes of carbon and oxygen in the ocean. *Nature (London)* **318**, 55–58.

Platzman, G. W. (1972). Two dimensional free oscillations in natural basins. *J. Phys. Oceanogr.* **2**, 117–138.

Pomeroy, L. R. (1974). The ocean's food web, a changing paradigm. *BioScience* **24**, 499–504.

Pomeroy, L. R. (1979). Secondary production mechanisms of continental shelf communities. *In* "Ecological Processes in Coastal and Marine Systems" (R. J. Livingston, ed.), pp. 163–186. Plenum, New York.

Pomeroy, L. R., and D. Deibel (1986). Temperature regulation of bacterial activity during the spring bloom in Newfoundland coastal waters. *Science* **233**, 359–361.

Postma, H. (1978). The nutrient content of North Sea water: changes in recent years, particularly in the Southern Bight. *Rapp. P.-V. Reun. Cons. Int. Explor. Mer* **172**, 350–352.

Powell, T. M., P. J. Richerson, T. M. Dillon, B. A. Agee, B. J. Dozier, D. A. Godden, and L. O. Myrup (1975). Spatial scales of current speed and phytoplankton biomass fluctuations in Lake Tahoe. *Science* **189**, 1088–1090.

Preller, R., and J. J. O'Brien (1980). The influence of bottom topography on upwelling off Peru. *J. Phys. Oceanogr.* **10**, 1377–1398.

Premuzic, E. T., C. M. Benkovitz, J. S. Gaffney, and J. J. Walsh (1982). The nature and distribution of organic matter in the surface sediments of world oceans and seas. *Org. Geochem.* **4**, 63–77.

Prior, D. B., J. M. Coleman, and E. H. Doyle (1984). Antiquity of the continental slope along the middle Atlantic margin of the United States. *Science* **233**, 926–928.

Pritchard, D. W., R. O. Reid, A. Okubo, and H. H. Carter (1971). Physical processes of water movement and mixing. *In* "Radioactivity in the Marine Environment," pp. 90–136. Natl. Acad. Sci., Washington, D.C.

Purnell, D. K. (1976). Solution of the advective equation by upstream interpolation with a cubic spline. *Mon. Weather Rev.* **104**, 42–48.

Qasim, S. Z. (1970). Some problems related to the food chain in a tropical estuary. *In* "Marine Food Chains" (J. H. Steele, ed.), pp. 45–51. Univ. of California Press, Berkeley.

Qasim, S. Z. (1973). Productivity of backwaters and estuaries. *In* "The Biology of the Indian Ocean" (B. Zeitzschel, ed.), pp. 143–154. Springer-Verlag, Berlin and New York.

Quay, P. D., and M. Stuiver (1980). Vertical advection-diffusion rates in the oceanic thermocline determined from ^{14}C distributions. *Radiocarbon* **22**, 607–625.

Quinn, W. H. (1974). Monitoring and predicting El Niño invasions. *J. Appl. Meteorol.* **13**, 825–830.

Rafter, T. A. (1955). ^{14}C variations in nature and the effect on radiocarbon dating. *N. Z. J. Sci. Technol.* **37**, 20–38.

Rainer, S. F., and R. C. Fitzhardinge (1981). Benthic communities in an estuary with periodic deoxygenation. *Aust. J. Mar. Freshwater Res.* **32**, 227–243.

Raiswell, R., and R. A. Berner (1985). Pyrite formation in euxinic and semi-euxinic sediments. *Am. J. Sci.* **285**, 710–724.

Ralston, A. (1965). "A First Course in Numerical Analysis." McGraw-Hill, New York.

Ramage, C. S. (1975). Preliminary discussion of the meteorology of the 1972–1973 El Niño. *Bull. Am. Meteorol. Soc.* **56**, 234–242.

Ramp, S. R., R. J. Schlitz, and W. R. Wright (1980). Northeast Channel flow and Georges Bank nutrient budget. ICES C.M. 1980/C:35.

Rasmussen, R. A., and M. A. Khalil (1981). Atmospheric methane (CH_4): trends and seasonal cycles. *J. Geophys. Res.* **86**, 9826–9832.

Rasmussen, R. A., and M. A. Khalil (1986). Atmospheric trace gases: trends and distributions over the last decade. *Science* **232**, 1623–1624.

Ratowsky, D. A., R. K. Lowry, T. A. McMeekin, A. N. Stokes, and R. E. Chandler (1983). Model for bacterial culture growth rate through the entire biokinetic temperature range. *J. Bacteriol.* **154,** 1222–1226.

Rau, G. H., R. E. Sweeney, I. R. Kaplan, A. J. Mearns, and D. R. Young (1981). Differences in animal ^{13}C, ^{15}N, and D abundance between a polluted and an unpolluted coastal site: likely indications of sewage uptake by a marine food web. *Estuarine Coast. Shelf Sci* **13,** 701–707.

Raup, D. M., and J. J. Sepkoski (1982). Mass extinctions in the marine fossil record. *Science* **215,** 1501–1503.

Redfield, A. C. (1936). An ecological aspect of the Gulf Stream. *Nature (London)* **138,** 1013.

Redfield, A. C. (1941). The effect of the circulation of water on the distribution of the calanoid community in the Gulf of Maine. *Biol. Bull. (Woods Hole, Mass.)* **80,** 86–110.

Redfield, A. C., and A. Beale (1940). Factors determining the distribution of populations of chaetognaths in the Gulf of Maine. *Biol. Bull. (Woods Hole, Mass.)* **79,** 459–487.

Redfield, A. C., B. H. Ketchum, and F. A. Richards (1963). The influence of organisms on the composition of seawater. *In* "The Sea" (M. N. Hill, ed.),.Vol. 2, pp. 26–77. Wiley (Interscience), New York.

Reeside, J. (1957). Paleoecology of the Cretaceous seas of the western interior of the United States. *In* "Treatise on Marine Ecology and Paleoecology, Vol. 2, Paleoecology" (H. S. Ladd, ed.), *Mem. Geol. Soc. Am.* No. 67, 505–542.

Reeve, M. R., J. E. Raymont, and J. K. Raymont (1970). Seasonal biochemical composition and energy sources of *Sagitta hispida*. *Mar. Biol. (Berlin)* **6,** 357–364.

Reeves, R. R., J. G. Mead, and S. Katona (1978). The right whale, *Eubalaena glacialis* in the western North Atlantic. *Rep. Int. Whaling Comm.* **28,** 303–312.

Reid, J. L. (1965). "Intermediate Waters of the Pacific Ocean." Johns Hopkins Press, Baltimore, Maryland.

Reid, P. C. (1975). Large scale changes in North Sea phytoplankton. *Nature (London)* **257,** 217–219.

Reid, P. C. (1977). Continuous plankton records: changes in the composition and abundance of the phytoplankton of the northeastern Atlantic Ocean and North Sea, 1958–1974. *Mar. Biol. (Berlin)* **40,** 337–339.

Reid, P. C. (1978). Continuous plankton records: large-scale changes in the abundance of phytoplankton in the North Sea from 1958 to 1973. *Rapp. P.-V. Cons. Int. Explor. Mer* **172,** 384–389.

Reid, R. O., and A. C. Vastano (1966). Orthogonal coordinates for analysis of long gravity waves near islands. Santa Barbara Specialty Conference in Coastal Engineering. *Proc. Am. Soc. Civ. Eng.,* pp. 1–20.

Reid, R. O., A. C. Vastano, R. E. Whitaker, and J. J. Wanstrath (1977). Experiments in storm surge simulation. *In* "The Sea" (E. D. Goldberg, I. N. McCave, J. J. O'Brien, and J. H. Steele, ed.), Vol. 6, pp. 145–168. Wiley (Interscience), New York.

Reish, R. L., R. B. Dosiro, D. Ruppert, and R. J. Carroll (1985). An investigation of the population dynamics of Atlantic menhaden *(Brevoortia tyrannus)*. *Can. J. Fish. Aquat. Sci.* **42,** 147–157.

Reitan, C. H. (1974). Frequencies of cyclones and anticyclones for North America, 1951–1970. *Mon. Weather Rev.* **102,** 861–868.

Reyssac, J. (1966). Quelques donnees sur la composition et l'evolution annuelle du phytoplankton au large d'Abidjan (mai 1964–mai 1965). *Doc. Sci. Prov. ORSTOM* No. 3, 1–31.

Reyssac, J. (1969). Measures de la production primaire par la methode du ^{14}C au large de la Cote d'Ivoire. *Doc. Sci. Prov. ORSTOM* No. 35, 1–16.

Richards, F. A. (1952). The estimation and characterization of plankton populations by pigment analyses. I. The absorption spectra of some pigments occurring in diatoms, dinoflagellates, and brown algae. *J. Mar. Res.* **11,** 147–155.

Richards, F. A. (1965). Anoxic basins and fjords. *In* "Chemical Oceanography" (J. P. Riley and G. Skirrow, eds.), Vol. 1, pp. 611–645. Academic Press, New York.

Richards, F. A., and T. G. Thompson (1952). The estimation and characterization of plankton populations by pigment analyses. II. A spectrophotometric method for the estimation of plankton pigments. *J. Mar. Res.* **11**, 156–172.

Richards, F. A., and A. C. Redfield (1954). A correlation between the oxygen content of sea water and the organic content of marine sediments. *Deep-Sea Res.* **1**, 279–282.

Richardson, S. L., and W. G. Pearcy (1977). Coastal and oceanic fish larvae in an area of upwelling off Yaquina Bay, Oregon. *Fish. Bull.* **75**, 125–145.

Richerson, P. J., T. M. Powell, M. R. Abbott, and J. A. Coil (1978). Spatial heterogeneity in closed basins. *In* "Spatial Pattern in Plankton Communities" (J. H. Steele, ed.), pp. 239–276. Plenum, New York.

Richey, J. E., J. T. Brock, R. J. Naiman, R. C. Wissmar, and R. F. Stallard (1980). Organic carbon. Oxidation and transport by the Amazon River. *Science* **207**, 1348–1351.

Riggs, S. R. (1984). Paleoceanographic model of Neogene phosphorite deposition, U.S. Atlantic continental margin. *Science* **223**, 123–131.

Riley, G. A. (1937). The significance of the Mississippi River drainage for biological conditions in the northern Gulf of Mexico. *J. Mar. Res.* **1**, 60–74.

Riley, G. A. (1939). Correlations in aquatic ecology with an example of their application to problems of plankton productivity. *J. Mar. Res.* **2**, 56–73.

Riley, G. A. (1941). Plankton studies. IV. Georges Bank. *Bull. Bingham Oceanogr. Coll.* **7**, 1–73.

Riley, G. A. (1942). The relationship of vertical turbulence and spring diatom flowerings. *J. Mar. Res.* **5**, 67–87.

Riley, G. A. (1946). Factors controlling phytoplankton populations on Georges Bank. *J. Mar. Res.* **6**, 54–73.

Riley, G. A. (1947). A theoretical analysis of the zooplankton population of Georges Bank. *J. Mar. Res.* **6**, 104–113.

Riley, G. A. (1951). Oxygen, phosphate, and nitrate in the Atlantic Ocean. *Bull. Bingham Oceanogr. Coll.* **13**, 1–125.

Riley, G. A. (1956). Oceanography of Long Island Sound. II. Physical oceanography. *Bull. Bingham Oceanogr. Coll.* **15**, 15–46.

Riley, G. A. (1963). Theory of food chain relations in the ocean. *In* "The Sea" (M. N. Hill, ed.), Vol. 2, pp. 438–463. Wiley, New York.

Riley, G. A. (1967a). Mathematical model of nutrient conditions in coastal waters. *Bull. Bingham Oceanogr. Coll.* **19**, 72–80.

Riley, G. A. (1967b). The plankton of estuaries. *In* "Estuaries" (G. H. Lauff, ed.), Publ. 83, pp. 316–326. Am. Assoc. Adv. Sci., Washington, D.C.

Riley, G. A. (1970). Particulate organic matter in sea water. *Adv. Mar. Biol.* **8**, 1–118.

Riley, G. A., and D. F. Bumpus (1946). Phytoplankton-zooplankton relationships on Georges Bank. *J. Mar. Res.* **6**, 33–46.

Riley, G. A., H. Stommel, and D. F. Bumpus (1949). Quantitative ecology of the plankton of the western North Atlantic. *Bull. Bingham Oceanogr. Coll.* **12**, 1–169.

Roache, P. J. (1972). On artificial viscosity. *J. Comput. Phys.* **10**, 169–184.

Roache, P. J. (1976). "Computational Fluid Dynamics." Hermosa Publ., Albuquerque, New Mexico.

Robinson, G. A. (1970). Continuous plankton records: variation in the seasonal cycle of phytoplankton in the North Atlantic. *Bull. Mar. Ecol.* **6**, 333–345.

Rochford, D. J. (1962). Hydrology of the Indian Ocean. II. The surface waters of the southeast Indian Ocean and Arafura Sea in the spring and summer. *Aust. J. Mar. Freshwater Res.* **13**, 226–251.

Rochford, D. J. (1975). Nutrient enrichment of east Australian coastal waters. II. Laurieton upwelling. *Aust. J. Freshwater Res.* **26**, 233–243.

Rodinova, K. F., E. A. Romankevich, T. N. Duzhikova, and M. S. Taklova (1973). Geochemistry of organic matter in recent marine sediments off the coast of Peru. *Vnigni, Tr. 138, Geochim. Sb.* **5**, 118–129.

Roithmayr, C. M. (1963). Distribution of fishing by vessels for Atlantic menhaden, 1955–59. *U.S. Fish. Wildl. Spec. Sci. Rep.* No. 434, 1–22.

Roman, M. R. (1984). Ingestion of detritus and microheterotrophs by pelagic marine zooplankton. *Bull. Mar. Sci.* **35**, 477–494.

Roman, M. R., A. L. Gauzens, and T. J. Cowles (1985). Temporal and spatial changes in epipelagic microzooplankton and mesozooplankton biomass in warm-core Gulf Stream ring 82-B. *Deep-Sea Res.* **32**, 1007–1022.

Romankevich, E. A. (1977). "Geochemistry of Organic Matter in the Ocean." Nauka, Moscow.

Rossby, C. G. (1938). On the mutual adjustment of pressure and velocity distribution in certain simple current systems. *J. Mar. Res.* **1**, 239–263.

Rossignal-Strick, M. (1983). African monsoons, an immediate climate response to orbital insolation. *Nature (London)* **303**, 46–49.

Rossignal-Strick, M. (1985). A marine anoxic event on the Brittany coast, July 1982. *J. Coast. Res.* **1**, 11–20.

Rossignal-Strick, M., W. Nesteroff, P. Olive, and C. Vergnaud-Grazzini (1982). After the deluge: Mediterranean stagnation and sapropel formation. *Nature (London)* **295**, 105–110.

Round, F. E., and R. M. Crawford (1981). The lines of evolution of the Bacillariophyta. I. Origin. *Proc. R. Soc. London Ser. B* **211**, 237–260.

Rowe, G. T. (1971). Benthic biomass in the PISCO, Peru upwelling. *Invest. Pesq.* **35**, 127–136.

Rowe, G. T., P. T. Polloni, and G. W. Horner (1974). Benthic biomass estimates from the northwestern Atlantic Ocean and the northern Gulf of Mexico. *Deep-Sea Res.* **21**, 641–650.

Rowe, G. T., C. H. Clifford, K. L. Smith, and P. L. Hamilton (1975). Benthic nutrient regeneration and its coupling to primary productivity in coastal waters. *Nature (London)* **255**, 215–217.

Rowe, G. T., and K. L. Smith (1977). Benthic-pelagic coupling in the mid-Atlantic Bight. *In* "Ecology of Marine Benthos" (B. C. Coull, ed.), pp. 55–66. Univ. of South Carolina Press, Columbia.

Rowe, G. T., and W. D. Gardner (1979). Sedimentation rates in the slope water of the northwest Atlantic Ocean measured directly with sediment traps. *J. Mar. Res.* **37**, 581–600.

Rowe, G. T., R. Theroux, W. Phoel, H. Quimby, R. Wilke, D. Koschoreck, T. Whitledge, P. Falkowski, and C. Fray (1987). Benthic carbon budgets for the continental shelf south of New England. *Cont. Shelf Res.* (in press).

Rubinson, H., and R. N. Clayton (1969). Carbon 13 fractionation between aragonite and calcite. *Geochim. Cosmochim. Acta* **33**, 997–1004.

Rumyantsev, A. I., and M. A. Darda (1972). Summer herring in the eastern Bering Sea. *In* "Soviet Fisheries Investigations in the Northeastern Pacific" (P. A. Moiseev, ed.), Vol. 70, pp. 409–434. VNIRO, Trudy, Moscow.

Russell, F. S., A. J. Southward, G. T. Boalch, and E. J. Butler (1971). Changes in biological conditions in the English Channel off Plymouth during the last half century. *Nature (London)* **234**, 468–470.

Ryan, W. B. (1972). Stratigraphy of late Quaternary sediments in the eastern Mediterranean. *In* "The Mediterranean Sea" (D. J. Stanley, ed.), pp. 149–169. Dowden, Hutchinson & Ross, Stroudsburg, Pennsylvania.

Ryan, W. B., and M. B. Cita (1977). Ignorance concerning episodes of oceanwide stagnation. *Mar. Geol.* **23**, 197–215.

Ryther, J. H. (1956). Photosynthesis in the ocean as a function of light intensity. *Limnol. Oceanogr.* **1**, 72–84.

Ryther, J. H. (1969). Photosynthesis and fish production in the sea. *Science* **166**, 72–76.

Ryther, J. H., and C. S. Yentsch (1957). The estimation of phytoplankton production in the ocean from chlorophyll and light data. *Limnol. Oceanogr.* **2**, 281–286.

Ryther, J. H., and C. S. Yentsch (1958). Primary production of continental shelf waters off New York. *Limnol. Oceanogr.* **3**, 327–335.

Ryther, J. H., J. R. Hall, A. K. Pease, and M. M. Jones (1966). Primary organic production in relation to the chemistry and hydrography of the western Indian Ocean. *Limnol. Oceanogr.* **11**, 371–380.

Ryther, J. H., D. W. Menzel, and N. Corwin (1967). Influence of the Amazon River outflow on the ecology of the western Tropical Atlantic. I. Hydrography and nutrient chemistry. *J. Mar. Res.* **25**, 69–83.

Ryther, J. H., and W. M. Dunstan (1971). Nitrogen, phosphorus, and eutrophication in the coastal marine environment. *Science* **171**, 1008–1013.

Ryther, J. H., D. W. Menzel, E. M. Hulburt, C. J. Lorenzen, and N. Corwin (1971). The production and utilization of organic matter in the Peru coastal current. *Invest. Pesq.* **35**, 43–60.

Sackett, W. M. (1986). $\delta^{13}C$ signatures of organic carbon in southern high latitude deep sea sediments; paleotemperature implications. *Org. Geochem.* **9**, 63–68.

Sackett, W. M., and R. R. Thompson (1963). Isotopic organic carbon composition of recent continental derived clastic sediments of the eastern Gulf of Mexico. *Am. Assoc. Pet. Geol. Bull.* **47**, 525–531.

Sackett, W. M., W. R. Eckelmann, M. L. Bender, and A. H. Be (1965). Temperature dependence of carbon isotope composition in marine plankton and sediments. *Science* **148**, 235–237.

Sackett, W. M., and W. S. Moore (1966). Isotopic variations of dissolved inorganic carbon. *Chem. Geol.* **1**, 323–328.

Saino, T., and A. Hattori (1980). ^{15}N natural abundance in oceanic suspended particulate matter. *Nature (London)* **283**, 752–754.

Sambrotto, R. N., J. J. Goering, and C. P. McRoy (1984). Large yearly production of phytoplankton in the western Bering Strait. *Science* **225**, 1147–1150.

Sancetta, C. (1983). Effect of Pleistocene glaciation upon oceanographic characteristics of the North Pacific Ocean and Bering Sea. *Deep-Sea Res.* **30**, 851–869.

Santander, H. (1976). The Peruvian Current. Part 2, Biological aspects. *In* "Proc. Workshop El Niño Phenom., Guayaquil, Ecuador," pp. 217–228. UNESCO, Geneva.

Santander, H., and O. S. de Castillo (1969). "El desove de la anchoveta *(Engraulis ringens J.)* en los periodos reproductivos de 1961, a 1968," Ser. Inf. Espec. 40, pp. 1–10. Inst. Mar, Callao, Peru.

Santander, H., and O. S. de Castillo (1977). Ichthyoplankton from the Peruvian coast. *In* "Proceedings of the symposium on warm water zooplankton" (S. Z. Qasim, ed.), pp. 105–123. Natl. Inst. Oceanogr., Goa.

Santos, S. L., and J. L. Simon (1980). Response of a soft bottom benthos to annual catastrophic disturbance in a South Florida estuary. *Mar. Biol. Prog. Ser.* **3**, 347–355.

Santschi, P. H., Y.-H. Li, J. J. Bell, R. M. Trier, and K. Kawtaluk (1980). Pu in coastal marine environments. *Earth Planet. Sci. Lett.* **51**, 248–265.

Sarkisyan, A. S. (1977). The diagnostic calculation of a large-scale oceanic circulation. *In* "The Sea" (E. D. Goldberg, I. N. McCave, J. J. O'Brien, and J. H. Steele, eds.), Vol. 6, pp. 363–458. Wiley, New York.

Saville, A. (1978). The growth of herring in the northwestern North Sea. *Rapp. P.-V. Reun. Cons. Int. Explor. Mer* **172**, 164–171.

Savin, S. M. (1977). The history of the Earth's surface temperature during the last 100 million years. *Annu. Rev. Earth Planet. Sci.* **5**, 319–355.

Schell, I. I. (1965). The origin and possible prediction of the fluctuations in the Peru Current and upwelling. *J. Geophys. Res.* **70**, 5529–5540.

Schell, W. R. (1977). Concentrations, physico-chemical states, and mean residence times of ^{210}Pb and [210]Po in marine and estuarine waters. *Geochim. Cosmochim. Acta* **41**, 199–1031.

Schell, W. R. (1982). Dating recent (200 years) events in sediments from lakes, estuaries and deep ocean environments using lead-210. *In* "Nuclear and Chemical Dating Techniques, Interpreting the Environmental Record" (L. A. Currie, ed.), pp. 331–361. Am. Chem. Soc., Washington, D.C.

Schlitz, R. J., and E. B. Cohen (1981). A nitrogen budget for the Gulf of Maine and Georges Bank. ICES C.M. 1981/L24, pp. 1–14.

Schmitt, R. W., and D. B. Olson (1985). Wintertime convection in warm core rings: thermocline ventilation and the formation of mesoscale lenses. *J. Geophys. Res.* **90**, 8823–8837.

Schneider, D. C., and G. L. Hunt (1982). Carbon flux to seabirds in waters with different mixing regimes in the southeastern Bering Sea. *Mar. Biol. (Berlin)* **66**, 332–344.

Schneider, D. C., G. L. Hunt, and N. M. Harrison (1986). Mass and energy transfer to seabirds in the southeastern Bering Sea. *Cont. Shelf Res.* **5**, 241–257.

Schoeninger, M. S., M. J. Deniro, and H. Tauber (1983). Stable nitrogen isotope ratios of bone collagen reflect marine and terrestrial components of prehistoric diet. *Science* **220**, 1381–1383.

Schopf, T. J. (1974). Permo-Triassic extinctions: relation to sea-floor spreading. *J. Geol.* **82**, 129–143.

Schopf, T. J. (1980). "Paleoceanography." Harvard Univ. Press, Cambridge, Massachusetts.

Schulenberger, E., and J. L. Reid (1981). The Pacific shallow oxygen minimum, deep chlorophyll maximum, and primary productivity, reconsidered. *Deep-Sea Res.* **28**, 901–919.

Scott, J. T., and G. T. Csanady (1976). Nearshore currents off Long Island. *J. Geophys. Res.* **81**, 5401–5409.

Scott, M. R., P. F. Salter, and J. E. Halverson (1983). Transport and deposition of plutonium in the ocean: evidence from Gulf of Mexico sediments. *Earth Planet. Sci. Lett.* **63**, 202–222.

Scott, W., and D. H. Miner (1936). Sedimentation in Winoma Lake. *Proc. Indiana Acad. Sci.* **45**, 275–286.

Sears, M. (1954). Notes on the Peruvian coastal current. I. An introduction to the ecology of Pisco Bay. *Deep-Sea Res.* **1**, 141–169.

Sears, M., and G. L. Clarke (1940). Annual fluctuations in the abundance of marine zooplankton. *Biol. Bull. (Woods Hole, Mass.)* **79**, 321–328.

Segar, D. A. (1985). Contamination of polluted estuaries and adjacent coastal oceans—a global review. *MESA N.Y. Bight Spec. Rep. Ser.* (in press).

Segar, D. A., and G. A. Berberian (1976). Oxygen depletion in the New York Bight apex: causes and consequences. *ASLO Symp.* No. 2, 220–239.

Seitz, R. (1967). Thermostad, the antonym of thermocline. *J. Mar. Res.* **25**, 203.

Seitzinger, S., S. Nixon, M. E. Pilsen, and S. Burke (1984). Denitrification and N$_2$O production in nearshore marine sediments. *Geochim. Cosmochim. Acta* (in press).

Seki, H., T. Tsuji, and A. Hattori (1974). Effect of zooplankton grazing on the formation of the anoxic layer of Tokyo Bay. *Estuarine Coast. Mar. Sci.* **2**, 145–151.

Seliger, H., J. A. Boggs, and W. H. Biggley (1985). Catastrophic anoxia in the Chesapeake Bay in 1984. *Science* **228**, 70–73.

Senum, G. I., and J. S. Gaffney (1985). A reexamination of the tropospheric methane cycle: geophysical implications. *In* "The Carbon Cycle and Atmospheric CO$_2$: Natural Variations, Archean to Present" (E. T. Sundquist and W. S. Broecker, eds.), Monogr. 32, pp. 61–69. Am. Geophys. Union, Washington, D.C.

Serobaba, I. I. (1974). Spawning ecology of the walleye pollock (*Theragra chalcogramma* Pallas) in the Bering Sea. *J. Ichthyol. (Engl. Transl.)* **14**, 544–552.

Sette, O. E. (1955). Consideration of midocean fish production as related to oceanic circulatory systems. *J. Mar. Res.* **14**, 398–416.

Shackleton, N. J. (1967). Oxygen isotope analyses and paleotemperatures reassessed, *Nature (London)* **215**, 15–17.

Shackleton, N. J. (1977). Carbon-13 in *Uvigerina:* Tropical rain forest history and the Equatorial Pacific carbonate dissolution cycles. *In* "The Fate of Fossil Fuel CO_2 in the Oceans" (N. R. Anderson and A. Malahoff, eds.), pp. 401–428. Plenum, New York.

Shackleton, N. J. (1985). Oceanic carbon isotope constraints on oxygen and carbon dioxide in the Cenozoic atmosphere. *In* "The Carbon Cycle and Atmospheric CO_2: Natural Variations, Archean to Present" (E. T. Sundquist and W. S. Broecker, eds.), Monogr. 32, pp. 412–417. Am. Geophys. Union, Washington, D.C.

Shackleton, N. J., M. A. Hall, J. Line, and C. Shuxi (1983). Carbon isotope data in core V19–30 confirm reduced carbon dioxide concentration in the ice age atmosphere. *Nature (London)* **306**, 319–322.

Shah, N. M. (1973). Seasonal variation of phytoplankton pigments and some of the associated oceanographic parameters in the Laccadive Sea off Cochin. *In* "The Biology of the Indian Ocean" (B. Zeitzschel, ed.), pp. 175–185. Springer-Verlag, Berlin and New York.

Shanks, A. L., and J. D. Trent (1980). Marine snow: sinking rates and potential role in vertical flux. *Deep-Sea Res.* **27**, 137–143.

Shannon, L. V., P. Schlittenhardt, and S. A. Mostert (1984). The NIMBUS-7 CZCS experiment in the Benguela Current region off southern Africa, February 1980. *J. Geophys. Res.* **89**, 4968–4976.

Shapiro, J. (1973). Blue-green algae: why they become dominant. *Science* **179**, 382–384.

Sharma, G. D. (1974). Contemporary sedimentary regimes of the eastern Bering Sea. *In* "Oceanography of the Bering Sea with Emphasis on Renewable Resources" (D. W. Hood and E. J. Kelley, eds.), Occas. Publ. Univ. Alaska Inst. Mar. Sci. No. 2, pp. 517–540.

Sharma, G. D. (1979). "The Alaskan Shelf." Springer-Verlag, Berlin and New York.

Sharp, J. H., C. H. Culberson, and T. M. Church (1982). The chemistry of the Delaware estuary. General considerations. *Limnol. Oceanogr.* **27**, 1015–1028.

Sheldon, R. W., A. Prakash, and W. H. Sutcliffe (1972). The size distribution of particles in the ocean. *Limnol. Oceanogr.* **17**, 327–340.

Sheldon, R. W., W. H. Sutcliffe, and A. Prakash (1973). The production of particles in surface waters of the ocean with particular reference to the Sargasso Sea. *Limnol. Oceanogr.* **18**, 719–733.

Shenker, J. M., and J. M. Dean (1979). The utilization of an intertidal marsh creek by larval and juvenile fishes: abundance, diversity, and temporal variation. *Estuaries* **2**, 154–163.

Shepard, F. S. (1963). "Submarine Geology," Harper, New York.

Sheppard, J. C., H. Westburg, J. F. Hopper, K. Ganesan, and P. Zimmerman (1982). Inventory of global methane sources and their production rates. *J. Geophys. Res.* **87**, 1305–1312.

Sherman, K. (1978). MARMAP, a fisheries ecosystem study in the NW Atlantic: fluctuations in icthyoplankton-zooplankton components and their potential for impact on the system. *In* "Workshop on Advanced Concepts in Ocean Measurements," pp. 1–38. Univ. of South Carolina, Columbia.

Sherman, K., C. Jones, L. Sullivan, W. Smith, P. Berriem, and L. Ejsymont (1981). Congruent shifts in sand eel abundance in western and eastern North Atlantic ecosystems. *Nature (London)* **291**, 486–489.

Shokes, R. F. (1976). Rate-dependent distributions of lead-210 and interstitial sulfate in sediments of the Mississippi River delta. Ph.D. Thesis, Texas A&M Univ., College Station.

Shuman, F. R., and C. J. Lorenzen (1975). Quantitative degradation of chlorophyll by a marine herbivore. *Limnol. Oceanogr.* **20**, 580–586.

Shushkina, E. A., M. E. Vinagradov, Y. I. Sorokin, and V. N. Mikheev (1978). The peculiarities of functioning of plankton communities in the Peruvian upwelling. *Oceanology* **18**, 886–902.

Silva, A., and C. D. Hollister (1973). Geotechnical properties of ocean sediments recovered with Giant Piston Core, 1, in the Gulf of Maine. *J. Geophys. Res.* **78**, 3597–3626.

Silver, M. W., and A. L. Alldredge (1981). Bathypelagic marine snow: deep sea algal and detrital community. *J. Mar. Res.* **39**, 501–530.

Simenstad, C. A., J. A. Estes, and K. W. Kenyon (1978). Aleuts, sea otters, and alternate stable state communities. *Science* **200**, 403–411.

Simons, T. J. (1980). Circulation models of lakes and inland seas. *Can. Bull. Fish. Aquat. Sci.* **203**, 34–51.

Simpson, J. H., and J. R. Hunter (1974). Fronts in the Irish Sea. *Nature (London)* **250**, 404–406.

Simpson, J. H., and R. D. Pingree (1978). Shallow sea fronts produced by tidal stirring. *In* "Oceanic Fronts in Coastal Processes" (M. J. Bowman and W. E. Esaias, eds.), pp. 29–42. Springer-Verlag, Berlin and New York.

Sindermann, C. J., and R. L. Swanson (1979). Historical and regional perspective. *In* "Oxygen Depletion and Associated Benthic Mortalities in the New York Bight, 1976" (C. J. Sindermann and R. L. Swanson, eds.), *NOAA Prof. Pap.* No. 11, 1–17.

Sklarew, R. C., A. J. Fabrich, and S. E. Pruger (1971). A Particle-in-Cell Method for Numerical Solution of the Atmospheric Diffusion Equation and Applications to Air Pollution Problems," Rep. 3SR-844. Div. Meteorol. Nat. Environ. Res. Cent., Research Triangle, North Carolina.

Slobodkin, L. B. (1961). "Growth and Regulation of Animal Populations." Holt, New York.

Sly, P. G. (1976). Lake Erie and its Basin. *J. Fish. Res. Board Can.* **33**, 355–370.

Small, L. F., and D. A. Ramberg (1971). Chlorophyll a, carbon and nitrogen in particles from a unique coastal environment. *In* "Fertility of the Sea" (J. Costlow, ed.), Vol. 2, pp. 475–492. Gordon & Breach, New York.

Small, L. F., H. Curl, and W. A. Glooschenko (1972). Estimates of primary production off Oregon using an improved chlorophyll light technique. *J. Fish. Res. Board Can.* **29**, 1261–1267.

Smayda, T. J. (1969). Some measurements of the sinking rate of fecal pellets. *Limnol. Oceanogr.* **14**, 621–625.

Smayda, T. J. (1970). The suspension and sinking of phytoplankton in the sea. *Oceanogr. Mar. Biol. Annu. Rev.* **8**, 353–414.

Smetacek, V. S. (1984). The supply of food to the benthos. *In* "Flows of Energy and Materials in Marine Ecosystems: Theory and Practice" (M. J. Fasham, ed.), pp. 517–548. Plenum, New York.

Smetacek, V. S. (1985). Role of sinking in diatom life-history cycles: ecological, evolutionary, and geological significance. *Mar. Biol. (Berlin)* **84**, 239–251.

Smetacek, V. S., K. von Brockel, B. Zietzschel, and W. Zenk (1978). Sedimentation of particulate matter during phytoplankton spring bloom in relation to the hydrographical regime. *Mar. Biol. (Berlin)* **47**, 211–226.

Smith, A. E., and I. Morris (1980). Pathways of carbon assimilation in phytoplankton from the Antarctic Ocean. *Limnol. Oceanogr.* **25**, 865–872.

Smith, D. J., G. Eglinton, and R. J. Morris (1983). Interfacial sediment and assessment of organic input from a highly productive water column. *Nature (London)* **304**, 259–262.

Smith, E. L. (1936). Photosynthesis in relation to light and carbon dioxide. *Proc. Natl. Acad. Sci. USA* **22**, 504–511.

Smith, G. B. (1981). The biology of the walleye pollock. *In* "The Eastern Bering Sea Shelf: Oceanography and Resources"(D. W. Hood and J. A. Calder, eds.), Vol. 2, pp. 527–552. Univ. of Washington Press, Seattle.

Smith, J. N., and C. T. Schafer (1979). Bioturbation of surficial sediments on the continental slope, east of Newfoundland. *Proc. NEA Semin. Radioecol., 3rd, Tokyo* pp. 225–235.

Smith, J. N., and A. Walton (1980). Sediment accumulation rates and geochronologies measured in the Saguenay Fjord using the Pb-210 dating method. *Geochim. Cosmochim. Acta* **44**, 225–240.

Smith, K. L. (1973). Respiration of a sublittoral community. *Ecology* **54**, 1065–1075.

Smith, K. L. (1978). Benthic community respiration in the N.W. Atlantic Ocean: *in situ* measurements from 40 to 5200 m. *Mar. Biol. (Berlin)* **47**, 337–347.

Smith, K. L., and J. M. Teal (1973). Deep-sea benthic community respiration: an *in situ* study at 1850 m. *Science* **179**, 282–283.

Smith, K. L., G. T. Rowe, and J. A. Nichols (1973). Benthic community respiration near the Woods Hole sewage outfall. *Estuarine Coast. Mar. Sci.* **1**, 65–70.

Smith, K. L., G. T. Rowe, and C. H. Clifford (1974). Sediment oxygen demand in an outwelling and upwelling area. *Tethys* **6**, 223–230.

Smith, P. E., and R. W. Eppley (1982). Primary production and the anchovy populations in the Southern California Bight: comparison of time series. *Limnol. Oceanogr.* **27**, 1–17.

Smith, R. C., and K. S. Baker. (1978a). The bio-optical state of ocean waters and remote sensing. *Limnol. Oceanogr.* **23**, 247–259.

Smith, R. C., and K. S. Baker. (1978b). Optical classification of natural waters. *Limnol. Oceanogr.* **23**, 260–267.

Smith, R. C., and W. H. Wilson (1981). Ship and satellite bio-optical research in the California Bight. *In* "Oceanography from Space" (J. R. Gower, ed.), pp. 281–294. Plenum, New York.

Smith, R. C., and K. S. Baker (1982). Oceanic chlorophyll concentrations as determined using Coastal Zone Color Scanner imagery. *Mar. Biol. (Berlin)* **66**, 269–279.

Smith, R. C., R. W. Eppley, and K. S. Baker (1982). Correlations of primary production as measured aboard ship in southern California waters and as estimated from satellite chlorophyll images. *Mar. Biol. (Berlin)* **66**, 281–288.

Smith, R. L. (1976). Waters of the sea: the ocean's characteristics and circulation. *In* "The Ecology of the Seas" (D. H. Cushing and J. J. Walsh, eds.), pp. 23–58. Blackwell, Oxford.

Smith, S. H. (1968). Species succession and fishery exploitation in the Great Lakes. *J. Fish. Res. Board Can.* **25**, 667–693.

Smith, S. L., and L. A. Codispoti (1980). Southwest monsoon of 1979, chemical and biological response of Somali coastal waters. *Science* **209**, 597–600.

Smith, S. L., and J. Vidal (1986). Variations in the distribution, abundance, and development of copepods in the southeastern Bering Sea in 1980 and 1981. *Cont. Shelf Res.* **5**, 215–239.

Smith, S. L., and P. V. Lane (1987). Grazing of the spring diatom bloom in the New York Bight by the calanoid copepods, *Calanus finmarchicus, Metridia lucens* and *Centropages typicus. Cont. Shelf Res.* (in press).

Smith, S. V. (1981). Marine macrophytes as a global carbon sink. *Science* **211**, 838–840.

Smith, W. G., J. D. Sibunka, and A. Wells (1978). Diel movements of larval yellowtail flounder, *Limanda ferruginea*, determined from discrete depth sampling. *Fish. Bull.* **76**, 167–178.

Smith, W. O., G. W. Heburn, R. T. Barber, and J. J. O'Brien (1983). Regulation of phytoplankton communities by physical processes in upwelling ecosystems. *J. Mar. Res.* **41**, 539–556.

Soderlund, R., and B. H. Svensson (1976). The global nitrogen cycle. *Ecol. Bull. Stoch.* **22**, 23–73.

Sorensen, J., B. B. Jorgensen, and N. P. Revsbech (1979). A comparison of oxygen, nitrate, and sulfate respiration in coastal marine sediments. *Microb. Ecol.* **5**, 105–115.

Sorokin, Y. I. (1973). Microbiological aspects of the productivity of coral reefs. *In* "Biology and Geology of Coral Reefs" (R. Endeau and R. Jones, eds.), pp. 17–95. Academic Press, New York.

Sournia, A. (1972). Une periode de poussees phytoplanctoniques pres de Nosy-Be (Madagascar) en 1971. II. Production primaire. *Cah. ORSTOM Ser. Oceanogr.* **10**, 289–300.

Sournia, A. (1977). Cycle annual du phytoplankton et de la production primaire dans les mers tropicales. *Mar. Biol. (Berlin)* **3**, 287–303.

Soutar, A., and J. D. Isaacs. (1974). Abundance of pelagic fish during the 19th and 20th centuries as recorded in anaerobic sediment off the Californias. *Fish. Bull.* **72**, 257–273.

Southward, G. M. (1967). Growth of Pacific halibut. *Rep. Int. Pac. Halibut Comm.* **43**, 1–40.

Spencer, D. W., P. G. Brewer, A. Fleer, S. Honjo, S. Krishaswami, and Y. Nozaki (1978). Chemical fluxes from a sediment trap experiment in the deep Sargasso Sea. *J. Mar. Res.* **36**, 493–523.

Spencer, D. W., M. P. Bacon, and P. G. Brewer (1980). The distribution of ^{210}Pb and ^{210}Po in the North Sea. *Thalassia Jugoslav.* **16**, 125–154.

Spiker, E. C. (1980). The behaviour of ^{14}C and ^{13}C in estuarine water: effects of *in situ* CO_2 production and atmospheric exchange. *Radiocarbon* **22**, 647–654.

Springer, A. M., D. G. Roseneau, E. C. Murphy, and M. I. Springer (1984). Environmental controls of marine food webs: food habits of seabirds in the eastern Chukchi Sea. *Can. J. Fish. Aquat. Sci.* **41**, 1202–1215.

Springer, A. M., and D. G. Roseneau (1987). Copepod-based food webs: auklets and oceanography in the Bering Sea. *Mar. Biol. (Berlin)* (submitted).

Staresinic, N., J. Farrington, R. E. Gagosian, C. H. Clifford, and E. M. Hulburt (1983). Downward transport of particulate matter in the Peru coastal upwelling: role of the southern anchoveta, *Engraulis ringens. In* "Coastal Upwelling: Its Sediment Record" (E. Suess and J. Thiede, eds.), pp. 225–240. Plenum, New York.

Stauffer, B., A. Neftel, H. Oeschger, and J. Schwander (1985a). CO_2 concentration in air extracted from Greenland ice samples. *In* "Greenland Ice Core: Geophysics, Geochemistry, and the Environment" (C. C. Longway, H. Oeschger, and W. Dansgaard, eds.), Monogr. 33, pp. 85–89. Am. Geophys. Union, Washington, D.C.

Stauffer, B., G. Fischer, A. Neftel, and H. Oeschger (1985b). Increase of atmospheric methane recorded in Antarctic ice core. *Science* **229**, 1386–1388.

Steele, J. H. (1962). Environmental control of photosynthesis in the sea. *Limnol. Oceanogr.* **7**, 137–150.

Steele, J. H. (1974). "The Structure of Marine Ecosystems." Harvard Univ. Press, Cambridge, Massachusetts.

Steele, J. H. (1978). Some comments on plankton patches. *In* "Spatial Pattern in Plankton Communities" (J. H. Steele, ed.), pp.1–20. Plenum, New York.

Steele, J. H., and D. W. Menzel (1962). Conditions for maximum primary production in the mixed layer. *Deep-Sea Res.* **9**, 39–49.

Steemann-Nielsen, E. (1934). "Untersuchungen über die Verbreitung, Biologie und Variation der Ceratien in Südlichen Stillen Ozean," DANA Rep. No. 4.

Steemann-Nielsen, E. (1958). A survey of recent Danish measurements of the organic productivity in the sea. *Rapp. P.-V. Reun. Cons. Int. Explor. Mer* **144**, 92–95.

Steemann-Nielsen, E., and A. Jensen (1957). Primary oceanic production, the autotrophic production of organic matter in the ocean. *Gal. Rep.* **1**, 47–135.

Stefansson, U. (1985). Oceanographic variations in recent decades and their impact on the fertility of the Iceland Sea. *In* "North Atlantic Deep Water Formation," *NASA Conf. Publ.* No. 2367, 19–22.

Stefansson, U., and F. A. Richards (1964). Distributions of dissolved oxygen, density, and nutrients off the Washington and Oregon coasts. *Deep-Sea Res.* **11**, 355–380.

Steinbeck, J. (1945). "Cannery Row." pp. 1–2. Viking Press, New York.

Steinbeck, J. (1954). "Sweet Thursday." pg. 3. Viking Press, New York.

Steinen, R. P., R. S. Harrison, and R. K. Mathews (1973). Eustatic low strand of sea level between 125,000 and 105,000 BP: evidence from the sub-surface of Barbados, West Indies. *Geol. Soc. Am. Bull.* **84**, 63–70.

Stenhouse, M. J., and M. S. Baxter (1976). Glasgow University Radiocarbon measurements. VIII. *Radiocarbon* **18**, 161–171.

Stephens, K. (1968). Data Record. Primary production data from the northeast Pacific Ocean, January 1966 to December 1967. *Rep. Sci. Fish. Res. Board Can.* No. 957.

Steven, D. M. (1971). Primary productivity of the tropical western Atlantic Ocean near Barbados. *Mar. Biol. (Berlin)* **10**, 261–264.

Stevenson, M. R., and H. R. Wicks (1975). Bibliography of El Niño and associated publications. *Bull. Inter-Am. Trop. Tuna Comm.* **16**, 451–501.

Stigebrandt, A. (1981). Cross thermocline flow on continental shelves and the location of shelf fronts. *In* "Ecohydrodynamics" (J. C. Nihoul, ed.), pp. 51–66. Elsevier, Amsterdam.

Stoddard, A. (1983). Mathematical model of oxygen depletion in the New York Bight: an analysis of biological, chemical and physical factors in 1975 and 1976. Ph.D. Thesis, Univ. of Washington, Seattle.

Stoddard, A., and J. J. Walsh (1986). Modeling oxygen depletion in the New York Bight: the water quality impact of a potential increase of waste inputs. In "Urban Wastes in Coastal Marine Environments" (D. A. Wolfe, ed.), *NOAA Prof. Pap.* (in press).

Stoecker, D. K., and J. J. Govoni (1984). Food selection by young larval gulf menhaden *(Brevoortia patronus)*. *Mar. Biol. (Berlin)* **80**, 299–306.

Stoker, S. (1981). Benthic invertebrate macrofauna of the eastern Bering/Chukchi continental shelf. *In* "The Eastern Bering Sea Shelf: Oceanography and Resources" (D. W. Hood and J. A. Calder, eds.), pp. 1069–1091. Univ. of Washington Press, Seattle.

Stokes, G. G. 1856. On the effect of the internal friction of fluids on the motion of pendulums. *Trans. Cambridge Philos. Soc.* **9**, 48–57.

Stommel, H. (1948). The westward intensification of wind-driven ocean currents. *Trans. Am. Geophys. Union* **29**, 202–206.

Stommel, H. (1960). "The Gulf Stream, a Physical and Dynamical Description." Univ. of California Press, Berkeley.

Stommel, H., and A. Leetma (1972). Circulation on the continental shelf. *Proc. Natl. Acad. Sci.* **69**, 3380–3384.

Strickland, J. H. (1958). Solar radiation penetrating the oceans. A review of requirements, data, and methods of measurement, with particular reference to photosynthetic productivity. *J. Fish. Res. Board Can.* **15**, 453–493.

Strickland, J. H. (1965). Production of organic matter in the primary stages of the marine food chain. *In* "Chemical Oceanography" (J. P. Riley and G. Skirrow, eds.), pp. 477–610. Academic Press, New York.

Stuiver, M. (1978). Atmospheric carbon dioxide and carbon reservoir changes. *Science* **199**, 253–270.

Stuiver, M., and H. A. Polach (1977). Reporting of ^{14}C data. *Radiocarbon* **19**, 355–363.

Stuiver, M., and H. G. Ostlund (1980). GEOSECS Atlantic radiocarbon. *Radiocarbon* **22**, 1–24.

Stuiver, M., R. L. Burk, and P. D. Quay (1984). ^{13}C/^{12}C ratios and the transfer of biospheric carbon to the atmosphere. *J. Geophys. Res.* **89**, 11731–11748.

Sturges, W., and J. P. Blaha (1976). A western boundary current in the Gulf of Mexico. *Science* **192**, 367–369.

Subba Rao, D. V. (1965). The measurement of total carbon dioxide in dilute tropical waters. *Aust. J. Mar. Freshwater Res.* **16**, 273–280.

Subba Rao, D. V., and T. Platt (1984). Primary production of arctic waters. *Polar Biol.* **3**, 191–201.

Suess, E. (1980). Particulate organic carbon flux in the oceans—Surface productivity and oxygen utilization. *Nature (London)* **288**, 260–263.

Suess, H. R. (1955). Radiocarbon concentrations in modern wood. *Science* **122**, 415–417.

Sverdrup, H. U. (1947). Wind-driven currents in a baroclinic ocean: with application to the equatorial currents of the eastern Pacific. *Proc. Natl. Acad. Sci. USA* **33**, 318–326.

Sverdrup, H. U. (1953). On conditions for the vernal blooming of phytoplankton. *J. Cons. Cons. Int. Explor. Mer* **18**, 287–295.

Sverdrup, H. U., M. W. Johnson, and R. H. Fleming (1942). "The Oceans." Prentice-Hall, Englewood Cliffs, New Jersy.

Swanson, R. L., H. M. Stanford, J. S. O'Connor, S. Chanesman, C. A. Parker, P. A. Eisen and G.

F. Mayer (1978). June 1976 pollution of Long Island ocean beaches. *J. Environ. Eng.* **104**, 1067–1085.

Sweeney, R. E., K. K. Liu, and I. R. Kaplan (1978). Oceanic nitrogen isotopes and their uses in determining the source of sedimentary nitrogen. *In* "Stable Isotopes in the Earth Sciences," *Bull. N.Z. Dep. Sci. Ind. Res.* No. 220, 9–26.

Sweeney, R. E., and I. R. Kaplan (1980). Natural abundances of ^{15}N as a source indicator for near-shore marine sedimentary and dissolved nitrogen. *Mar. Chem.* **9**, 81–94.

Swift, D. J., J. W. Kofoed, P. J. Saulsburg, and P. Sears (1972). Holocene evolution of the shelf surface, central and Atlantic shelf of North America. *In* "Shelf Sediment Transport: Process and Pattern" (D. J. Swift, D. B. Duane, and O. H. Pilkey, eds.), pp. 499–574. Dowden, Hutchinson & Ross, Stroudsburg, Pennsylvania.

Systems and Applied Sciences Corp. (1984). "Users Guide for the Coastal Zone Color Scanner Compressed Earth Gridded Data Sets of the Northeast Coast of the United States for February 28 through May 27, 1979," SASC-T-5-5100-0002-0008-84, pp. 1–15, Vienna, Virginia.

Taguchi, S. (1976). Relationship between photosynthesis and cell size of marine diatoms. *J. Phycol.* **12**, 185–189.

Takahashi, T., and A. E. Azevedo (1982). The ocean as a CO_2 reservoir. *In* "Interpretation of Climate and Photochemical Models, Ozone, and Temperature Measurements" (R. A. Reckand and J. R. Hammell, eds.), pp. 83–109. Am. Inst. Phys., New York.

Talling, J. F. (1957). Photosynthetic characteristics of some fresh-water plankton diatoms in relation to underwater radiation. *New Phytol.* **56**, 29–50.

Tamiya, H., E. Hase, K. Shibata, A. Mituya, T. Iwamura, T. Nihei, and T. Sasa (1953). Kinetics of growth of *Chlorella*, with special reference to its dependence on quantity of available light and temperature. *In* "Algal Culture from Laboratory to Pilot Plant" (J. S. Burlew, ed.), *Carnegie Inst. Wash. Publ.* No. 600, 204–232.

Tan, F. C., and P. M. Strain (1979). Organic carbon isotope ratios in recent sediments in the St. Lawrence estuary and the Gulf of St. Lawrence. *Estuarine Coast. Mar. Sci.* **8**, 213–225.

Taniguchi, A. (1973). Phytoplankton-zooplankton relationships in the western Pacific Ocean and adjacent seas. *Mar. Biol. (Berlin)* **21**, 115–121.

Tappan, H. (1968). Primary production, isotopes, extinction, and the atmosphere. *Palaeogeogr. Palaeoclimatol. Palaeoecol.* **4**, 187–210.

Tappan, H., and A. R. Loeblich (1970). Geobiologic implications of fossil phytoplankton evolution and time-space distribution. *In* "Palynology of the Late Cretaceous and Early Tertiary" (R. M. Kosanke and S. T. Cross, eds.), pp.247–340. Michigan State Univ., Lansing.

Taylor, C. C., H. B. Bigelow, and H. W. Graham (1957). Climatic trends and the distribution of marine animals in New England. *Fish. Bull.* **57**, 293–345.

Tenore, K. R. (1977). Food chain pathways in detrital feeding benthic communities, a review, with new observations on sediment resuspension and detrital recycling. *In* "Ecology of Marine Benthos" (B. C. Coull, ed.), pp. 37–54. Univ. of South Carolina Press, Columbia.

Tenore, K. R. (1985). Seasonal changes in soft bottom macroinfauna of the U.S. South Atlantic Bight. *In* "Oceanography of the Southeastern U.S. Continental Shelf" (L. P. Atkinson, D. W. Menzel, and K. A. Bush, eds.), *Coast. Estuarine Sci.* **2**, 130–140.

Thayer, G. W., D. E. Hoss, M. A. Kjelson, W. F. Hettler, and M. W. Lacroix (1974). Biomass of zooplankton in the Newport River estuary and the influence of postlarval fishes. *Chesapeake Sci.* **15**, 9–16.

Thiede, J., and T. H. Van Andel (1977). The paleoenvironment of anaerobic sediments in the Late Mesozoic South Atlantic. *Earth Planet. Sci. Lett.* **33**, 301–309.

Thomas, J. P. (1966). The influence of the Altamaha River on primary production beyond the mouth of the River. M.S. Thesis, Univ. of Georgia, Athens.

Thomas, J. P. (1981). Large Area Marine Productivity-Pollution Experiments (LAMPEX)—a series of studies being developed to hasten the operational use of remote sensing for living marine resources and environmental quality. *In* "Oceanography from Space" (J. F. Gower, ed.), pp. 403–409. Plenum, New York.

Thomas, J. P., W. C. Phoel, F. W. Steimle, J. E. O'Reilly, and C. E. Evans (1976). Seabed oxygen consumption—New York Bight Apex. *ASLO Spec. Symp.* No. 2, 354–369.

Thomas, J. P., J. E. O'Reilly, C. N. Robertson, and W. C. Phoel (1978). Primary production and respiration over Georges Bank during March and July 1977. ICES C.M. 1978/L:37.

Thomas, J. P., J. E. O'Reilly, A. Draxler, J. A. Babinchak, C. N. Robertson, W. C. Phoel, R. Waldhauer, C. A. Evans, A. Matte, M. Cohn, M. Nitkowski, and S. Dudley (1979). Biological processes: productivity and respiration. *In* "Oxygen Depletion and Associated Benthic Mortalities in the New York Bight, 1976" (C. J. Sindermann and R. L. Swanson, eds.), *NOAA Prof. Pap.* No. 11, 231–262.

Thomas, W. H. (1970a). On nitrogen deficiency in tropical Pacific oceanic phytoplankton: photosynthetic parameters in poor and rich water. *Limnol. Oceanogr.* **15,** 380–385.

Thomas, W. H. (1970b). Effect of ammonium and nitrate concentration on chlorophyll increases in natural tropical Pacific phytoplankton populations. *Limnol. Oceanogr.* **15,** 380–394.

Thomas, W. J., and E. G. Simmons (1960). Phytoplankton production in the Mississippi Delta. *In* "Recent Sediments, Northwest Gulf of Mexico" (F. P. Shepard, F. B. Phleger, and T. H. van Andel, eds.), pp. 103–116. Am. Assoc. Pet. Geol., Tulsa, Oklahoma.

Thompson, J. D. (1974). The coastal upwelling cycle on a betaplane: hydrodynamics and thermodynamics. Ph.D. Thesis, Florida State Univ., Tallahassee.

Thordardottir, T. (1973). Successive measurements of primary production and composition of phytoplankton at two stations west of Iceland. *Nor. J. Bot.* **20,** 257–270.

Tilseth, S., and B. Ellertsen (1984). The detection and distribution of larval Arcto-Norwegian cod, *Gadus morhua,* food organisms by an *in situ* particle counter. *Fish. Bull.* **82,** 141–156.

Timonin, A. G. (1971). The structure of plankton communities of the Indian Ocean. *Mar. Biol. (Berlin)* **9,** 281–289.

Tingle, A. G., D. A. Dieterle, and J. J. Walsh (1979). Perturbation analysis of the New York Bight. *In* "Ecological Processes in Coastal and Marine Systems" (R. J. Livingston, ed.), pp. 395–435. Plenum, New York.

Tissot, B. P., and D. H. Welte (1978). "Petroleum Formation and Occurrence." Springer-Verlag, Berlin and New York.

Tissot, B. P., G. Deroo, and A. Hood (1978). Geochemical study of the Uinta Basin: Formation of petroleum from the Green River formation. *Geochim. Cosmochim. Acta* **42,** 1469–1485.

Tranter, D. J., R. R. Parker, and G. C. Cresswell (1980). Are warm-core eddies unproductive? *Nature (London)* **284,** 540–542.

Tranter, D. J., G. S. Leech, and D. Airey (1983). Edge enrichment in an ocean eddy. *Aust. J. Mar. Freshwater Res.* **34,** 665–680.

Trask, P. D. (1953). The sediments of the Western Gulf of Mexico. II. Chemical Studies of sediments of the Western Gulf of Mexico. *Pap. Phys. Oceanogr. Meteorol.* No. 12, 47–120.

Tsuji, T., H. Seki, and A. Hattori (1973). Results of red tide formation in Tokyo Bay. *J. Water Pollut. Control Fed.,* **46,** 165–172.

Turner, J. T. (1977). Sinking rates of fecal pellets from the marine copepod *Pontella meadii. Mar. Biol. (Berlin)* **40,** 245–259.

Turner, R. W., S. W. Woo, and H. R. Jitts (1979). Estuarine influences on a continental shelf plankton community. *Science* **206,** 218–220.

Tyler, M. A., and H. H. Seliger (1978). Annual subsurface transport of a red tide dinoflagellate to its bloom area: water circulation patterns and organism distribution in the Chesapeake Bay. *Limnol. Oceanogr.* **23,** 227–246.

United States Court of Appeals (1979). *National Sea Clammers Association and Lovgren, G.* v. *City of New York et al* ., No. 79–1360, U.S. Court of Appeals for the District of New Jersey, Newark.

Urey, H. C. (1947). The thermodynamic properties of isotopic substances. *J. Chem. Soc., London* 562–581.

Urrere, M. A., and G. A. Knauer (1981). Zooplankton fecal pellet fluxes and vertical transport of particulate organic matter in the pelagic environment. *J. Plankton Res.* **3**, 369–387.

Vail, P. R., R. M. Mitchum, and S. Thompson (1978). Seismic stratigraphy and global changes in sea level. *In* "Seismic Stratigraphy: Application to Hydrocarbon Exploration" (C. E. Payton, ed.), *Mem. Am. Assoc. Pet. Geol.* **26**, 83–97.

Vail, P. R., and J. Hardenbol (1979). Sea level changes during the Tertiary. *Oceanus* **22**, 71–79.

Valdivia, J. (1976). Biological aspects of the 1972–73 El Niño. Part 2, The anchovy population. *In* "Proc. Workshop El Niño Phenom., Guayaquil, Ecuador," pp. 73–82. UNESCO, Geneva.

Valiela, I. (1984). "Marine Ecological Processes." Springer-Verlag, Berlin and New York.

Van Bennekom, A. J., W. W. Gieskes, and S. B. Tijssen (1975). Eutrophication of Dutch coastal waters. *Proc. R. Soc. London Ser. B* **189**, 359–374.

Van Bennekom, A. J., G. W. Berger, W. Helder, and R. T. deVries (1978). Nutrient distribution in the Zaire estuary and river plume. *Neth. J. Sea Res.* **12**, 296–323.

Van Bennekom, A. J., and W. Salomons (1981). Pathways of nutrients and organic matter from land to ocean through rivers. *In* "River Inputs to Ocean Systems," pp. 35–51. UNESCO, Geneva.

Varga, R. S. (1962). "Matrix Iterative Analysis." Prentice-Hall, Englewood Cliffs, New Jersey.

Vargo, G. (1968). Studies of phytoplankton ecology in tropical and subtropical environments of the Atlantic Ocean. II. Quantitative studies of phytoplankton distribution in the Straits of Florida and its relation to physical factors. *Bull. Mar. Sci.* **18**, 5–60.

Vedernikov, V. I., and A. A. Solov'yeva (1972). Primary production and chlorophyll in the coastal waters of the Barents Sea. *Oceanology* **6**, 557–563.

Veeh, H. H. (1966). Th^{230}/U^{238} and U^{234}/U^{238} ages of Pleistocene high sea level stand. *J. Geophys. Res.* **71**, 3379–3386.

Veizer, J., W. T. Holser, and C. K. Wilgus (1980). Correlation of $^{13}C/^{12}C$ and $^{34}S/^{32}S$ secular variations. *Geochim. Cosmochim. Acta* **44**, 579–587.

Venkatesan, M. I., M. Sandstrom, S. Breener, E. Ruth, J. Bonilla, I. R. Kaplan, and W. E. Reed (1981). Organic geochemistry of surficial sediments from eastern Bering Sea. *In* "The Eastern Bering Sea Shelf: Oceanography and Resources" (D. W. Hood and J. A. Calder, eds.), Vol. 1, pp. 389–410. Univ. of Washington Press, Seattle.

Vidal, J. (1978). Effects of phytoplankton concentration, temperature, and body size on rates of physiological processes and production efficiency of the marine planktonic copepods, *Calanus pacificus* Brodsky and *Pseudocalanus* sp. Ph.D. Thesis, Univ. of Washington, Seattle.

Vidal, J., and T. E. Whitledge (1982). Rates of metabolism of planktonic copepods as related to body weight and temperature of habitat. *J. Plankton Res.* **4**, 77–84.

Vidal, J., and S. L. Smith (1986). Biomass, growth and development of populations of herbivorous zooplankton in the southeastern Bering Sea during spring. *Deep-Sea Res.* **33**, 523–556.

Vildoso, A. C. (1976). Biological aspects of the 1972–73 El Niño. Part 1, Distribution of the fauna. *In* "Proc. Workshop El Niño Phenom., Guayaquil, Ecuador," pp. 63–72. UNESCO, Geneva.

Viollier, M., and B. Sturm (1984). CZCS data analysis in turbid coastal waters. *J. Geophys. Res.* **189**, 4977–4985.

Voituriez, B., and A. Herbland (1979). The use of the salinity maximum of the Equatorial Undercurrent for estimating nutrient enrichment and primary production in the Gulf of Guinea. *Deep-Sea Res.* **26**, 77–83.

Vollenweider, R. A. (1965). Calculation models of photosynthesis-depth curves and some implications regarding day rate estimates in primary production measurements. *In* "Primary Productivity in Aquatic Environments" (C. R. Goldman, ed.), pp. 425–457. Univ. of California Press, Berkeley.

Vollenweider, R. A. (1968). Scientific fundamentals of the eutrophication of lakes and flowing waters, with particular reference to nitrogen and phosphorus as factors in eutrophication. OECD Tech. Rep. DAS/CS1/68-27, pp. 1–95.

Vollenweider, R. A., M. Munawar, and P. Stadlemann (1974). A comparative review of phytoplankton and primary production in the Laurentian Great Lakes. *J. Fish. Res. Board Can.* **31,** 739–762.

Volterra, V. (1926). Varazioni e fluttuazione del numero d'individuli in specie animali conviventi. *Mem. Acad. Linza* **6,** 31–113.

Von Arx, W. S. (1962). "Introduction to Physical Oceanography." Addison-Wesley, Reading, Massachusetts.

Voorhis, A. O., D. C. Webb, and R. C. Millard (1976). Current structure and mixing in the shelf/slope water front south of New England. *J. Geophys. Res.* **81,** 3695–3708.

Voronina, N. M. (1972). The spatial structure of interzonal copepod populations in the Southern Ocean. *Mar. Biol. (Berlin)* **15,** 336–343.

Vukovitch, F. M., B. W. Crissman, M. Bushnell, and W. J. King (1979). Some aspects of the oceanography of the Gulf of Mexico using satellite and *in situ* data. *J. Geophys. Res.* **84,** 7749–7768.

Wajih, S., S. Naqui, R. J. Noronha, and C. V. Reddy (1982). Denitrification in the Arabian Sea. *Deep-Sea Res.* **29,** 459–469.

Waldron, K. D. (1978). Ichthyoplankton of the eastern Bering Sea: 11 February to 16 March 1978. Natl. Mar. Fish. Serv. N.W. Fish. Cent., Seattle.

Walker, J. C. (1974). Stability of atmospheric oxygen. *Am. J. Sci.* **274,** 193–214.

Walsh, J. J. (1971). Relative importance of habitat variables in predicting the distribution of phytoplankton at the ecotone of the antarctic upwelling ecosystem. *Ecol. Monogr.* **41,** 291–309.

Walsh, J. J. (1972). Implications of a systems approach to oceanography. *Science* **176,** 969–975.

Walsh, J. J. (1975). A spatial simulation model of the Peru upwelling ecosystem. *Deep-Sea Res.* **22,** 201–236.

Walsh, J. J. (1976). Herbivory as a factor in patterns of nutrient utilization in the sea. *Limnol. Oceanogr.* **21,** 1–13.

Walsh, J. J. (1977). A biological sketchbook for an eastern boundary current. *In* "The Sea" (E. D. Goldberg, I. N. McCave, J. J. O'Brien, and J. H. Steele, eds.), Vol. 6, pp. 923–968. Wiley (Interscience), New York.

Walsh, J. J. (1978). The biological consequences of interaction of the climatic, El Niño, and event scales of variability in the eastern tropical Pacific. *Rapp. P.-V. Reun. Cons. Int. Explor. Mer* **173,** 182–192.

Walsh, J. J. (1980). Concluding remarks: marine photosynthesis and the global CO_2 cycle. *In* "Primary Production in the Sea" (P. G. Falkowski, ed.), pp. 497–506. Plenum Press, New York.

Walsh, J. J. (1981a). Shelf-sea ecosystems. *In* "Analysis of Marine Ecosystems" (A. R. Longhurst, ed.), pp. 158–196. Academic Press, New York.

Walsh, J. J. (1981b). A carbon budget for overfishing off Peru. *Nature (London)* **290,** 300–304.

Walsh, J. J. (1983). Death in the sea: enigmatic phytoplankton losses. *Prog. Oceanogr.* **12,** 1–86.

Walsh, J. J. (1984). The role of the ocean biota in accelerated ecological cycles: a temporal view. *BioScience* **34,** 499–507.

Walsh, J. J., and D. A. Dieterle (1986). Simulation analysis of plankton dynamics in the northern Bering Sea. *In* "Marine Interfaces Ecohydrodynamics" (J. J. Nihoul, ed.), pp. 401–428. Elsevier, Amsterdam.

Walsh, J. J., and R. C. Dugdale (1971). A simulation model of the nitrogen flow in the Peruvian upwelling system. *Invest. Pesq.* **35,** 309–330.

Walsh, J. J., and S. O. Howe (1976). Protein from the sea: a comparison of the simulated nitrogen and carbon productivity of the Peru Upwelling Ecosystem. *In* "Systems Analysis and Simulation in Ecology" (B. C. Patten, ed.), Vol. 4, pp. 47–61. Academic Press, New York.

Walsh, J. J., and C. P. McRoy (1986). Ecosystem analysis in the southeastern Bering Sea. *Cont. Shelf Res.* **5**, 259–288.

Walsh, J. J., J. C. Kelley, R. C. Dugdale, and B. W. Frost (1971). Gross biological features of the Peruvian upwelling system with special reference to possible diel variation. *Invest. Pesq.* **35**, 25–42.

Walsh, J. J., J. C. Kelley, T. E. Whitledge, J. J. MacIsaac, and S. A. Huntsman (1974). Spin-up of the Baja California upwelling ecosystem. *Limnol. Oceanogr.* **19**, 553–572.

Walsh, J. J., T. E. Whitledge, J. C. Kelley, S. A. Huntsman, and R. D. Pillsbury (1977). Further transition states of the Baja California upwelling ecosystem. *Limnol. Oceanogr.* **22**, 264–280.

Walsh, J. J., T. E. Whitledge, F. W. Barvenik, C. D. Wirick, S. O. Howe, W. E. Esaias and J. T. Scott (1978). Wind events and food chain dynamics within the New York Bight. *Limnol. Oceanogr.* **23**, 659–683.

Walsh, J. J., T. E. Whitledge, W. E. Esaias, R. L. Smith, S. A. Huntsman, H. Santander and B. R. DeMendiola (1980). The spawning habitat of the Peruvian anchovy, *Engraulis ringens. Deep-Sea Res.* **27**, 1–27.

Walsh, J. J., G. T. Rowe, R. L. Iverson, and C. P. McRoy (1981). Biological export of shelf carbon is a neglected sink of the global CO_2 cycle. *Nature (London)* **291**, 196–201.

Walsh, J. J., E. T. Premuzic, J. S. Gaffney, G. T. Rowe, W. Balsam, G. Harbottle, R. W. Stoenner, P. R. Betzer, and S. A. Macko (1985). Storage of CO_2 as organic carbon on the continental slopes off the mid-Atlantic Bight, Southeastern Bering Sea, and the Peru coast. *Deep-Sea Res.* **32**, 853–883.

Walsh, J. J., C. P. McRoy, T. H. Blackburn, L. K. Coachman, J. J. Goering, J. J. Nihoul, P. L. Parker, R. B. Tripp, A. M. Springer, T. E. Whitledge, and C. D. Wirick. (1986a). The role of Bering Strait in the carbon/nitrogen fluxes of polar marine ecosystems. *In* "Marine Living Systems of the Far North" (L. Rey and V. Alexander, eds.), E. A. Brill, Leiden, Netherlands. In press.

Walsh, J. J., T. E. Whitledge, J. E. O'Reilly, W. C. Phoel, and A. F. Draxler (1986b). Nitrogen cycling within seasonally stratified and tidally mixed regions of the mid-Atlantic Bight: the New York shelf and Georges Bank. *In* "Georges Bank and Its Surrounding Waters" (R. H. Backus, ed.), MIT Press, Cambridge, Massachusetts. In press.

Walsh, J. J., D. A. Dieterle, and W. E. Esaias (1987a). Satellite detection of phytoplankton export from the mid-Atlantic Bight during the 1979 spring bloom. *Deep-Sea Res.* **34**, 675–703.

Walsh, J. J., C. D. Wirick, L. J. Pietrafesa, T. E. Whitledge, and F. E. Hoge (1987b). High frequency sampling of the 1984 spring bloom within the mid-Atlantic Bight: synoptic shipboard, aircraft, and *in situ* perspectives of the SEEP-I experiment. *Cont. Shelf Res.* (in press).

Walsh, J. J., D. A. Dieterle, and M. B. Meyers (1987c). A simulation analysis of the fate of phytoplankton within the mid-Atlantic Bight. *Cont. Shelf Res.* (in press).

Wangersky, P. J. (1965). The organic chemistry of sea water. *Am. Sci.* **53**, 358–374.

Ward, B. B., R. J. Olson, and M. J. Perry (1982). Microbial nitrification rates in the primary nitrite maximum off southern California. *Deep-Sea Res.* **29**, 247–255.

Waterbury, J. B., S. W. Watson, R. L. Guillard, and L. E. Brand (1979). Widespread occurrence of a unicellular, marine planktonic cyanobacterium. *Nature (London)* **277**, 293–294.

Watson, A. J., and M. Whitfield (1985). Composition of particles in the global ocean. *Deep-Sea Res.* **32**, 1023–1039.

Watts, J. C. (1958). The hydrology of a tropical West African estuary. *Bull. Inst. Fr. Afr. Noire* **20**, 697–752.

Weaver, F. M., and S. W. Wise (1974). Opaline sediments of the southeastern coastal plane and horizon A: Biogenic origin. *Science* **184**, 899–901.

Weaver, S. S. (1979). *Ceratium* in Fire Island Inlet, Long Island, New York (1971–1972). *Limnol. Oceanogr.* **24**, 553–558.

502 References

Webster, F. (1969). Turbulence spectra in the ocean. *Deep-Sea Res.* **16**, 357–368.

Wefer, G., E. Suess, W. Balzar, G. Liebezeit, P. J. Muller, C. A. Ungerer, and W. Zenk (1982). Fluxes of biogenic components from sediment trap deployment in circumpolar waters of the Drake Passage. *Nature (London)* **299**, 145–147.

Weinstein, M. P. (1981). Plankton productivity and the distribution of fishes on the southeastern U.S. continental shelf. *Science* **214**, 351–354.

Weiss, R. F. (1981). The temporal and spatial distribution of tropospheric nitrous oxide. *J. Geophys. Res.* **86**, 7185–7195.

Weiss, R. F., and H. Craig (1976). Production of atmospheric nitrous oxide by combustion. *Geophys. Res. Lett.* **3**, 751–753.

Weiss, R. F., J. L. Bullister, R. H. Gammon, and M. J. Warner (1985). Atmospheric chlorofluoromethanes in the deep equatorial Atlantic. *Nature (London)* **314**, 603–610.

Welander, P. (1957). Wind action on a shallow sea: some generalizations of Ekman's theory. *Tellus* **9**, 45–52.

Welch, W. R. (1968). Changes in abundance of the green crab, *Carcinus maenas*, in relation to recent temperature changes. *Fish. Bull.* **67**, 337–345.

Wellman, R. P., E. D. Cook, and H. R. Krouse (1968). Nitrogen-15: Microbiological alteration of abundance. *Science* **161**, 269–270.

Welschmeyer, N. A., and C. J. Lorenzen (1985). Chlorophyll budgets: zooplankton grazing and phytoplankton growth in a temperate fjord and the central Pacific gyres. *Limnol. Oceanogr.* **30**, 1–21.

Wespestad, V. G., and L. H. Barton (1981). Distribution, migration, and status of Pacific herring. *In* "The Eastern Bering Sea Shelf: Oceanography and Resources" (D. W. Hood and J. A. Calder, eds.), Vol. 1, pp. 509–525. Univ. of Washington Press, Seattle.

Wharton, W. G., and K. H. Mann (1981). Relationship between destructive grazing by the sea urchin, *Strongylocentrotus droebachiensis*, and the abundance of American lobster, *Homarus americanus* on the Atlantic coast of Nova Scotia. *Can. J. Fish. Aquat. Sci.* **38**, 1339–1349.

Whitcomb, V. L. (1970). Oceanography of the Mid-Atlantic Bight in support of ICNAF, September–December 1967. *USCG Oceanogr. Rep.* No. 35.

Whitledge, T. E. (1978). Regeneration of nitrogen by zooplankton and fish in the northwest Africa and Peru upwelling ecosystems. *In* "Upwelling Ecosystems" (R. Boje and M. Tomczak, eds.), pp. 90–100. Springer-Verlag, Berlin and New York.

Whitledge, T. E., and C. D. Wirick (1983). Observations of chlorophyll concentrations off Long Island from a moored *in situ* fluorometer. *Deep-Sea Res.* **30**, 297–309.

Whitledge, T. E., and C. D. Wirick (1986). Development of a moored *in situ* fluorometer for phytoplankton studies. *In* "Tidal Mixing and Plankton Dynamics" (M. J. Bowman, C. S. Yentsch, and W. J. Petersen, eds.), Springer-Verlag, Berlin and New York. In press.

Whitledge, T. E., W. S. Reeburgh, and J. J. Walsh (1986). Seasonal inorganic nitrogen distributions and dynamics in the southeastern Bering Sea. *Cont. Shelf Res.* **5**, 109–132.

Whitmore, F. C., K. O. Emery, H. B. Cooke, and D. J. Swift (1967). Elephant teeth from the Atlantic continental shelf. *Science* **156**, 1477–1481.

Wickett, W. P. (1967). Ekman transport and zooplankton concentrations in the North Pacific Ocean. *J. Fish. Res. Board Can.* **24**, 581–594.

Wiebe, P. H., S. H. Boyd, and C. Winget (1976). Particulate matter sinking to the deep-sea floor at 2000 m in the tongue of the ocean, Bahamas, with a description of a new sedimentation trap. *J. Mar. Res.* **34**, 341–354.

Wiebe, P. H., V. A. Barber, S. H. Boyd, C. S. Davis, and G. R. Flierl (1985). Macrozooplankton biomass in Gulf Stream warm-core rings: spatial distribution and temporal changes. *J. Geophys. Res.* **90**, 8885–8901.

Wigley, R. L., and A. D. McIntyre (1964). Some quantitative comparisons of offshore meiobenthos and macrobenthos south of Martha's Vineyard. *Limnol. Oceanogr.* **9**, 485–493.

Wigley, R. L., and R. B. Theroux (1979). Macrobenthic invertebrate fauna of the middle Atlantic Bight region: faunal composition and quantitative distribution. *U.S. Geol. Surv. Prof. Pap.* No. 10.

Wilkinson, W. B., and L. A. Greene (1982). The water industry and the nitrogen cycle. *Philos. Trans. R. Soc. London Ser. B* **296**, 459–475.

Williams, P. M., H. Oeschger, and P. Kinney (1969). Natural radiocarbon activity of dissolved organic carbon in the Northeast Pacific Ocean. *Nature (London)* **224**, 256–258.

Williams, P. M., J. A. McGowan, and M. Stuiver (1970). Bomb carbon-14 in deep-sea organisms. *Nature (London)* **227**, 375–376.

Williams, P. M., M. C. Stenhouse, E. M. Druffel, and M. Koide (1978). Organic-^{14}C activity in an abyssal marine sediment. *Nature (London)* **276**, 698–701.

Winokur, M. (1948). Photosynthesis relationship of *Chlorella* species. *Am. J. Bot.* **35**, 207–214.

Winter, D. F., K. Banse, and G. C. Anderson (1975). The dynamics of phytoplankton blooms in Puget Sound, a fjord in the northwestern United States. *Mar. Biol. (Berlin)* **29**, 139–176.

Wirick, C. D. (1981). Marine herbivores and the spatial distributions of phytoplankton. Ph.D. Thesis, Univ. of Washington, Seattle.

Witkamp, M. (1966). Decomposition of leaf litter in relation to environment, microflora, and microbial respiration. *Ecology* **47**, 194–201.

Wolf, T. (1979). Macrofaunal utilization of plant remains in the deep sea. *Sarsia* **64**, 117–136.

Wollast, R. (1983). Interactions in estuaries and coastal waters. *In* "The Major Biogeochemical Cycles and Their Interactions" (B. Bolin and R. B. Cook, eds.), pp. 385–407. Wiley, New York.

Wolman, M. G. (1971). The nation's rivers. *Science* **174**, 905–918.

Wong, C. S. (1978). Atmospheric input of carbon dioxide from burning wood. *Science* **200**, 197–200.

Wong, W. (1976). Carbon isotope fractionation by marine phytoplankton. Ph.D. Thesis, Texas A&M Univ., College Station.

Wood, B. J. (1974). Fatty acids and saponifiable lipids. *In* "Algal Physiology and Biochemistry" (W. D. Stewart, ed.), pp. 236–265. Univ. of California Press, Berkeley.

Woodwell, G. M., R. H. Whittaker, W. A. Reiners, G. E. Likens, C. C. Delwiche, and D. B. Botkin (1978). The biota and the world carbon budget. *Science* **199**, 144–146.

Wooster, W. S., and O. Guillen (1974). Characteristics of El Niño in 1972. *J. Mar. Res.* **32**, 387–404.

Worsley, T. (1974). The Cretaceous-Tertiary boundary event in the ocean. *In* "Studies in Paleo-Oceanography" (W. W. Hay, ed.), *Spec. Publ. Soc. Econ. Paleontol. Mineral.* No. 20, 94–125.

Wroblewski, J. S. (1977). A model of phytoplankton plume formation during variable Oregon upwelling. *J. Mar. Res.* **35**, 357–394.

Wroblewski, J. S. (1984). Formulation of growth and mortality of larval northern anchovy in a turbulent feeding environment. *Mar. Ecol. Prog. Ser.* **20**, 13–22.

Wroblewski, J. S., and J. G. Richman (1987). The nonlinear response of plankton to wind mixing events—implications for larval fish survival. *J. Plankton Res.* **9**, 103–123.

Wyatt, T. (1971). Production dynamics of *Oikopleura dioica* in the southern North Sea, and the role of fish larvae which prey on them. *Thalassia Jugosl.* **7**, 435–444.

Wyatt, T. (1972). Some effects of food density on the growth and behavior of plaice larvae. *Mar. Biol. (Berlin)* **14**, 210–216.

Wyatt, T. (1974). The feeding of plaice and sand eel larvae in the Southern Bight in relation to the distribution of their food organisms. *In* "The Early Life History of Fish" (J. H. Blaxter, ed.), pp. 245–251. Springer-Verlag, Berlin and New York.

Wyatt, T. (1976). Food chains in the sea. *In* "The Ecology of the Seas" (D. H. Cushing and J. J. Walsh, eds.), pp. 341–360. Blackwell, Oxford.

Wyrtki, K. (1981). An estimate of equatorial upwelling in the Pacific. *J. Phys. Oceanogr.* **11,** 1205–1214.

Wyrtki, K., E. Stroup, W. Patzert, R. Williams, and W. Quinn (1976). Predicting and observing El Niño. *Science* **191,** 343–346.

Yentsch, C. S., and J. H. Ryther (1959). Relative significance of the net phytoplankton and nanoplankton in the waters of Vineyard Sound. *J. Cons. Cons. Int. Explor. Mer* **24,** 231–238.

Yentsch, C. S., and D. W. Menzel (1963). A method for the determination of phytoplankton chlorophyll and phaeophytin by fluorescence. *Deep-Sea Res.* **10,** 221–231.

Yingst, J. Y., and R. C. Aller (1982). Biological activity and associated sedimentary structures in HEBBLE-area deposits, western North America. *Mar. Geol.* **48,** M7-M15.

Yoder, J. A. (1983). Statistical analysis of the distribution of fish eggs and larvae on the southeastern U.S. continental shelf with comments on oceanographic processes that may affect larval survival. *Estuarine Coast. Shelf Sci* **17,** 637–650.

Yoder, J. A., L. P. Atkinson, S. S. Bishop, E. E. Hofmann, and T. N. Lee (1983). Effect of upwelling on phytoplankton productivity of the outer southeastern United States continental shelf. *Cont. Shelf Res.* **1,** 385–404.

Yoder, J. A., L. P. Atkinson, S. Bishop, J. O. Blanton, T. N. Lee, and L. J. Pietrafesa (1985). Phytoplankton dynamics within Gulf Stream intrusions on the southeastern United States continental shelf during summer 1981. *Cont. Shelf Res.* **4,** 611–635.

Yoder, J. A., S. S. Bishop, L. P. Atkinson, C. R. McClain, T. Paluszkiewicz, and A. D. Mahood (1987). Effects of the Loop Current on phytoplankton production of the outer southwest Florida continental shelf. *J. Plankton Res.* (submitted).

Yoshida, K., and H.-L. Mao (1957). A theory of upwelling of large horizontal extent. *J. Mar. Res.* **16,** 40–54.

Zaret, T. M., and T. R. Paine (1973). Species introduction in a tropical lake. *Science* **182,** 449–455.

Zeitzschel, B., P. Dieckmann, and L. Uhlmann (1978). A new multi-sample sediment trap. *Mar. Biol. (Berlin)* **45,** 285–288.

Zimmerman, P. R., and J. P. Greenberg (1983). Termites and methane. *Nature (London)* **302,** 354.

Zscheile, F. P. (1934). A quantitative spectrophotometric analytical method applied to solutions of chlorophylls, a and b. *J. Phys. Chem.* **38,** 95–102.

Glossary

a	Time of sunrise
$a_0, a_1, \ldots, a_n$	Coefficients of a Fourier series
A	Acceleration of a fluid
A_c	Sediment carbon accumulation rate
α	Constant bottom velocity in linearized bottom stress
b	Product of specific gravity, percent carbon, and percent volume of sediment
$b_0, b_1, \ldots, b_n$	Coefficients of either multiple regressions or Fourier series
$B_{x,y}$	Components of the bottom friction
β	$\partial f / \partial y$, change of Coriolis parameter with latitude
c	Phase velocity, $\sqrt{gH}$
C	Drag coefficient
CZCS	Coastal zone color scanner
d	Julian day
dw	Dry weight
D	Depth of surface Ekman layer
DN	Detrital nitrogen
DOC	Dissolved organic carbon
DON	Dissolved organic nitrogen
δ	Ratio of photosynthetically active radiation (PAR) to total solar radiation
$\delta\,^{13}C$	Isotopic ratio of ^{13}C to ^{12}C
$\delta\,^{14}C$	Isotopic ratio of ^{14}C to ^{12}C
$\delta\,^{15}N$	Isotopic ratio of ^{15}N to ^{14}N
$\delta\,^{18}O$	Isotopic ratio of ^{18}O to ^{16}O
$\delta\,^{34}S$	Isotopic ratio of ^{34}S to ^{32}S
Δ	Sampling interval
e	Elevation of the sea surface
E	Einstein, unit of 5×10^4 g-cal of visible radiation
ε	Resultant algal growth rate (hr^{-1})
ε_m	Maximum algal growth rate (hr^{-1})
f	Coriolis parameter, or planetary vorticity
F	Fraction of surface kinetic energy available for mixing

505

$F_{x,y}$	Components of the surface friction
FPN	Fish particulate nitrogen
g	Acceleration of gravity, 980 cm sec^{-2}
G	Herbivore
γ	Grazing rate of herbivores
h	Depth of bottom Ekman layer
h_c	Critical depth
h_e	Depth of euphotic zone
h_m	Mixed layer depth of water
h_s	Secchi depth
H	Depth of the water column
H_m	Mixed layer depth of sediment
$H(f)$	Transfer function of a digital filter
i	Day length, or photoperiod
I_c	Compensation light intensity (PAR)
I_h	Average irradiance of a mixed water column
I_k	Half-saturation light intensity at which $\mu = \mu_m/2$
I_0	Total incident solar irradiance (g-cal cm^{-2} hr^{-1})
$I_0^{s,w}$	Incident radiation at summer (s) and winter (w) solstices
I_m	Maximum incident radiation at local apparent noon (LAN)
I_s	Saturation light intensity for maximum algal photosynthesis
$I(z)$	Subsurface light field of photosynthetically active radiation (PAR) at depth z
j	Chemical reaction rate of a homogeneous gas
J	Number of collisions of molecules in an ideal gas
k	Diffuse attenuation coefficient
k_w	Attenuation due to water
k_p	Attenuation due to phytoplankton
k_s	Attenuation due to dissolved substances
K_f	Numerical dispersion coefficient
$K_{x,y,z}$	Eddy coefficients of the Reynolds stress
K_z	Sediment mixing coefficient
ℓ	Excretion rate of herbivores
l	Lag of the autocovariance or crosscovariance
L	Length scale
L_c	Critical length, above which algal patch forms
L_0	Wave number
λ	Rate of respiration
λ_C	Decay constant of ^{14}C (0.000121 yr^{-1})
λ_{Pb}	Decay constant of ^{210}Pb (0.0311 yr^{-1})
m	Mass flux of nutrient (μg-at. l^{-1} hr^{-1})
M	Microalgae, or phytoplankton
μ	Resultant rate of photosynthesis (mg C mg chl^{-1} hr^{-1})
μ_m	Maximum rate of light-saturated photosynthesis, or assimilation index

n	Half-saturation constant (μg-at. 1^{-1}) of nutrient uptake, at which $\chi = \chi_m/2$
N	Nutrient concentration (μg-at. 1^-)
N_0	Threshold nutrient concentration
N^2	Brunt–Väisälä frequency
η	Molecular viscosity
o	Consumption rate of carnivores
O	Carnivores
θ	Half-saturation constant of internal nutrient pool (pg-at. $cell^{-1}$)
θ_p	Half-saturation grazing constant
Ω	Angular velocity of the earth
$P_{1,2,3}$	Photosynthetic parameters
P	Pressure field
PP	Particulate phosphorus
PN	Particulate nitrogen
PSi	Particulate silicon
P_0	Zooplankton grazing threshold
P_z	Fish grazing threshold
χ	Resultant rate of nutrient uptake (hr^{-1})
χ_m	Maximum rate of nutrient uptake (hr^{-1})
q	Minimum cell quota (pg-at $cell^{-1}$)
Q	Maximum cell quota at which cell division occurs
Q_{10}	Temperature coefficient of metabolic activity, usually 2, such that a rate doubles for every 10°C increment of temperature
r	Bottom resistance coefficient
r^2	Coefficient of determination in regression analysis
$.r_1$	Cross-correlation coefficient
r_p	Particle radius
R	Depth-integrated buoyancy field, $\int_{-H}^{e} \rho g dz$
Ri	Richardson number
Ro	Rossby number
ρ	Density of seawater
ρ_p	Density of particle
s_1	Light-limitation parameter of photosynthesis
s_2	Light-inhibition parameter of photosynthesis
S	Activation energy of a chemical reaction
S_a	Apparent sedimentation rate
S_t	True sedimentation rate
Sv	Sverdrup unit of transport (10^6 m^3 sec^{-1})
σ_t	$(\rho - 1) \times 10^3$
t	Time
T	Temperature in degrees Celsius
T_k	Temperature in kelvins
$T_{l,u}$	Minimum and maximum temperatures at which photosynthesis stops

T_0	Period of a fundamental harmonic
τ_s	Surface wind stress
τ_b	Bottom drag stress
ϕ	Veer angle of bottom Ekman layer
u	x component of the flow field
u_E	x component of flow in the Ekman layer
U	x component of the depth-integrated transport
v	y component of the flow field
v_E	y component of flow in the Ekman layer
V	y component of the depth-integrated transport
w	z component of the flow field
w_i	Weight of a digital filter
w_s	Sinking velocity of phytoplankton, or other particles
W	Wind speed
ω	Relative vorticity
x	Cartesian coordinate, positive northward
$X_i, \ldots, X_n$	Independent variables of a regression equation or Fourier series
ξ	Gas constant, 2 cal $^{\circ}T_k^{-1}$ g-at^{-1}
y	Cartesian coordinate, positive eastward
$Y_1, \ldots, Y_n$	Dependent variables of a regression equation
z	Cartesian coordinate, positive downward
Z	Layer depth of vertically segmented model
ZPN	Zooplankton particulate nitrogen

Subject Index